Centrifugal Pump Clinic

MECHANICAL ENGINEERING

A Series of Textbooks and Reference Books

Editor: L.L. FAULKNER Columbus Division, Battelle Memorial Institute, and Department of Mechanical Engineering, The Ohio State University, Columbus, Ohio

Associate Editor: S.B. MENKES Department of Mechanical Engineering, The City College of the City University of New York, New York

1. Spring Designer's Handbook, *by Harold Carlson*
2. Computer-Aided Graphics and Design, *by Daniel L. Ryan*
3. Lubrication Fundamentals, *by J. George Wills*
4. Solar Engineering for Domestic Buildings, *by William A. Himmelman*
5. Applied Engineering Mechanics: Statics and Dynamics, *by G. Boothroyd and C. Poli*
6. Centrifugal Pump Clinic, *by Igor J. Karassik*
7. Computer-Aided Kinetics for Machine Design, *by Daniel L. Ryan*
8. Plastics Products Design Handbook, Part A: Materials and Components; Part B: Processes and Design for Processes, *edited by Edward Miller*
9. Turbomachinery: Basic Theory and Applications, *by Earl Logan, Jr.*
10. Vibrations of Shells and Plates, *by Werner Soedel*
11. Flat and Corrugated Diaphragm Design Handbook, *by Mario Di Giovanni*
12. Practical Stress Analysis in Engineering Design, *by Alexander Blake*
13. An Introduction to the Design and Behavior of Bolted Joints, *by John H. Bickford*
14. Optimal Engineering Design: Principles and Applications, *by James N. Siddall*
15. Spring Manufacturing Handbook, *by Harold Carlson*
16. Industrial Noise Control: Fundamentals and Applications, *edited by Lewis H. Bell*
17. Gears and Their Vibration: A Basic Approach to Understanding Gear Noise, *by J. Derek Smith*

18. Chains for Power Transmission and Material Handling: Design and Applications Handbook, *by the American Chain Association*
19. Corrosion and Corrosion Protection Handbook, *edited by Philip A. Schweitzer*
20. Gear Drive Systems: Design and Application, *by Peter Lynwander*
21. Controlling In-Plant Airborne Contaminants: Systems Design and Calculations, *by John D. Constance*
22. CAD/CAM Systems Planning and Implementation, *by Charles S. Knox*
23. Probabilistic Engineering Design: Principles and Applications, *by James N. Siddall*
24. Traction Drives: Selection and Application, *by Frederick W. Heilich III and Eugene E. Shube*
25. Finite Element Methods: An Introduction, *by Ronald L. Huston and Chris E. Passerello*
26. Mechanical Fastening of Plastics: An Engineering Handbook, *by Brayton Lincoln, Kenneth J. Gomes, and James F. Braden*
27. Lubrication in Practice, Second Edition, *edited by W. S. Robertson*
28. Principles of Automated Drafting, *by Daniel L. Ryan*
29. Practical Seal Design, *edited by Leonard J. Martini*
30. Engineering Documentation for CAD/CAM Applications, *by Charles S. Knox*
31. Design Dimensioning with Computer Graphics Applications, *by Jerome C. Lange*
32. Mechanism Analysis: Simplified Graphical and Analytical Techniques, *by Lyndon O. Barton*
33. CAD/CAM Systems: Justification, Implementation, Productivity Measurement, *by Edward J. Preston, George W. Crawford, and Mark E. Coticchia*
34. Steam Plant Calculations Manual, *by V. Ganapathy*
35. Design Assurance for Engineers and Managers, *by John A. Burgess*
36. Heat Transfer Fluids and Systems for Process and Energy Applications, *by Jasbir Singh*
37. Potential Flows: Computer Graphic Solutions, *by Robert H. Kirchhoff*
38. Computer-Aided Graphics and Design, Second Edition, *by Daniel L. Ryan*
39. Electronically Controlled Proportional Valves: Selection and Application, *by Michael J. Tonyan, edited by Tobi Goldoftas*
40. Pressure Gauge Handbook, *by AMETEK, U.S. Gauge Division, edited by Philip W. Harland*
41. Fabric Filtration for Combustion Sources: Fundamentals and Basic Technology, *by R. P. Donovan*
42. Design of Mechanical Joints, *by Alexander Blake*
43. CAD/CAM Dictionary, *by Edward J. Preston, George W. Crawford, and Mark E. Coticchia*

44. Machinery Adhesives for Locking, Retaining, and Sealing, *by Girard S. Haviland*
45. Couplings and Joints: Design, Selection, and Application, *by Jon R. Mancuso*
46. Shaft Alignment Handbook, *by John Piotrowski*
47. BASIC Programs for Steam Plant Engineers: Boilers, Combustion, Fluid Flow, and Heat Transfer, *by V. Ganapathy*
48. Solving Mechanical Design Problems with Computer Graphics, *by Jerome C. Lange*
49. Plastics Gearing: Selection and Application, *by Clifford E. Adams*
50. Clutches and Brakes: Design and Selection, *by William C. Orthwein*
51. Transducers in Mechanical and Electronic Design, *by Harry L. Trietley*
52. Metallurgical Applications of Shock-Wave and High-Strain-Rate Phenomena, *edited by Lawrence E. Murr, Karl P. Staudhammer, and Marc A. Meyers*
53. Magnesium Products Design, *by Robert S. Busk*
54. How To Integrate CAD/CAM Systems: Management and Technology, *by William D. Engelke*
55. Cam Design and Manufacture, Second Edition; with cam design software for the IBM PC and compatibles, disk included, *by Preben W. Jensen*
56. Solid-State AC Motor Controls: Selection and Application, *by Sylvester Campbell*
57. Fundamentals of Robotics, *by David D. Ardayfio*
58. Belt Selection and Application for Engineers, *edited by Wallace D. Erickson*
59. Developing Three-Dimensional CAD Software with the IBM PC, *by C. Stan Wei*
60. Organizing Data for CIM Applications, *by Charles S. Knox, with contributions by Thomas C. Boos, Ross S. Culverhouse, and Paul F. Muchnicki*
61. Computer-Aided Simulation in Railway Dynamics, *by Rao V. Dukkipati and Joseph R. Amyot*
62. Fiber-Reinforced Composites: Materials, Manufacturing, and Design, *by P. K. Mallick*
63. Photoelectric Sensors and Controls: Selection and Application, *by Scott M. Juds*
64. Finite Element Analysis with Personal Computers, *by Edward R. Champion, Jr. and J. Michael Ensminger*
65. Ultrasonics: Fundamentals, Technology, Applications, Second Edition, Revised and Expanded, *by Dale Ensminger*
66. Applied Finite Element Modeling: Practical Problem Solving for Engineers, *by Jeffrey M. Steele*
67. Measurement and Instrumentation in Engineering: Principles and Basic Laboratory Experiments, *by Francis S. Tse and Ivan E. Morse*

68. Centrifugal Pump Clinic, Second Edition, Revised and Expanded, *by Igor J. Karassik*

Additional Volumes in Preparation

Mechanical Engineering Software

Spring Design with an IBM PC, *by Al Dietrich*

Mechanical Design Failure Analysis: With Failure Analysis System Software for the IBM PC, *by David G. Ullman*

Centrifugal Pump Clinic
Second Edition, Revised and Expanded

Igor J. Karassik
Chief Consulting Engineer
Dresser Pump Company
Harrison, New Jersey

Taylor & Francis
Taylor & Francis Group
Boca Raton London New York

CRC is an imprint of the Taylor & Francis Group,
an informa business

To my coauthors, that is,
those engineers and centrifugal pump operators
whose questions I answered and who sometimes took issue
with the answers I gave to others

Library of Congress Cataloging-in-Publication Data

Karassik, Igor J.
    Centrifugal pump clinic / Igor J. Karassik. -- 2nd ed., rev. and expanded.
       p.   cm.
    Bibliography: p.
    Includes index.
    ISBN 0-8247-8072-8
    1. Centrifugal pumps.   I. Title.
TJ919.K29   1989
621.6'7--dc20   89-1423
                                                          CIP

This book is printed on acid-free paper.

Copyright © 1989 by MARCEL DEKKER, INC. All Rights Reserved

Neither this book nor any part may be reproduced or transmitted in any form or by any means, electronic or mechanical, including photocopying, microfilming, and recording, or by any information storage and retrieval system, without permission in writing from the publisher.

MARCEL DEKKER, INC.
270 Madison Avenue, New York, New York 10016

Current printing (last digit):
10 9 8 7 6

Preface to the Second Edition

The fact that man has learned to handle water in any appreciable quantities and move it from one place to another is nothing short of miraculous. By all rights, man should have developed a strong aversion toward having anything to do with water. It erodes, corrodes, abrades, and causes wire-drawing. It rusts, busts, wears, tears, dilutes, and shocks. It flashes into steam at the slightest provocation, then collapses back into its liquid state with much accompanying damage to metal surfaces and noise. It expands and contracts with temperature changes, creating unexpected stresses in the piping that carries it. It tempts pump designers into believing that it can act as a lubricant for so-called water-lubricated bearings and then betrays their confidence by failing to adequately carry out this assigned task. Occasionally, it develops a tendency to get into places where it has no business to be, such as oil-lubricated bearings and oil reservoirs.

Not content with dealing with water, man went on messing around with a whole batch of other liquids, each of them even more unpredictable than water. When he found that he had to move both water and these other liquids from one place to another in greater quantities than he could carry on his shoulders, he had to invent a mechanism to do so in his stead. That is how pumps were invented.

As I said, why anybody would want to have anything to do with such unpleasant elements as liquids, I do not know, unless it is because neither man nor any of his technological processes can exist without them. In order to develop some means of coexistence, it has become necessary for us to learn—I almost said to guess—what liquids will do under a number of peculiar circumstances.

Every year, young engineers join the staffs of pump companies that make it their principal business to produce equipment for pumping water and other liquids. Others join companies which, in turn, use these pumps. It is to the latter that I have addressed myself throughout my own career in an attempt to help them endure the indignities that pumping liquids visits upon them.

While my entire career has been spent with a manufacturer of pumps, my sympathy and my concerns have remained on the side of the users of these pumps. This is why I have enjoyed the opportunity to expand and update the first edition of *Centrifugal Pump Clinic* and thus continue my dialogue with these users and help them solve the problems they encounter. During the years that have elapsed since the publication of the first edition, I have written many more Clinic articles. Their inclusion in this second edition will, I hope, alleviate even more problems that face pump users.

For they certainly face many problems. And if we pump engineers do not try to help solve these problems, I am afraid that there will be more truth than poetry in a doggerel I wrote just the other day:

> They frequently say, when things don't look right:
> 'There's a light at the end of the tunnel, alright.'
> But once in a while, I'm afraid that there might
> Be a tunnel that looms at the end of the light.

Igor J. Karassik

Preface to the First Edition

At first glance, the question-and-answer format of this book may appear somewhat primitive and artificial. I feel therefore that I owe an explanation to its readers, an explanation which is really quite simple. The material presented in this book appeared first in the form of articles in a number of technical magazines. It was taken from actual correspondence dealing with various aspects of centrifugal pump application and operation.

Many years ago, when engineers and users of centrifugal pumps first began to write me about problems they encountered in the field, I noticed that a number of my correspondents had the same problems in common. I decided to save this portion of my correspondence for the purpose of publishing some of it in the form of articles. The first of these appeared in 1956.

As soon as the articles started to appear, they generated a host of new questions from the readers. In almost all cases, these readers permitted me to use their questions and my answers in new articles. The questions I chose to utilize were mainly intended to acquaint the readers with various ways in which they could improve the operation of their equipment or modify this equipment in order to save operating or maintenance costs. In many cases they were also intended to illustrate the diagnosis of various operating difficulties, which, for one reason or another, arise to plague pump operators.

The articles were later issued in reprint form and distributed by Worthington Corporation. It rapidly became apparent that the material had great interest because an increasing number of engineers requested to be placed on the mailing list for these reprints. Time and again, readers would write me that one or another of these articles would appear providentially in time to solve some particular problem that had been puzzling them. Further interest was evidenced when technical magazines in India, Argentina, Spain, Italy, and Japan started republishing these articles abroad.

Many of my readers had been writing me with the suggestion that I choose the most significant of these articles and publish them in a book form. I had already written a book, *Centrifugal Pumps, Selection, Operation, and Maintenance,* published by the F. W. Dodge Division of the McGraw-Hill Book Company, but I knew that there would be no conflict between this book and the Centrifugal Pump Clinic article series. If anything, the two books would be supplementary.

When I embarked on the project of choosing the questions to be used in this book, I realized that some material from another series of my articles, published under the title of "Steam Power Plant Clinic" in the magazine *Combustion* could be incorporated very usefully in the same book. Once the material had been selected, I proceeded with the task of rearranging the questions into a logical sequence and of grouping similar topics to facilitate easy reference.

During the years that followed the 1964 publication of *Engineers' Guide to Centrifugal Pumps,* I continued to write the "Centrifugal Pump Clinic" and the "Steam Power Plant Clinic" series of articles on which the book was based. These new clinics included material which would make a new edition of the book much more complete and, therefore, more useful to the reader. As this new material nearly doubles the information contained in the original text, it is indeed a new book.

The questions treated in this book touch on practically all the pertinent problems faced by engineers who are involved in one way or another with pumping equipment. While heavy emphasis is placed on operating and maintenance problems, the book embodies some general information that should be considered by engineers selecting centrifugal pumps for fluid-handling systems. The first chapter deals with application problems and is followed by a chapter on design and construction questions. This latter chapter is considerably shorter than all the others. This is rather natural, since I did not want to include material which can be found in my book *Centrifugal Pumps.* The third chapter deals with pump installation and the fourth with operation. Proper attention to these areas will serve to reduce considerably the cost and annoyance of service interrruptions and pump repairs. The fifth chapter deals with questions on preferred methods of pump maintenance. Finally, the last chapter answers questions on field difficulties. This, unfortunately, is a chapter which cannot ever be completed. Certainly my own experience has taught me that, while certain types of difficulties recur only too frequently, one constantly meets completely new and sometimes mysterious difficulties which must be diagnosed and solved without benefit of previous experience.

I have hopes that this material will be found useful by three basic groups of engineers. First, the plant engineering and operating group, facing many of the problems discussed in this book in their daily job, may find this a good working tool. Second, design engineers in consulting engineering firms and in the central engineering staffs of large companies should find it a useful reference when tackling pump application problems. Finally, recent engineering graduates entering into their careers should take advantage of previous experience to avoid many of the pitfalls in the field of pumping problems that others have already encountered and solved.

I feel certain that *Centrifugal Pump Clinic* will be a useful contribution to the literature available to those personnel who apply, operate, and maintain centrifugal pumps in our constantly more complex technological civilization.

I wish to thank my colleagues at Worthington Pump, W. C. Krutzsch, W. H. Fraser, and J. E. C. Valentin, who coauthored several of the clinics which have been included.

Likewise, I wish to thank the magazines in whose pages the various clinics appeared as articles and who have kindly permitted me to incorporate this material in my book: Chemical Processing, Combustion, National Engineer, The Plant, Plant Engineering, Power Engineering, Southern Engineering, Southern Power & Industry, and World Pumps.

Finally, I would like to thank Graham Garratt, Vice President of Marcel Dekker, Inc., who encouraged me to undertake the task of preparing this volume.

Igor J. Karassik

Contents

Preface to the Second Edition   *iii*
Preface to the First Edition   *v*

1. Application   1

1.1	How Far to Go in Preparing Pump Specifications	1
1.2	Reducing Impeller Diameter to Correct Motor Overload	5
1.3	Effect of Pump Speed Change	6
1.4	Calculation of RPM for Different Conditions of Service	7
1.5	More on the Effect of Pump Speed Change	11
1.6	Pump Capacity Versus Impeller Diameter	11
1.7	Narrow and Wide Impellers	13
1.8	Impeller Cutdown for Multistage Pumps	16
1.9	How to Cut and Trim Impellers	18
1.10	Field Test for Required NPSH	22
1.11	NPSH Tests	27
1.12	How to Measure Suction Lift and Suction Head	28
1.13	Effect of Altitude on Atmospheric Pressure	33
1.14	NPSH Versus Impeller Eye Diameter	33
1.15	Recommended Suction and Discharge Nozzle Velocities	36
1.16	Parable About NPSH and Hard Cash	38
1.17	More About NPSH and Velocity Head	41
1.18	More About Velocity Head and Entrance Loss	43
1.19	Effect of Changing Heater Pressure on NPSH	44
1.20	Inert Gas Pressurization of Reservoir at Pump Suction	45

1.21	Effect of Temperature on Required NPSH	46
1.22	More on Temperature and NPSH	50
1.23	NPSH Required When Handling Hydrocarbons	51
1.24	Injection of Cold Water to Boiler Feed Pump Suction	57
1.25	Submergence Control with 15-ft NPSH	61
1.26	"Uncontrolled" Operation or Operation in the Break	65
1.27	Suction Specific Speed	69
1.28	More About Suction Specific Speed	76
1.29	Required NPSH at Reduced Flows	79
1.30	Suppression Test	84
1.31	Effect of Speed on the Required NPSH	88
1.32	Effect of Entrained Air on NPSH	88
1.33	NPSH Required for Cryogenic Pumps	92
1.34	Head and Brake Horsepower of High Specific Speed Pumps at Shutoff	94
1.35	Efficiency of Pump with Inducer	97
1.36	Retrofitting a Pump with an Inducer	97
1.37	Effect of Viscosity on Centrifugal Pump Performance	98
1.38	Formulas, Symbols, and Abbreviations	99
1.39	How to Develop Approximate System-Head Curves	101
1.40	More about System-Head Curves	104
1.41	Discharge Piping Configurations	107
1.42	Friction Losses in Fittings	109
1.43	Altering Design Capacity of a Pump	113
1.44	Effect of Oversizing Centrifugal Pumps	118
1.45	Operation in a Closed Circuit	121
1.46	Parallel or Series Operation?	127
1.47	Is Pump Performing at Capacity?	130
1.48	"Drooping" Head-Capacity Curves	133
1.49	Parallel Operation of Steam Turbine-Driven Pumps	135
1.50	Variable-Speed Operation of Multiple-Pump Installation	138
1.51	Constant- and Variable-Speed Pumps in Parallel	142
1.52	Rated Speeds for High-Pressure Boiler Feed Pumps	144
1.53	Main Shaft Drive of Feed Pumps	149
1.54	Paralleling Centrifugal and Reciprocating Pumps	150
1.55	Effect of Water Temperature on Pump bhp	150
1.56	Design Capacity of Condensate Pumps	151
1.57	Should Circulating Pump Handle Cold or Warm Brine?	153
1.58	Testing Pumps at Reduced Speed	154
1.59	Pressurizing Suction of Hydraulic Pumps	158

	1.60	Correct Discharge Pressure for Boiler Feed Pumps	160
	1.61	Trends in Feedwater Cycles: Open or Closed?	164
	1.62	Split Feedwater Pumping Cycle	166
	1.63	Condensate Booster Pumps	170
	1.64	Simplified Feed System	174
	1.65	Variable-Speed Condensate Booster Pumps in a Closed Feedwater Cycle	176
	1.66	Variable Speed for Condenser Circulating Pumps	177
	1.67	Source of a Small Quantity of High-Pressure Feedwater	179
	1.68	Bid Evaluation	181
	1.69	Attainable Pump Efficiencies	183
	1.70	Optimum Specific Speed Selection	188
2.	**Pump Construction**		**190**
	2.1	Packed Boxes Versus Mechanical Seals	190
	2.2	Further Comments on Mechanical Seals	191
	2.3	Extra Deep Stuffing Boxes	193
	2.4	Sleeve Bearings Versus Antifriction Bearings	193
	2.5	Vertical Turbine Pump Bearings	197
	2.6	Coupling End Float	200
	2.7	Use of Cast Steel Casings for Boiler Feed Pumps	202
	2.8	Taper Machining of Axially Split Casing Flanges	205
	2.9	Single Suction Versus Double Suction	206
	2.10	Two Pumps in Series with Single Driver	210
	2.11	Balancing Holes in Impellers	214
	2.12	Impeller Pump-out Vanes	217
	2.13	Balancing Axial Thrust of Multistage Pumps	220
	2.14	Single or Double Rings	225
	2.15	Reverse Threads in Wearing Rings	228
	2.16	Shaft Deflection of Multistage Pumps	228
	2.17	Shaft Sleeves	230
	2.18	Water-Flooded Stuffing Box	231
3.	**Installation**		**235**
	3.1	Location of Discharge Nozzles on End-Suction Pumps	235
	3.2	Raised Faces on Pump Nozzles	237
	3.3	Eccentric Reducers at Pump Suction	238
	3.4	Reduction of Suction and Discharge Piping Size	240
	3.5	Are Baseplates Needed?	243

3.6	Grouting of Baseplates	245
3.7	Shims Under Pump Baseplates	246
3.8	Motor Doweling	248
3.9	Expansion Joints	249
3.10	More on Expansion Joints	250
3.11	Does Axial Thrust of Vertical Turbine Pumps Act on the Foundations?	252
3.12	Further Comments on Axial Thrust of Vertical Turbine Pumps	258
3.13	Shaft Elongation of Vertical Turbine Pumps	260
3.14	Saving Stuffing-Box Leakage	263
3.15	Sealing Water for Stuffing Boxes	267
3.16	Sealing Stuffing Boxes with Brine	270
3.17	Vortex Formation in Reservoir at Pump Suction	271
3.18	More on Vortex Formation	277
3.19	Air Entrainment into Deep-Well Pumps	279
3.20	Bypass to Drain Instead of to Suction	280
3.21	Temperature-Rise Control of Recirculation Bypass	282
3.22	Modulating Bypass Valves and Single Flow Orifice	285
3.23	Common Recirculation Line for Several Pumps	289
3.24	Location of Recirculation Control Valve	290
3.25	Location of Flowmeter for Bypass Control	294
3.26	Balancing Device Leak-off Return	298
3.27	Check Valve in Balancing Device Leak-off Line	301
3.28	Balancing Device Leak-off from a Lean Oil Pump	303
3.29	Relief Valve in Boiler Feed Pump Suction Line	305
3.30	Interconnected Spare Boiler Feed Pump	308
3.31	Cleaning Boiler Feed Pump Suction Lines	314
3.32	Suction Strainers	315
3.33	Differences in Adjacent Steam Power Plant Units	317
3.34	Location of Suction Bell of Vertical Pump	318
3.35	Extended Storage of Boiler Feed Pumps	319
3.36	Turning the Shaft of Pumps in Storage	323
	Bibliography on Wet-Pit Pump Intakes	325

4. Operation — 327

4.1	Testing Motors for Rotation	327
4.2	Warm-up of Boiler Feed Pumps	331
4.3	Should Suction or Discharge Be Throttled?	333
4.4	Can a Centrifugal Pump Be Operated Against a Closed Discharge?	334

4.5	Starting Boiler Feed Pumps	335
4.6	What Are the Best Means to Prime Centrifugal Pumps	336
4.7	Priming Large Horizontal Pumps	337
4.8	Protection Against Loss of Prime	338
4.9	Centrifugal Pumps in Series	342
4.10	Frequent Starting of Electric Motors	343
4.11	Run-out Capacity of Parallel Pumps When Run Singly	344
4.12	Keep Flashing Pumps Running	347
4.13	Minimum Operating Speed	349
4.14	Basis for Minimum Flow	351
4.15	Elimination of Recirculation Bypass	351
4.16	Temperature Rise for Liquids Other than Water	355
4.17	Continuous Versus Intermittent Operation	357
4.18	Cooling and Sealing Water for Standby Pumps	360
4.19	Bearing and Gland Cooling	362
4.20	Hand-Operated Auxiliary Oil Pumps	363
4.21	Lubricating Oil Characteristics	364
4.22	Rules for Operation and Maintenance of Centrifugal Pumps	369
4.23	Automation of Boiler Feed Pump Operation	373
4.24	Pumping Through Idle Feed Pumps to Fill the Boiler	382
4.25	Testing Pumps at Different Temperatures	383
4.26	Minimum Flow of Condensate Pumps	385
4.27	Excess of Total Head in a Boiler Feed Pump	387
4.28	Tripping the Boiler Feed Pump During Transient Operating Conditions	389
4.29	Monitoring Pump Characteristics	394
4.30	Checking Pump Performance Without Dismantling or Special Test Setup	398
4.31	Using Temperature Rise to Measure Pump Efficiency	401
4.32	Testing Boiler Feed Pumps for Wear	407
4.33	Field Tests When Proper Instrumentation Is Not Available	409
4.34	Testing for Shutoff Head	412
4.35	Testing Vertical Pumps Horizontally	416
4.36	Measuring Bearing Temperature	420
4.37	Turning Gear Operation	421
4.38	Changes in Shaft Deflection	424
4.39	Impeller Cutdown	425
4.40	More About Impeller Cutdown	425
4.41	Use of Oversize Clearances During Early Stages of Operation	426

	4.42	Minimum Start-up Time for Motor-Driven Standby Pumps	427
	4.43	How to Retrofit Inadequate Minimum Flow By-passes	428
	4.44	Flashing in Boiler Feed Pump Suction Header When Booster Pumps Are Tripped	433
5.	**Maintenance**		**436**
	5.1	Frequency of Complete Overhauls	436
	5.2	Life Between Overhauls for Boiler Feed Pumps	438
	5.3	Replacement with a More Efficient Pump	440
	5.4	Maintenance Tools	442
	5.5	Spare Parts for Boiler Feed Pumps	443
	5.6	Repair or Replace?	445
	5.7	Wearing Ring Clearances	447
	5.8	More About Wearing Ring Clearances	449
	5.9	Measuring Internal Clearances (Axially Split Casing Pumps)	450
	5.10	Measuring Internal Clearnaces (Radially Split Casing Pumps)	451
	5.11	Checking Internal Clearnaces Without Opening Pump	454
	5.12	Replacement of Casing Gaskets	456
	5.13	Further Comments on Gaskets of Axially Split Casing Pumps	457
	5.14	More Comments on Casing Gaskets	462
	5.15	Sprayed-on Casing Gaskets	466
	5.16	Straightening Pump Shafts	466
	5.17	Eroded Casing Volute Tongue	468
	5.18	How to Pack a Pump	468
	5.19	Hands-on or Hands-off?	470
	5.20	Diagnosis from Worn Stuffing-Box Packing	473
	5.21	Can Shaft Sleeves Be Reground?	474
	5.22	Use of Molybdenum Disulfide on Shrink Fits	475
	5.23	Locking Wearing Rings in Place	476
	5.24	Draining Pumps Handling Corrosive Liquids	479
6.	**Field Troubles**		**480**
	6.1	Documentation of Field Troubles	480
	6.2	Suction Lift at 5200 ft Elevation	484
	6.3	Speeding Up a Pump	484
	6.4	Why Do These Pumps Lose Suction?	490
	6.5	Loss of Capacity After Starting	493

Contents   xv

6.6	Further Comments on Loss of Capacity After Starting	495
6.7	Failure of a Pump to Deliver Any Capacity	497
6.8	Unwatering a Basin	499
6.9	NPSH Problem When Pumping from a Reservoir or How Best to Drain a Tank	501
6.10	Oxygen in Feedwater	504
6.11	Admitting Air into Pump Suction	508
6.12	Float-Controlled Valve in Pump Discharge	509
6.13	Float Switches Versus Float-Controlled Valves	512
6.14	Is Engineering an Exact Science?	513
6.15	Reversed Impeller	515
6.16	Transient NPSH Conditions	517
6.17	More on Transient Conditions in Open Feedwater Cycles	520
6.18	More About Antiflash Baffling	524
6.19	Horizontal Runs in the Suction Piping of Boiler Feed Pumps	529
6.20	Suction Piping Vibrations	530
6.21	Effect of Transients in Closed Feedwater Cycles	537
6.22	Effect of Reverse Flow	543
6.23	Reverse Rotation of Vertical Pumps	544
6.24	Zero-Speed Indicator	546
6.25	Use of Two Check Valves in Series	547
6.26	Effect of Reverse Rotation on Boiler Feed Pumps	548
6.27	Frozen Leak-off Line	551
6.28	Diagnosing Pump Troubles by Type of Noise	556
6.29	More on Centrifugal Pump Noise	557
6.30	Noisy Jet Pump Causes Concern	559
6.31	Vibration Caused by Operation at Low Flow	563
6.32	More About Operation at Reduced Flow	566
6.33	Standby Boiler Feed Pump Common to Two Units	569
6.34	Comments on Question 6.33	571
6.35	Short Ball Bearing Life	573
6.36	More About Short Bearing Life	577
6.37	Rapid Wear of Wearing Rings	580
6.38	Casing Wear	581
6.39	Wear Caused by Sand in River Water	582
6.40	Galvanic Corrosion	583
6.41	Corrosion of Cast-Iron Casings of Boiler Feed Pumps	584
6.42	Pumps Handling Seawater	586
6.43	Further Discussion of Materials for Seawater Pumps	588
6.44	Corrosion of Bronze-Fitted Cast-Iron Bowl Vertical Turbine Pumps	588

6.45	Unequal Damage to the Two Sides of a Double-Suction Impeller	589
6.46	Leakage at Boiler Feed Pump Gasket	592
6.47	Shaft Sleeve and Packing Wear	594
6.48	Source for Condensate Injection Sealing	595
6.49	Possible Contamination of Feedwater at Condensate Injection Seals	596
6.50	Maintenance Problems with Mechanical Seals	597
6.51	Centrifugal Separators	601
6.52	Short Life of Mechanical Seals	605
6.53	Acid-Handling Chemical Pumps	608
6.54	How to Limit Bearing Abrasion	610
6.55	Iron Compound Deposit on Impellers	614
6.56	Elbow Erosion in Bypass Piping	616
6.57	Noise in Boiler Feed Pump Recirculating Lines	616
6.58	Excessive Vibration in a Vertical Pump Installation	619
6.59	Effect of Frequency Reduction on Capability of Steam Power Plant Pumps	621
6.60	Overload of Motor Drivers	624
	Bibliography	627

Index   *629*

1
Application

Question 1.1   How Far to Go in Preparing Pump Specifications

We are in the process of overhauling our purchasing practices for plant equipment. In the course of each year, we place orders for a large number of centrifugal pumps of many different types and sizes. Is it sound practice to prepare very complete and strictly detailed specifications, or is it better to outline the required conditions of service and let each pump manufacturer submit a quotation on the equipment he believes will best serve our requirements?

Answer.   The question very carefully avoids giving a clue as to the present practices in this repect. Thus, it is impossible to guess whether the decision must be to liberalize and simplify the present system of dictating every detail of construction and choice of material or to tighten the reins on loose methods of purchasing whatever manufacturers' representatives recommend as suitable or desirable.

   Whatever the present policies, it appears that changes are contemplated. But this is an extremely controversial question and rather difficult to answer quite objectively. I prefer, therefore, to permit my personal feelings on this subject to color my answer to some extent, and since I preface it with this warning, it should not cause anyone to take violent exception. The points covered here are basically those contained on the manufacturer's typical data sheet (Fig. 1.1), the data entered there being the minimum amount of information required to select and build a pump.

WORTHINGTON	CENTRIFUGAL PUMP DATA			OPERATIONS NO. _____
				SALES OFFICE _____
QUANTITY	PUMP TYPE	TYPE DRIVER	RATING CURVE	S.O. NUMBER _____
				ITEM. NO. _____ PG. _____ OF _____ ADD'L NO. _____

CONDITIONS OF SERVICE	MATERIALS OF CONSTRUCTION
SERVICE _____	CHECK ONE ☐ STANDARD CLASS _____
*LIQUID _____	☐ ALL IRON ☐ ALL BRONZE
*TEMP. °F _____	☐ SPECIAL (SPECIFY)
*SP. GR. @ P.T. _____	SHAFT _____
*OTHER: PH, SOLIDS, VISCOSITY, ETC. _____	SHAFT SLEEVES _____
	IMPELLER _____
☐ EXACT DUPLICATE	IMP'L RINGS _____
☐ SIMILAR TO ORDER	CASING _____
SER. NO. _____	CASING RINGS _____

	RATED	▲MAX.	▲MIN.	GLAND _____
U.S. GPM				STAGE PIECES _____
TOTAL HEAD FT.				OTHER _____
RPM				
EFFICIENCY %				
BHP				REF. SPEC. NO. _____
NPSH - AVAIL/REQ'D				BASEPLATE UNDER: ☐ PUMP ☐ MOTOR ☐ COMMON
SUCTION HEAD FT.				BEARINGS: ☐ BALL ☐ SLEEVE & BALL ☐ SLEEVE & KINGSBURY
SUCTION LIFT FT.				LUBRICATION: ☐ GREASE ☐ OIL ☐ CIRCULATING OIL
SUCT./DISCH. PRESS. (PSIG)				COUPLING: ☐ STD ☐ L.E.F. ☐ SPACER
*IF NOT GIVEN, ASSUME COLD CLEAR WATER		▲WHEN ADD'L. C.O.S. ARE SPECIFIED		COUPLING GUARD: ☐ STD ☐ HINGED ☐ MESH
				OTHER (AUX. PIPING, ETC.)

DRIVER DATA	
☐ MOTOR ☐ TURBINE ☐ ENGINE ☐ GEAR	☐ STD BOX
☐ FURNISHED BY WORTHINGTON ☐ CUSTOMER	☐ WATER COOLED BOX
MAKE _____	PACKING:-TYPE _____
HP _____ RPM _____	MECHANICAL SEAL:
	MANUFACTURER: _____ MATERIALS OF CONSTRUCTION
· TYPE ENCLOSURE ELEC. CHAR	TYPE : _____ METAL PARTS : _____
☐ INDUCTION ☐ ODP PHASE ___	SEAT : _____
☐ SYNCHRONOUS ☐ TEFC	MFG. CODE _____ ROTARY FACE: _____
☐ VARIABLE SPEEDS* ☐ EXPROOF HERTZ ___	SEAT MOUNTINGS: _____ GLAND : _____
	VOLTAGE ___	API CODE NO. _____ API SEAL PIPE PLAN _____
*TYPE _____ BEARINGS ___ ☐ BALL ☐ SLEEVE	COOLER: SIZE _____ MANUFACTURE _____
___°C RISE AMBIENT _____ SERV. FACTOR _____	
	ROTATION FROM COUPLING END OF PUMP
	☐ CLOCKWISE ☐ COUNTERCLOCKWISE
	NOZZLE POSITION _____

ADD'L DATA: INSUL, HAZARDOUS DUTY, EFF., ETC.	CUSTOMER REQUIREMENTS
	DRAWING APPROVAL REQ'D: ☐ YES ☐ NO
	TIME _____ COPIES _____
	INSTRUCTION & PARTS LIST COPIES _____
MOUNTED BY ☐ WORTH. ☐ CUSTOMER	TYPICAL PERF. CURVE COPIES _____
CUSTOMER'S PRINT ATTACHED ☐ YES ☐ LATER**	SEND TO ☐ DIRECT "SOLD TO ADDRESS"
**NEED PRINT OR MAKE, ENCLOSURE AND FRAME BEFORE ELEVATION DRAWINGS CAN BE SUBMITTED.	ATT'N. _____
	☐ OTHER (SPECIFY)

VERTICAL PUMPS	
MTR. STAND NO. _____ Distance MTR. Floor to ℄ Casing _____	☐ PRICED RECOMMENDED SPARES LIST
VERT. SHAFTING: "A" SECT. TYPE _____ SIZE _____	
"B" SECT. NO. _____ TYPE _____ SIZE _____	TESTS: HYDRO: ☐ WITNESS ☐ NON-WITN'S ☐ NONE
FLANGE SIZES: MOTOR _____ PUMP _____	RUNNING: ☐ WITNESS ☐ NON-WITN'S ☐ NONE
GUIDE BEARINGS: DISTANCE MTR. FLOOR TO EA. BRG.	FIELD (ACCEPT.): ☐ BY WORTH. ☐ BY CUST. ☐ NONE
NO. _____ TYPE _____ 1. _____ 2. _____ 3. _____ 4. _____	OTHER _____
MISCELLANEOUS	CERT. SHOP PERF. CURVE - COPIES _____ APPROVAL ☐ YES
	REQ'D PRIOR TO SHIPPING: ☐ NO
	INSPECTION (DESCRIBE IN DETAIL)

Figure 1.1 Data sheet from the author's company listing the information required to select and build a pump.

To begin with, I do not believe that centrifugal pump specifications can ever be too explicit or too detailed where the operating conditions are concerned. There are frequently circumstances that appear unessential to the purchaser that can affect the ultimate life of the equipment very materially. I shall cite one specific instance: The specifications of a general-service vertical wet-pit pump that once came to my attention had been extremely detailed in all respects except in indicating the range of operating capacities. The inference had been that the pump would operate constantly at or about the design capacity. After the pump had been installed and very shortly after its initial operation, it was discovered that a serious shaft whip prevented satisfactory behavior of the stuffing-box packing. The pump shaft broke right under the impeller 2 months from initial operation. Investigation disclosed that 80% of the time the pump was operated in a range of flows from 25 to 10% of design conditions. This particular style of pump had a decided radial thrust in this range of capacities and was not suitable for operation except in the general range of 75 to 120% of rated flow. Had attention of the manufacturer been drawn to the actual type of operation, a pump suitable for this service could easily have been selected at very little higher cost to the user and the serious difficulties avoided.

Although we cannot treat in too great detail all the items that must be thoroughly investigated and made known to the prospective bidders, it may be advisable to list some of the essential information that should be stipulated in the specifications:

1. What is the nature of the liquid to be pumped?
    a. Is it fresh or salt water; acid or alkali; oil, gasoline, or slurry; or other?
    b. Is it cold or hot, and if hot, at what temperature? What is the vapor pressure of the liquid?
    c. What is its specific gravity?
    d. Is it viscous or nonviscous?
    e. Is it clear and free from suspended foreign matter or dirty and gritty? If the latter, what are the size and nature of the solids, and are they abrasive? If the liquid is of a pulpy nature, what is the consistency expressed in percentage or in pounds per cubic foot of liquid? What is the suspended material?
    f. What is the chemical analysis, pH value, and so on? And what variations are expected in this analysis?
2. What is the required capacity as well as the minimum and maximum amounts of liquid the pump will be called upon to deliver?
3. What are the suction conditions? Is there a suction lift or a suction head? What variations are expected in these conditions?

4. What are the discharge conditions?
   a. What is the static head? Is it constant or variable?
   b. What is the friction head?
   c. What is the maximum discharge pressure against which the pump must deliver the liquid?
5. Is the service continuous or intermittent?
6. Is the pump to be installed in a horizontal or vertical position? And if the latter,
   a. In a wet pit?
   b. Or in a dry pit?
7. What type of power is available to drive the pump, and what are the characteristics of this power?
8. What space, weight, or transportation limitations are involved?
9. Where is the installation located? This should include reference to elevation above sea level, geographic location with its effect on recommended spare parts, and immediate surroundings that might affect accessibility.
10. Are there special requirements or marked preferences with respect to the design, construction, or performance of the pump?

It is the last item that can easily become a very controversial issue. I believe that these special requirements or preferences should be listed but that the purchaser avoid too rigid an attitude toward a manufacturer who deviates from them in his quotation, provided that a sound explanation is given for the deviation. Remember that some of these preferences may be based on insufficient knowledge of the most modern practice or designs.

On the other hand, they may as likely originate from experience with pumps operating under the same conditions that the new pumps will have to meet. Much valuable information may be made available to the pump manufacturer, which will enable him to furnish that type of equipment that will give the longest and most reliable service.

But I firmly believe that customers' recommendations should be limited to their experience with pumps operating under similar conditions lest their preferences result in the purchase of very special equipment. Whenever the manufacturer's standard construction can be used, it is preferable to specially build units, both from the point of view of initial cost and that of repair parts later.

To be sure, there are circumstances that may force the customer to write very "tight" specifications. This is the case, for instance, with municipal or federal agencies. The legal problems that arise with the purchase of any equipment by these agencies dictate the use of such specifications lest the agency be constrained to purchase unsatisfactory equipment merely because the bidder's price is the lowest of all those submitted. But when such a situation does not exist, as in the case of private concerns who are not forced to buy the lowest bid, excess zeal in circumscribing the possible offerings into very narrow limits and in demanding special construction, special materials, and special tests when these are not needed does not lead to the selection of the most economical equipment.

Many customers, such as consulting engineering firms or large-volume buyers, find it advisable to develop "prefabricated" specifications for a variety of centrifugal pump services, such as boiler feed, heater drain, condensate, ash sluicing, and circulating water. This practice generally leads to a reduction in the cost of specification preparation, especially if customers frequently avail themselves of the help that can be given by pump manufacturers in reviewing these standard specifications in the light of the latest developments and experience.

In summary, I believe that centrifugal pump specifications should be as complete and detailed as possible with regard to "what must be performed by the pump" and as general as possible in restricting the manufacturer within the framework of unnecessary preferences or special treatment.

Question 1.2 Reducing Impeller Diameter to Correct Motor Overload

We have a centrifugal pump that is required to pump 20 GPM of water at 90 psi discharge pressure. The pump has a 7 in. diameter impeller and is driven by a 5 hp, 3500 RPM motor. When tested, it pumped 20 GPM at 100 psi pressure. The motor horsepower developed on this test was 6 hp. Evidently the motor is somewhat overloaded, and our discharge pressure is too high. What rules of proportion must be applied to reduce the impeller diameter to proper size?

Answer. A centrifugal pump is a velocity machine. A change in impeller diameter will change the peripheral speed of the impeller directly in proportion to this change. Furthermore, the velocities in the pump impeller and in the casing for *similar points on the characteristic curve* will vary directly in the same proportion. Therefore, the pump capacity, which is a direct function of the velocities, will vary directly as the impeller diameter ratio. The total head, which is a function of the square of the peripheral speed, will vary as the square of the diameter ratio. Finally, since the power consumption varies as the product of the head and of the capacity, the power will vary as the cube of the diameter ratio.

However, it must be remembered that, with a given impeller design, a reduction in impeller diameter will result in a change in the basic design of the impeller and thus may affect the pump characteristics. For this reason, the basic rule can be applied over only a limited range, depending on the type of the impeller. Furthermore, if the impeller diameter is reduced excessively, the pump efficiency will be somewhat affected, so that the power consumption will no longer follow the cube of the diameter ratio relationship.

In the particular case at hand and presuming that the suction lift or suction head is negligible (you give no value for this item), the efficiency of the pump will probably change very little if the impeller diameter is reduced to give 90 psi discharge pressure instead of 100 psi. The shape of the head-capacity curve at 20 GPM is probably very flat at the present operating point, so that the diameter ratio can be *approximated* from the square root of the two pressures. This results in an impeller diameter of about 6 5/8 in. After the impeller is cut, the vane tips should be filed very carefully to approximate the finish of the vanes as they now appear. Presuming that the impeller cutdown does not exceed reasonable limits for this pump (and this would not be likely), the power consumption at 20 GPM would vary directly with the pressure. (Note that in this case the power does not vary as the cube of the impeller diameter, since we do not change the capacity at the operating point.) Thus, the power consumption at 20 GPM after the cutdown can be estimated at 5.4 hp.

Note that when larger pumps and operating conditions farther to the right of zero flow are dealt with this approximation is not quite correct. It becomes neessary to determine by trial and error, on the slide rule, the diameter ratio that will step up the desired capacity (directly as the diameter ratio) and the desired total head (as the square of the ratio) to a set of Q and H conditions right on the test curve of the pump. Having established this ratio, it becomes quite easy to plot the resulting head-capacity curve directly from the original one. One simply sets the ratio on the lower scale of the slide rule, and several points on the test curve having been selected, the capacities are stepped down directly on the same scale and the total heads on the square scale.

Question 1.3 Effect of Pump Speed Change

If the pump in the preceding question were belt driven, could the speed be changed to obtain the same result?

Answer. Since it is the peripheral speed of the impeller that determines the total head and the capacity of the pump, obviously it is immaterial whether the peripheral speed is changed by cutting down the impeller diameter or the pump speed. In this particular case, if the pump developed 100 psi net pressure at 3500 RPM, the speed would have had to be reduced in the same proportion as we found it necessary to cut the impeller diameter, in other words to 3320 RPM.

Chapter 1

The important formulas to remember are, therefore,

$$\frac{GPM_2}{GPM_1} = \frac{D_2}{D_1} = \frac{\sqrt{H_2}}{\sqrt{H_1}} = \frac{\sqrt[3]{bhp_2}}{\sqrt[3]{bhp_1}}$$

and

$$\frac{GPM_2}{GPM_1} = \frac{RPM_2}{RPM_1} = \frac{\sqrt{H_2}}{\sqrt{H_1}} = \frac{\sqrt[3]{bhp_2}}{\sqrt[3]{bhp_1}}$$

Question 1.4 Calculation of RPM for Different Conditions of Service

How does one calculate the speed at which a given pump must be operated to meet a head and capacity condition other than those corresponding to the performance at the speed for which test curve data are available? Is this done by a trial-and-error approach, or is there a simpler method to derive the desired answer? Can a computer be used to calculate such speed changes?

Answer. These calculations are required quite frequently and can be carried out very simply through the use of the so-called affinity laws. As stated in question 1.3, when the speed is changed (1) the capacity for any given point on the pump characteristic curve varies directly as the speed, *and at the same time*, (2) the head varies as the square of the speed while (3) the brake horsepower varies as the cube of the speed. In other words, if subscript 1 is given to the conditions under which the characteristics are known and subscript 2 denotes the conditions at some other speed, then

$$\frac{Q_2}{Q_1} = \frac{n_2}{n_1}, \frac{H_2}{H_1} = \left(\frac{n_2}{n_1}\right)^2$$

and

$$\frac{P_2}{P_1} = \left(\frac{n_2}{n_1}\right)^3$$

These relations can be used safely for moderate speed changes such as those that occur when a pump is operated at variable speed. They may not be as accurate in large speed changes, when the ratio of speeds approaches 2:1 or higher.

To illustrate the process involved in calculating the required speed for some set of capacity and head conditions that does not fall right on the available H-Q curve, let us assume that we are dealing with the known pump performance as given on Fig. 1.2 at a speed of 1800 RPM. We want to determine the speed at which this pump would have to run to deliver 2000 GPM and a total head of 150 ft. The steps required are as follows:

1. Select an arbitrary capacity greater than the 2000 GPM, such that it will be located on the parabola defined by the affinity laws:

$$\frac{Q_3}{Q_2} = \frac{n_3}{n_2}, \frac{H_3}{H_2} = \left(\frac{n_3}{n_2}\right)^2 = \left(\frac{Q_3}{Q_2}\right)^2$$

Say, for instance, that $Q_3 = 2500$ GPM. Then,

$$H_3 = H_2 \left(\frac{Q_3}{Q_2}\right)^2 = 150 \left(\frac{2500}{2000}\right)^2 = 234.4 \text{ ft}$$

2. Draw the portion of the parabola defining the affinity laws between 2000 GPM, 150 ft and 2500 GPM, 234.4 ft. It can be assumed that this is essentially a straight line.
3. The intersection of this straight line with the head-capacity curve of the pump at 1800 RPM corresponds to 2395 GPM and 215 ft. We can now determine the speed required to meet the desired conditions of 200 GPM and 150 ft:

$$n_2 = 1800 \sqrt{\frac{150}{215}} = 1503 \text{ RPM}$$

In practice, I would round this off to 1505 RPM.

Of course, I could have calculated the required speed by multiplying 1800 RPM by the ratio between 2000 and 2395 GPM (the capacity at which the section of the parabola I constructed intersects the head-capacity curve at 1800 RPM). The reason I prefer to use the square root of the ratio between the heads is that this reduces by about one-half the possible error in constructing the parabola or in reading the intersection point.

The method I have described is used in drawing a curve of speed and power consumption for a pump that, when operated at variable speed, is made to meet the requirements of a system-head curve. Several points on that curve are used to calculate the necessary data, and the results may be used in the evaluaton of the power savings that would accrue from variable-speed operation.

Of course, you must remember that the speed calculations so derived are not necessary for the control of the pump speed. This is generally done

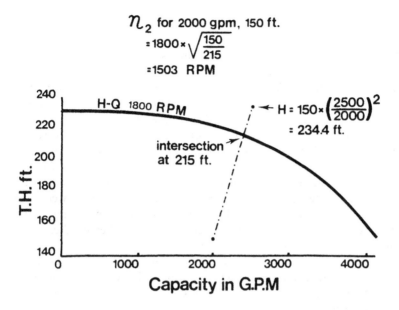

Figure 1.2 Calculation of speed change.

completely automatically, with the particular characteristic that dictates the speed requirement providing the trigger mechanism to the control mechanism.

Similar affinity laws exist for changes in impeller diameter, within reasonable limits of impeller cutdown. In other words,

$$\frac{Q_2}{Q_1} = \frac{D_2}{D_1}, \quad \frac{H_2}{H_1} = \left(\frac{D_2}{D_1}\right)^2$$

and

$$\frac{P_2}{P_1} = \left(\frac{D_2}{D_1}\right)^3$$

Some deviation occurs even with relatively modest cutdowns, and Fig. 1.3 shows the recommended cutdown related to the theoretical cutdown for radial flow impellers, expressed in percentage form.

In the example just given, assuming that the original impeller was 14.75 in., the new impeller would have to be cutdown to a ratio of $\sqrt{150/215}$, or 83.3% theoretically. Applying the correction indicated on

Fig. 1.3, however, we would cut the diameter to 85.5% of its original diameter, or 12.6 in.

It must be noted, however, that although speed variation can be used to meet a variety of temporarily required conditions of service, like most surgical procedures the cutdown of an impeller is irreversible. Therefore, this is used only when a permanent change in operating conditions is required.

Of course, a computer can be used for these calculations, except that this seems to be akin to using a sledgehammer to kill a gnat. Because there is a simpler way to find the answer, I have not given much thought to the exact nature of the computer program one would have to create for this purpose. But in general terms, I assume that the computer would be instructed first to create a mathematical model for the head-capacity curve at whatever speed the pump will have been tested. Then the computer would be given a set of instructions to construct a series of other head-capacity curves using the affinity laws and then continue superimposing these newly generated curves over the grid until one of the curves would go right through the point of head and capacity at which one wishes to determine the speed. Now, at that moment the computer would apparently stop and give you the answer that in order to meet that head and capacity this particular pump has to run at such and such a speed. The process might be really fairly rapid as far as the computer is concerned, but is this necessary? It seems to be rather complex, but the means that I've shown for determining the speed for any given head and capacity are really not that complicated.

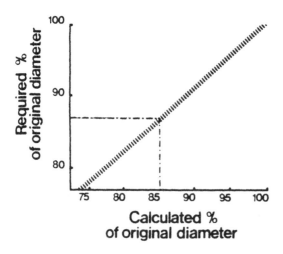

Figure 1.3 Corrections for impeller cutdown.

Chapter 1

On the other hand, when the process of correcting capacity, head, and power readings to a preselected speed is involved on a repetitive basis, as in the case of tests conducted at the manufacturer's test stand, a computer is nowadays frequently used for the conversion of readings taken at various points of the pump curve. Thus, one can instantaneously produce a complete pump curve at the rated constant speed, even though the individual readings may have been taken at speeds deviating somewhat from this rated speed.

Question 1.5   More on the Effect of Pump Speed Change

Do the affinity laws for changes in speed hold over any and all speed ratios?

Answer.   Obviously, they do not. There are a number of individual pump characteristics, particularly those having to do with various pump losses, that are responsive to such parameters as the Reynolds number. Therefore, very extreme reductions in pump speed will not result in a strict adherence to the affinity laws. On the other hand, for all practical purposes—either in the carrying out of pump tests or in using speed changes to alter the pump performance—the speed changes one normally encounters are such that to all intents and purposes you can assume that the affinity laws are not seriously violated.

There is one qualifying statement that I should like to make about this: when testing large horsepower pumps at reduced speeds, one should consider the use of correction factors in predicting the pump efficiency at full speed. Parenthetically, I should like to note a paradox in the recommendations of the Hydraulic Institute in this connection. Its Standards recognize the effect of the changing the Reynolds number when tests are carried out at temperatures other than the design conditions and provide a formula for correcting test efficiencies for this effect. But they do not give any such recognition to the effect of speed changes that *do* change velocities and, therefore, the Reynolds number. Yet this number does not "know" whether it has changed because of a change in the kinematic viscosity of the liquid or because of a change in velocity. Whether this paradox is eventually recognized and eliminated is anybody's guess.

Question 1.6   Pump Capacity Versus Impeller Diameter

I would like to refer to a statement that has appeared in a number of articles on centrifugal pumps that deals with the effect of cutting down the impeller diameter on the capacity and head produced by a centrifugal pump. It is stated that for similar points on the characteristic curve the pump capacity will vary directly as the impeller diameter, but the head will vary as the square of this impeller.

This statement seems to be contrary to general belief. I should like to point out that the discharge (or capacity) varies as both the area and the velocity. Reducing the impeller diameter would reduce both the area at the impeller exit and the velocity. Should not the capacity, therefore, vary as the square of the impeller diameter instead of directly as this diameter?

Answer. The functional relationship between a change in impeller diameter and the resulting change in capacity may be readily established in either of the two following ways:

1. Empirically, by analysis of actual test results
2. Theoretically, by comparison of discharge velocity triangles or other analytical means

In either case the results substantiate the fact that capacity varies directly with change in diameter, subject to slight variations that will be discussed later. Considering the empirical approach first, the substantiating evidence lies in literally thousands of tests that have been made and are still being regularly conducted by both pump manufacturers and pump users. Although a review of any substantial number of such tests covering a wide range of different pump types would undoubtedly turn up some designs that would indicate a capacity change greater than that expected, it would also reveal some showing a smaller change than expected and on the whole would indicate that the large majority performed very closely in accordance with the rule that capacity varies directly with diameter.

Insofar as theoretical proof is concerned, there would probably be little value in and even less need for a complete theoretical explanation in this book, since this can readily be found in textbooks on pump design. However, I would like to comment briefly on your statement regarding reduction of discharge area with reduction of impeller diameter.

Assuming for purposes of discussion a radial-flow closed impeller with parallel shrouds, it is obvious that *circumferential* area is reduced in direct proportion to any reduction in impeller diameter. However, this area does not represent the effective discharge area of the impeller and therefore does not exert the principal controlling effect on pump capacity. If such an impeller is considered as a series of rotating flow channels (each channel being formed by the upper side of one vane, the underside of an adjacent vane, and the inside surfaces of the shrouds), it is then apparent that the *effective* discharge area is equal to the normal area between any two vanes times the number of vanes. Referring to Fig. 1.4, this area is represented by A × B for full diameter and by A × B' at the reduced diameter. In most impellers these two areas will remain very nearly or exactly the same for the

Chapter 1

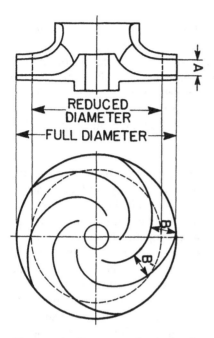

Figure 1.4 Geometry of reducing impeller diameter.

range of diameters over which the impeller is normally applied. Since this area remains constant, capacity then varies only as velocity or, in other words, directly with impeller diameter.

The fact that variations from this rule are sometimes observed during pump tests does not in any way detract from its validity but rather points up the variations and limitations encountered in practical pump designs. For example, the necessity for obtaining special characteristics of pump performance, such as an unusually flat or unusually steep characteristic curve, may sometimes force the designer to distort the impeller areas or configuration from the optimum, thus affecting similarity relationships. In addition, internal losses in a pump prevent us from measuring the performance of the impeller itself and influence overall performance to vastly different degrees in different designs. These and a number of other similar factors all serve to illustrate why deviations from the general rule are to be expected.

Question 1.7 Narrow and Wide Impellers

I have read somewhere that the capacity of a centrifugal pump varies approximately with the width of the impeller. That is, if the impeller is 3/4 in.

wide and the head capacity is like that shown in Fig. 1.5, then if the impeller were made only 5/8 in. wide on the outside diameter, the head curve would fall off in the ratio of 5/8:3/4, or 5/6. Now I would like to know if the efficiency curve for the 3/4 in. wide impeller is as indicated on the sketch, what will it be for the 5/8 in wide impeller? Will the whole efficiency curve shift to the left, or will it remain the same as for the 3/4 in. wide impeller but fall of in the higher capacity range as I have indicated (see Fig. 1.5)?

Answer. The problem involved here is somewhat complex, and it is difficult to make general rules that would apply to any and all designs. Although theoretically the head-capacity curve would fall off in the ratio of the impeller widths as you indicate, this would be true only within a narrow range of widths. Should too narrow an impeller be used in a given pump casing, there could take place an excessive amount of turbulence and shock losses, which not only would tend to reduce the capacity of the pump but also may react unfavorably on the head generated by the pump. Therefore, the head at shutoff or zero delivery may be less than that produced by the full-width impeller. If the reduction in width is not excessive, however, the head at shutoff will be approximately the same, as it can be controlled within certain limits by proper design of vane exit angles and other design factors.

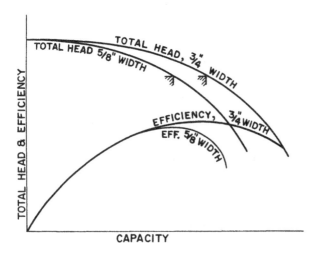

Figure 1.5 Assumed effect of narrowing impeller width on pump performance.

The best efficiency point of a narrow impeller will move to a lower capacity approximately in the same proportion as the head-capacity curve, *provided* that other portions of the impeller are properly adjusted in design. In other words, it is also necessary to reduce the inlet area between the vanes. Another way of expressing this is that a factor less than 1.0 is applied to the entire impeller, after which the impeller is extended to its original diameter. The casing design may also affect the behavior of the impellers. If, for instance, the wider impeller is being severely throttled in the casing itself, the narrow one may show very little change in performance.

Except for very small changes in width, the best efficiency will be decreased somewhat by narrowing the impeller. This decrease is caused by three separate factors:

1. Increasing turbulence and shock losses.
2. Increase in the proportion between the disk horsepower and the useful water horsepower.
3. Increase in the proportion between the leakage and mechanical losses and the useful water horsepower.

The disk horsepower is that power required to drag the impeller through the liquid surrounding it and is caused by the friction of the liquid against the shrouds (or walls) of the impeller. This loss will remain essentially constant, regardless of the width of the impeller, assuming that the speed and the impeller diameter remain constant. However, the decrease in the pump capacity for any given total head will decrease the net output of the pump, and therefore the proportion of the losses to the net output will increase, lowering the pump efficiency.

In the same manner, with the same differential pressure across the wearing rings, the leakage losses remain constant. So do the mechanical losses in the bearings and at the stuffing boxes. The overall effect can best be visualized by examining the formula for pump efficiency:

$$e = \frac{\text{water hp}}{\text{brake hp}}$$

where brake hp is made up of the sum of the following elements:

Water horsepower
Hydraulic losses
Disk horsepower
Leakage losses
Mechanical losses

If we reduce the water horsepower but keep the disk horsepower, leakage losses, and mechanical losses constant, the efficiency will obviously be reduced.

The aforementioned should not be interpreted to mean that narrow impellers should not be or are not used. As a matter of fact, in many cases a standard line of pumps will include two groups of impellers: One (called the 100% impellers) is designed to utilize the pump casings fully, and the second group (called the 80% impellers, for instance) is used to move the best efficiency point to a lower capacity. In this manner, greater coverage is obtained from a line of pumps.

Assuming that the narrower impeller has been designed with proper attention to the necessary details, the performance of the pump will be similar to that indicated in Fig. 1.6.

Question 1.8   Impeller Cutdown for Multistage Pumps

We have a two-stage centrifugal pump that we intend applying in another location and on different service. The requirements of head and capacity are lower than the original pump rating, and we plan to cut the impeller diameter. Should the cut be made on both impellers equally, or can a bigger cutdown be made on one impeller only?

Answer.   This depends in part on the particular two-stage pump in question and in some cases on the type of service for which the pump is used. Considering these factors, our answer to this question could be any of the three following:

1. Impellers must be cut down equally.
2. It is preferred but not essential that impellers be cut down equally.
3. It is preferred that the entire cut be made on the second-stage impeller.

Probably the simplest way to illustrate when each of these answers is applicable is to show sectional assembly drawings of three different pumps and to discuss each in turn. Thus, Fig. 1.7 is typical of a style of pump widely used for capacities up to approximately 2000 GPM and heads up to 1000 ft. In this design, each of the impellers is axially unbalanced, since the back shroud is subject to a pressure nearly equal to that in the volute over its full extent, but the front shroud is subject to an equal pressure only down to the diameter of the wearing ring. This leaves a net unbalanced load acting toward the impeller eye from the center of the pump. However, since the two impellers are mounted back to back, the unbalanced loads offset each other, and the pump remains in axial balance provided that the head developed by each impeller is equal. If one impeller is smaller than the other, the axial balance is destroyed, and since the pressures developed by each stage may be fairly high, an axial load of several hundred pounds

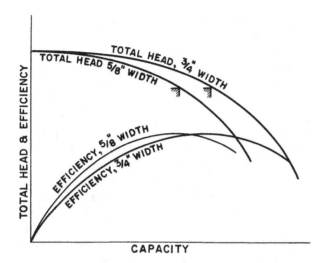

Figure 1.6 Actual effect of narrowing impeller width on pump performance.

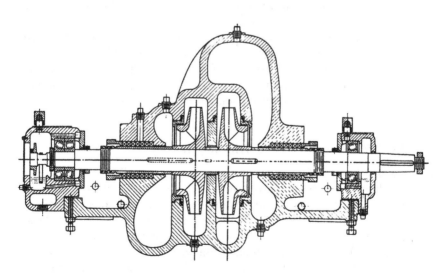

Figure 1.7 Both impellers of this two-stage pump are axially unbalanced, but as they face in opposite directions their individual unbalanced axial thrusts balance each other.

may be imposed on the thrust bearing. Thus, with a pump of this type, impellers must be cut down equally.

The two-stage pumps shown in Fig. 1.8 is a type generally used for larger capacities, and since a double-suction impeller is used in each stage, neither creates any axial unbalance. Thus, there is no mechanical restriction here on the procedure to be used. It should be noted, however, that when an impeller is cut down, it generally becomes less efficient. Furthermore, the rate of efficiency reduction also increases with decreasing diameter. Thus, for a cut of any significant amount, the pump will perform more efficiently if both impellers are cut equally. This is therefore the preferred procedure.

Figure 1.9 shows a typical two-stage condensate pump. In this design the impellers are mounted face to face in order to subject the stuffing boxes to the discharge pressures of the respective stages, thus minimizing the possibility of air leakage through the stuffing boxes into the pump. Because of the breakdown in total head that occurs in a condensate pump when operating on submergence control, the first-stage impeller may be partially (and in some cases completely) ineffective at low capacities. To avoid a complete and rather lengthy description of the principles of condensate-pump application, we shall simply state here that under these circumstances it is sometimes desirable to maintain the first-stage impeller at or near maximum diameter. It is therefore preferable that the entire cut be made on the second-stage impeller. The resulting axial unbalance is no problem here, since the thrust bearing must be designed to take the entire unbalance of one stage anyway owing to the possibility of complete breakdown of head in the first stage.

The examples provided cover the conditions most frequently encountered in cutting impellers in two-stage pumps but are by no means the only possibilities. They do, however, suffice to show that no one procedure will be best in all cases and that when any doubt exists, the manufacturer of the particular pump in question should be consulted.

Question 1.9 How to Cut and Trim Impellers

I have a question relative to cutting or trimming impellers.

We have a number of large two-stage centrifugal pumps in crude-oil pipeline service operating at approximately 900 ft head and 7500 GPM. From time to time impellers must be cut either to meet reduced head-capacity conditions without throttling or to prevent motor overload. Specifically, I would like your opinion on whether (1) impellers should be "undercut," that is, just the vanes turned down to the new diameter and the shrouds left at the original diameter, or (2) both the shrouds and vanes should be cut to the new diameter.

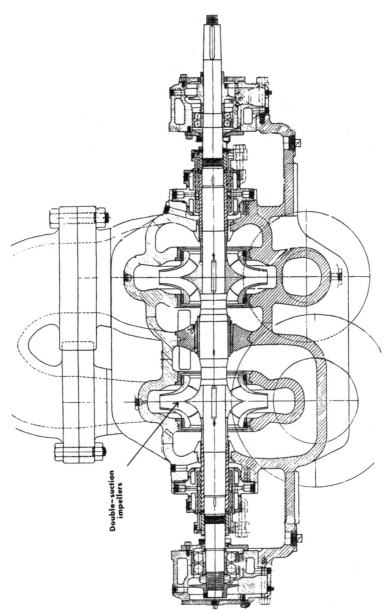

Figure 1.8 Two-stage pump with double-suction impellers for large capacities.

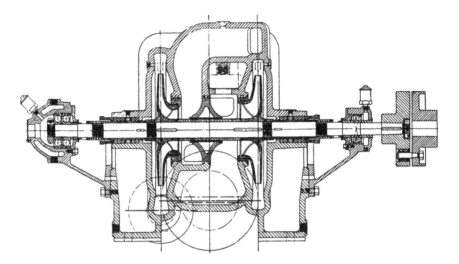

Figure 1.9 Typical two-stage condensate pump.

Answer. It is definitely recommended that when impellers must be cut down to reduce the head-capacity curve of a centrifugal pump, the impeller shrouds be reduced to the same diameter as the impeller vanes in order to get the maximum benefit in power reduction from the cutdown.

Should only the impeller vanes be cut down in diameter, the useful work of the impeller would be reduced as desired, but the disk friction horsepower would remain unchanged, and the overall pump efficiency would be unnecessarily reduced.

This reduction can be quite significant because the disk horsepower varies with the fifth power of the impeller diameter, following the relation

$$hp_{disk} = Kn^3D^5$$

where K = experimental factor
n = pump speed in RPM
D = impeller diameter in inches

Let us examine the significance of the possible reduction in efficiency in relation to the particular problem you have outlined. Assuming that the crude oil handled by your two-stage pumps has a specific gravity of 0.80 and that the pump efficiency is 85%, the power consumption at 7500 GPM and 900 ft head is 1600. Your pumps probably operate at 1750 RP, and each impeller has a diameter of 22 3/4 in. For this specific speed of impeller (impeller type) the disk horsepower is approximately 5% of the total power input, or 80 bhp.

Let us assume that it is desired to reduce the pump head-capacity curve so that we can meet an operating condition of 7000 GPM at 800 ft. The impeller can be cut down to approximately 21 1/2 in., as shown in Fig. 1.10. If the impeller shrouds were cut down with the vanes, the disk horsepower would be reduced as the fifth power of the impeller diameter and

$$\text{New disk hp} = 80 \times \left(\frac{21.5}{22.75}\right)^5 = 60$$

Failing to cut down the impeller shrouds would therefore cause an increase in disk horsepower of 20. Assuming that at the new conditions the efficiency was still 85%, the power consumption would have been 1330 hp, and a 20 hp increase in power consumption is equivalent to a loss of 1.5% in efficiency.

After an impeller has been cut down, it is necessary to taper the vanes at their discharge tips so as to restore the tip widths to their original dimensions. This is generally done by *overfiling*, that is, filing the impeller vanes on their leading edge, as in Fig. 1.11. In some cases it may be advantageous to *underfile*, that is, remove metal from the underside of the vane, as in Fig. 1.12. The latter method is frequently used when a slight increase in capacity is desired. The filing should be blended into the impeller waterway to a minimum distance as indicated in Table 1.1.

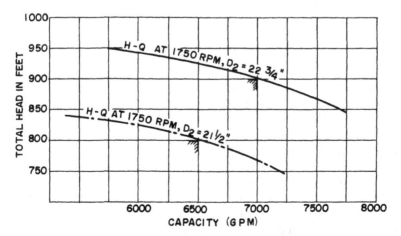

Figure 1.10 Effect of trimming impeller diameter on pump head-capacity curve.

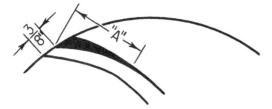

Figure 1.11 Overfiling impeller tips after cutdown.

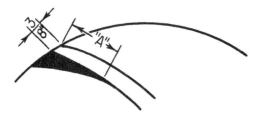

Figure 1.12 Underfiling impeller tips after cutdown.

It is recommended to rebalance the impellers after they have been cut down. Mount the impeller on an arbor, placing the ends of the arbor on parallel and level knife edges. If the impeller is out of balance, it will turn the arbor, and the heavier portion of the impeller will come to rest at the bottom. Metal must be removed from this portion of the impeller by mounting it off center in a lathe and taking a cut from the shroud, deepest at the periphery as in Fig. 1.13.

Question 1.10 Field Test for Required NPSH

In the near future, I shall be checking out a centrifugal pump that, to date, has not been meeting the capacity at the desired head. Part of the test will be to determine the required NPSH. My question at this time concerns this phase of the test.

Is there a published procedure for this part of the test, or could you give me a step-by-step explanation? Assume that the suction and discharge piping are correct as well as the valves and the instrumentation and the pump is below the liquid supply level. Also assume that we have the head-capacity curve established as well as the horsepower required.

I would greatly appreciate any information you could supply to enlighten me on this phase of the test.

Table 1.1 Blending of a Filing After Cutdown

Impeller diameter (in.)	A, distance of blend (in.)
≤ 10	1½
10 1/16–15	2½
15 1/16–20	3½
20 1/16–30	5
≥ 30	6

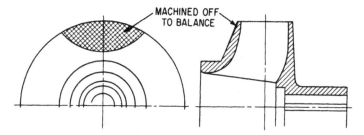

Figure 1.13 Out-of-balance impeller can be corrected by mounting off center in lathe and taking cut from shroud.

Answer. If I understand your question correctly, you have run a regular head-capacity-power test on this pump already and are satisfied that your measurements are reasonably accurate, or you still have to run such a test but, again, consider that the test will yield sufficiently accurate results. Your problem, therefore, remains to set up a test procedure that will give you equally accurate information on the effect of available NPSH on the pump performance.

The Standards of the Hydraulic Institute include a section outlining recommended methods for conducting *cavitation* tests, in other words, for establishing the minimum NPSH (net positive suction head) for satisfactory operation.

The available NPSH is a characteristic of the system in which a centrifugal pump works and is the difference between the absolute suction head and the vapor pressure of the liquid at the pumping temperature. Figure 1.14 illustrates the method of calculating the available NPSH for a pump that takes its suction from a supply above the pump centerline. If the supply is open to atmosphere, the value of P_s is, of course, equal to the atmospheric pressure.

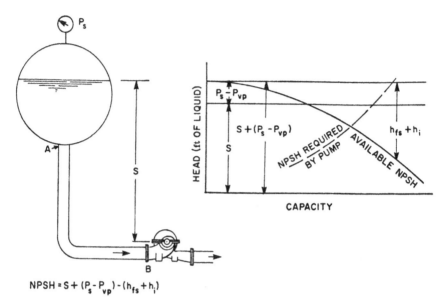

NPSH $= S + (P_s - P_{vp}) - (h_{fs} + h_i)$

Figure 1.14 Calculations of available NPSH; h_{fs} = friction loss in suction line from A to B; h_1 = entrance loss at A; P_{vp} = vapor pressure of liquid at pumping temperature. All heads and pressures must be expressed in feet of liquid at the pumping temperature with proper algebraic signs. Although P_s and P_{vp} can be in either gauge or absolute values, they must both be measured above the same datum.

If the available NPSH is less than the minimum required by the pump at the desired capacity, the pump will be unable to meet its head-capacity conditions. A typical group of performance curves for a pump operating under varying suction conditions is shown in Fig. 1.15.

You will note that the reduction in head for any specific suction limitation is not abrupt; in other words, the head-capacity curve does not coincide with the curve with ample excess NPSH up to some capacity and then break off suddenly. Partial cavitation starts at some capacity lower than the complete breakdown, and the head-capacity curve starts to depart slightly, then more and more, from its normal shape. Thus, operation at point A, for instance, may result in some reduction in head from the head developed with greater NPSH.

The determination of critical suction conditions in accordance with the Hydraulic Institute Test Code is tied with the use of cavitation coefficient σ (sigma), defined as

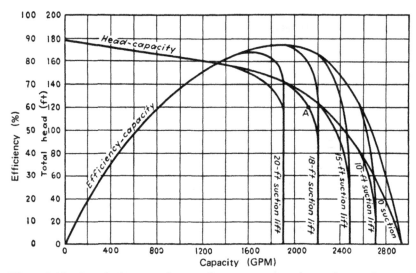

Figure 1.15 A typical group of curves for a pump operating under varying suction conditions.

$$\sigma = \frac{h_{sv}}{H}$$

where h_{sv} = net positive suction head
H = total pump head per stage

The critical value of sigma at which cavitation begins is found by operating the pump at constant speed, correcting all values to constant RPM conditions, and plotting the corrected head and the effficiency against sigma as shown in Fig. 1.16.

In the higher range of sigma values, both the efficiency and the head will remain substantially constant. But as sigma is reduced, a point will be reached where the curves break away from a substantially horizontal line. This departure indicates the beginning of cavitation.

Such a test is relatively easy to carry out in a laboratory. It may be somewhat more difficult for an existing installation. For instance, it may not be possible to vary the static elevation between the supply and the pump centerline, to install a stilling chamber in the suction as recommended by the Hydraulic Institute, or to arrange a closed circuit for the pump.

However, if we understand the meaning of "available NPSH," and if we take certain precautions in varying it during the test, we can still obtain a reasonably accurate plot similar to that illustrated in Fig. 1.16. You can

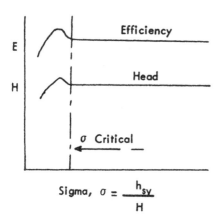

Figure 1.16 Critical value of sigma.

establish the available NPSH by direct measurement once the vapor pressure of the liquid at the pumping temperature is known. The method recommended for measuring the suction head is outlined in the Hydraulic Institute Standards, Test Code Section.

If you are limited in your ability to vary the static elevation, you may be in the position of varying the vapor pressure of the liquid. This will be the case, however, only if you take suction from a tank open to atmosphere, so that the value of the P_s-P_v in Fig 1.14 can be varied. In the case of a boiler feed pump taking its suction from a deaerating direct-contact heater, P_s is always equal to P_v. Thus, changing the operating temperature here does not change the available NPSH, which is equal to the static submergence less friction losses.

Failing to accomplish your ends by varying the static elevation or the vapor pressure, you will be forced to throttle the suction, even though the Hydraulic Institutes does not recommend this procedure unless a stilling chamber can be provided downstream of the throttling valve. If this throttling takes place as far as possible from the pump, the results will not be affected to a significant degree, although they cannot be termed absolutely accurate.

The test should be run at the rated or guaranteed capacity, adjusting this capacity as necessary. Both efficiency and head should be plotted against sigma, as shown in Fig. 1.16.

I have assumed that you are dealing with water. The problem of cavitation and determination of adequate limits when liquids are being handled that are mixtures of several compounds each having different vapor pressures and latent heats, such as hydrocarbons, is considerably more complex. For more detailed information on the effect of nonhomogeneous liquids on cavitation, I refer you to Question 1.23.

Question 1.11 NPSH Tests

I would like to present to you a problem that arose recently in our plant.

It was decided to remove a number of centrifugal pumps of a laid-up plant and to reuse them for service in one of our new plants now under construction. It has become necessary, therefore, to find out the required NPSH of each pump in order to determine its suitability for its intended service.

Apart from instrumentation, the equipment available for pump testing consists of a test basin divided into two parts and connected by a sluice gate. The test procedure that I decided upon will be described. However, I am not fully satisfied with the accuracy that can be expected from such a setup (see Fig. 1.17).

Test Procedure

1. The inlet valve is slightly open.
2. The pump is started against a closed discharge valve.
3. The discharge valve will be opened gradually, until the pump loses suction.
4. At this point, the following readings will be recorded: (a) pressure (or head) in the discharge pipe, (b) capacity delivered (GPM), (c) suction pressure, (d) barometric pressure, (e) water temperature.
5. The setting of the inlet valve will be altered, and the whole procedure will be repeated at five or six different inlet valve settings.
6. The NPSH will be calculated from the readings recorded in step 4, and the NPSH curve can be drawn for the entire range of capacities.

I am not fully satisfied with this arrangement and would appreciate it very much if you could give me your comments and/or suggestions.

Answer. In general, the test setup you indicate is probably as satisfactory as you can achieve under the circumstances. I would suggest, however, replacing the wire gauze (item 9 in Fig 1.17) by some form of stilling chamber.

I assume that the purpose of the wire gauze is to provide a measure of *stilling* in the suction pipe so as to reduce the turbulence created by valve 2. I am not certain that you will achieve the purpose in this manner. I believe it would be preferable to use an enlargement in the pipe between 2 and 4 to

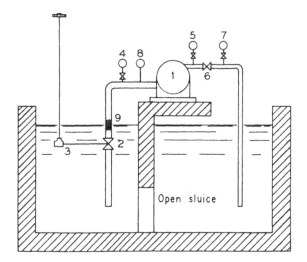

Figure 1.17 Proposed test setup: (1) pump to be tested, (2) gate valve in suction line (below water level), (3) linkage and extended valve spindle, (4) vacuum gauge in suction line, (5) pressure gauge in discharge line, (6) throttling valve, (7) flowmeter, (8) temperature indicator, (9) wire gauze.

form a stilling chamber as described in the Standards of the Hydraulic Institute (see Test Code Section). This enlargement in the suction pipe should contain screens or baffles to dissipate the turbulence from the throttle valve and to distribute the flow evenly to the end that the pump takes a flow free from undue turbulence.

Your test procedure could likewise be improved. Complete loss of suction, as you intend to establish in step 3, is not a proper means of establishing required NPSH, since it corresponds to an excessive reduction of NPSH.

I have described a recommended test procedure for required NPSH in Question 1.10. In brief, by accepted definition, the beginning of cavitation is stated to have taken place when, operating the pump at constant speed, efficiency and/or total head plotted against sigma departs materially from substantially horizontal lines. Thus, instead of changing the flow by throttling valve 6, you should attempt to establish the value of critical sigma at several different flows by manipulating valve 2, adjusting valve 6 correpondingly to attempt to maintain a fixed capacity for the point being tested.

Question 1.12 How to Measure Suction Lift and Suction Head

We shall shortly be running an acceptance test on a new pump and, in attempting to set up the instrumentation for this test, have raised some questions on which there seems to be some difference of opinion among various people in our own organization. We would appreciate any comments you would be willing to offer on the following points:

1. In establishing suction lift, is it necessary to add velocity head to the gauge reading to determine the actual lift?
2. If this pump were operating with head on suction instead of with a suction lift, what effect would this have on the answer to the preceding question?

Answer. That a certain degree of confusion is likely to exist in connection with the question you have asked is adequately demonstrated by the frequency with which it recurs. This confusion undoubtedly stems largely from the natural and accepted practice of using atmospheric pressure as the reference point for other pressure measurements. Thus, with pressures above atmospheric, we deal always with positive values, and no confusion arises, since our thinking proceeds along the lines of straightforward arithmetical addition and subtraction. In dealing with pressures below atmospheric, however, we are confronted with negative values and must resort to algebraic addition and subtraction. Although this is undeniably elementary, the transformation of the physical concepts into mathematical quantities seems to introduce just enough complication to confuse the issue.

This being the case, the obvious solution would be to use only absolute pressures. This is a useful concept, and if it helps in keeping things straight, there is no objection to its use. It is not customary, however, and since it is by no means essential, the following may serve to clarify the problem.

In discussion suction lift (or suction head), it is useful to keep in mind the basic purpose of any pump in any system, which is simply to add energy to the fluid stream. Thus, the work done by the pump is represented by the increase in the energy of the fluid stream as it passes through the pump or, in other words, by subtracting its energy at the inlet from its energy at the outlet.

These measurements of energy, when applied to a unit of mass of the fluid, are referred to as measurements of *head*, expressed in feet of the liquid pumped. Thus, the *total head* of a pump is simply the difference between the *total discharge head* and the *total suction head*. If all pressure measurements are expressed as absolute values, these two terms are always positive, and total head is determined by simple subtraction.

When pressures are expressed in gauge values, however, the total suction head is frequently negative. Under these circumstances, we must be careful to subtract algebraically in order to determine total head as follows:

$$H = H_d - (-H_s) \tag{1.1}$$

where H is total head and the subscripts d and s refer to discharge and suction, respectively. Numerically, of course, this is equivalent to

$$H = H_d + H_s \tag{1.2}$$

Thus, in effect, the total suction lift (negative suction head) is added to the total discharge head to obtain total head.

From the foregoing, it should be evident that suction head and suction lift are treated in exactly the same fashion but that in dealing with a lift we must be careful to observe correct usage of the algebraic sign conventions. Having established this, we are faced now with only one remaining necessity, which is to examine the individual components of *any* head measurement.

These components are simply the three forms of energy that were represented by Bernoulli in his well-known equation of potential energy, pressure energy, and velocity energy. In pump test work, the first of these is accounted for by referring all head measurements to a common elevation. Pressure energy is measured by means of a Bourdon gauge, mercury manometer, or some other form of pressure-sensing device. Velocity energy is usually obtained by calculation.

Since gauges will generally remain in fixed positions during any test, the gauge elevation corrections (potential energy) can be worked out in advance and will remain constant for all readings. Pressure and velocity energy, on the other hand, will vary with capacity. Thus, for each test point, our head readings must include both terms. In the case of suction head,

$$H_s = H_{sp} + H_{sv} \tag{1.3}$$

where the subscripts p and v refer to pressure and velocity, respectively (see also Fig. 1.18).

In the case of a suction lift, we have a negative suction head, but since the velocity head component must always be positive, this expression becomes

$$-H_s = -H_{sp} + H_{sv} \tag{1.4}$$

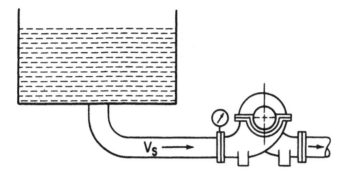

Figure 1.19 Total suction head. Gauge reading (corrected to pump centerline) = Sp, $V_s^2/2_g$ (at point of gauge attachment) = H_{sv}. Total suction head $S = Sp - H_{sv}$.

This equation indicates simply that the suction gauge reading shows a negative head that is lower than the true (or total) negative suction head by the amount of the velocity head. However, in reading a suction gauge or in speaking of suction lift, we are referring to a measurement below atmospheric, and hence the gauge indication must show a higher numerical value to denote this lower absolute value. Thus, if we were to indicate suction lift as S, we would in effect be multiplying Eq. (1.4) by -1, and it becomes

$$S = S_p - H_{sv} \tag{1.5}$$

(See Fig. 1.19) The answer to your first question, since it is expressed in terms of suction lift, must be that it is necessary to subtract the velocity head rather than add it to obtain the actual total suction lift of the pump. Expressed in other words, the total suction lift is lower, numerically, than the gauge reading since the fluid contains some energy in the form of velocity head that does not show on the gauge.

In the case of suction head, it should be evident from the preceding discussion that the velocity head must be added to obtain the actual total suction head. This is no different from the case of suction lift, except that in the former instance we gave cognizance to the fact that we were measuring a negative quantity.

To summarize the effect of these two situations on pump total head measurements, we may say that, when we refer to suction lift, our total head is equal to the total discharge head plus the total suction lift:

$$H = H_d + S \tag{1.6}$$

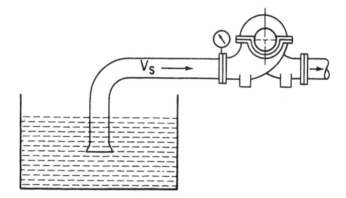

Figure 1.19 Total suction lift. Gauge reading (corrected to pump centerline) = Sp, $V_s^2/2_g$ (at point of gauge attachment) = H_{sv}. Total suction lift $S = S_p - H_{sv}$.

or

$$H = H_{dp} + H_{dv} + S_p - H_{sv}$$

When we speak of suction head, be it negative or positive relative to atmosphere, our total head is equal to the total discharge head minus the total suction head. For positive suction head,

$$\begin{aligned} H &= H_d - H_s \\ &= (H_{dp} + H_{dv}) - (H_{sp} + H_{sv}) \\ &= H_{dp} + H_{dv} - H_{sp} - H_{sv} \end{aligned} \quad (1.7)$$

For negative suction head,

$$\begin{aligned} H &= (H_{dp} + H_{dv}) - (-H_{sp} + H_{sv}) \\ &= H_{dp} + H_{dv} + H_{sp} - H_{sv} \end{aligned} \quad (1.8)$$

Note that, numerically, Eq. (1.8) will produce the same results as Eq. (1.6), which, of course, is as it should be.

We might suggest that prior to running the acceptance test, you refer to the Standards of the Hydraulic Institute, Test Code Section, which will give you considerable information on various other phases of centrifugal pump testing.

Question 1.13 Effect of Altitude on Atmospheric Pressure

Is there a handy reference in table or chart form for the effect of altitude on atmospheric pressure to provide corrections for permissible suction lifts in installations made in locations considerably above sea level?

Answer. Indeed such charts are available, and two such charts are reproduced here. Figure 1.20 shows the approximate barometer reading for altitudes from sea level to 15,000 ft, with the corresponding atmospheric pressures expressed both in feet of water at 62 °F and in pounds per square inch. (The atmospheric pressure at sea level is equal to 33.95 ft of 62 °F water, but a 62 °F water barometer would only read 33.36 ft as the vapor pressure of water at that temperature is 0.256 lb/in.^{2} or 0.59 of water.) The reason I say that Fig. 1.20 shows approximate values is that there will always be some variation in the pressure at any location due to atmospheric conditions.

Reduction in atmospheric pressure at higher elevations makes it necessary to consider this carefully in the location of a pump in regard to the suction water level. For example, if a pump is installed at 10,000 ft elevation, the atmospheric presure is only 23.3 ft of water or 10.6 ft less than at sea level. If the pump is operated with a suction lift of 4 ft, the pump design would have to be such that the pump could operate on a 14.6 ft lift at sea level.

Figure 1.21 shows the approximate reduction in atmospheric pressure with altitude, plotted for 62 °F (18.7 °C) water, but the values on this chart can be used for cold water of any temperature between 32 and 80 °F. For liquids other than cold water, divide the reduction in atmospheric pressure indicated on the chart by the specific gravity (62 °F water = 1.0) of the liquid at the pumping temperature.

Figure 1.22 gives the specific gravity of water at different temperatures.

Question 1.14 NPSH Versus Impeller Eye Diameter

Can the adequacy of an impeller from the point of view of suction conditions be estimated from information on the eye diameter and on the area at the eye? In other words, are the impeller velocity and the peripheral speed at the impeller eye suitable criteria to judge the ability of an impeller to handle a certain capacity under given suction conditions without cavitation?

Answer. The two design factors that you mention, namely, impeller eye diameter and eye area, enter to a certain degree into the determination of a given impeller to handle a certain flow volume under specific suction

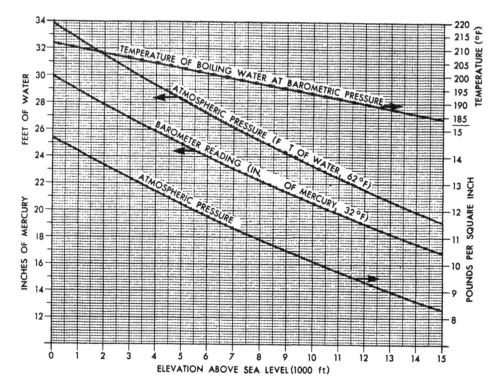

Figure 1.20 Approximate atmospheric pressure at altitudes from sea level to 15,000 ft. *Note*: Various authorities do not agree on the exact change in atmospheric pressure with change in altitude. There is also some variation in the pressure at any location due to atmospheric conditions, so the values shown on this chart should be considered approximate.

conditions. However, they are not the only factors or necessarily the most important ones. Other information might be needed to estimate the performance of the impeller, such as area between the vanes, vane angle for screw-vane impellers, and location of the leading edge of the vanes within the impeller profile.

But the most important fact is that it is highly dangerous for people not too familiar with centrifugal pump design to set themselves as the judge of the relative excellence of several impeller designs strictly on the basis of a dimensional comparison among these impellers. Different designers use different methods to produce an impeller that will perform satisfactorily under the specific operating conditions that it must meet. Without the knowledge of these methods, it would be extremely rash—if I am permitted to be frank—to assume that the manipulation of a few

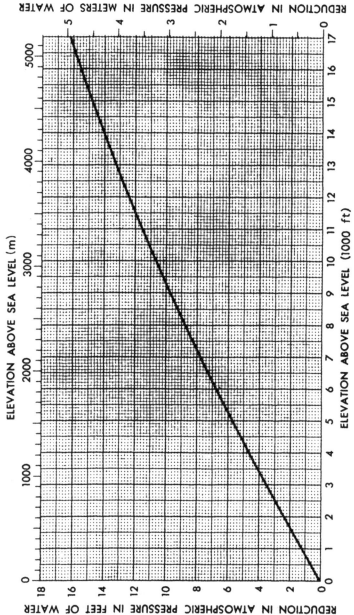

Figure 1.21 Approximate reduction in atmospheric pressure with altitude. *Note:* The values on this chart were plotted for 62 °F (18.7 °C) water but can be used for cold water of any temperature between 32 and 80 °F. For liquids other than cold water, divide the head in feet or meters by the specific gravity (62 °F water = 1.0) of the liquid at the pumping temperature to obtain the head in feet or meters of liquid. Values are approximate. See note with Fig. 1.20.

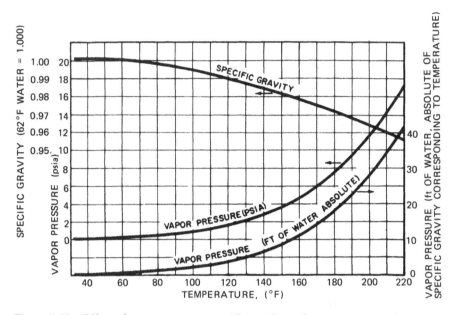

Figure 1.22 Effect of temperature on specific gravity and vapor pressure of water.

selected numbers and dimensions will yield sufficient information on the ability of an impeller to perform as required.

Question 1.15 Recommended Suction and Discharge Nozzle Velocities

Our engineering department has prepared a standard set of specifications for pumps that will be used by our purchasing department in obtaining quotations from pump manufacturers. I have noted that these specifications limit discharge and suction nozzle velocities to 15 and 10 ft/sec, respectively. I would like to know whether it is advisable to specify these velocities and, if so, whether the figures listed fall within reasonable limits.

Answer. Most unfortunately, my answer to both your questions has to be "no." I do not believe that such pump design parameters as discharge and suction nozzle velocities should be part of the specifications prepared by the purchaser of pumps. Nor are the velocities you have mentioned reasonable maximum limits. Note, for instance, that the API-610 Standards covering centrifugal pumps for general refinery service and issued by the American Petroleum Institute do not specify any limitations at either the suction or discharge flanges.

Chapter 1　　37

ANSI (American National Standards Institute) Standards do not directly specify velocities, but because they indicate capacity ranges for different sizes of pumps, they imply certain velocities. Thus, using ANSI Standards, discharge velocities range from 25.2 ft/sec to as high as 37.1 ft/sec.

Important to remember is that, generally speaking, a pump installation involves a reducer at the suction and an increaser at the discharge, as illustrated in a typical installation (see Fig. 1.23). This is intended to reduce friction losses in the suction and discharge piping for reasons of economy. The water does not know where the piping ends and the pump begins. As a matter of fact, once the water has entered the pump at the suction flange, it continues to accelerate further as it approaches the impeller because the velocity at the eye of the impeller is higher than at the suction nozzle.

The reverse is true on the discharge size: The water velocities in the casing are higher than they are at the discharge nozzle.

Let us examine the specific case of an 8 in. double-suction single-stage pump designed to deliver 4700 GPM when operating at its best efficiency point and at 1775 RPM. The case throat area of this 8 in. pump is 21.1. in.^{2}, so that the casing throat velocity is 72 ft/sec. This velocity, incidentally, is necessary to the development of the pump total head and cannot be significantly altered without affecting pump performance, regardless of any restrictions imposed at the pump suction or discharge. This velocity of 72 ft/sec is reduced at the discharge flange to approximately 30 ft/sec.

To accomplish even this reduction requires a discharge nozzle about 20 in. long. To get the velocity down to 15 ft/sec, we would have to extend it at least another 20 in. even if the angle of divergence were maintained at the maximum for efficient energy conversion.

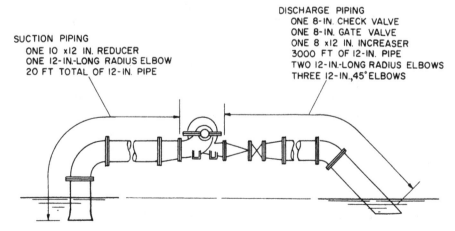

Figure 1.23 Typical pump installation showing use of reducer and increaser in suction and discharge piping.

Providing this added nozzle length as a part of the pump casing would have the following unfavorable results:

1. This would be more costly than accomplishing the same purpose by the use of standard increasers, since it would affect the cost of almost all machining operations on the pump casing. We would be forced to use bigger machines because of the larger overall dimensions, and the hourly costs are higher on larger, more expensive machines. In addition, this would undoubtedly lead to wider base plates, higher shipping weights, and so on.
2. Most freqently, horizontal double-suction pump installations require up or down elbows at both suction and discharge. The longer nozzle would preclude the possibility of obtaining the required velocity reduction in a long radius increasing elbow, thus adding in many cases to the floor space required for the installation and thus to the cost of building space required.
3. The longer nozzle would impose these potential penalties even on those applications for a given pump when it is operating at lower speed. For instance, if the same 8 in. pump were applied at a speed of 1175 RPM, with an extended nozzle to limit the discharge velocity to 15 ft/sec at 1775 RPM, it would then have a discharge velocity of only 10 ft/sec, which is certainly unnecessarily low.
4. Finally, one should remember that it is generally the practice to place check valves and gate valves between the pump and the increaser in order to reduce the cost of these valves. The use of larger discharge nozzles would preclude such economies.

Remember, as I have stated, that the water does not care whether it gets slowed down in the pump or in the piping. It makes better sense to do it in the piping on the basis that to do so allows greater flexibility in pump application and, in most cases, a more economic solution.

Question 1.14 dealt with the danger of specifying impeller eye diameters and eye velocities. The same holds true for pump nozzle dimensions. The imposition of arbitrary velocity restrictions at the pump suction or discharge is not in the customer's best interest.

Question 1.16 Parable About NPSH and Hard Cash

When figuring available NPSH at the suction of a centrifugal pump, is it permissible to include the velocity head in addition to the static pressure available, or is the available NPSH calculated as exclusive of the velocity head?

Answer. Many years ago, when I was just beginning my career in the field of centrifugal pumps, I became involved in an argument over this very question with one of my colleagues. Try as I would to prove to him by conventional means that the velocity head at the pump suction forms a legitimate component of the available NPSH (net positive suction head over the vapor pressure), I could not convince him. In desperation, I resorted to writing a parable on kinetic and pressure energy. The parable succeeded where I had failed.

Grown men seldom play at whimsical games unless they are under the mollifying influence of alcoholic beverages. Nevertheless, to assure the ultimate success of this parable, we shall have to assume that the little game I shall describe is to be played in cold sobriety, as only under such conditions shall we be able to wend our way through the cerebral gyrations that this game involves.

It is first supposed that our victim receives a definite sum of money, all in silver coins, after which the rules of the game are carefully explained to him. He is to start on a journey during which no additional moneys will be made available to him. He will, however, be permitted to make change all along the way, that is, to trade his silver for copper pennies and vice versa. But even this trading process will be regulated. The number of pennies in his possession must at all times bear a definite relation to the speed and the mode of traveling he is employing. The higher his speed, the more pennies he must have, the number of pennies to increase in geometric progression. Of course, should he slow down in his journey, for instance, step from an express train to a horse and buggy, he is enjoined to reverse the process, that is, to change a commensurate number of pennies back into silver. In order not to complicate our game unnecessarily, we shall assume that our traveler can always find accommodating strangers willing to act as a second party to these little transactions.

We shall impose a second rule on our traveler in the form of a penalty for traveling fast. As he goes along, he will hand out payment to the attendants of the conveyances he uses. The higher the speed, the greater the cost, again in geometric progression.

However, he must pay careful attention to the fact that, whether he pays in pennies or in silver, the number of pennies left in his possession must at all times fulfill the requirements of our first rule.

We almost forgot that, in this game, altitude, just like speed, can be bought or sold. If our traveler decided to climb on the way to his final destination, he pays out some silver coins as he climbs, in proportion to the heights he reaches. Of course, we must also reward him for availing himself of the better mode of traveling, that is, downhill. Thus, if he travels down instead of up, we shall give him an extra sum of silver as a bonus.

As a final rule, we shall extract a promise from our traveler, overburdened as he already is with dictatorial restrictions. At no time must be find himself with less than some previously fixed sum of money in silver. Should he at any times travel so fast, pay out so much, and trade so much silver for pennies that his reserve of silver falls below the prescribed minimum, he has lost.

Arriving at his destination with silver equivalent to or in excess of the minimum, he wins. He had better not dawdle on the way, however, as he will be expected to reach the end at a certain time.

Did you like the game? Well, as all good games, this one has a meaning. The silver that our man received at the onset of his journey is obviously the pressure energy at the source where a centrifugal pump takes its suction. The bonus for sliding downhill and the penalty for climbing are the static elevation between the liquid level and the pump centerline. If silver is pressure or potential energy, copper must be the kinetic energy, and the trading of silver for the baser metal is illustrative of the transformation of pressure energy into kinetic energy as the liquid velocity increases. Of course, as the velocity decreases, some of the kinetic energy is transformed back into pressure energy, and we get some silver back for our pennies. Paying for our traveling in geometric proportion to the rapidity of our conveyance? Friction losses, of course, increase approximately as the square of the fluid velocity. Finally, the predetermined amount of silver we are to hold in reserve is the vapor pressure of the liquid at the pumping temperature. Fall below this minimum, and you lose.

This parable should shed some light on a question that frequently puzzles engineers, that is, whether the kinetic energy (velocity head) at the pump suction nozzle can be considered available energy in computing the available net positive suction head over and above the vapor pressure. Obviously it can, and it should. When he arrives at the pump suction nozzle, our traveler has a defnite amount of silver and some pennies. Now when he enters the rocket ship that will take him on the last leg of his journey into the inner sanctum, between the impeller entrance vanes, all he has to do is to trade as much silver into pennies as will make up the difference between the pennies he has to carry at these high speeds and the pennies he already has. The more copper he has on hand as he reaches the pump suction nozzle, the fewer additional pennies he needs.

Once the traveler reaches the inner sanctum without mishaps, he really does get a h____ of a surprise. They pile so much silver onto him that he does not have much time left to count pennies, even though he still has some. As a matter of fact, he probably gets a lot of pennies in there, which some kind soul trades for silver so that he will not have too much of a load

Chapter 1 41

to carry. Generally, prosperity goes to his head, and he starts spending both silver and pennies like a drunken sailor. But that is another story.

Question 1.17   More About NPSH and Velocity Head

I have just finished reading your interesting parable about NPSH and hard cash. I was particularly drawn to this discussion because I had remembered reading some centrifugal pump application articles you authored several years ago.

While reading this parable, I did not recall that you had included the velocity head in determining the NPSH in the aforementioned articles. On checking back it appears that my memory was correct because an earlier article, your formula for NPSH did not include the velocity head term.

I would appreciate it if you would clarify this apparent discrepancy.

Answer. It is always a pleasure for an author to be made aware that his readers study his writing in detail and that some of them take time to write him whenever they find a statement that is not clear or with which they disagree. This is particularly flattering when, as in this case, a reader reaches back so many years in memory to carry out a comparison between two apparently conflicting statements I have made.

The apparent contradiction might have been avoided had I defined my terms somewhat more completely in both cases. In saying that the velocity head is a legitimate component of the available NPSH in addition to the static pressure available, I should have elucidated further that the static pressure referred to is a reading of this pressure as measured on a gauge at the pump suction nozzle and corrected to the elevation at the pump centerline. If, however, the formula for NPSH is based on static elevation differences between the level of the supply at the suction and the pump centerline, the velocity head cannot and should not be included. The reasons for this is that the velocity head is "created" by this static elevation difference. It is this difference in head plus the pressure at the suction source that causes flow to take place. Therefore, you cannot count the velocity head energy component twice.

To clarify this entire matter further, I am reproducing as Fig. 1.24 the drawing that accompanied that article, and I shall quote the formula to which you make reference:

$$\text{NPSH} = S - h_{fs} - h_i + P_s - P_{vp}$$

where S = static head between the suction liquid level and the pump centerline
h_{fs} = friction loss in suction line
h_i = entrance loss in suction line
P_s = pressure on suction liquid surface
P_{vp} = vapor pressure of liquid

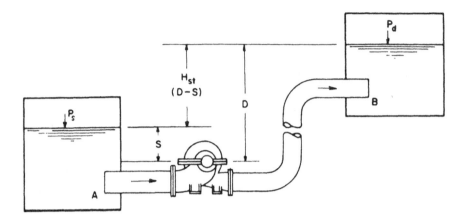

Figure 1.24 Diagram of pump installation.

All pressures and heads must be expressed in feet of the liquid at the pumping temperature with the proper algebraic sign. Although P_s and P_{vp} can be either in gauge or in absolute values, they must both be measured under the same conditions.

On the other hand, a gauge located at the pump suction nozzle and exactly at the pump centerline would read a static pressure h_g, which lacks the velocity head component to be a measure of the total energy available at the pump suction. The relation between h_g and the other items shown in Fig. 1.24 is

$$h_g = P_s + (S - h_{fs} - h_i) - h_v$$

where h_v is the velocity head. Therefore, the *total energy* at the pump suction, or suction head, is

$$h_s = h_g + h_v = P_s + (S - h_{fs} - h_i)$$

Since NPSH is defined as the difference between the total energy at the pump suction h_s and the vapor pressure of the liquid, we can develop it as follows:

$$\begin{aligned} \text{NPSH} &= h_s - P_{vp} \\ &= P_s + (S - h_{fs} - h_i) - P_{vp} \end{aligned}$$

as stated in the article you have referred to.

Chapter 1   43

Summarizing the velocity head component *must* be included if you are determining the NPSH from a gauge reading at the pump suction but is already included if the NPSH is established from a difference in elevations.

Question 1.18   More About Velocity Head and Entrance Loss

Reference is made to one of your centrifugal pump clinics that concerns the determination of available NPSH. You give the formula

NPSH = S − h_{fs} − h_i + P_s − P_{vp}

Since velocity head does not appear in the formula, the questions that arise are the following: (1) Is it intended to include velocity head in the value for entrance loss, or (2) was velocity head deliberately excluded from the formula?

In an actual problem that I now have, the 14 in. suction of a vertical sewage pump connects to the wet well through a 14 × 36 in. increaser and an existing 36 in. wall pipe. In the event that it is necessary to include velocity head, should it be computed on the basis of the 36 in. tank connection or the 14 in. pump suction connection?

Your assistance in this problem will be very much appreciated.

Answer.   To understand the manner in which available NPSH should be calculated, it is best to start with the basic concept of NPSH. In the pumping of liquids, the pressure at any point in the suction line or within the pump impeller must never be reduced to a value below the vapor pressure of the liquid. The available energy that can be utilized to let the liquid through the suction piping and suction waterways of the pump into the impeller is thus the total suction head less the vapor pressure of the liquid at the pumping temperature. This available energy—measured at the suction opening of the pump and corrected to the centerline datum line of the pump—has been named *net positive suction head* or NPSH available. It must be equal to or greater than the required NPSH of the pump, the latter value expressing the amount of energy required to get the liquid between the impeller vanes where additional energy will be transmitted to the liquid pumped. The required NPSH is a function of the pump design and is generally established empirically by tests.

Whether velocity head will or will not enter into the calculations of available NPSH depends entirely on whether gauge measurements or static measurements are used in determining total suction head. Since we are dealing with *total energy* available at the pump suction nozzle, before deducting vapor pressure, we need consider the various forms in which this energy may be available. The three components of energy are (1) pressure energy, (2) potential energy, and (3) velocity of kinetic energy.

The first of these is represented by the value of P_s in the equation given. In the case illustrated in Fig. 1.24, the velocity energy (or velocity head) is created by the static elevation (S), and thus both potential and velocity energy are represented by the value of S.

The energy at the pump suction is reduced by the losses at the entrance to the suction piping (at point A) and by the friction losses in the suction piping. All the foregoing explains the structure of the equation and the reason velocity head does not enter into it but the entrance losses do.

When the energy at the pump suction is measured by means of a pressure gauge as in Fig. 1.18, the gauge reading will express the static energy at the pump suction nozzle and therefore includes both the pressure energy at the liquid surface and the potential energy created by the static elevation S. The reading also reflects the fact that losses will have occurred between point A (entrance to the piping) and the suction nozzle. The gauge reading, on the other hand, does not include a certain portion of the true total energy at the pump suction, namely the velocity head. Thus, to express the total energy at the suction before deducting the vapor pressure so as to determine available NPSH, it becomes necessary to add to the gauge reading the value of the velocity head at the pump suction nozzle.

To return to the second portion of your question, if you were dealing with an existing installation and if you were determining available NPSH from a gauge reading, you should include the velocity head at the pump nozzle that is based on the velocity at the 14 in. opening. However, from the wording of your question, I assume that you intend to calculate available NPSH from static elevation readings and by calculating entrance and friction losses. In this event, the entrance loss must be calculated on the basis of a 36 in. opening.

Entrance losses are generally expressed as percentages of the velocity head, these percentages varying with the tye of entrance provided. If a plain end pipe projects through the well wall into the suction well, the entrance loss will be approximately 80% of the velocity head. This will be reduced to 50% of the velocity head if the pipe has a plain end flush with the wall. Finally, if a bell mouth is provided, the entrance loss will be 25% of the velocity head (with the bell mouth slightly rounded) and as little as 10% if the bell mouth is well rounded.

Question 1.19 Effect of Changing Heater Pressure on NPSH

Our boiler feed pumps take their suction from a deaerating heater operating at 5 psig. We would like to revamp this installation and operate

Chapter 1

the heater at 15 psig so as to tie it into our process steam line. The heater is suitable for this pressure, but we are concerned over the boiler feed pump requirements. Will the heater have to be raised to compensate for the higher temperature water that the pump will handle?

Answer. Assuming that the boiler feed pumps are now operating satisfactorily and the existing submergence is sufficient, there will be no need to alter the installation.

A centrifugal pump requires a certain amount of energy in excess of the vapor pressure of the liquid pumped to cause flow into the impeller. By definition, the net positive suction head (or NPSH) represents this net energy referred to the pump centerline *over and above* the vapor pressure of the liquid. This definition makes the NPSH automatically independent of any variations in temperature and vapor pressure of the feedwater.

The feedwater in the storage space of a direct-contact heater from which the boiler feed pump takes its suction is under a pressure corresponding to its temperature. Therefore, the energy available at the first-stage impeller over and above the vapor pressure is the static submergence between the water level in the storage space and the pump centerline less the friction losses in the suction piping.

Question 1.20 Inert Gas Pressurization of Reservoir at Pump Suction

We are installing a closed-cycle process in which a 1500 GPM centrifugal pump takes its suction from a 5000 gal vessel and discharges through a heat exchanger and the pumped hydrocarbon returns to the vessel at the suction of the pump. The hydrocarbon (specific gravity = 0.8) is made up of volatile fractions, the lightest of which are at the boiling point at the surface of the vessel under atmospheric pressure and at the prevailing temperature, which is about 45°F.

The static submergence to the pump centerline is only 4 to 5 ft, so that assuming about 1 ft friction loss in the suction, the system provides only 3 to 4 ft available NPSH. The pump rating curve indicates that 10 ft NPSH are required. We cannot raise the vessel itself, nor is it practical to lower the pump elevation. It has been suggested that we could solve our problem by slightly pressurizing the vessel with an inert gas, such as carbon dioxide. But I remember reading in one of your articles that changing the deaerating heater pressure in a boiler feed pump installation has no effect on the available NPSH since increasing this pressure automatically increases the vapor pressure of the feedwater. Isn't our case similar, in that increasing the pressure in the vessel we would automatically increase the vapor pressure and thus leave the available NPSH unaffected?

Answer. The two cases are not exactly the same, and you can safely count on increasing the available NPSH of your installation by pressurizing the vessel with carbon dioxide (as an inert gas).

If you refer to Fig. 1.14, you will note that the available NPSH is essentially the difference between the suction pressure at the pump centerline and the vapor pressure of the liquid at the pumping temperature. If the liquid pumped is boiling at the surface of the vessel, in other words, if P_s if equal to P_v, the available NPSH is equal to the static submergence over the pump centerline less the losses in the suction piping. If you pressurize the vessel, that is, if you increase P_s without changing the temperature of the liquid pumped, you will not be changing the liquid vapor pressure, and the available NPSH will be increased by the increase in P_s. In your particular case, the desired increase in available NPSH is 10 ft (required) less the 3 ft (available) or 7 ft. Expressed in psi, this is

$$\text{Desired increase in pressure} = \frac{(10 - 3) \times 0.8}{2.31} = 2.4 \text{ psi}$$

To be on the safe side, I would pressurize the vessel by 3 or 4 psig. From your question, I infer that the slight amount of carbon dioxide that will be dissolved in the hydrocarbons being pumped will have no ill effects. If you have any doubts on this score, you may consider pressurizing the vessel with nitrogen.

If we return to the question of a boiler feed pump taking its suction from a deaerating heater (Fig. 1.25), the situation is quite different. The heater pressure can be increased only by increasing the pressure of the steam supplied to the heater. This, in turn, increases the feedwater temperature to correspond to the new operating pressure; the vapor pressure P_v remains equal to the heater pressure P_s. Thus, increasing the heater pressure does not affect the available NPSH of a boiler feed pump installation.

Question 1.21 Effect of Temperature on Required NPSH

I have heard conflicting statements regarding the effect of temperature on the required NPSH for boiler feed pumps. Some engineers claim that more NPSH is required at higher temperatures and produce as evidence a chart published in the Standards of the Hydraulic Institute that shows margins to be added to the recommended NPSH-margins that increase with increasing temperatures. On the other hand, I have heard recently that less NPSH is required when handling hot water than with cold water. Which of these statements is correct?

Chapter 1

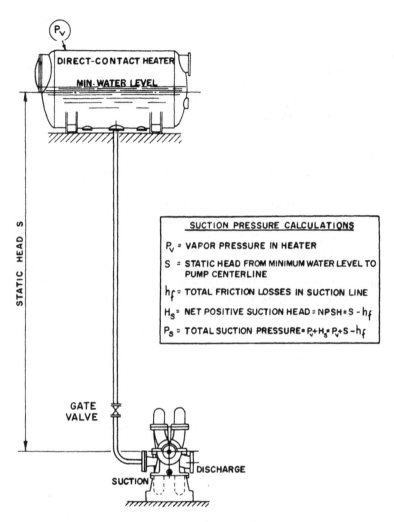

Figure 1.25 Method of determining total suction pressure.

Answer. Oddly enough, both statements are correct if they are properly interpreted. If we refer strictly to steady-state conditions, the required NPSH is reduced with increasing temperatures. The theory underlying this effect is fairly simple but need not be discussed in detail here. It is based on the fact that mild and partial cavitation can take place in a pump without causing extremely unfavorable effects. The higher the temperature, the less volume will be occupied by the steam into which a small amount of water will flash if NPSH is reduced until cavitation just begins. Thus, the same amount of pressure reduction below the vapor pressure will liberate less steam volume and will reduce the pump effective capacity by a lesser amount.

This effect can also be demonstrated by strictly logical considerations. Suppose that we are considering the effect of temperature on the performance of centrifugal pumps, operating at a given speed and supplied with a fixed NPSH, when handling

1. Cold water at 70°F
2. Hot water at 705.4°F (critical temperature for water)

It is obvious that in the case of cold water at 70°F, the break in the head-capacity curve will occur at some capacity determined by the geometry of the impeller and by the speed at which the pump is operating as indicated in Fig. 1.26.

When it comes to handling water at 705.4°F, that is, the critical temperature at which no evaporation takes place (as long as the pressure is maintained at 3206.2 psia), the volume occupied by the water is the same as that occupied by steam. Under these conditions, a slight reduction in the available NPSH can have no appreciable effect on the performance characteristics of the pump (when expressed in foot-pounds per pound plotted against volume), since no true cavitation can take place. This statement is predicated on the fact that a change from water to steam takes

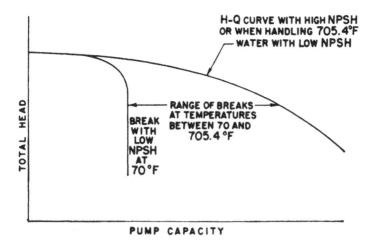

Figure 1.26 Effect of temperature on maximum capacity at fixed pump speed and with fixed NPSH.

place without a change in volume occupied by either fluid. In other words, the head-capacity curve of a pump handling 705.4 °F water with less than the theoretically required NPSH will coincide with the head-capacity curve with cold water and ample NPSH, as shown in Fig. 1.26.

Having established the two limits of performance characteristics at 70 and 705.4 °F, we can assume that the relationship between the location of the break and the pumping temperature is a continuous function. Therefore, all breaks at temperatures between these two extreme temperatures must take place between the two limits of capacity indicated.

This would indicate that the practice of assuming that the required NPSH is independent of the pumping temperature is conservative and unrealistic and that this NPSH is definitely reduced with increasing temperatures.

On the other hand, the margin that used to be recommended to be added by the Standards of the Hydraulic Institute (this has been eliminated in the latest edition of the standards) took into consideration the actual conditions that prevail in a steam power plant, which differ considerably from the "steady-state" conditions assumed in establishing the relation between NPSH requirements and operating temperature. It is well known that "transient conditions," such as sudden load reductions, will introduce unfavorable effects on the suction from a deaerating direct-contact heater. A very severe reduction in available NPSH follows the sudden load reduction.

Therefore, means must be employed to take care of the time lag that exists between the instantaneous reduction of pressure in the heater that follows a sudden load drop and the ultimate reduction of temperature at the pump suction after the feedwater already in the suction piping will have been pumped out into the discharge header. The most logical solution is to provide a factor of safety for the installation through the addition of some arbitrary amount to the required NPSH under steady-state conditions.

The degree of NPSH reduction caused by load reduction is intimately connected with the operating pressure in the heater and hence with the operating temperature: the higher this pressure, the more severe the NPSH reduction.

The chart in the Standards of the Hydraulic Institute was therefore intended to compensate, at least partially, for this effect. The margin indicated was purely empirical and did not necessarily cover all cases, since in addition to the operating temperature several other factors enter into establishing the amount by which the available NPSH will be reduced after a sudden load reduction. Some of these factors are the storage volume in the heater, the configuration and internal volume of the suction piping, and the behavior of the feedwater regulator following the sudden load reduction. Thus, in some cases, the recommended margin was insufficient, but in others it could be excessive. The merit of the chart was in bringing attention to the importance of giving this problem consideration.

Returning now to the thought that under steady-state conditions required NPSH values are reduced with increased temperatures, it is necessary to introduce a word of caution: It would be highly unwise to rush headlong to the conclusion that the values of recommended NPSH for boiler feed pumps should be drastically reduced. We must remember that boiler feed pumps are frequently required to operate over wide ranges of temperature and that a given installation cannot very well take advantage of a permissible reduction in NPSH requirements at the top operating temperature if operation at lower temperatures will wipe out any assistance gained by the effect we have described. The effect is also very useful in affording us additional protection during the transient conditions that prevail when severe load reduction occurs in a steam power plant. Let us preserve this protection, not give in to the temptation to cut corners.

Question 1.22 More on Temperature and NPSH

I have heard it stated that temperature has a major effect on the NPSH required by a centrifugal pump and that pumps handling hot water are much less subject to cavitation and its consequences than cold-water pumps. Is this true, and if so, what are the reasons for this?

Answer. The statement is quite true, and I have discussed it at some length in Question 1.21. The basic reason is that a slight amount of cavitation can take place in a pump without necessarily leading to very unfavorable effects. This should not be interpreted, incidentally, as recommending that pump installations should always be designed in that marginal area where cavitation is sure to take place but merely as an observation of a fact.

The degree of interference with the proper operation of the pump caused by such minor cavitation will bear a definite relation to the temperature of the liquid handled by the pump. Remember that when we say that a pump is cavitating, we mean that, somewhere within the confines of the pump, the pressure will have fallen below the vapor pressure of the liquid at the prevailing temperature. Thus, a small portion of the liquid handled by the pump will vaporize, and this vapor will occupy considerably more space within the impeller than the equivalent mass of liquid before vaporization.

If the pump is handling water at normal temperatures, the volume of a bubble of steam is tremendously larger than the volume of the original quantity of the water. For instance, at 40°F, 1 lb of water occupies 0.016 ft^3, but steam at the same temperature occupies 2441 ft^3. The ratio of the two volumes is 152,500! The rapidity with which this ratio diminishes as water temperature increases is illustrated in Fig. 1.27, which presents a plot of the ratio of the volume of steam to the volume of the equivalent mass of water for temperatures between 50 and 705.4 °F (the critical temperature of water at which steam occupies the same volume as water). At 212 °F, 1 lb of water occupies 0.0167 ft^3 and 1 lb of steam 26.81 ft^3, so that the ratio of volumes is only 1605-almost 100 times less than at 50 °F.

Since at 705.4 °F the ratio of steam to water volume is 1.0, no true cavitation can take place, since a slight reduction in the available NPSH can have no appreciable effect on the performance characteristics of the pump. This is illustrated by Fig. 1.26, which shows the effect of temperature on the maximum capacity of a pump at a fixed pump speed and with a given fixed value of NPSH. You will see that the head-capacity curve of a pump handling 705.4 °F water and installed with less than the theoretically required NPSH will coincide with the head-capacity curve of the same pump when handling cold water and ample NPSH.

Question 1.23   NPSH Required When Handling Hydrocarbons

I understand that centrifugal pumps handling certain liquids, such as hydrocarbons, have different suction performance than they do with water. Is this correct, and why is this so?

Answer.   You are correct. In most cases, a pump handling hydrocarbons does require less NPSH than when it handles water, and I will try to explain why this is so.

Cavitation occurs when the absolute pressure within an impeller falls below the vapor pressure of the liquid, and bubbles of vapor are formed; these bubbles collapse further within the impeller when they reach a region of higher pressure. The minimum required NPSH for a given capacity and at a given pump speed is defined as that difference between the absolute suction head and the vapor pressure of the liquid pumped at the pumping temperature that is necessary to prevent cavitation.

The fact that a pump is cavitating manifests itself by one or more of the following signs: noise, vibration, drop in the head-capacity and efficiency curves, and-with time-damage to the impeller by pitting and erosion. The minimum NPSH is determined in the pump manufacturer's laboratory by a test in which both total head and efficiency are measured at a given speed and capacity with varying NPSH conditions. At the higher values of NPSH, the values of head and efficiency remain substantially constant. As the NPSH is reduced, a point is finally reached where the curves break, showing the impairment of pump performance caused by cavitation.

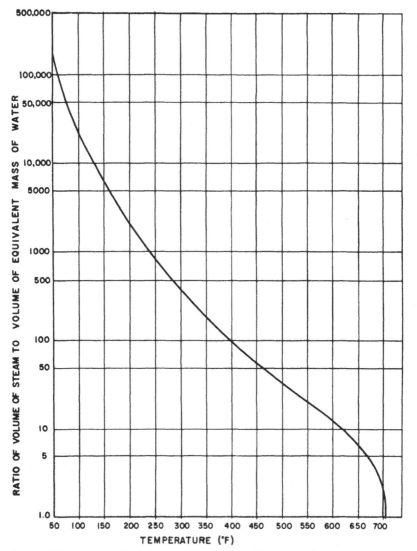

Figure 1.27 Effect of temperature on ratio of volume of steam to volume of equivalent mass of water.

NPSH tests of centrifugal pumps are normally carried out on cold water, and both the Hydraulic Institute Standards curves and pump manufacturers' rating curves indicate NPSH requirements on cold water. Thus, it might be assumed that the NPSH required by a centrifugal pump for satisfactory operation is independent of the liquid vapor pressure at the pumping temperature. This is not true. It is merely an oversimplification used to illustrate the definition that NPSH is a measurement of the energy in the liquid at the pump suction over the datum line of its vapor pressure.

At the same time, both laboratory and field tests run on pumps handling a wide variety of liquids and over a range of temperature have always shown that the NPSH required for a given capacity and with a given pump apparently vary appreciably. For example, the required NPSH when handling some hydrocarbons is frequently much less than that required when the pump handles cold water. Even when pumping water, there is definite evidence that required NPSH decreases when the water temperature increases. Altogether it became evident quite a number of years ago that the reduction in the required NPSH must be a function of the vapor pressure and of the characteristics of the liquid handled by the pump.

Because in the case of hydrocarbons this reduction in NPSH could play a most important role in the relative costs of a refinery installation, it was in connection with hydrocarbons that most efforts were directed at understanding the phenomena involved. Thus, it was thought that rules could be developed to predict the effect of liquid characteristics on the required NPSH to take advantage of this phenomenon without the risk of overoptimistic assumptions.

Such rules have been developed by members of the Hydraulic Institute and incorporated in its standards.

Briefly, what was the historical development of these efforts? As already noted, comparison of laboratory and field tests had for a long time demonstrated that both liquid characteristics and operating temperatures affected the required NPSH.

In the early and middle 1940s, it was thought that these variations were only apparent and that if "true vapor pressures" or "bubble point" pressures were to be used in the calculations of test NPSH, the discrepancies would disappear, and complete correlation with water cavitation data would exist. Corrections for NPSH with hydrocarbons were nevertheless used, as a matter of policy rather than based on accepted theoretical deductions. It was believed that a reduced NPSH could be justified for the following reasons:

1. Oil companies' specifications generally called for a maximum capacity and head at a minimum NPSH. In practice, it was unlikely that these two requirements would be imposed simultaneously. In fact, some of the field conditions are self-regulating. For instance, low capacity occurs at low NPSH, as a result of a reduced flow in the system. Under these conditions, even if the pump capacity falls off, NPSH is increased, and equilibrium is eventually attained.

2. The effect of cavitation with hydrocarbons was noted to be not as severe as with water; that is, the head-capacity curve does not break off suddenly for two reasons:
   a. Only the lighter fractions will boil first.
   b. The specific volume of hydrocarbon vapors is very small in comparison with that of water vapor.

Obviously, these facts do not tell the whole story, as many other factors affect the behavior of a pump handling hydrocarbons with low NPSH. Thus, attempts to arrive at a more reasoned understanding continued while some interim correction factors of an approximate nature were being used.

These efforts centered on the accumulation and comparison of many tests, using a variety of pumps and handling many different hydrocarbons. These tests, in turn, helped generate a variety of correction curves for NPSH. Some of these charts would on occasion lead to rather impractical conclusions, and additional rules were then introduced to avoid this situation. Sometimes, the user of the chart was cautioned not to exceed a 50% correction for NPSH based on water. In other cases, certain arbitrary limits were imposed, such as a 3 ft minimum NPSH for pumps under 4 in. and 5 ft minimum for larger pumps.

In 1951, the Hydraulic Institute Standards incorporated a conversion chart for hydrocarbons (see Fig. 1.28), which has since been updated. It provided an estimate of the NPSH required by a centrifugal pump handling hydrocarbons of various gravities and vapor pressures in percentages of that required by the same pump when handling cold water. These curves were derived from an accumulation of experimental data and do not pretend to be arrived at by analytical means.

In using this chart, one entered Fig. 1.28 at the specific gravity at pumping temperature and proceeded vertically to the inclined line corresponding to the absolute vapor pressure at this pumping temperature. The percentage of the NPSH on water that will be required for this particular hydrocarbon was read on the left scale.

It may be interesting to examine one particular aspect of the first Hydraulic Institute Standards correction chart. Since the reduction in the required NPSH is probably affected most by the liquid characteristics (vapor pressure, temperature, and composition), it would seem more logical to express this reduction in absolute terms rather than in terms of a percentage of the NPSH value on cold water. For the same reason, it would seem that this reduction should be essentially independent of the pump capacity. In other words, if it can be predicted that when handling a particular hydrocarbon at a given temperature, the required NPSH will be reduced by, say, 3 ft, this same 3 ft reduction should probably apply not only at the best efficiency but also at all other capacities.

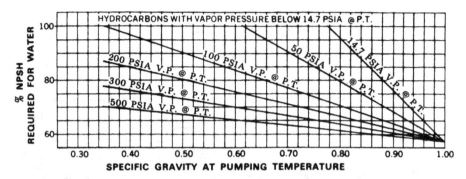

Figure 1.28 NPSH corrections in Hydraulic Institute Standards of 1951.

This approach was incorporated in the 1975 edition of the Hydraulic Institute Standards, and the latest revised correction chart is shown in Fig. 1.29 in which the NPSH reduction is expressed in absolute terms of "so many feet" rather than as a percentage.

To use this chart, enter at the bottom of the chart with the pumping temperature in °F and proceed vertically upward to the vapor pressure in psia. From this point, follow along or parallel to the sloping lines to the right side of the chart, where the NPSH reductions in feet of liquid may be read on the scale provided. If this value is greater than one-half of the NPSH required for cold water, deduct one-half of the cold-water NPSH to obtain the corrected NPSH required. If the value read on the chart is less than one-half of the cold-water NPSH, deduct this chart value from the cold-water NPSH to obtain the corrected NPSH required.

Because of the absence of available data demonstrating NPSH reductions greater than 10 ft, the chart has been limited to that extent, and extrapolation beyond that limit is not recommended.

Warnings are included in the Hydraulic Institute Standards regarding the effect of entrained air or gases. This circumstance can cause serious deterioration of the head-capacity curve, of the efficiency, and of the suction capabilities even when relatively small percentages of air or gas are present.

The fact remains that there is insufficient correlation at this moment among the many tests cited in the technical literature. It appears rather probable that the very characteristics of a pump—that is, its specific speed and its actual design—play some role in the actual reduction in NPSH right along with the characteristics of the hydrocarbon. This role may be minor, but it probably does exist.

I believe that an exhaustive analysis of the phenomena that take place in a pump handling hydrocarbons is beyond the scope of our discussion. As

56   *Application*

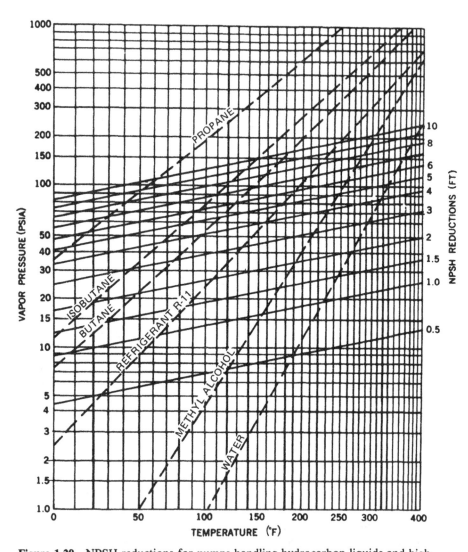

Figure 1.29 NPSH reductions for pumps handling hydrocarbon liquids and high-temperature water (Courtesy Hydraulic Institute Standards of 1975.)

a matter of fact, such an analysis would at best be open to argument, because several somewhat conflicting interpretations still exist with respect to what actually takes place.

Whether a more rigorous theoretical deviation of NPSH reduction is ever developed is really immaterial. The important fact remains that as further experience is gained and more and more experimental data are accumulated, the validity of correction charts will be even greater.

However, I wish to give you a word of caution: it is probably best to use the correction factor as an additional safety factor rather than as a license to reduce the available NPSH. This is a personal opinion, but one I share with a number of rotating machinery specialists of some of the major petroleum and petrochemical companies.

I should not fail to mention that all the above is predicated on the present definition of NPSH required, which is stipulated in the Hydraulic Institute Standards to be that NPSH available that at a given speed and capacity does not produce more than a 3% reduction in total head. There is a great deal of agitation to change this definition so that NPSH required would correspond to only a 1% reduction in total head, or even in just 0% reduction. It is my opinion that the redefinition may take place at some time in the future. If this occurs, the correction factors curve will disappear, as in my opinion at 0% drop in head, the NPSH required for hydrocarbons will prove to be the same as that for cold water.

If you are interested in a more complete discussion of this question of 3% versus 0% in head, you may wish to locate a copy of the Proceedings of the First International Pump Symposium, Houston, Texas, May 1984. It was sponsored by Texas A&M University, and you can probably get a copy of the Proceedings from them. It contains an article of mine, "A Map of the Forest . . . Understanding Pump Suction Behavior: Where Do We Go From Here?" that deals specifically with all these facts.

Unfortunately, one cannot claim that our knowledge of suction conditions and of cavitation is an exact science. One could say, instead, that if all the papers that have dealt with this subject were to be laid end to end, they would not reach a conclusion. There are at least as many sacred and immovable opinions on this matter as there are authors of such papers. What I want to say is that there is truth in the statement that if you search far enough, you will always find some authoritative paper that will confirm whatever opinion you happen to favor. And it is to illustrate these thoughts that I once drew a map, which is reproduced as Plate 1.1. References on the map are obvious to pump users: one of the rivers and the lake are named for incipient cavitation; the sigma for which the Confusion River delta is named is, of course the ratio $NPSH_r/H$, which also carries the name "Thoma-Moody" cavitation parameter. Finally, the cliffs bear the name of 3% drop in head used to define NPSH required. But I refuse to explain the reference to the dragons. For those who know pumps, no explanation is needed; for those who do not, no explanation will make sense.

Question 1.24 Injection of Cold Water to Boiler Feed Pump Suction

When our steam power plant was revamped and modernized 15 years ago, a deaerating heater operating at 95 psia constant pressure was added to the

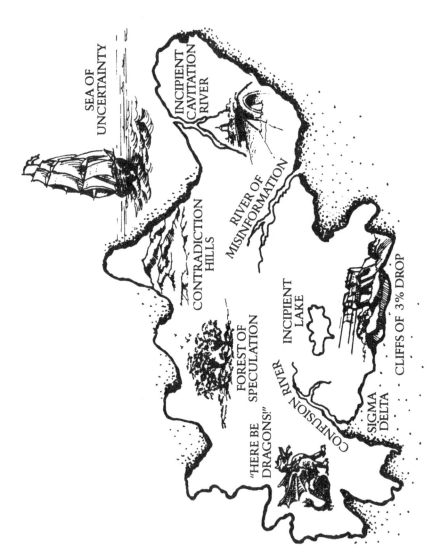

Plate 1.1 "A Map of the Forest".

installation. It was impossible to locate it an an elevation sufficient to provide the required NPSH to the boiler feed pumps, and a line was provided to inject some 227 °F water into the suction line. Could you explain the manner in which this injection overcomes the lack of sufficient NPSH? Does this injection have any harmful effects? How can we determine the minimum amount of injection required?

Answer. The required NPSH, or net positive suction head, is the energy in feet of the liquid pumped that must be made available over and above the vapor pressure of the pumped liquid in order to avoid flashing and cavitation in the centrifugal pump impeller. On the other hand, the available NPSH of a system is the difference between the suction pressure and the vapor pressure and, as shown in Question 1.10, can be expressed as

$$\text{Available NPSH} = \frac{(P_s - P_v) \times 2.31}{\text{sp gr}} + (S - h_{fs} - h_i)$$

where P_s = pressure in suction vessel
 P_v = vapor pressure at pumping temperature
 S = static elevation from suction vessel liquid level to pump centerline
 h_{fs} = friction losses in suction piping
 h_i = entrance loss
 h_f = total friction in suction piping = $h_{fs} + h_i$

Then, if H_s is the required NPSH for the boiler feed pump, proper operation of the pump will be assured only if

$$\frac{(P_s - P_v) \times 2.31}{\text{sp gr}} + (S - h_f - h_i)$$

is greater than or equal to H_s. Inasmuch as in a direct-contact deaerating heater the pressure P_s is substantially equivalent to the vapor pressure P_v, the only available NPSH is represented by the difference between the static head S and the friction losses h_f, as illustrated in Fig. 1.25. It is obvious that, in many cases, physical limitations will reduce the value of S to a nominal amount, possibly insufficient to provide sufficient NPSH.

One of the solutions employed in some cases to overcome this difficulty is to inject colder water into the pump suction. This has the effect of lowering the vapor pressure and therefore of artificially increasing the available NPSH. The method is sometimes referred to as *subcooling*.

Two charts are appended to illustrate the effect of colder water injection upon the available NPSH. The first, in Fig. 1.30, illustrates the effect

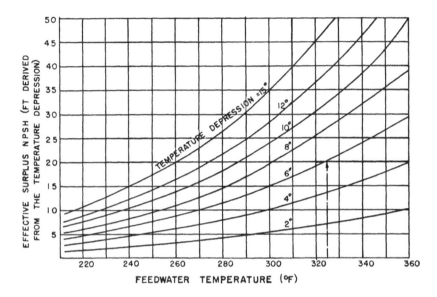

Figure 1.30 Effect of subcooling on available NPSH at various initial feedwater temperatures.

of subcooling (or temperature depression) upon the available NPSH at various initial feedwater temperatures. The second, in Fig. 1.31, shows the temperature depression resulting from cold-water injection plotted against the difference in temperature between the feedwater and the injection stream and given for varying ratios of injection flows.

For instance, if it were desired to provide 20 ft additional NPSH to the pump in question, which handles 325 °F feedwater, the required temperature depression is 6 °F. The injection water temperature is 227 °F, giving us a 98 °F difference between feedwater and injection water temperatures. From Fig. 1.31 we can see that the injection flow must be 6.2% of the total feedwater flow.

The colder injection flow must obviously be taken from the upstream side of the feedwater cycle, ahead of the deaerating heater. To introduce nondeaerated condensate into the boiler feed pump suction is not recommended, especially in the case of high-pressure boilers, where the possible effects of oxygen contamination may be quite severe. On the other hand, it may be that the deaeration provided in the condenser hotwell is reasonably good, in which case the introduction of a little over 6% water that has bypassed the deaerating heater may not have serious results. The fact that this installation has operated in this manner for 15 years would indicate that no serious troubles have occurred.

Chapter 1   61

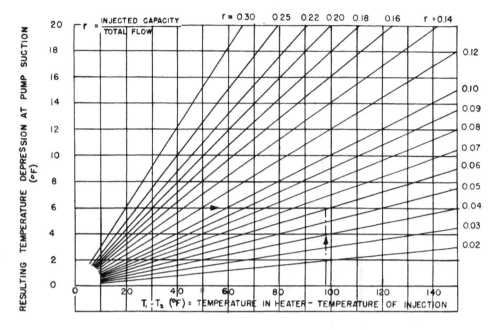

Figure 1.31 Required amount of cold-water injection for a given temperature depression.

Of course, if it were desired to eliminate entirely any possible oxygen contamination, it would be possible to install a heat exchanger in the suction piping so as to subcool the feedwater without introducing any contamination there. The friction loss through the heat exchanger should be added to the desired increase in available NPSH and the amount of subcooling necessary determined from Fig. 1.30.

To avoid wasting the heat by the subcooling process, the cooling medium employed can be the condensate itself on the way to the deaerating heater, as shown in Fig. 1.32.

Question 1.25 Submergence Control with-15 ft NPSH

We are considering the application of a centrifugal pump that will be very similar to the *tail pump* serving a barometric condenser. In other words, it will draw its suction from a vessel under vacuum, and the available NPSH will be the actual submergence to the pump centerline. For the particular conditions encountered, we have been advised that the pump will require approximately 15 ft of NPSH. Because of our desire for the simplest possible installation, we have considered the pump service to be suitable for

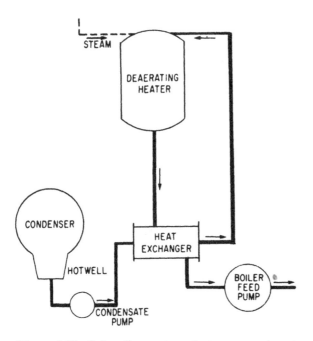

Figure 1.32 Subcooling system that uses condensate of the plant as cooling medium.

submergence control, that is, operation of the pump in the break, with the inflow to the vessel regulating the available submergence and hence the pump delivery, just as in the case of typical condensate installations.

We have been discouraged from using submergence control in this case, and the manufacturer has recommended that a throttling valve be used in the pump discharge, operated by means of a float control.

We would like to know why submergence control is suitable for normal condensate service but not applicable here.

Answer. NPSH control, or submergence control as it is commonly called, is inherently automatic, and since it requires no control equipment, it has been used very widely in steam power plants to regulate the delivery of condensate pumps. A typical performance curve for a pump operated on submergence control is shown in Fig. 1.33, which also shows the system-head curve. It will be noted that, at capacities B, C, and D, the available NPSH is exactly equal to the required NPSH. This will always be the case for any capacity smaller than that at which the "uncontrolled" head-capacity curve intersects the sytem-head curve, since in this region the level

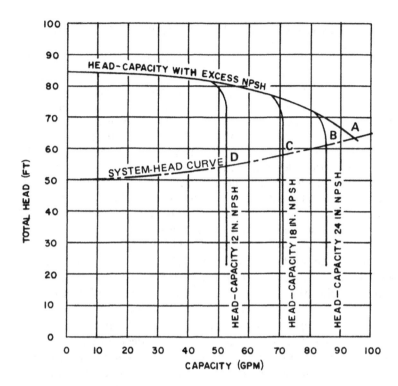

Figure 1.33 Condensate pump performance curve showing how capacity is controlled by means of submergence control.

in the hotwell will be pumped down until the available NPSH will have been reduced to just equal the NPSH required by the pump, after which it cannot be reduced any further.

Thus, when a pump is operated under submergence control, the pump is operating *in the break*, that is, cavitates, at all capacities less than A. The pump may be more or less noisy at all capacities below A, depending on a large number of factors involving design and operating conditions. However, the cavitation that occurs is generally not severely destructive in nature in the case of condensate pumps.

To understand the effect of submergence control on pumps operating with considerably higher values of NPSH, it is necessary to analyze just what the mechanics of cavitation consists of. Cavitation is a generic term that covers the entire scale of phenomena taking place in and about an impeller when it is operated under conditions in which local pressures fall close to or below the vapor pressure of the liquid pumped. In other words, the term covers not only the actual formation of vapor bubbles but also the limitation on capacity, the incidental noise, and the destructive effect on the impeller metal.

All these effects, I repeat, originate in the formation of vapor bubbles when liquid pressure drops just below the vapor pressure. This is followed by the recollapse of these bubbles to liquid when pressure increases to a point just above the vapor pressure. The cause of cavitation damage is the shock wave set up by the collapse, or *implosion*, of these bubbles.

In a pump, these bubbles form at the inlet to the impeller when the available NPSH becomes equal to or slightly less than the required NPSH. The bubbles are carried along in the liquid stream as it passes through the impeller until they reach a point where sufficient head has been generated just to exceed the vapor pressure. At this point, the bubbles collapse. It is at this point where damage becomes evident. It cannot occur upstream of this point, since in that region there exists a mixture of liquid and vapor, and the energy of the shock wave is dissipated in alternate compression and expansion of the vapor. But in the pure liquid phase, the shock wave is propagated through the liquid until it is arrested by the surface of the impeller. If at the point of arrest the shock wave possesses sufficient energy, it actually displaces a minute particle of material from the surface of the metal. Frequent and repeated occurrence of this process produces the pitting that is the usual symptom of cavitation. The degree of pitting for a given degree of cavitation is, of course, related to the relative *toughness* of the metal used in producing the impeller. In addition, there may take place a side effect involving a chemical reaction of the metal to any gas held in solution in the liquid and liberated by the reduction in pressure.

That serious pitting occurs in pumps that cavitate at what may be referred to as "usual" levels of NPSH, such as 15 or 20 ft, is widely known. But it is a curious and much less known fact that condensate pumps operating on submergence control work under essentially continuous cavitating conditions and yet do not usually suffer damage from this cavitation. There must, of course, be an explanation for this observed fact, and the most logical explanation lies in the concept of *energy level* of the fluid stream. Although this is by no means the only possible explanation for this phenomenon and in itself may be only one of several contributing factors, it does lend itself to a rather ready explanation and is understandable without any detailed knowledge of pump design theory.

This energy level is indicated by the NPSH available. Although this is generally expressed simply in feet of liquid, its actual units are foot-pounds per pound. In other words, the NPSH available is a measure of the available energy per pound of the liquid pumped. Thus, in a condensate pump operating at 3 ft NPSH, the collapse of the vapor bubbles occurs at an energy level of 3 ft-lb/lb, and the intensity of the shock wave of which I spoke is determined by this energy level.

In a pump operating at 15 ft. NPSH, the energy of the fluid stream is five times as great, and the intensity of the shock wave similarly greater. In view of this 5:1 ratio, it is not surprising that one of these instances causes no damage, but the other one does. The difficulty, however, arises in establishing the limit of NPSH for nondestructive cavitation.

Despite the extensive literature on the subject of cavitation accumulated in recent years, there is still a great scarcity of positive knowledge regarding the limiting factors of damage-free operation of centrifugal impellers under cavitating conditions. Most of the technical papers that treat this subject deal with the fluid dynamics involved and develop various theoretical relations that, with the help of certain empirical coefficients, will yield approximate conditions under which cavitation may start. But whether operation under cavitating conditions will or will not have an abnormally destructive effect on the impeller remains a matter of experience. We must consider that even the words "abnormally destructive" are subject to very wide interpretation. What constitutes satisfactory life to one operator may be far from satisfactory to another. As a result, it becomes necessary to accept the fact that any limiting recommendations that may be presented here are strictly qualitative and subjective. They must be reevaluated by the reader in the light of his personal experience and opinion.

The limits are undoubtedly different in every installation because of minor differences in pump design, materials of construction, character of the liquid, and general characteristics of the system in which the pump is operating. Admittedly, in a rather arbitrary fashion, it has been the practice of the writer to select a limit of 5 ft NPSH for satisfactory life of a pump on submergence control. This limit may be considered as somewhat conservative but will generally ensure the installation against undesirable and unexpected difficulties.

Question 1.26 "Uncontrolled" Operation or Operation in the Break

In purchasing and designing pump systems, we frequently run into an application in which we can allow the pump to operate "uncontrolled"; that is, we let the suction operate on free level and the level regulate the capacity of the pump. An example of this is in the green liquor transfer pumps carrying 200 °F liquor (SG. 1.1) from the kraft recovery smelt tanks to the green liquor clarifier. Figure 1.34 shows the arrangement. The green liquor in the tank is open to the atmosphere.

We designed a pump installation without controls to handle rejects from the tertiary cleaners and secondary screen of a large newsprint machine. The pump suction is connected to a tank about 6 ft high, and the

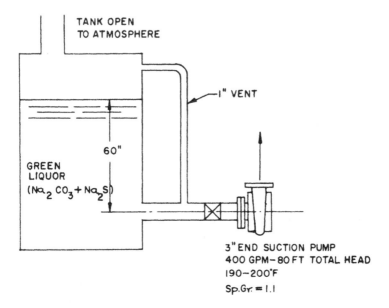

Figure 1.34 This pump arrangement has been used for at least 10 years.

pump casing is vented by a 1 in. line back to the tank. Tank level varies a couple of feet between the air-bound and flooded suction conditions. Neither accurate level control nor uniform flow were important. A control valve on the discharge side would probably have plugged frequently under low flow conditions. The pump wear plate has shown excessive wear after 8 months' service, but this is attributed more to the abrasive fluid than to cavitation. The mill is satisfied that, all things considered, this is a good solution to the level or flow control problem in this particular service.

We also have many applications in which the fluid (usually steam condensate) is at some saturated pressure in a flash tank or in a level-controlled surge tank. In this case, we put a level controller acting on a control valve in the pump discharge to avoid blow-through should the pressure downstream of the pump be reduced (for instance when 50 psig condensate becomes contaminated and is drained directly to sewer). (See Fig. 1.35.) When the level is controlled, we always make sure that the available NPSH is greater than the required NPSH.

In many applications in which blow-through is no problem or the tank is at atmospheric pressure, we run into level controllers on the tanks with level control valves, past the pump. In such applications, is the level control justified? If the pump will stand up to operating continuously on free suction level, even with some increased maintenance, it seems that the

Chapter 1

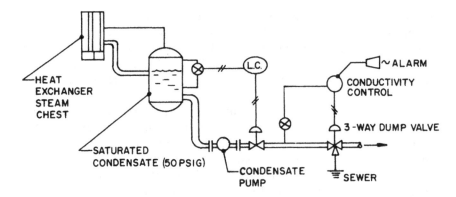

Figure 1.35 Level-controlled condensate pumping system.

cost savings and simplicity of operation would easily justify the removal of the level control. One application that comes to mind is the turbogenerator condenser condensate removal system. I would appreciate hearing your comments on pump serviceability and maintenance when operating as described.

Answer. Your observations on the operation of the green liquor pumps are very interesting and serve to illustrate the fact that there generally are several ways to operate a particular system. The preferable method will always depend on local circumstances and on the personal preferences of the plant operators, since the relative weight given to initial cost and simplicity on one hand and to the life of the components on the other varies widely.

The "uncontrolled" operation that you describe is variously spoken of as *operation in the break* or *submergence control*. It was originally developed and applied to condensate pump service in electric steam power plants many years ago, as described in Question 1.25. The condensate pump is installed with no throttling valve in its discharge piping and feeds the deaerating heater downstream of the pump controlled strictly by the variation in the hotwell level.

The actual mechanics of this submergence control are illustrated in Fig. 1.33. When the NPSH available exceeds that required, the pump operates on its normal head-capacity curve. The capacity delivered by the pump will then correspond to the intersection of this head-capacity curve and the system head curve. In this particular case, this would be somewhere near 90 GPM.

If at this particular moment, the amount of steam being condensed is only 70 GPM, the excess of capacity being pumped out will reduce the level in the hotwell. When this level is reduced to, say, 24 in., the condensate pump will cavitate, and its head-capacity curve will be altered to that shown in Fig. 1.33 for 24.-in. NPSH. The flow will be reduced to 85 GPM. Since this is still in excess of the 70 GPM that is being condensed, the level in the hotwell will continue to fall until it reaches a value of only 18 in.

At this point, the pump can only handle 70 GPM, and an equilibrium will have been reached until such time that the condensed steam falls below 70 GPM or increases beyond it. The former would further reduce the level in the hotwell, simultaneously reducing the pump's ability to deliver the condensate, or, conversely, the level would increase, permitting the pump to handle more condensate. In either case, equilibrium would again be reestablished.

The operation of the pump in the vertical portion of the head-capacity curve (at the intersection with the system-head curve) is called operation in the break. Special impeller designs are used for this type of service to operate on low NPSH and with a special physical configuration to give reasonably quiet operation and long life even though cavitating all the time that they are in operation. Such pumps generally are limited to 100 ft in the first stage, and when higher heads are encountered, a multistage construction is employed.

It has been found that although operation in the break can be reasonably satisfactory with low values of required NPSH, of the order of 2 ft to at most 5 ft, pumps requiring higher NPSH values tend to be noisy, vibrate, and wear rather rapidly. Thus, operation in the break is recommended only for pumps in the former category.

Submergence control was very popular some time ago for condensate service in steam power plants, but several circumstances have united to make it less practical at this moment.

1. The operating pressures of the deaerators have been rising considerably. Since the deaerator pressure varies almost directly with the operating load, the system-head curve between the condenser and the deaerator has been steepened. This means that the reduction in required total head between full load and light loads is so large that cavitation must extend to several stages of the pump. This is not conducive to good operation since, generally, only the first-stage impeller is properly designed to operate in the break.
2. In many cases, the desire to use vertial can condensate pumps at higher speeds with NPSH values as high as 10 to 15 ft leads to the situation I mentioned earlier, in which the type of pump required is not suitable for submergence control.

Chapter 1 69

You will notice that I have sometimes referred to *submergence control* and sometimes to *operation in the break*. The two terms are not really synonymous since submergence control is accomplished by the operation in the break. They are, however, used interchangeably by pump users and pump designers.

Returning to your specific case, however, rather than to present-day large steam power plants, I am certain that there are a great many applications both in process work and in pumping condensate in which submergence control is both feasible and sound practice. It is necessary to examine each case individually, and most important of all, operators must realize that they may be trading lower initial cost and simplicity of operation for a slighty more frequent maintenance. They, therefore, are the ones who must be the ultimate judge in this choice. They should make it with full knowledge of what the choice implies.

Question 1.27 Suction Specific Speed

Because of the frequent use of the expression *specific speed* by pump designers in articles on centrifugal pumps that appear in the technical literature, this expression and its meaning are probably quite will understood by most pump users. On the other hand, I have recently seen references to *suction specific speed*, and I am not entirely sure of its meaning. Would you care to discuss this subject?

Answer. You are quite right in saying that the meaning of specific speed is probably well understood by most pump users, I shall therefore limit myself to a very brief discussion of it, mainly to provide the background for an explanation of suction specific speed.

The early technology of centrifugal pumps was strongly influenced by the experience in the construction of water turbines. The latter were exploited in commercial applications on a greater scale than the centrifugal pump in the late nineteenth and twentieth centuries. As a matter of fact, in continental Europe, it was the turbine builders who branched out most widely into the construction of centrifugal pumps and who became the leading pump manufacturers in the early 1900s. Thus the desire and, I might say, the need to classify the characteristics of centrifugal pumps became most easily satisfied by borrowing techniques developed in classifying water turbines. The specific speed-an index number that identifies the pump type-was first applied to centrifugal pumps around 1905 and was an extension and a modification of the specific speed classification principle developed earlier for water turbines.

Mathematically, this index number is expressed as

$$N_s = \frac{n\sqrt{Q}}{H^{3/4}} \tag{1.9}$$

When English units are used, the rotative speed n is expressed in revolutions per minute, the capacity Q in gallons per minute, and the total head H in feet.

It should be well understood that the specific speed is an index number that can be used to classify the general outline of impeller profiles, the typical performance characteristics of centrifugal pumps, and the range of efficiencies attainable with a given pump "type." For example, Fig. 1.36 illustrates the approximate impeller shapes, the optimum efficiencies, and the shape of the head, power, and efficiency curves of pumps over a wide range of specific speeds.

The development of the specific speed concept was a most important one in the history of centrifugal pump development. At once, it became possible to utilize test and design data on existing pumps to develop new designs of dimensionally similar pumps but of larger or smaller size because the specific speed of a pump remains independent of size. And when centrifugal pump builders found that they were encountering excessive difficulties centering about the suction conditions of centrifugal pump installations, it was to the concept of specific speed classification that they first turned to codify reasonable limitations for these suction conditions.

Codifying Suction Conditions

Every piece of machinery is preordained to have its Achilles' heel. That of the centrifugal pump can generally be found in its suction. This fact must have been discovered quite early in the commercial application of centrifugal pumps and, probably, accepted as an inescapable penalty exacted in return for the advantages that were made available by this newer means of raising water at a more reasonable cost than the slow, cumbersome direct-acting steam pumps in use at that time. Limitations on permissible suction lifts were imposed on strictly empirical grounds. And since the understanding of the phenomenon of cavitation was imperfect, these limitations were sometimes overconservative and sometimes quite optimistic. The attendant difficulties could not have failed to stimulate a considerable amount of head scratching by centrifugal pump designers.

The need for a more exact understanding of the phenomena that take place in the suction area of a centrifugal pump became more pressing as the requirements imposed on these pumps grew in magnitude. Here again, the thinking of the designers must have been heavily influenced by the experience of water turbine builders, since cavitation is a problem of no mean

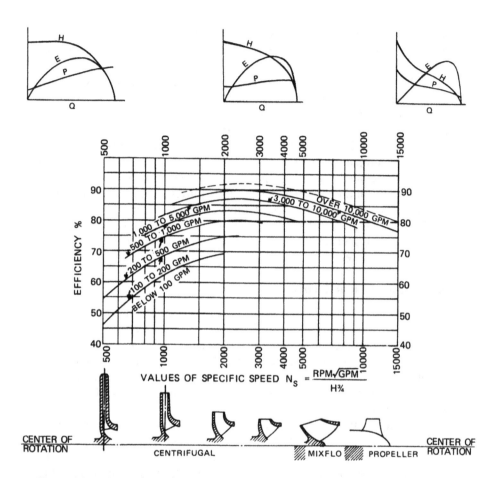

Figure 1.36 Approximate impeller shapes, optimum efficiencies, and the shape of the head, power, and efficiency curves of pumps over a wide range of specific speeds.

magnitude with this type of machinery. In 1922, at the Hydroelectric Conference held in Philadelphia, H. B. Taylor and L. F. Moody first presented the concept of a parameter *sigma* to facilitate the description of the conditions under which cavitation occurs. Sigma was defined as

$$\sigma = \frac{H_s}{H} \tag{1.10}$$

where H_s = net positive suction head
H = total head

At about the same time, D. Thoma was developing the same concept in Germany, and therefore sigma has since then been known in centrifugal pump terminology as the Thoma-Moody parameter.

Means were then available to relate the operating conditions of a centrifugal pump—its capacity, head, and rotating speed—to the minimum net positive suction head required for satisfactory operation. But commerical pressures of a very severe competitive situation seem to have outweighed sound engineering judgment much too often in the 1920s. The number of companies manufacturing centrifugal pumps had proliferated without necessarily a corresponding increase in the number of knowledgeable and experienced engineers. Spurred on by the advantage of offering a higher operating speed than competition or of guaranteeing satisfactory operation with higher suction lifts, some comapnies made installations that had disastrously expensive consequences for user and manufacturer alike.

Specific Speed Limit Charts

An organization for the discussion and solution of technical problems in the realm of pumping machinery had been formed some years before by the older, larger, and most reputable pump manufacturers under the name of the Hydraulic Institute. Standards had been developed and published by this institute to codify sound practices, including testing and guarantee practices. A committee was appointed to investigate centrifugal pump suction problems.

The committee proceeded to collect information on centrifugal pump installations in which cavitation troubles had been experienced as well as on satisfactory installations. It was found that in order to avoid difficulties, the specific speed of the pump should be kept below a certain value for any given total head and suction lift conditions. The conclusions of the committee were published in October 1932 in the Hydraulic Institute Standards in the form of charts that have since become commonly known among centrifugal pump engineers and users as *specific speed limit charts*.

For the record, it is important to remember that these charts were strictly empirical. They did not indicate that pumps built for the limits allowed were necessarily the best design or that pumps built to lower limits were not more economical in certain cases or finally that pumps could not be designed and built for higher limits. All that these charts were intended to indicate was that, for a given set of head, capacity, and suction conditions, a certain maximum rotative speed should give assurance that the pump would be capable of giving satisfactory service.

As experience was accumulated on better designs than described in these first charts of 1932, revised charts were prepared and published by

Chapter 1

the Hydraulic Institute. Figure 1.37 illustrates one such chart issued in 1975 giving specific speed limits for double-suction pumps having the shaft passing through the eye of the impeller.

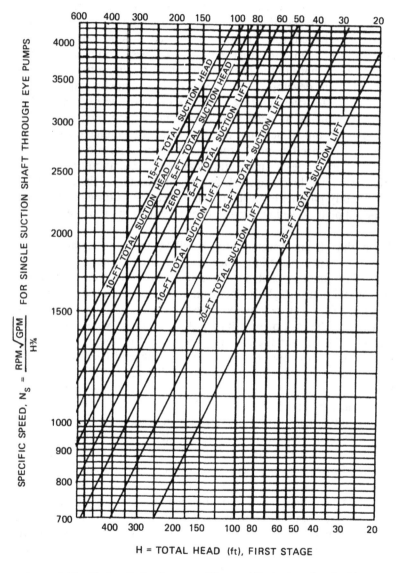

Figure 1.37 Hydraulic Institute specific speed limit chart for double-suction pumps with the shaft passing through the eye of the impeller.

Search for an Improved Classification

But the application of the Thoma-Moody parameter concept or of the specific speed limit charts as they were originally developed had a very important shortcoming: the fact that satisfactory suction conditions were tied directly to the total head developed by the pump. And yet it must be obvious that the performance of an impeller from the point of view of cavitation cannot be affected too significantly by conditions existing at its discharge periphery. On the other hand, it is these conditions at the discharge periphery that are the prime factor in determining the total head that the impeller will develop. In other words, if an impeller exhibits certain suction characteristics, cutting down its diameter within reasonable limits and thus reducing its head should have no influence on its suction capabilities. At the same time, since the total head H is changed, a strict interpretation of the specific speed limit charts would indicate that, unless the suction lift were to be commensurately altered, the maximum permissible specific speed must be changed. Likewise, to maintain a fixed value for the Thoma-Moody parameter, a reduction in head by cutting the impeller diameter should be followed by a proportional reduction in the net positive suction head H_s.

I hope that I am forgiven for dwelling in some detail on the developments that took place in connection with the elimination of this inconsistency. The fact is that I was intimately connected with these developments and that it might be interesting to have on record the manner in which the problem was eventually solved. All this took place in 1937. I had been assigned the task of searching out a theoretically defensible relationship among the various factors that affect the "suction capability" of centrifugal pumps. Unknown to me, the same task had been assigned to two of my colleagues at Worthington, G. F. Wislicenus and R. M. Watson. Both of these men, incidentally, have since distinguished themselves in the area of centrifugal pump theory and of fluid dynamics.

My own investigation was not fruitful in results, and I was about ready to claim that no logical relationship could be demonstrated to exist when I learned of the parallel investigation being carried on by Wislicenus and Watson. They had progressed considerably further than I had, and they had developed reasonable evidence of the fact that the Thoma-Moody parameter H_s/H must be a function of the specific speed $nQ^{1/2}/H^{3/4}$. But the head factor was refusing stubbornly to disappear from the relationship, and they and I knew that, at least for a certain range of specific speeds, conditions at the impeller discharge could not be affecting suction conditions.

In retrospect, the steps that finally led us to the solutions were childishly simple. All that was required was a mere algebraic manipulation.

We suddenly saw that if instead of trying to relate sigma to the specific speed we looked for a relationship between the specific speed and the three-quarter power of sigma, the total head H disappeared very conveniently from the relation. The algebraic steps that were followed are indicated in Fig. 1.38. Thus was born the concept of suction specific speed, to which we gave that name because of its similarity of structure with specific speed.

Of course, if you look up the first two papers that were presented on this new method of representing cavitation results, you will find that a much more sophisticated derivation of the suction specific speed was developed by us.* It can be derived using either similarity considerations or dimensional analysis. But the interesting fact is that it was first stumbled upon by much simpler means.

Specific speed $n_s = \dfrac{nQ^{1/2}}{H^{3/4}}$

Cavitation parameter $\sigma = \dfrac{H_s}{H}$

$\qquad\qquad\qquad = $ **Function of** N_s**?**

Let

$\sigma^{3/4} = \dfrac{H_s^{3/4}}{H^{3/4}} = \dfrac{1}{S} \times n_s$

$\dfrac{H_s^{3/4}}{H^{3/4}} = \dfrac{1}{S} \times \dfrac{nQ^{1/2}}{H^{3/4}}$

and

$S = \dfrac{nQ^{1/2}}{H_s^{3/4}}$

$S = $ **suction specific speed**

Figure 1.38 Derivation of suction specific speed concept.

To say that this concept was received with unanimous approval would be an exaggeration. Many years later I will not be criticized for divulging the fact that a very bitter controversy sprang up immediately over the validity of the concept. But the controversy was short-lived, and the suction specific

---

*"Some Notes on a New Method of Representing Cavitation Results" by Dr. G. F. Wislicenus, R. M. Watson, and I. J. Karassik, presented at the Hydraulic Institute Meeting in New York, Dec. 6, 1937, and "Cavitation Characteristics of Centrifugal Pumps Described by Similarity Considerations" by G. F. Wislicenus, R. M. Watson, and I. J. Karassik, presented at the Spring Meeting of the ASME at Los Angeles, Calif., March 23-25, 1938 (*Transactions of the ASME*, Jan. 1939).

speed has been accepted as the most convenient parameter for describing the suction conditions of centrifugal pumps.

Finally, the 14th edition of the Hydraulic Institute Standards (June 1983) eliminated the earlier shortcomings of the specific speed limit charts of earlier editions. They are now based on a suction specific speed of 8500 for both single- and double-suction impellers. (See Figs. 1.39 and 1.40.) The total head factor has finally been eliminated from these charts. In addition, the recommended values are now expressed in terms of NPSH instead of suction lifts or suction heads, significantly simplifying their use.

Question 1.28 More About Suction Specific Speed

What are the capacities and NPSH values used in the formula for suction specific speed when analyzing pump performance in relation to the internal recirculation at the impeller suction?

Answer. In my answer to the previous question, I failed to make it clear that in using the formula for suction specific speed,

$$S = \frac{n\sqrt{Q}}{H_s^{3/4}}$$

the values for Q and H_s must represent the conditions corresponding to the capacity at best efficiency. Values calculated for any other flow do not have any relevancy and do not describe the suction characteristics of that pump. The guaranteed conditions of service may or may not correspond to this best efficiency flow—they rather seldom do.

In addition, it must be remembered that the suction speed represents the pump performance at the inlet to the impeller. Therefore, for single-suction impellers, Q is the total flow, but for double-suction impellers, Q is taken as one-half the total flow.

What may be less obvious is that the suction specific speed value for any pump must be calculated on the basis of the pump performance with the maximum impeller diameter for which it was designed. This stricture will become apparent when one considers that the internal recirculation at the pump suction occurs because of certain conditions that arise in and around the inlet of the impeller, conditions that are not necessarily affected by cutting down the impeller diameter.

What I mean to say is that cutting down the impeller diameter moves the best efficiency point to a lower flow value but does not reduce the capacity at which suction recirculation will occur. As a matter of fact, in certain cases (in which the ratio between the eye diameter and the impeller diameter exceeds 0.5), the onset of suction recirculation may be triggered by the discharge side recirculation. Here, cutting down the impeller

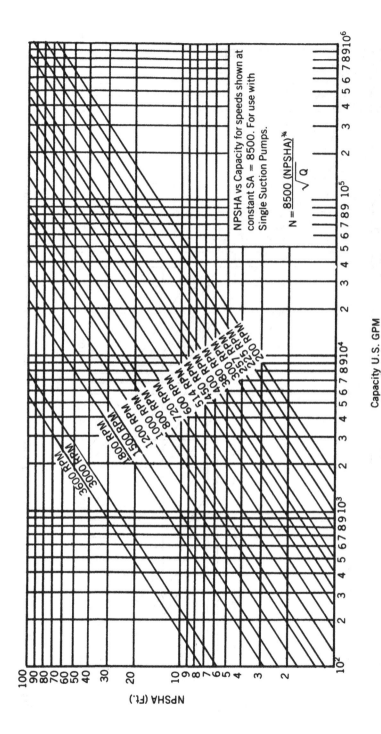

Figure 1.39 Recommended maximum operating speeds for single-suction pumps. (From H.I. Standards, 1983).

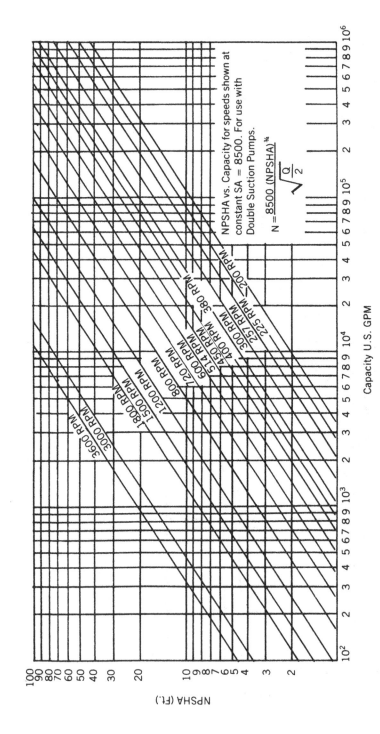

Figure 1.40 Recommended maximum operating speeds for double-suction pumps. (From H.I. Standards, 1983).

diameter may actually increase the actual flow at which suction recirculation takes place.

For this reason, whenever the impeller configuration is such that the ratio of D_1 to D_2 may be close to 0.5, it is recommended to check this ratio and, if applicable, to calculate the capacity at the discharge side recirculation since this value will now govern the onset of suction recirculation.

Question 1.29 Required NPSH at Reduced Flows

In designing a centrifugal pump installation we always check that the available NPSH exceeds the required NPSH for the pump. These data are supplied by the pump manufacturer in the form of a curve of required NPSH plotted against flow. However, this curve usually covers a limited zone, say from 50 to 100% of flow. Although this curve is very useful in designing the sytem for normal operation, the transient operations create some problems. During abnormal conditions, the pump may run at varying capacities (as low as 10% and as high as 150% of rated conditions), and NPSH requirements will naturally change. Is there any safe and conservative method of extrapolating the required NPSH figures beyond the data furnished by the pump manufacturer?

Recently, I was told by a pump manufacturing company representative that when run at 20 to 25% of their rated flow centrifugal pumps require more NPSH than at the rated condition. In other words, the required NPSH versus flow characteristic curve looks like a parabola, concave side up. He referred to this as *incipient cavitation*. I would like to have your comments on this aspect, too.

Answer. Let me first address myself to the question of NPSH requirements at flows higher than rated, or rather higher than the capacity at best efficiency. In general, all the test results that I have examined for radial-flow and mixed-flow centrifugal pumps indicate that it is perfectly safe to assume that the NPSH required from 100% flow on up to as much as 150% varies as the square of the capacity.

With extremely few exceptions, all deviations from this general rule indicated NPSH values just slightly lower than those calculated on the basis of the square law. Even the few exceptions showed NPSH values of only a couple percent higher than calculated on the square of the capacity basis. Thus, although the theoretical value of the NPSH at 150% flow would be 225% of that at 100% flow, the overall range of NPSH values will run from 220 to 230% of that NPSH , with values below 225% predominating statistically.

When it comes to flows below 100% of rated conditions, the situation becomes quite complex and, I might even say, controversial. Because a centrifugal pump can be designed only for one capacity at a given head and speed, the geometry of the impeller can be ideal only for these rated conditions; at all flows *below* this rated capacity, the distribution of velocities and the flow angles are distorted. The required NPSH no longer follows the "square of the flow" rule but decreases at a lesser rate. As a matter of fact, at a certain reduced flow—which varies with each impeller design—a new phenomenon arises, that of *incipient internal recirculation*. This phenomenon leads to disturbances in performance, hydraulic pulsations, and even destructive action on the impeller metal.

But even before we discuss these effects, let us address ourselves to another complex question: that of determining cavitation limits. Over the years, various test techniques have been used to determine and define the cavitation characteristics of centrifugal pumps. But we must realize that regardless of the fact that once ground rules are established we can determine quantitative results, the very nature of these ground rules makes the results we obtain qualitative rather than quantitative.

For instance, although the noise and vibration that accompany cavitation may be used as a means of detection, this is a most inaccurate and inexact method of defining cavitation limits. Two somewhat more specific approaches have therefore been used, with the *cavitation limit* of a pump being defined as (1) an upper limit of capacity for a given NPSH and speed of rotation (Fig. 1.41) or (2) a lower limit of NPSH at a given capacity and speed of rotation (Fig. 1.42). A modified refinement of the second definition has generally been preferred. For instance, the Hydraulic Institute Standards suggest that *suppression tests* be performed over a range of capacities and that plots of head, efficiency, and power input be made for various capacities against a range of values of NPSH. A change in performance, such as a drop in head or in efficiency, is considered to be an indication of cavitation. Because of the difficulty of determining the exact condition when this change takes place, it is the practice to define *required NPSH* as that value at which a drop of 3% in head will have taken place. This is illustrated on Fig. 1.43. The required NPSH at capacity Q and speed n is H_{s2}, which is that NPSH at which the head will have been reduced by 3%.

There are certain problems introduced by such a definition. In certain cases, the use of a 3% drop in head to define the cavitation limit may lead to NPSH values with insufficient margins of safety to prevent ultimate cavitation damage to the impeller. This is referred to in the Hydraulic Institute Standards and the use of 1% drop in head as a limit recommended for certain services.

Another complication is introduced by the fact that at certain reduced flows, centrifugal pumps may exhibit pressure pulsations as shown in Fig. 1.44. The presence of these pulsations prevents obtaining any accuracy in the use of the accepted definition of cavitation limits, since neither the total head developed by the pump nor the suction pressure can be determined accurately in the capacity region where this phenomenon occurs. For this reason, I do not

Chapter 1

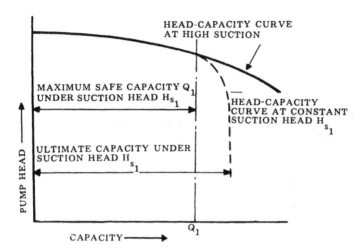

Figure 1.41 Use of head-capacity curve to define cavitation limit.

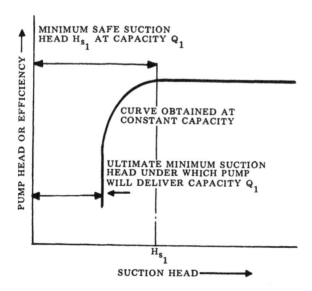

Figure 1.42 Use of change in head to define cavitation limit.

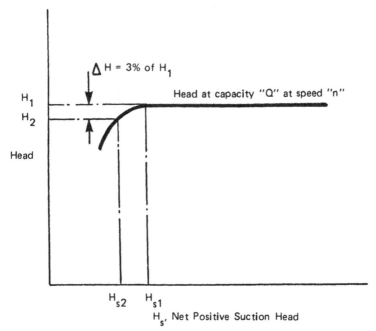

Figure 1.43 Definition of cavitation limit using a 3% drop-off in head.

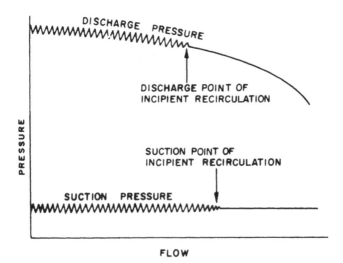

Figure 1.44 Characteristics of internal recirculation.

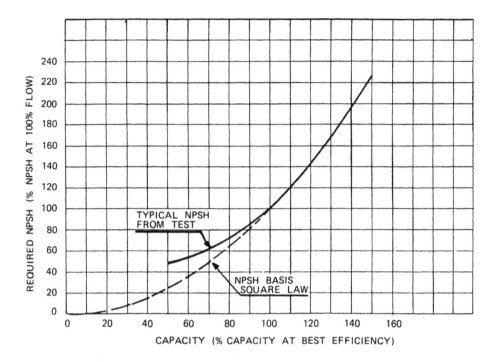

Figure 1.45 Test NPSH values

consider NPSH curves extended below 50% of best efficiency flow to be too accurate. In general, however, these curves will resemble those shown in Fig. 1.45, where the deviation from the square law is clearly indicated.

A few words about your reference to NPSH curves resembling concave-side-up parabolas. At the capacity at which incipient internal recirculation occurs at the suction eye of an impeller, the flow at the outer eye diameter has a tendency to reverse itself (see Fig. 1.46) and develops a vortex. It is the collapse of these vortices that creates a form of cavitation leading to impeller damage. Although it is impossible to calculate or measure the exact value of NPSH required to prevent this cavitation, it can be indicated qualitatively as shown in Fig. 1.47, which shows conditions prevailing in two different impellers: (A), an impeller with very low NPSH and an extremely oversized eye diameter, and (B), an impeller with moderate NPSH requirements. It becomes evident that impeller (B) develops an internal recirculation at flows considerably lower than impeller (A) and is less apt to be subject to cavitation damage.

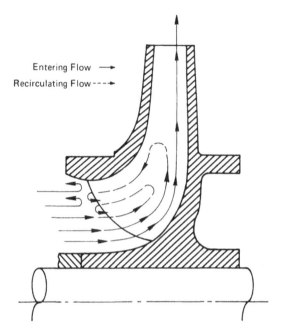

Figure 1.46 Section through a single-suction impeller indicating schematically the recirculation of liquid at the inlet during operation at low capacities.

Question 1.30 Suppression Test

I have several questions with regard to suction specific speed, as follows:

1. When purchasing a new pump, when is it advisable to request a suppression test to determine NPSH?
2. Is the guideline of keeping the suction specific speed S below 10,000 to 12,000 still valid; is this value applicable for any liquid?

Answer. As stated in the answer to the previous question, the minimum NPSH required is determined by a test in which the total head is measured at a given speed and capacity with varying NPSH available conditions. Preferably this test is conducted in a loop such as that described in Fig. 1.48, and it is then referred to as a "suppression test." The pump takes its suction from a closed vessel in which the pressure level can be adjusted and the vapor pressure of the liquid can be varied by adjusting the temperature level. The difference between the pressure level at the pump centerline and the vapor pressure of the liquid controls the available NPSH.

The results of such a test, plotted against NPSH, appear in a form similar to that in Fig. 1.43. At the higher values of NPSH available, the

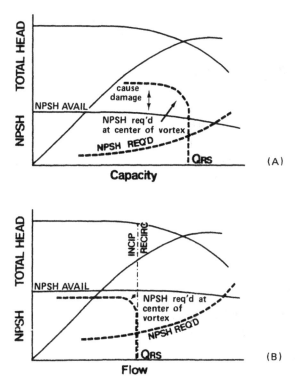

Figure 1.47 Relation between NPSH, suction recirculation, and cavitation: (A) low NPSH design; (B) moderate NPSH design.

values of head remain substantially constant. As the available NPSH is reduced, a point is reached where the curves break, showing the impairment of performance caused by cavitation. The exact value of NPSH where cavitation starts is, as mentioned earlier, difficult to pinpoint. It is for this reason that the Hydraulic Institute Standards define required NPSH as that NPSH available at which a drop of 3% in head has been demonstrated.

The Hydraulic Institute Standards permit two simpler forms of cavitation tests. In the first arrangement (Fig. 1.49), the pump takes its suction from a constant-level sump through a throttle valve that is followed by a section of pipe containing a screen and straightening vanes. The operation of the throttle valve is used to vary the available NPSH. In the second arrangement (Fig. 1.50), the pump takes it suction from a relatively deep sump in which the level can be varied to establish the desired available NPSH.

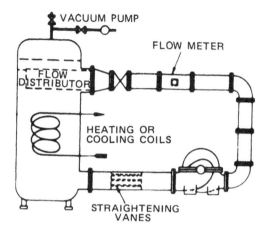

Figure 1.48 Closed-loop setup horizontal pump.

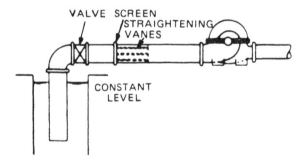

Figure 1.49 Constant-level sump supply.

It is rather difficult to give a cut-and-dried answer to your first question unless one were to interpret the meaning of the words "a new pump" to be that your are buying a pump that is the first of its kind, that is, a brand-new design that has not been manufactured before. In that case, I would prefer to have the NPSH characteristics determined by a suppression test, assuming that we are not dealing with such a large pump that the suppression test is impractical. In such an event, calculations of NPSH required by the prototype can be determined by the methods indicated in the Hydraulic Institute Standards (see 14th edition, 1983).

If, however, we are dealing with pumps that have been manufactured previously, it is more than probable that the manufacturer will have carried out suppression tests on at least one such pump and you would probably

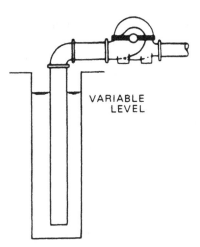

Figure 1.50 Deep sump supply.

save considerable expense by waiving the suppression test and accepting cavitation tests as described in Figs. 1.49 or 1.50. You will note the comments given in the Standards regarding possible problems encountered if the liquid contains dissolved air or gas.

I should add that the decision varies among various users. For instance, I know that several oil companies specify suppression tests whenever the NPSH available exceeds the NPSH required by less than 4 ft.

API-610, 6th edition, hedges: paragraph 4.3.4.2 says: "A vacuum tank suppression test is preferred. NPSH testing by suction valve throttling may be used when mutually agreed upon."

Guidelines on suction specific speed values vary considerably. They depend on a wide range of considerations. Probably the foremost question involves whether the pump will be operated over a wide range of capacities. Another major consideration is the size and horsepower of the pump in question. Certainly a 12,000 S value is excessive if we are dealing with a 5,000 hp boiler feed pump, and I wouldn't exceed 8500 to 9500 for such service. On the other hand, an S value of 12,000 will probably not create any major problems for a small pump. The reason for different guidelines for different pump horsepowers is the following: higher values of suction specific speed cause internal recirculation to occur at higher percentages of the best efficiency capacity; at low energy levels, suction recirculation is not particularly harmful, which explains why large pumps operating at low speed (and therefore low heads) can be run successfully at high suction specific speeds. When dealing with high speeds (high rotating speeds and fluid velocities), suction recirculation becomes harmful, shortening impeller life and—in some cases—producing extremely high pressure pulsations. I might add that even high values of S can be used satisfactorily, provided you set up reasonable minimum flow conditions.

Question 1.31 Effect of Speed on the Required NPSH

How does the NPSH required vary with the speed of the pump? Is the following relation correct?

$$\frac{NPSH_1}{NPSH_{r2}} = \left(\frac{N_1}{N_2}\right)^2$$

Answer. We have conducted many NPSH tests on pumps of different specific speeds and at different speeds, but we have never found a consistent deviation from the affinity laws, which say that as the capacities vary directly with the speed, NPSH required values vary with the square of the speed. There are two factors that may cause apparent deviations in the affinity or modeling laws:

1. Particularly at low values of required NPSH, the dissolved air content of the water gives an apparent higher NPSH value. Then, when these results are stepped up to higher speeds and higher NPSH values at which the air content has less effect, the NPSH as measured on test appears to be less than the theoretically stepped-up value.
2. Low values of NPSH at reduced speeds are more difficult to measure accurately than the higher values at higher speeds. In other words, we are stepping up readings with a greater margin of error to a higher speed with a smaller margin of error.

Some authorities have said that they have data indicating that the NPSH increases with some exponent somewhat less than 2.0, but they also refer to the fact that the effect may be only apparent and caused by a more readily liberated air or gas in solution. I personally prefer to consider that using an exponent of 2.0 is a safer approach when stepping up results. If, as we step up the speed, the NPSH required ends up being slightly less than predicted, nothing will be lost and the user will be pleased that the results are better than anticipated.

Question 1.32 Effect of Entrained Air on NPSH

What is the effect of entrained air bubbles on the NPSH requirements of a centrifugal pump?

Answer. This is a fairly complex phenomenon and, as of this moment, cannot be described by an exact mathematical relation that could be applied indiscriminately to any given pump. As a matter of fact, I find myself compelled to discuss this subject in a much broader context than strictly in connection with the required NPSH, for reasons that will become evident presently.

To start with, we shall assume that the original NPSH tests for the pump in question have been conducted with deaerated cold water, as specified in the Hydraulic Institute Standards. Let us then consider what happens when the pump is again tested with cold water that contains a certain amount of entrained air. The air bubbles will pass from the suction toward the entrance to the impeller, in other words into a zone where the ambient pressure is even lower than at the pump suction flange. Air being a compressible fluid, the bubbles will expand in volume inversely as the absolute ambient pressure. In other words, they will occupy a volume even greater by percentage than at the suction flange.

But whatever that volume by percentage may be, once these bubbles reach the entrance to the impeller, they will occupy space that would normally be filled with liquid. It follows that for the same net capacity of liquid being handled by the pump, the liquid velocities will be higher than if there were no air present.

These higher velocities, in turn, will require a greater transformation of static pressure into kinetic energy, and consequently the presence of the entrained air will have led to a greater lowering of the ambient static pressure at the entry to the impeller. Thus, I am led to conclude that the presence of entrained air (or gas) tends to increase the required NPSH for all rates of flow.

The key question remains "by how much?" And this is a difficult question to answer because the effect on the NPSH will be masked by the effect of the entrained air on the head-capacity performance of the pump. Our problem stems from the fact that we are confronted by two separate but simultaneous phenomena:

1. The currently accepted definition of required NPSH is that it is that value of NPSH that, for a given capacity and at a given speed, produces a 3% drop in total head, as illustrated in Fig. 1.43.
2. On the other hand, it is a well-known and documented fact that if as little as 1% by volume air or gas is entrained with the liquid pumped, the head-capacity curve is noticeably reduced. As this percentage by volume increases, the reduction becomes even more drastic, until at 6% by volume we reach a condition in which most pumps cease to perform satisfactorily.

Let us examine the performance of a specific pump under conditions such that the NPSH available does not affect the head-capacity curve, in other words, conditions such that what we are seeing is solely the effect of entrained air. (See Fig. 1.51). You will note that at 750 GPM, the total head is 83.5 ft with 0% air and falls to 78 ft with 2% entrained air, a loss of 6.5%. Obviously, this loss exceeds the 3% used to define the NPSH required.

It seems to me that there might be a way out of our quandary, although I am not fully satisfied that the results would be fully valid. We could conduct our tests for NPSH, not with deaerated water, but rather with whatever percentage of entrained air we wish to use. The head-capacity curve with ample NPSH would be used as our benchmark. Then, a regular suppression test would be conducted and the required NPSH with that percentage of entrained air established. Finally, this last value would be compared with the required NPSH at the tested capacity under deaerated water conditions. I repeat that the results may not completely isolate the effect of the entrained air on the required NPSH.

But what is more important is that in my opinion the results may not necessarily be "generic," that is, applicable to all pumps of that general type. Certainly much more analysis and many more tests would be required before one could answer your question with precision.

There is one more caveat to enter here: I know of no 100% accurate method for determining the percentage of entrained air or gas under any given conditions. The data cited are most often estimated rather than measured.

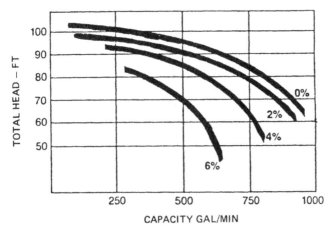

Figure 1.51 Effect of entrained air on head-capacity of a centrifugal pump.

Chapter 1

But this is not the entire story, and we should look at a few additional matters. Oddly enough, for instance, the presence of entrained air does have some beneficial effects as well as the unfavorable ones I have cited. Under inadequate suction conditions, when flashing occurs in the entrance passages of the impeller, the vapor bubbles are swept along until they reach an area of higher pressure between the vanes, such that the vapor bubbles are recondensed. It is this collapse of the vapor bubbles and the consequent high velocities of the liquid rushing to fill the voids that cause both the noise of cavitation and the damage sustained at the metal surfaces of the impeller.

But if the liquid contains a certain amount of entrained air or gas, the air bubbles *do not* collapse because they are noncondensable. They are only slightly compressed, proportionally to the ambient pressures. And wherever they are, they act to absorb the energy of the liquid rushing in to fill the voids created by the condensation of the vapor bubbles. In so doing, they tend to reduce the noise of cavitation.

This is a fact that has been well known and documented for, I imagine, the last 60 or 70 years, and has been frequently resorted to on many occasions known to me. In other words, a very slight amount of air is introduced *on purpose* at the pump suction, and the characteristic cavitation noise either disappears or is satisfactorily reduced. Of course, the process served by the pump in question must be such that the presence of air in the liquid (generally water) can be tolerated. Such is the case for instance with sewage pumps that may be discharging into aeration basins. On the other hand, one wouldn't dream of quieting condensate or boiler feed pumps in this manner.

The effect on the damage to the impeller metal is a much more complex matter. Theoretically, assuming a liquid and metal combination such that corrosion is not a possible factor, the damage may also be reduced by the effect of some of the air bubbles acting as buffers to absorb the impact of the recondensed liquid on the metal. On the other hand, in many cases the reverse has been observed, with the damage appearing to have been intensified by the presence of entrained air. This is probably because the presence of air may accelerate the process of corrosion sufficiently to mask the beneficial effect of reducing the mechanical impact.

One such example involves the use of cast-iron impellers in pumps in cooling tower service. Cast iron is not particularly corrosion resistant, but it is even more susceptible to damage in the presence of the highly oxygenated water that exists in this particular service. As a matter of fact, there is ample evidence of severe pitting damage to cast-iron impellers even in the absence of suction condition deficiencies or of pump operation at excessively reduced flows. And once the corrosion process, accelerated by the presence of oxygen, takes place, the high velocities within the impeller remove any protective film that may have formed on the surface, exposing new virgin metal to the same corrosive attack. For these reasons, it is my firm opinion that cast-iron impellers should *never* be used in cooling tower service. On the other hand, the problem does not seem to exist with respect to cast-iron casings used in this service, provided that the impellers are made of 11 to 13% chrome-steel (or of aluminum-bronze, if the water analysis permits this combination). The probable reason for the acceptability of cast iron for the casing material is that the prevailing velocities are much lower and that no significant flow separation occurs near the casing metal surfaces.

Question 1.33 NPSH Required for Cryogenic Pumps

It is found that many pumps used in liquid propane gas (LPG) transport have NPSH characteristics similar to those shown in Fig. 1.52. Why does the required NPSH increase with reduced flows between A and B?

Answer. It has been frequently claimed by users that the required NPSH curve of cryogenic pumps has a marked increase as the capacity falls to some 20 or 30% of design conditions. I think that we may unravel this mystery if we examine the facts without any preconceived opinion.

The question involves us with some ambiguity introduced by our semantics. If I consider your question on strictly its literal interpretation, the answer is, "No, a pump handling cryogenic liquids does not require any greater NPSH at reduced capacities than would be the case if the pump were to handle water." But had you instead worded your question, "Should a pump handling cryogenic liquids be provided with appreciably more available NPSH than it would require when handling water, if the pump is expected to operate at reduced flows?" my answer would be an unequivocal "yes."

The problem of semantics arises from our accepted definition of NPSH, be it required or available. The only practical means to define

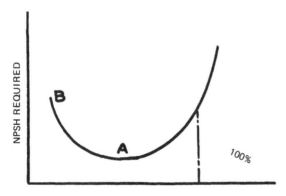

Figure 1.52 NPSH required for cryogenic pumps.

NPSH is to refer to conditions prevailing at the pump suction flange and corrected to the pump centerline (assuming for the sake of simplifying this explanation that we dealing with a horizontal pump). The energy at that location less the vapor pressure of the liquid, expressed in ft-lb/lb or feet, is defined as the NPSH available, but the NPSH required by the pump at a given speed and for a given capacity is defined as the NPSH available when the total head of the pump is reduced by exactly 3%. (See Fig. 1.43.)

The observed facts are strictly apparent, not real. What one sees in a test for NPSH of a cryogenic pump is an error in measuring the NPSH available, not an increase in NPSH required.

After all, remember, one does not measure the NPSH required, one measures the available NPSH and then determines the required NPSH by observing the test values and calling that NPSH available at which a 3% drop in head occurs the required NPSH. The available NPSH is stated to be equal to the energy over and above the vapor pressure at the pumping temperature *at the pump suction flange.*

In most cases, the fact that the vapor pressure at the impeller inlet differs to some extent from the vapor pressure at the suction flange can be neglected because this difference is negligible. But in the case of cryogenic pumps this difference can become so large under certain conditions that the effect is quite dramatic.

At or near the best efficiency point this cannot introduce any significant error since the temperature rise in the pump is negligible and the flow past the wearing ring of the first stage is but a diminutive fraction of the flow into the pump. Thus, the temperature at the eye of the impeller does not change appreciably from the temperature at the suction flange and the assumed vapor pressure is essentially correct.

But as the capacity is reduced, the temperature rise increases while the leakage flow increases as a percentage of the suction flow. Of course, the calculated temperature rise takes place in the discharge passages. But some of the liquid flows back into the suction through the clearance joints and mixes with the incoming liquid.

The net effect is that the temperature at the eye of the impeller is no longer the same as at the suction flange, nor of course is the vapor pressure.

Consider, for instance, the effect of an increase in liquid temperature of 1 °F on the vapor pressure of water at 80 °F and, say, of methane at the usual pumping temperature of -240 °F. For water,

Temperature	Vapor pressure
80 °F	0.507 psia
79 °F	0.490 psia
$\Delta t = $ 1 °F Difference	0.017 psi
or	0.04 ft

and for methane,

Temperature	Vapor pressure
−240 °F	33 psia
−220 °F	64 psia
Δt = 20 °F Difference	31 psi
Δt = 1 °F Difference	1.5 psi
or, at specific gravity of 0.4 =	8.95 ft

In other words, an increase in temperature of 1 °F increases the vapor pressure of 80 °F water by 0.04 ft and that of −240 °F methane by 8.95 ft.

If we were to imagine that at some low flow the effect I have described raises the liquid temperature by 0.5 °F at the eye of the impeller, the result is to increase the vapor pressure by a negligible amount if the liquid is 80 °F water but by as much as almost 4.5 ft if it is methane at −240 °F.

This increase in vapor pressure is not normally taken into account when running the NPSH test, and therefore the real NPSH available is 4.5 ft less than the apparent NPSH available, if we use the temperature rise I have assumed. Since by definition the NPSH required is that NPSH available that will not cause a drop in total head of over 3%, *It appears that the NPSH Required Has gone up.* But it hasn't, really.

Question 1.34 Head and Brake Horsepower of High Specific Speed Pumps at Shutoff

Given the capacity, design head, and operating speed of a vertical pump having a specific speed within the range of 6000 to 9000 (U.S. units),

1. Is it possible to predict the shutoff head of the pump? Could this be different for different pump manufacturers?
2. Is it possible to have an impeller design for such a pump that would have a nonoverloading power curve?
3. Since most pumps in this range of specific speeds have a rapidly rising power curve toward shutoff, what should be the basis for the selection of the motor rating? Should it just cover the duty point requirement with a margin of, say, 15%, or must it be adequate to meet the shutoff horsepower condition?
4. What is the safest zone of operation in terms of percentage of the duty point for such pumps?

Answer. 1. Given the capacity, head, and speed of the pump, it is not only possible but also relatively easy to predict the shutoff head of the pump, once certain design characteristics have been selected. What is meant by this statement is that for any given specific speed there is a rise to shutoff values that can be achieved by manipulating some of the impeller and/or casing design factors. Within this range there is a certain optimum or preferred value, and it is such an optimum that is indicated on such curves as that in Fig. 1.53. Deviations from these optimum values can and do occur, not only for different pump manufacturers but even for impellers developed for various pump lines by any single manufacturer. But these deviations are not very significant in magnitude.

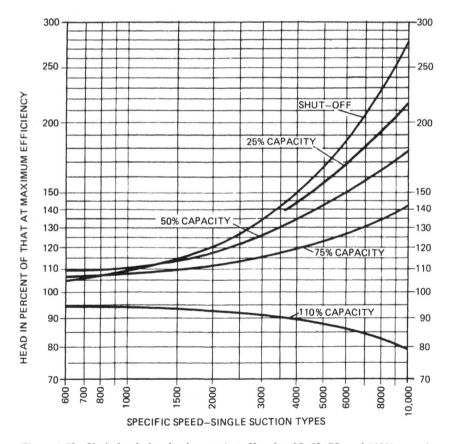

Figure 1.53 Variation in head values at shutoff and at 25, 50, 75, and 110% capacity with specific speed.

2. In the range of specific speeds of 6000 to 9000, it is not possible to develop impellers that would have a nonoverloading power curve. Optimum bhp values in terms of bhp at design are shown in Fig. 1.54. You will note that the following values are indicated:

Specific speed	bhp at shutoff in terms of bhp at design flow (%)
6000	130
9000	190

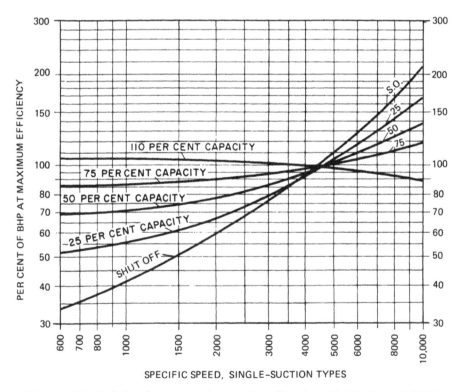

Figure 1.54 Variation in power values at shutoff and at 25, 50, 75, and 110%, capacity with specific speed.

The value of 130% could probably be reduced to 125% or even 120% by design manipulation, but this might have very unfavorable effects on the pump performance.

3. The motor rating for pumps having high shutoff bhp values need not be based on those values but rather should be chosen to cover the normal expected operating range of the pump. To avoid overloading the motor during start-up, the butterfly valve in the discharge is timed to open to 1/8 or 1/4 by the time the motor reaches half-speed. This, of course, requires the plant designer to work out the rate at which the valve opens and the pump accelerates and to properly balance these two factors. Of course, some margin should be provided in the motor rating, and the 15% you refer to should be adequate for this purpose. But no hard-and-fast standards can be established to correlate maximum pump bhp and motor rating. The final decision has to be based on whether overload capacity is built into the motor, on what the permissible motor temperature rise is, on the ambient temperature, and other factors.

4. Likewise, it is difficult to predict the range of the "safe operating zone" of pumps of 6000 to 9000 specific speed, as this depends on too many factors. The safest zone is, of course, right at the "best efficiency" capacity. However, I suggest that in most cases, operation between 80% and 120% of that capacity should be reasonably safe.

Question 1.35 Efficiency of Pump with Inducer

Does the use of an inducer affect pump efficiency?

Answer. Theoretically, the answer is yes: it will generally reduce the pump efficiency, but by an insignificant amount. Generally, rating curves do not differentiate in this respect between pumps with and without inducers as far as efficiencies are concerned. Consider that the inducer develops probably less than 10% of the total head of the pump. It would take a 10% reduction in efficiency between inducer and conventional impeller efficiencies to result in a 1% overall reduction. And remember that in the majority of cases the inducer head will be even less than the 10% I mentioned, sometimes as little as 5% or less.

Question 1.36 Retrofitting a Pump with an Inducer

Can inducers be retrofit to existing pumps?

Answer. An inducer is a low-head axial-flow impeller with few blades placed in front of a conventional impeller, frequently used as an option for

ANSI and API-610 pumps. It requires considerably less NPSH than a conventional impeller. Inducers are used for this reason to reduce the NPSH requirements of a given pump or to permit the pump to operate at higher speeds with a given NPSH available.

Whether a pump can be retrofit with an inducer will depend on a number of factors. For instance, if the pump in question is available from the manufacturer optionally with or without an inducer, the answer, obviously, is "yes." On the other hand, if this option does not exist as a standard, one would have to consult the manufacturer. An inducer must be carefully matched to the conventional impeller it precedes, and only the original manufacturer can determine whether such a matching inducer can be accommodated within the pump. Furthermore, if the inducer is not a standard optional feature, the procedure of designing and producing this inducer may make this solution extremely expensive and some other means should be considered to reduce the NPSH required or to increase the NPSH available.

Question 1.37 Effect of Viscosity on Centrifugal Pump Performance

What are the limits to the use of centrifugal pumps imposed by the viscosity of the liquid pumped?

Answer. The real limits are essentially imposed by the economics of pump selection. Two of the major losses sustained by a centrifugal pump are those created by the fluid friction of the liquid in the pump and the disk horsepower, that is, the power expended in dragging the impeller through the pumped liquid contained in the pump casing. Both these losses vary with the viscosity of the liquid, increasing very rapidly as the viscosity increases so that the head-capacity curve and the power consumption differ unfavorably from their values when handling water.

Considerable testing has been carried out to determine the effect of viscosity on the performance of centrifugal pumps, and the Hydraulic Institute Standards provides charts that can be used in predicting the performance of pumps handling liquids of varying viscosities from the knowledge of the pump performance when handling water. Figure 1.55 shows the effect of viscosities ranging from 32 SSU (that of water) through 4000 SSU. Although the pump had an efficiency of 76% at (BEP) when handling water, the pump efficiency is reduced to about 20% when handling liquids with a viscosity of 4000 SSU.

Obviously an evaluation must be made between centrifugal pumps and positive displacement pumps (screw, gear, piston, or plunger) to make

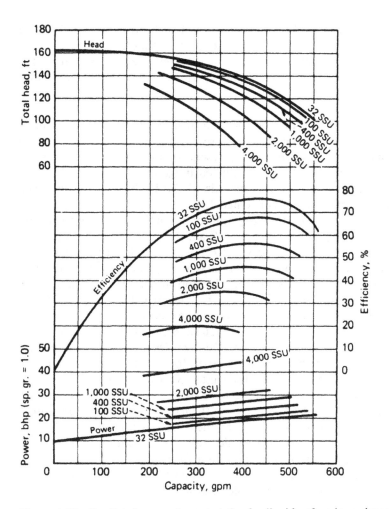

Figure 1.55 Predicted pump characteristics for liquids of various visocosities.

a decision justifiable on the economics of the case. But as a general rule it can be stated that in most cases 2000 SSU will prove to be the upper limit for centrifugal pumps.

Question 1.38 Formulas, Symbols and Abbreviations

I find frequent occasion to refer to a set of engineering data published by a pump manufacturing company. It contains several pages entitled "Formulas Useful in Pump Application." Each page lists a number of formulas and provides a tabulation of the abbreviations used in these formulas. On one of these pages I find the abbreviation H = head in feet, and on the succeeding page appears the abbreviation TH = total head in feet.

Which of these abbreviations is the accepted one, and is there such a thing as an "authorized" list of abbreviations to be used in connection with centrifugal pumps?

Answer. The fact that any specific item of information appears in print does not necessarily make it right or sacred. In this particular case the term H is the one normally accepted to represent head or total head. The most authentic and, one might say, authorized usage of symbols or abbreviations is that contained in the Standards of the Hydraulic Institute. The latest issue of these standards is dated 1983.

Your questions points up the typical confusion that arises too frequently from sloppy thinking among engineers. (I am sure that a similar ailment can be found in other professions as well.) An engineer will generally take particular care to verify the validity of formulas and equations used or presented to readers but will seldom spend as much effort in selecting nomenclature or symbols. The engineer seems to be saying, "Oh, well, I know perfectly well what I mean. It should be quite obvious to the reader."

To begin with, we should distinguish between the two words *abbreviation* and *symbol*. If we refer to *Webster's Dictionary*, we find that an abbreviation is a letter or a few letters used for a word or phrase, such as "Gen" for Genesis and "USA" for United States of America.

On the other hand, according to Webster, a symbol is "a letter or character which is significant; a sign, as the letters and marks representing things and operations in chemistry, mathematics, astronomy, etc."

In this particular case, TH would really be an abbreviation for total head, while H is a symbol to be used in a formula.

There are certain basic rules that should be observed in connection with symbols. Among these, I would cite two:

1. A single symbol should be used for a single concept.
2. Usage must be recognized.

If you refer to the Hydraulic Institute Standards, you will find that both these rules have been respected. Letter symbols have been assigned to certain concepts. Subscripts are used to distinguish between subsidiary concepts. For example, note the formulas presented in Table 1.2.

Note in the examples that although h is the symbol for net positive suction head, NPSH is the abbreviation for the same concept. This is probably one of the best examples that illustrate the difference between an abbreviation and a symbol.

It is interesting to note that usage changes among engineers as well as in the field of languages in general. However, in the case of engineers,

Table 1.2 Formulas Illustrating Distinction Between Abbreviations and Symbols

H = total head, in feet
h_s = total suction head, in feet
h_a = atmospheric pressure, in feet absolute
P_a = atmospheric pressure, in psia
h_{sv} = net positive suction head, in feet (NPSH)

Source: Standards of the Hydraulic Institute.

usage generally yields to logic rather than to the vagaries o the acceptance of once-ungrammatical practices into the spoken language.

A typical case in point is the disappearance of the expression *total dynamic head*, frequently abbreviated as TDH. It once referred to what is now called total head.

Question 1.39   How to Develop Approximate System-Head Curves

You and other authors frequently make use of the term *system-head curve*, and I am interested in learning how one goes about developing the head and flow information necessary to plot such a curve for an existing system.

A typical water system at our plant would consist of a combination of fixed- and variable-speed centrifugal pumps discharging into a common header. We attempt to control the minimum system pressure by varying the pump speed and/or the number of pumps operating.

None of our water is metered, at either the source or point of use. The elevation of the end of the system is approximately 100 ft above the pump discharge. Over 50% of the system flow is used at the higher elevations.

I will greatly appreciate your help and explanations on this subject.

Answer.   A pump operating in a system must develop a total head, which is made up of several components:

1. The static head between the source of supply and the point of delivery
2. The difference in pressures (if any) existing on the liquid at the source of supply and at the point of delivery
3. The frictional losses in the piping, valves, and so on, in the system
4. Entrance and exit losses at the source of supply and at the point of delivery, respectively

5. The difference in velocity heads at the pump discharge and at the pump suction

Because the first two components generally do not vary with flow, they can be lumped together into a single term and become the *total static head*. On the other hand, the remaining three components vary as a function (roughly as the square) of the flow through the system. They are frequently lumped together into a single term and considered as *friction losses*.

If the sum of the total static head and of the friction losses for a series of assumed flows is plotted against flow (or pump capacity), the resulting curve is called the *system-head curve*. Such a curve is illustrated in Fig. 1.56. If either the static head or the pressure difference varies under certain conditions, it is necessary to plot two or more such curves. At least one of these curves should correspond to the minimum total static head and one curve to the maximum total static head.

To determine the capacity that a given pump or a group of pumps will deliver into the system, it is merely necessary to superimpose the head-capacity curve of the pump or pumps on the system-head curve. The intersection of the head-capacity and system-head curves will indicate the flow that will take place through the system (see Fig. 1.57.)

The construction of system-head curves for installations in which water is taken from branch lines at different elevations or where several pumps that have considerable individual piping discharge into a common line is a fairly complex undertaking, and it would not be practical to present the procedure involved in a discussion such as this. I suggest that you refer to a book on centrifugal pumps and to their application.*

I am puzzled by your remark that none of the water is metered. It would be impossible to plot any kind of a curve without some knowledge of the flows that take place. Even an approximation would be better than nothing. Unless you have some information on the flows delivered, you can make no evaluation of the relative state of repair of your pumps. Nor can you determine whether it is better to use more or fewer pumps at any particular time.

You can probably install some form of metering orifices at various strategic points. With the readings obtained from these meters and head measurements obtained by pressure gauges, you will be able to plot the curves needed to evaluate your system.

---

*One possible reference is to Chap. 18 of *Centrifugal Pumps* by Igor Karassik and Roy Carter, McGraw-Hill, New York, 1960.

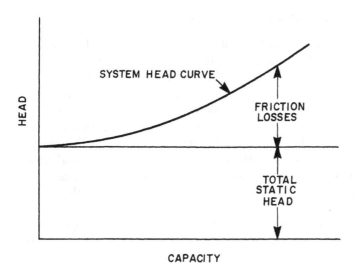

Figure 1.56 Construction of system-head curve.

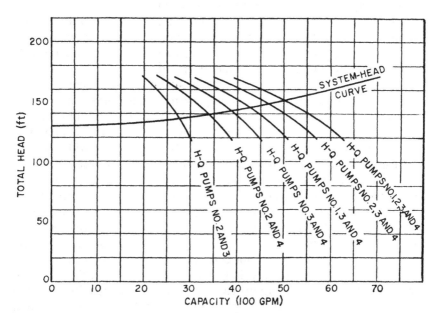

Figure 1.57 Head-capacity curves superimposed on system-head curve.

Question 1.40 More about System-Head Curves

The information contained in your answer to Question 1.39 was very thought provoking. I was also very pleased to have the reference to your book. It will be invaluable for present and future pump problems.

Our main distribution system consists of one 30, and 24, and two 16 in. underground cast-iron mains. Because of the size and location of these mains, the installation of orifice or venturi meters would be a project of considerable magnitude and cost and cannot be justified at this time.

I had considered tapping the mains for the temporary installation of pitometers. However, plant management is not in favor because of the possible danger of cracking these old cast-iron pipes and the resulting serious consequences.

In view of these apparent limitations on metering, it appears that I will have to forego, for the present, my plans to develop a system-head curve.

Answer. All is not lost. In simple words, there is always more than one way to skin a cat. You will remember that I said "Even an approximation would be better than nothing" with respect to the rates of flow through your system. Let us then address ourselves to the problem of developing a procedure by which such an approximation can be obtained.

Of course, the relative accuracy of the approximation will depend on the amount of information you have on the pumps serving your system. I shall therefore present several suggestions-each of these based on a different assumption as to the available information.

Our first assumption will be that the pumps are of recent "vintage" and that you have on hand the pump performance curves, in the form either of certified test curves or of proposal curves. If the pumps are fairly new, or if you have recently restored them to their original condition by renewing the internal wearing parts, today's performance will essentially duplicate the pump performance at the time of the shop test.

A reasonably close approximation of the pump capacity a any given operating condition can then be obtained by measuring as accurately as possible the total head developed by the pump and the power consumption at this condition. Referring to the pump performance curve, read the pump capacity at that total head and at that power consumption. If the two points of capacity so read do not coincide, average the two capacities.

In this manner, regardless how many pumps are operating at a given time and with a given condition, you can establish the capacity of each of these pumps. The sum of these capacities is the total flow delivered into the system under the particular conditions of the system at that moment, that is, with the amount of flow taken off at all the distribution points at that moment.

To be able to develop system-head curves for each of your branch lines, all that would be necessary would be to shut off delivery to *all* other branches while taking readings of the total head developed by the pump (which gives you the capacity approximation) and readings of pressures at each branching-off point in your piping. The combination of these readings will yield a series of points on each individual system-head curve. (Details on constructing these system-head curves are given in the reference in Question 1.39.)

Once you have constructed a family of system-head curves, you can develop as much information as you wish for any combination of operating conditions, since by measuring the total head of all pumps running, you obtain information on the total capacity delivered into the system. You can even approximate the flow into any delivery branch by taking pressure readings at the branching-off point and at the point of delivery: Reference to the individual sytem-head curves for each branch will establish flow into it from the head readings.

If the pumps are old and have not been recently restored to their original condition, the approximation will be somewhat less accurate because it can no longer be assumed that the present head-capacity curves coincide with the curves submitted to you initially by the manufacturer. However, even here, the error of the approximation will probably not exceed a few percentage points. Judging by the size of the piping in your system (which gives an idea of the capacities involved) and by the probable range of the total heads encountered (100 ft static elevation plus friction losses), these are probably single-stage double-suction horizontal pumps. In estimating the present hydraulic characteristics of these pumps, it is only necessary to note that the major potential cause of any reduction in capacity from the original condition of such pumps is the wear at the wearing rings, in other words, the increased leakage through the increased clearances at these rings.

The effect of an increased leakage on the pump head-capacity curve is illustrated in Fig. 1.58. The net capacity of a pump at any given head is reduced by the increase in leakage. Theoretically, the leakage varies with the pressure differential across the running joint (approximately as the square root of this differential) and is therefore not constant at all heads. However, such a nicety of calculation can be disregarded for our purpose. Therefore, we can assume that this increase in leakage remains constant to all heads. To plot the effect of the increase in leakage, Fig. 1.58 shows that a constant value representing this increase is deducted from the capacity at a series of total heads (H_A, H_B, H_C, H_D, and so on). The new head-capacity curve is obtained by joining the points thus obtained.

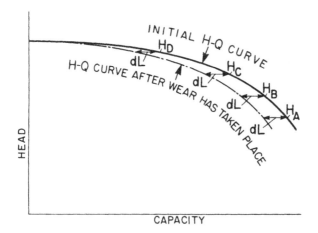

Figure 1.58 Effect of wear on head-capacity curve of centrifugal pumps.

The only assumption that remains to be made is the value of this increased leakage. For pumps in the specific speed range involved in your installation, the initial leakage losses will probably not exceed 1½ to 2%. Unless the pumps are very old and have not been repaired for some 20 years, it is doubtful that the leakage has increased to more than double or triple its initial value. Therefore this increase in leakage must be in the range of 3 to 6% of the design capacity. Depending on your knowledge of the state of repair of these pumps, you can assume a value within this range and plot an approximation of the present head capacity of your pumps.

Finally, we can assume the worst: You have no test curves, and you are unable to obtain any information from the pump manufacturer because the name plates have been lost or for any other reason. You can still obtain a reasonable approximation of the pump characteristics from physical measurements that you will have to make on these pumps. A detailed procedure is presented in the book to which I made reference earlier.* In brief, the procedure consists first of determining the pump specific speed from measurements of the impeller outside diameter and of its eye diameter. Various head and capacity design constants can then be established from the knowledge of the specific speed. These constants and the measurements of the width of the impeller, the number of impeller

---

Ibid, pp. 205-209.

vanes and their width, and the impeller diameter will yield the pump capacity and total head at its best efficiency point at any given operating speed. Finally, having head and capacity values as the best efficiency point and knowing the pump specific speed, you can reconstruct the entire set of pump curves, that is, head-capacity, power-capacity, and efficiency-capacity.

Question 1.41 Discharge Piping Configurations

Please permit me to make an inquiry about discharge piping installation for a centrifugal pump. This question refers to pp. 185–186 of the book *Centrifugal Pumps: Selection, Operation and Maintenance* by I. Karassik and R. Carter under the heading "Typical Discharge Systems."

System II of Fig. 17.14 on p. 185 (Fig. 1.59) presents a discharge piping installation in which discharge is directly at the bottom of an elevated open tank and systems IVa and IVb with discharge piping routed to the top of an open elevated tank with a siphon leg. On p. 186, you stated that systems IVb and V would theoretically be the most efficient among the six systems shown. However, this statement was not elaborated upon, and it is here where I request clarification.

My questions are these: (1) Will a centrifugal pump with discharge piping installed in accordance with system IVa and IVb be more efficient than the same type of pump installed as in system II in terms of (a) power consumption-that is, will system IVa or IVb consume less power than the system II configuration? Why?; (b) volume of water discharge to tank per unit of time-that is, will the pump having discharge piping as in system IVa or IVb pump more water to the tank in 1 hr than the same pump with discharge piping conforming to system II? Why? (2) Which of the three systems of discharge piping is most economical in the long run? Why?

Assume that the pumps are motor driven, are exactly the same, are to pump the same kind of liquid (water), were installed the same (i.e., same suction lift, same size and type of suction pipe, same number and type of fittings, and so on), and the only difference is in the discharge piping configuration (system II or either system IVa or IVb).

Answer. I regret that the illustration you refer to is somewhat ambiguous and can lead to confusion. This is caused by the fact that the discharge piping is indicated schematically rather than to scale. Thus, the dimension D can be assumed to be the same in systems, II, III, IVa, and IVb, which is not true if the level of the liquid above the pump centerline is the same in all four cases.

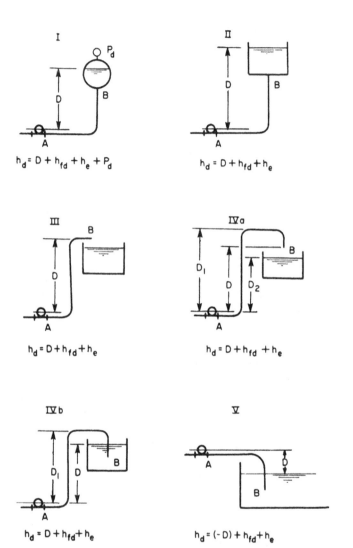

Figure 1.59 Determination of discharge heads for six typical discharge layouts: P_d = pressure deviation from atmospheric; h_e = exist loss at B; h_{fd} = friction loss from A to B (including any siphon loss).

Actually, the value of D is the same for cases II and IVb; it is slightly higher for case IVa and highest for case III. If we were to neglect the difference in friction losses among the four systems, the lowest discharge head requirement would occur in case II, followed by case IVb (if efficiency of siphon is not quite 100%). Case IVa requires a higher discharge head, because liquid exits the piping at some elevation above tank level. Finally, the highest discharge head requirement occurs in system III, because static elevation from pump centerline to exit from piping is highest.

Actual differences in friction losses may affect exact differences among discharge heads in all four cases but probably to a very slight degree. Case II should have the lowest friction loss; case III has one more elbow and longer piping; case IVa has two more elbows and still longer piping; finally, case IVb has the two extra elbows and longest piping.

Now to your questions. They are difficult to answer, because one can assume two different situations.

In the first situation, I shall assume that it is desired to deliver a given volume of water per unit of time to the tank. In such a case, the lowest discharge head and, hence, the lowest total head will result in lowest power consumption. As explained earlier, the most efficient arrangement is system II, followed by cases IVb, IVa, and III in that order.

If, on the other hand, I assume we are dealing with the same pump in all cases, the situation changes. The capacity delivered by a given pump into a system is determined by the intersection of its head-capacity curve with the system-head curve (Fig. 1.56). In turn, the system-head curve is made up of the sum of the static head and all the friction, entrance, and exit losses.

To determine exactly how much capacity will be delivered into the tank, it will be necessary to construct the actual system-head curves for cases II, III, IVa, and IVb. Since the static head is lowest for cases II and IVb and case IVb has slightly more friction losses than case II, the order in which we can rank the four cases in order of capacity delivered is as follows: case II, maximum capacity; case IVb; case IVa; and case III, lowest capacity.

This all presumes that piping has been selected to obtain maximum effectiveness of the siphon and that differences in friction losses in cases IVb, IVa, and III do not overcome differences in static head.

Of course, if we are dealing with a pump of such specific speed that power consumption increases with capacity, the case with highest capacity will take the maximum horsepower. If you want to be very precise in your evaluation, you could proceed with a calculation of power consumption per unit volume of liquid delivered to the tank. Except for very large installations, this is rather an academic exercise and hardly justifiable.

Question 1.42   Friction Losses in Fittings

I have found considerable discrepancy in frictional data given by various manufacturers for several types of valves and pipe fittings. The equivalent lengths, based on new steel pipe seem to vary considerably. For instance, I find the equivalent length of gate valves given as low as 3.2 ft for 16 in. pipe and as high as 17.3 ft. In the case of 90° long radius elbows, for the same 16 in. size, I find equivalent lengths as low as 10 ft and as high as 26.6 ft.

Which frictional data should be used? Why the discrepancy? These variations can make a great deal of difference in calculating the friction losses of a system and therefore in the required total head for pumps selected to serve these systems.

Answer. The discrepancies found in the data are relatively understandable and certainly easier to explain that it is to decide which of the values to use. I shall therefore deal with your second question first.

To begin with, a valve or fitting designed and produced by one manufacturer will rarely be identical to that produced by another, even though the material and the fit-up dimensions of both conform to the same standard. Such variations might be minimal in the case of a 45° weld-end elbow but could be significant in the case of a flanged globe valve or swing-check valve. Even something as simple as a standard reducer could be of either a conical or bell mouth design, with different head loss characteristics for each of the two.

Furthermore, the published values you have questioned are usually determined by tests, and this provides the opportunity for the introduction of additional variations. Presumably, these should be of a relatively minor nature, but they are over and above those previously suggested and however small do not help the situation. Such errors may arise from variations in the test setups, in the reduction of data, or in the interpretation of the results.

The existence of conflicting information in this area does indeed complicate the process of designing piping systems, but it is just one of those many complications we must all suffer as the price we pay for trying to improve the way we do things. As long as there are new valves and fittings being designed, there will be new values of head loss to be determined.

As a practical matter, however, it is obviously necessary to arrive at some conclusion regarding which values should be used for design purposes. This should not really be as difficult as it might at first seem, since the possibilities are relatively limited and since some variations in performance of a pumping system must be anticipated in any event. In fact, the last statement provides a fairly good basis for selecting a design philosphy.

In almost all cases, as soon as a pump goes into operation in the system to which it is applied, the process of gradual deterioration of both the pump and the system begins. However slowly, the pump begins to wear, and its output begins to diminish. Conversely, the system may start to experience increases in surface roughness or buildup of scale or products of corrosion, and its resistance begins to increase. The design philosophy selected should anticipate these changes.

If equivalent lengths for valves and fittings are chosen at or near the highest published values, the probability is that the actual pressure drop through the system, when new, will be less than the calculated value. This is illustrated by system-head curves 1 and 2 in Fig. 1.60, where curve 1 is the calculated curve and curve 2 shows a possible location for the actual curve. For some time, the system deterioration that may then occur will serve only to raise the actual system-head curve 2 to a level more nearly in line with that which was calculated and closer to the conditions for which the pump was selected.

This would be represented in the illustration by a gradual change in pump capacity from point A to point B. If concurrently, however, the pump output were decreasing, as would be indicated by a change from head-capacity curve 4 to 5, its capacity would then lie somewhere below the line AB in the horizontally shaded area ABCD. Since most of this area falls to the right of the design capacity, it should be apparent that there will be a period of time, which may be of considerable duration, during which the system's requirements and the pump's capabilities will tend to converge toward the design capacity before they ultimately begin again to diverge. As a result, the time during which the installation may be operated without the necessity for repair will have been prolonged.

In most situations the margin in capacity provided by this approach will be helpful in terms of operational life without creating any significant effect on other characteristics of the pump or system. This would generally be true when the NPSH available to the pump exceeds that which it requires by a reasonable margin, as illustrated by curves 6 and 7, which intersect at a capacity greater than that represented by point A.

When the available NPSH provided by the system just equals or barely exceeds that required by the pump, as indicated by curves 7 and 8, which intersect in this illustration at exactly the required capacity, the design philosophy should be altered.

In this case it would be preferable to use minimum values of equivalent lengths, thus tending to produce an actual system-head curve higher than calculated, as represented by curve 3. Initially, the pump would operate in this system somewhere along the line BE, which corresponds to that segment of its "as-new" performance curve that lies within the range of friction loss uncertainty arising from variations in available data on valves, fittings and even pipe friction.

During the period in which the system head remains between curves 1 and 3, but as the pump output begins to deteriorate from curve 4 to curve 5,

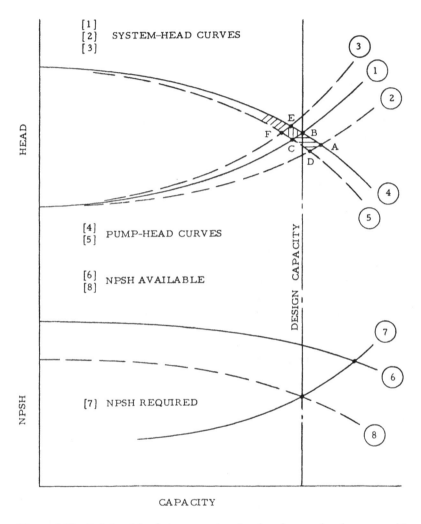

Figure 1.60 Relationships between system-head and pump-head curves, with consideration of NPSH available versus NPSH required.

the actual capacity will fall somewhere within the vertically shaded area BCFE.

As time passes, and should the system deteriorate further, output capacity will depart even more from the design value, gradually diminishing within the diagonally cross-hatched band lying to the left of line EF. It should be evident that because of the limited NPSH available, it has become necessary to design the system in such a way that the pump capacity cannot exceed its design value, since otherwise it would be subject to unsatisfactory operation and abnormal wear due to cavitation arising from insufficient NPSH.

As inferred above, this type of situation will probably be less frequently encountered than the one described earlier, but because it can be costly to correct the consequences of unexpected cavitation damage, it cannot be ignored.

Finally, I would like to comment briefly on the observation you made in the statement of this problem that these differences in frictional values can make a great deal of difference in determining the system head. There are cases in which this is indeed true, but fortunately they constitute a distinct minority. Most pump installations must accomplish at least one of the following:

1. Raise the pump liquid from one elevation to another
2. Increase the static pressure within the liquid
3. Transport the fluid over some distance

Any one or any combination of these is likely in most cases to result in head requirements far exceeding the friction losses represented by fittings and valves, thereby mitigating the effect of any uncertainties about them.

It should be further noted, however, that the design philosophy I have suggested may also be applied to variations that may be discovered in values for friction losses in the pipe itself whenever the application does include the last of the aforementioned three requirements.

Question 1.43 Altering Design Capacity of a Pump

In certain pump applications, as in refineries, it often happens that the flow through a pipe system is increased over the original conditions, sometimes by several hundred percent. In a conversation the other day, it was suggested that the pumps that serve a distribution sytem of this kind could be adapted to the throughput variations by providing suitable inserts in the housing. The impeller, or course, would remain unchanged.

I seem to recall that sometime, somewhere, I had read or heard of such a method but am unable to get the matter pinned down. Could you recall any such instance from your experience?

Answer. I regret to say that the method of increasing throughput in a pumping system that you describe seems a bit "faulty" to me for several reasons. If the piping system remains unchanged, even doubling the capacity (let alone increasing it several hundred percent) would make the pump incapable of developing the increased total head requirements. Let us assume, for instance, that, of the total head developed by a pump to meet initial service conditions, 80% goes to overcome static head differences and the terminal static pressure, and 20% is the friction loss in the piping and valves. Since friction losses vary approximately with the square of the flow, doubling the capacity would increase friction losses fourfold to 80% and the pump would now have to develop 160% of the total head against which it worked previously. Unless the impeller diameter or the pump speed were to be increased markedly, the pump obviously could not do this.

If I understand you correctly, the pump casing would be provided initially with some sort of filler pieces. Then, later, when a greater capacity was required, the filler pieces or portions of them would be removed. This would increase the pump casing area and, by this theory, increase the pump capacity correspondingly.

It is probable that some increase in capacity (at the same total head as before) would occur but nowhere nearly as significant an increase as you imply. The casing areas are only partially responsible in establishing pump capacity; the pump impeller areas are much more of a factor. Thus, unless the casing is throttling the effect of the impeller excessively to begin with, no significant increase will take place merely by enlarging the casing areas. Even if a throttling action is taking place, a capacity increase of some 10 to 20% at most (at the same head, remember) can be expected.

A somewhat different means of varying pump capacity (without materially affecting the head) is the use of so-called 100 and 80% impellers. Standard lines of pumps in the medium and larger sizes, say 6 in. and above, are frequently designed to accommodate either of two impeller designs: a 100% impeller that fully loads up the casing and a narrower wheel designed for, say 80% of the standard capacity. For instance, a pump designed to have its best efficiency at 5000 GPM and 200 ft may also have an impeller pattern that would have its best efficiency at 4000 GPM and 200 ft.

The basic purpose of this arrangement is to avoid building too many intermediate pump patterns and thus to reduce total development costs, which, after all, have to be passed on to the customer. On occasion, it is possible to take advantage of this if the pump was fitted with an 80% impeller initially by replacing it with a 100% impeller. However, a significant increase in capacity can be achieved only if the driver can carry the increased power requirements (or can be economically replaced) and if the head requirement of the system at the increased flow falls within the maximum head that can be developed with the maximum diameter in this casing.

It is also possible that you had heard of another arrangement that can be used to limit the capacity of a pump and pumping system combination so as to prevent severe driver overload. Centrifugal pumps of low specific speed types generally have an "overloading" type of bhp curve; in other words, their power consumption continues to increase as the capacity increases and the total head decreases. Even pumps whose bhp curve flattens out at some maximum capacity may overload their drivers if permitted to operate at capacities beyond their design capacities unless the driver is selected to match the maximum pump bhp at any capacity. For instance, referring to Fig. 1.61, if the pump were designed for 1000 GPM and 130 ft, the pump bhp would be 41. If the pump were never expected to operate beyond this 1000 GPM, a 40 hp motor could be used without incurring a very serious overload. But if, under certain conditions, the total head of the system were permitted to fall appreciably below 130 ft at capacities in excess of 1000 GPM, the pump would operate at these higher capacities, and the bhp might reach 50 or even higher, with serious effects on the motor life.

It is possible, however, to introduce a means of limiting the pump capacity whenever necessary to prevent possible operation at capacities at which motor overload would occur. I suggested such means in 1943 for standard close-coupled or frame-mounted pumps normally built for stock in standard combinations. This means consisted of an orifice to be located in the discharge of a pump whenever it was sold for conditions warranting the use of a motor sufficient for normal operation but that would be overloaded at reduced heads. The effect of such an orifice (see Fig. 1.62) is illustrated in Fig. 1.61.

Let us assume that the operating conditions call for 900 GPM at 115 ft. The normal selection for these conditions would have required the use of a 50 hp 1750 RPM motor. At 900 GPM this pump develops 133 ft and takes 39 hp. If 100 ft of the total required head was static and 15 ft represented the friction losses, we can construct our system-head curve and establish that the intersection of this system-head curve with the pump head-capacity curve would take place at 1170 GPM and 125 ft. The pump bhp would be 44.5, and a 50 hp motor would definitely be required. But if an orifice were installed in the pump discharge such that it would absorb 18 ft of friction at 900 GPM, the pump capacity could never exceed this 900 GPM value nor the pump bhp exceed 39, and a 40 hp motor would be sufficient.

The curve was drawn to indicate that the net pump head-capacity curve is reduced by the amount of 18 ft at 900 GPM and by a varying amount at other flows, proportional to the square of the capacity. Actually, the effect can be said to increase artificially the friction in the discharge piping by these same amounts, and we could have drawn a new system-head curve. In other words, the orifice does nothing that a manually positioned throttling valve would not do, except that it does it permanently or, at least, until it is removed. This eliminated the danger that an operator would

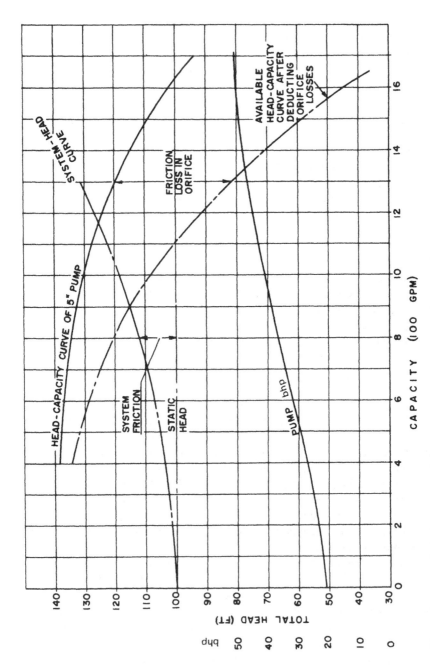

Figure 1.61 Performance of 5 in. pump at 1750 RPM with orifice in discharge.

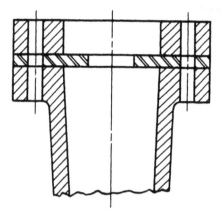

Figure 1.62 Arrangement of capacity-limiting orifice.

inadvertently open the valve, increase the flow through the system, and overload the driver.

The arrangement of the capacity-limiting orifices I had suggested is shown in Fig. 1.62. Each size pump was to have had a series of orifices of varying dimensions to be used for different operating conditions.

A more refined arrangement has recently been adopted by a refinery that is utilizing this principle to install a standardized pump on a variety of different services for different capacities and heads.* The pumps are provided with a plug-type discharge orifice to modify the head-capacity curve available to the pump system (see Fig 1.63).

Obviously, the power consumed by the pump at any given capacity is that required to develop the normal total head, and the friction loss through the orifice is a net loss. The user must evaluate this loss against the distinct advantage of using a single-size pump on a multitude of different services and the ability to convert a pump to different conditions of service with at most a change of drivers.

You will find that such an evaluation may justify the use of an orifice in the discharge in the range of small pumps only and generally only if the range between minimum and maximum capacities to which a given size pump is applied is not excessive.

---

*"Pump Standardization . . . . a User's Concept" by R. G. Jobe, Shell Chemical Corporation, Houston, TX., Paper 59-A-162, contributed by the Petroleum Division for presentation at the Annual Meeting of the ASME, Nov. 29 to Dec. 4, 1959, Atlantic City, N.J.

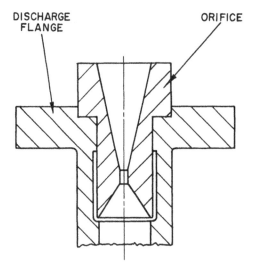

Figure 1.63 Plug-type discharge orifice used to modify the system head-capacity curve.

Question 1.44 Effect of Oversizing Centrifugal Pumps

Can oversizing a centrifugal pump create serious operating problems beyond the obvious fact that its purchase cost may be unnecessarily increased and that power consumption may also be adversely affected?

Answer. I would like to take this occasion to thank the magazines in which these questions and answers first appeared for having encouraged their readers to write me and present any problems they may have in connection with centrifugal pumps. Not only have many readers taken advantage of this invitation, but I have also received much interesting information to confirm observations I have personally made in the field. I might even say that, as in the case of some less serious (but more widely read) Broadway columnists, once in a while my column was written by the readers themselves.

Thus, for instance, I might cite a very interesting example of my advice to select a pump that exactly suits the job it should do rather than oversize the application. This example was supplied to me by Mr. Stocker of the Allison Division of General Motors.

At a gas plant coke-quenching station, about 15 tons of red hot coke were quenched at a time with a deluge of water from an overhead tank. The quenching water was returned into a reservoir for reuse. It was first run through decanters to get rid of the coke particles. A pump returned this water along with some makeup from the reservoir to the overhead tank. Of course, the decanters did not do a perfect job, and the water was quite dirty with small coke particles. In addition, since relatively little makeup was used and since little opportunity existed for getting rid of the heat picked up in the quenching process, the water was quite warm.

The original pump installed on the quenching job was inadequate and had been replaced by a pump retired from another application. This was a horizontal pump, installed above the reservoir and provided with a foot valve in the suction line.

The pump impellers were wearing out every few weeks. This was attributed to the abrasive coke particles in the water. The pump was quite noisy. In addition, the operators were concerned by the fact that the electrical supply lines to the pump motor were somewhat undersized and likely to cause trouble.

An investigation showed that the head developed by the pump was considerably in excess of that required. This permitted the pump head-capacity curve to intersect the system-head curve well beyond the design capacity of the pump. After some calculations, the impeller diameter was reduced by 1/2 in. to reduce the head-capacity curve and, incidentally, to reduce the load on the motor.

This reduced the power required by the pump, so that the electrical lines did not have to be replaced. The pump also ran considerably more quietly. Some spare impellers of the smaller diameter were ordered. These spare impellers, incidentally, may still be on the shelf, for when the pump was opened about a year later, the impeller showed no signs of wear.

No doubt, the original wear was caused by cavitation rather than by the abrasiveness of the coke particles. In turn, cavitation was caused by a combination of factors:

1. Operation at excess capacity, requiring greater NPSH
2. Operation at high capacity, causing a high pressure drop through the foot valve and suction line, increasing the suction lift
3. Warm water, with a higher vapor pressure and, therefore, reduction of the effect of atmospheric pressure

This cavitation had been indicated by the noise observed when the original larger diameter impeller was used, but it had not been recognized as such. Mr. Stocker's analysis of this problem is quite correct. The effect of oversizing a pump is illustrated in Figs. 1.64 and 1.65. The first shows qualitatively the effect of oversizing a pump on the capacity delivered to the system and on the power consumption. Obviously, the system-head curve will intersect the head-capacity curve of the pump at a capacity much in excess of that desired. The power consumption exceeds significantly that

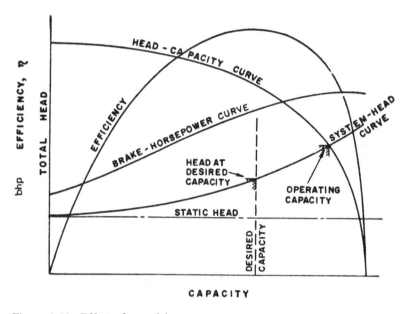

Figure 1.64 Effect of oversizing a pump.

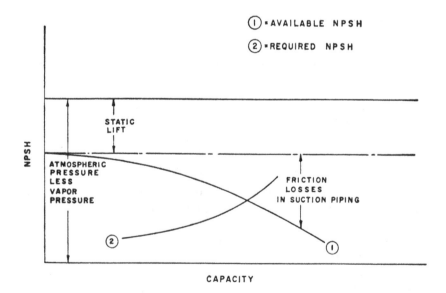

Figure 1.65 Available and required NPSH.

power that would have been required to handle the desired flow. At best the pump can be throttled on the discharge side and the power reduced somewhat. But if the pump head and capacity had been selected to match the system requirements, very important power savings would have been made available. In addition, operation of this pump so far out on the head-capacity curve may be unsatisfactory from the point of view of suction conditions.

Figure 1.65 illustrates the fact that the available NPSH for the installation Mr. Stocker describes is the difference between the atmospheric pressure and the vapor pressure less the sum of the static lift and the friction losses in the suction piping. Thus, the available NPSH decreases rapidly with an increase in capacity. In turn, the NPSH required by the pump increases with capacity. If the available NPSH cannot satisfy the requirements of the pump, the latter will cavitate and operate in the break. Obviously, as the water temperature increases, its vapor pressure increases too, and the baseline from which we start (atmospheric pressure less vapor pressure) is reduced, bringing us into the danger zone much more quickly.

Question 1.45 Operation in a Closed Circuit

We have recently modified an existing closed station heating system and have noted that the pump suction operates in a vacuum. Figure 1.66 is a schematic diagram of the system and Fig. 1.67 the head-capacity curve for the pump. Heat transfer calculations have established that the required total flow is 80 GPM. It would seem to me that, in a closed system, the total pressure loss through the system is equal to the total head across the pump. In such a system, what other means are available to reduce the pump capacity other than trimming the impeller or reducing pump RPM? Would a recirculating line from the pump discharge back to the suction be of any benefit? The system apparently operates satisfactorily, but we would prefer to operate with a positive suction to reduce air infiltration.

With regard to centrifugal head-capacity curves, which comes first, the chicken or the egg? For example, with a pump taking suction from an open tank (constant suction pressure), does throttling the discharge valve reduce the flow, and as a result does the pump then seek the discharge head necessary for that flow? Or does throttling the valve raise the discharge head and thereby establish the flow? It is obvious from the above that I do not understand all that I know about centrifugal pumps. Your comments will be most appreciated.

Answer. There are essentially two separate problems that interest you: (1) the relation between the pump performance characteristics and the system

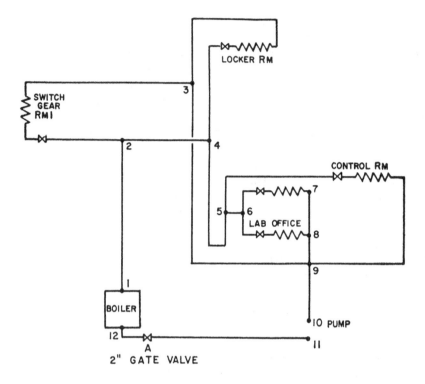

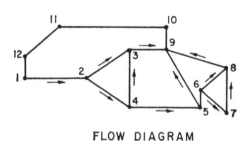

Figure 1.66 Schematic diagram of hot-water heating system.

characteristics and how to control one or the other to obtain the desired flow most economically and (2) how to avoid operation with vacuum at the pump suction.

When operating at constant speed, a centrifugal pump has a definite head-capacity characteristic. In other words, at any given flow it will

Chapter 1

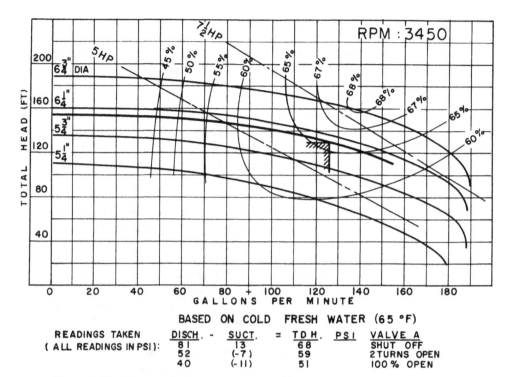

Figure 1.67 Pump performance curve on cold water.

develop a certain total head. This characteristic, when plotted with capacity as abscissas and total head as ordinates, is the pump head-capacity curve. Insofar as the system is concerned, it also has certain characteristics whereby to cause a certain flow to take place through the system, a given head must be made available to overcome static head differences and the friction losses through the system at that flow. With varying flow, the friction losses vary approximately as the square of the flow. By plotting the sum of any static head that exists in the system and of the friction losses at varying flows against these values of flow, one determines the so-called system-head curve. Now if the pump head-capacity curve is superimposed on the system-head curve, the intersection between the two curves determines the operating point. In other words, the capacity that will flow through the system will be the capacity at the intersection of the two curves, and the head required to cause this flow will correspond to the total head developed by the pump at that capacity.

I have frequently been asked to expand on this explanation so that one could visualize this "physically." To do so it may help to image what would happen if the flow were to vary from the flow at the intersection of the two curves. If the flow were reduced, the friction losses would be reduced with it, and the head required (on the system-head curve) would also be reduced, while the available head (the total head developed by the pump at this lower flow) would increase. Thus, there would be excess head available, and this excess head would cause the flow to increase back to a value where equilibrium exists, that is, to a value at which the required head is exactly balanced by the available head. On the other hand, no greater flow than that determined by the intersection of the two curves can exist, since the required head at this greater flow would exceed the head that is available from the pump.

To change the flow without changing the impeller diameter (and hence the head-capacity curve of the pump at the operating speed), it is necessary to alter either the pump head-capacity curve by varying the pump speed or the system-head-capacity curve by changing the friction loss component. This is done by manipulating a valve to increase the friction losses in the system artificially if the flow must be decreased or to decrease the friction losses by opening the valve if flow is to be increased.

There is a third means available to alter the resulting flow, and that is to bypass or recirculate some of the flow from the discharge back to the source of suction. This, essentially, reduces the capacity available to flow into the system proper by the amount that is being bypassed. This method of control is employed in certain special cases, generally when it is not desirable to reduce the pump capacity below a certain minimum value. (In your particular case, incidentally, this is not an attractive solution, as you will presently see.)

The installation you describe is a closed system, and therefore there is no static component to the system-head curve, this curve being made up of the friction losses in the system only. To determine the flows that will prevail in the system, I have replotted the head-capacity curve of your pump in Fig. 1.68 and also three "derived" system-head curves. These are probably only approximate, since I had no direct capacity readings corresponding to the net pressure data that appear on the curve of Fig. 1.67. Incidentally, your tabulation indicates that the suction and discharge pressures given refer to cold water at 65 °F. If this interpretation is correct, the total head data I have calculated can be used as is. If, on the other hand, the readings you give were taken at some higher temperatures, the total heads I indicate in Table 1.3 should be corrected for the specific gravity and would be slightly higher.

To construct the system-head curves with the valve fully open and with the valve open two turns, it is merely necessary to find on the pump curve the capacity that corresponds to the total head developed by the

Chapter 1

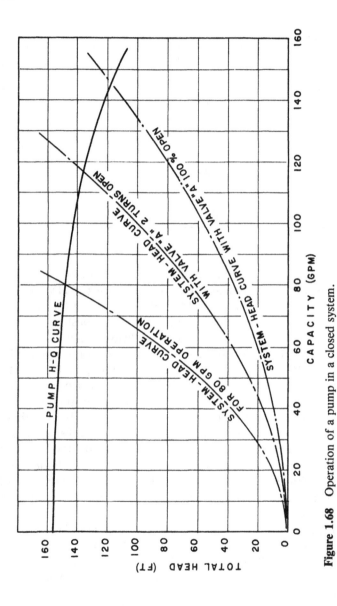

Figure 1.68 Operation of a pump in a closed system.

Table 1.3 Calculations of System-Head Curves

GPM	Head with valve 100% open (ft)	Head with valve open two turns (ft)
0	0	0
40	8.9	16
80	35.6	63.6
117	—	136
120	80	143
146	118	
155	133	

pump under these two conditions. For instance, with the valve fully open, the net pressure developed by the pump is 51 psi (or 118 ft). The pump develops 118 ft at a capacity of 146 GPM. Assuming that friction losses vary as the square of the capacity, it is possible to construct a system-head curve by calculating friction losses at flows below and above 146 GPM, as per Table 1.3. The same is done for the valve two turns open; 59 psi corresponds to 136 ft, which is developed by the pump at 117 GPM.

If you wish to hold the flow to 80 GPM, all that is necessary is to close down on valve A beyond the two-turns-open position. If you have no means to measure the flow accurately, you may hit it by checking out the total head developed by the pump, which should be in the neighborhood of 148 ft. This is the head developed by the pump at 80 GPM and would correspond to the intersection of the head-capacity curve with the new system-head curve with the valve in the desired position to obtain 80 GPM. We can construct this sytem-head curve exactly as we have constructed the other two. (See Table 1.4.)

It is preferable to throttle the pump discharge rather than to bypass some of the pump capacity to the suction. By looking at your curve you can estimate that at 80 GPM and 148 ft the pump takes 5 bhp. If, on the other hand, the pump runs at 146 GPM and 118 ft and you recirculate 60 GPM, the pump will take about 6.5 bhp.

You imply that you prefer not to trim the impeller. However, if you do not contemplate flow requirements in excess of 80 GPM, this would be the most economical solution. You may prefer the ability of providing some control of flow and therefore to operate at 80 GPM with the valve partially closed—let us say with the valve two turns open. Under these conditions, and unless the pressure at the boiler must be maintained at some

Table 1.4 System-Head Curve Points for a flow of 80 GPM Through the System

GPM	Head for 80 GPM operation in system (ft)
0	0
20	9.25
40	37
60	83.25
80	148
85	166

level, the pump need develop only 62.5 ft. This is less than the pump rating curve indicates as the minimum cutdown of the impeller. If we were to cut it to its minimum diameter, the total head at 80 GPM would be 98 ft and the pump power consumption only about 3.5 bhp.

As to the suction conditions, I assume that the pressure in the boiler (or rather hot-water heater) is set at some particular value by means of a thermostat on one hand and a relief valve on the other. Then, if the friction losses between the heater and the pump suction exceed the gauge reading at the heater, the suction pressure will fall below atmospheric. Whether this will no longer be the case at a flow of 80 GPM through the system, you can determine only by trial. However, if you wish to avoid air infiltration into the pump, you need only inject a small flow of water into the stuffing-box seal cage. This injection can probably be taken directly from the pump discharge through a small valve.

Comments

Your comments regarding our heating system problem and the data that you sent to us are very much appreciated. Your assumption as to the method of operation is correct: The operating pressure of our boiler is 30 lb. Operating at a higher pressure will cause the safety valve to lift.

Question 1.46 Parallel or Series Operation?

When two pumps are installed to take advantage of one- and two-pump operations for a varying capacity demand, is it preferable to use the two pumps in parallel or in series if a greater range in capacity is desired between single- and two-pump operation?

Answer. This will depend on the shape of the system-head curve as well as, to some extent, on the shape of the pump head-capacity curve. The system-head curve is constructed by adding the friction losses at all capacities to the static head that will prevail. The capacity that will be delivered into the system will correspond to the intersection of the system-head curve with the head-capacity curve of the pump or pumps used in the system. To construct the head-capacity curve of two pumps in parallel, it is necessary to add the capacities of the individual pumps for various total heads. Contrariwise, the head-capacity curve of two pumps in series is constructed by adding the individual pump heads for various capacities.

If the static component represents a large portion of the system head and the friction losses are low, parallel operation is preferable. On the other hand, if the system head is composed almost entirely of friction, operating the two pumps in series will give more flow through the system than if the pumps operate in parallel.

To illustrate this general statement, let us consider a few examples showing rather extreme characteristics for both pumps and systems. For this purpose I have prepared illustrations showing all variables in terms of percentage of rated values. This facilitates comparisons and provides general solutions. This will also avoid any possible connotations of particular pump sizes or types.

Figure 1.69 represents a type of pump having a flat head-capacity curve (A-B-C) and shows in addition the combined head-capacity curves of two pumps in series (curve D-E-F) and of two pumps in parallel (curve A-G-H). Superimposed on the pump curves are a flat system-head curve (J-B-H-F) consisting primarily of static head and a steep system-head curve (K-B-G-E) composed entirely of line friction.

For a system composed entirely of friction, it is immediately evident that two pumps in parallel would operate at point G and would provide only 105% of the capacity obtained with a single pump. The same two pumps in series, however, would operate at point E and provide 128% of single-pump capacity. Thus, in this instance, there is an obvious advantage in series operation. The particular circumstances used in this example are quite typical of pipeline service, and indeed pipeline pumps are generally installed precisely this way.

Consider the same pumps applied to the flat system-head curve. Two pumps operating in series at point F still show a slight advantage in capacity over parallel operation at point H. But against this slight advantage, we must weigh the following disadvantages:

1. Brake horsepower, which has been indicated by dashed lines in Figure 1.69, would total 230% of the rated horsepower for one

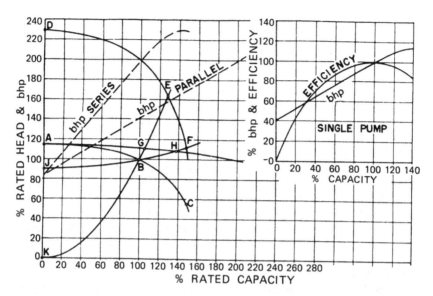

Figure 1.69 Series and parallel operation of pumps with flat head-capacity curves.

pump. For this power input we would obtain 148% of single-pump capacity, indicating a power capacity ratio of 1.55. With parallel operation, the bhp is 164% and capacity is 138%, yielding a power capacity ratio of 1.19. Thus, there is a decided disadvantage to series operation in terms of cost per gallon of liquid pumped.

2. With pumps operating in series, the stuffing-box pressures of the second-stage pump will be increased by an amount equal to the total head developed by the first-stage pump unless special features are incorporated in the pump to alleviate this condition. These special features add to the initial pump cost, but on the other hand their absence may lead to increased maintenance cost for packing and shaft sleeve replacement. This disadvantage exists for series operation under any circumstances, but in the case of a steep system-head curve it might be accomodated in return for a more significant improvement in capacity.

3. With two pumps in series operation at point F, each pump must handle 148% of normal capacity. This may well result in a serious NPSH problem at the first-stage pump, since the required NPSH increases very rapidly beyond the design capacity of 100%. For parallel operation at point H, each pump would handle only 69% of normal capacity, and the problem of NPSH is completely eliminated.

4. If the pump driver had been selected without any margin over the power requirement at rated capacity, operating at point F will overload the driver.

Considering the same system-head curves but a type of pump having a steep head-capacity curve, the picture (as shown in Fig. 1.70) changes radically. The identification of the curves and of the operating points corresponds to that given in Fig. 1.69.

In this case, the capacity advantage of series operation on the steep system-head curve is greatly diminished compared with the previous illustration. Here, series operation yields only 122% of single-pump capacity compared with 115% for parallel operation. Unlike the previous case, however, the total power consumption is actually less for the greater capacity obtained from series operation than it would be for the smaller capacity that would be realized by parallel operation. This is caused by the fact that the power consumption curve of such pump types actually diminishes instead of increasing as capacity increases.

On a flat system-head curve, however, operating this type of pump in parallel would show a definite advantage over series operation, yielding 160% of single-pump capacity as against 134%, respectively. This is indicated by comparison of points H and F in Fig. 1.70.

These illustrations have been selected to substantiate the general statement made in the first paragraph of the answer, but the figures and discussion provided should be adequate to permit the construction of a similar set of curves for any specific installation. This would then provide a definite answer in any particular case under consideration.

Question 1.47 Is Pump Performing at Capacity?

We have two two-stage centrifugal pumps rated at 250 GPM and 400 ft total head, driven by 50 hp electric motors. They pump cold water into an overhead tank against a 273 ft head. Each pump has its own intake under a 10 ft positive suction head. The two pump discharges are connected to a single header.

When running alone, each pump discharges 330 GPM at a pressure of 150 psi, but when the pumps are operating in parallel, the total capacity delivered is only 465 GPM at 180 psi. Why do these pumps deliver so much less capacity in parallel, and what must be done to them to restore them to their rated condition?

Answer. A careful analysis will show that there is nothing radically wrong with these pumps and that they are practically in the same condition as when they were new.

Chapter 1   131

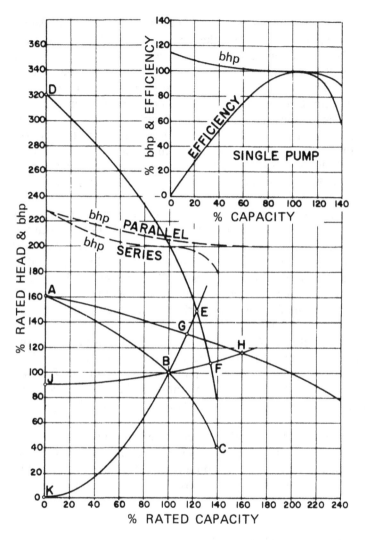

Figure 1.70 Series and parallel operation of pumps with steep head-capacity curves.

Two pumps operated in parallel will deliver against any given total head the sum of their individual capacities at that head. Thus, if each pump was designed for 250 GPM and 400 ft, two such pumps in parallel will deliver 500 GPM at 400 ft, assuming that the pumps have met their guarantees and have not lost capacity through wear.

This does not mean, however, that two pumps in parallel will deliver through a given piping system twice the capacity that either of them can deliver alone. It is a characteristic of centrifugal pumps operating in a hydraulic system that the capacity delivered will correspond to the intersection of the pump head-capacity curve with the system-head curve.

We can attempt to construct both these curves from the information given in the question. In carrying out these calculations, it has been assumed that the 273 ft head referred to is the static discharge head, that the friction losses in the suction piping are negligible, and that the 150 and 180 psi referred to are discharge pressures. It has been further assumed that, unless proved otherwise in this analysis, the pumps meet their rated design guarantees of 250 GPM at 400 ft head.

On this basis, we can develop two points on the pump head-capacity curve, namely, the design point and the total head at 330 GPM, since this last is equal to $(150 \times 2.31) - 10 = 336.5$ ft. We can likewise establish a system-head curve, since we know that the total static head is $273 - 10 = 263$ ft and therefore that the friction losses at 330 GPM are $336.5 - 263 = 73.5$ ft. Since friction losses can be assumed to vary as the square of the capacity, we can tabulate data at several arbitrary flows and plot a system-head curve, as shown in Table 1.5.

It will be seen (Fig. 1.71) that the head-capacity curve of the two pumps operating in parallel intersects the system-head curve at 465 GPM and a total head of 410 ft, with each pump delivering one-half of 465 GPM, or 232.5 GPM. This seems to be quite probable, since a pump designed for 250 GPM at 400 ft can be expected to deliver 232.5 GPM at 410 ft.

If the total head is 410 ft and the suction head is 10 ft, the discharge pressure will be $(410 + 10)/2.31$ or 181.5 psi if we neglect the difference between the velocity heads at the suction and discharge nozzles. Actually, we should also have assumed a slight increase in suction piping friction

Table 1.5 Calculation of Total System-Head

Capacity (GPM)	Fraction losses (ft)	Total system head (ft)
250	42	305
300	61	324
330	73.5	336.5
400	108	371
500	168	431

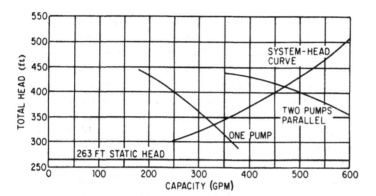

Figure 1.71 Based on the facts in the question, head-capacity and system-head curves have been plotted on this graph.

losses and hence a reduction in the suction head and therefore in the discharge pressure. It would therefore appear that the slight discrepancy between the observed and calculated discharge pressures at 465 GPM can be easily accounted for without assuming that the pumps are appreciably worn. In other words, the two pumps operating in parallel cannot be expected to deliver more than 465 GPM through this system unless the friction losses in the piping are reduced.

The friction losses observed appear to be on the high side for this type of installation, and it might be advisable to analyze this system with the view in mind of reducing these losses. If this is done, two choices are available to the operator: the pumps may be left as they are and a greater capacity will be available, or the impellers can be cut down slightly so that the present capacities are handled by the pumps but with some reduction in power consumption.

Question 1.48 "Drooping" Head-Capacity Curves

I have a question concerning the so-called "stable" and "unstable" types of performance characteristics. I have an application involving the circulation of hot water through a radiant heating system. A certain manufacturer has recommended a pump that has a head-capacity curve with a "drooping characteristic." It has always been my thought that a drooping characteristic is unstable but that it may not cause any difficulties in a system in which the total head is comprised of approximately one-half-elevation and one-half friction losses. Could you enlighten me on this matter?

Answer. You are quite correct in your understanding of pumps with drooping characteristics. Such curves are considered to be unstable. On the other hand, the installation in question and the conditions surrounding the application of any given pump have a great deal of bearing on whether or not difficulties may be encountered if the pump has a drooping characteristic. For instance, if paralleling problems are not encountered, that is, if the pump is never to operate in parallel with a second pump, a drooping characteristic will not necessarily make it impossible to use the pump, assuming that the system head is made up as you state of approximately one-half static head and one-half friction head.

On the other hand, if two pumps are operated in parallel, there is always the possibility that one of the pumps will carry the major portion of the load and the second pump a lesser portion, both operating at the same head but at different capacities. For instance, in Fig. 1.72, one pump may be operating at point A while the second pump handles capacity B.

As the demand is reduced by means of throttling the discharge, the pump that is operating at the right of the maximum head point C will start climbing up on the curve and develop a pressure greater than that developed by the second pump. This will close the check valve in the discharge line from this second pump, making it operate against complete shutoff. If no provision is made to protect the pumps against this situation by means of a recirculating bypass, the pump will overheat and become damaged.

In addition, if two pumps are to be operated in parallel and only one pump is running at some particular time, it may be impossible to start the second pump, since the pressure it will develop at shutoff is less than the pressure developed by the pump that is running. Referring to Fig. 1.72, this will happen whenever the pump running is handling a capacity less than E. Under these conditions, when the idle pump is started it develops insufficient pressure to lift its check valve.

In some systems, a drooping characteristic may cause problems even if the paralleling system does not exist. This will occur if a great portion of the system-head curve is made up of static elevation or pressure and a very small portion of it by friction. In such a case, operation of the pump at or near the maximum total head it can develop might cause the pump to go into swings. If a reduction in capacity is required, a valve in the discharge would be throttled. Theoretically, this should reduce the capacity by increasing the frictional component. However, this throttling action would result in operating the pump at a point at which it develops somewhat less head than it did prior to the throttling operation. At that very moment, the pressure in the system is still equal to the pressure that existed prior to throttling, and the pump suddenly develops less pressure (at the left of its

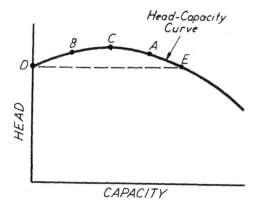

Figure 1.72 Drooping head-capacity pump curve.

maximum head) than exists in the system. A tendency to cause backflow arises, and the pump might go into swings.

However, as I have stated previously, there are many applications in which the use of a pump with drooping characteristics does not necessarily introduce excessive problems.

Question 1.49   Parallel Operation of Steam Turbine-Driven Pumps

We have installed two steam turbine-driven single-stage 5 in. centrifugal chilled-water pumps, each rated at 1080 GPM, 125 ft head, at 2240 RPM. We plan to operate both pumps in parallel at speeds from about 40 to 100% by controlling the steam admission at the turbines by means of a differential-pressure impulse across the remotest coil in the chilled-water distribution system. The idea is to achieve variable chilled-water flow in response to changing system demand. In most air-conditioning installations, the designers call for constant flow. This we consider a waste of energy and prefer the variable-speed turbine-driven pump when steam is available and can be employed economically.

The question, if two pumps are used in parallel operation and there occurs a slight difference in amount and pressure of steam to each turbine owing to piping arrangement (125 psig nominal), causing a slight difference in speed between the two pumps, what will be the effect on the combined pump discharge? Will the pump that operates at a slightly higher speed hog the flow, or will each pump discharge according to its speed?

Answer. I do not intend to dispute your reasoning with regard to variable versus constant-flow operation of the chilled water. However, if design chilled-water velocities are maintained even at part load, a more favorable evaporating temperature will exist at this load than if the flow were also reduced. Since the power consumption of the chilled-water pumps is only 5 to 10% of the power consumption of the refrigerant compressor, the power savings for the compressor will be considerably more significant than the potential savings in reducing the chilled-water pump power. This is the reason that in most installations the chilled-water flow is maintained constant regardless of the load.

Now, as to the hydraulics involved here, unless the circumstances surrounding this particular installation are very unusual and one of the pumps operates at a very much reduced speed below the other, they may be expected to share the load within reasonable limits. Certain precautions may be necessary, however, to eliminate the danger that would arise if one of the pumps were to hog the entire load.

It is necessary first to realize that the head-capacity curves of two pumps intended to operate in parallel do not necessarily have to be exact duplicates of each other under certain circumstances.

To construct the combined head-capacity curve of two pumps operating in parallel, the individual capacities of each pump are added at various values of total head. The total flow that these two pumps will deliver into a system will correspond to the intersection of this combined head-capacity curve with the system-head curve. In turn, the capacity contributed by each individual pump will be that capacity on its own H-Q curve corresponding to the head at the intersection point mentioned above.

The curves in Fig. 1.73 illustrate this. I have assumed a certain head-capacity curve for pump A when it operates at 2240 RPM, and I have further assumed that pump B can be operated at a maximum speed of only 95% of that pump A, or 2128 RPM.

We first plot the combined head-capacity curves of pumps A and B. Until pump A delivers 1140 GPM and develops 124 ft head, pump B can deliver no flow into the common header, and its check valve remains closed. At lower heads, pump B starts delivering into the system and adding its flow to that of pump A. Thus, the combined head-capacity curve starts at 0 capacity and 136 ft, continuing along curve 1 until the flow reaches 1140 GPM and both pumps deliver their combined flow along curve 3.

To determine what each pump will contribute, I have also assumed that the pump rating conditions were selected to correspond to the system-head curve (to simplify matters, I have neglected the probability that some margin was included) and, therefore, that the system-head curve is made

Chapter 1

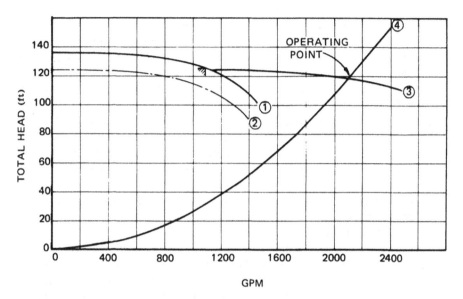

Figure 1.73 To construct the combined head-capacity curve for two pumps in parallel, the individual capacities of each pump are added at various total heads.

up of friction only and passes through 2160 GPM (twice 1080) and 125 ft. Under these conditions, the two pumps will deliver a total flow of 2100 GPM at a total head of 118 ft, which corresponds to the intersection of curves 3 and 4. You will notice that, at this total head of 118 ft, pump A, operating at 2240 RPM, delivers 1260 GPM, and pump B, operating at 2128 RPM, delivers only 840 GPM, making up the total of 2100 GPM.

Obviously, the difference in the flows from each pump will depend on the difference in operating speeds and on the shape of the pump curves and of the system-head curve. If the two speeds are closer together than we have assumed, the difference between the two individual flows will be less marked. In the contrary case, pump A will carry an even greater share of the total load. It it were desired to establish exactly the degree to which a particular speed difference will affect the distribution of flow between the two pumps, it would only be necessary to plot a family of curves at different speeds over the curves that I have drawn.

Although this is not probable in the particular case analyzed here, a point can finally arrive because of a very considerable difference in speed at which one pump will hog the entire flow because the second pump does not develop sufficient head even at shutoff to discharge into the system. Such a possibility is much more likely to occur if the system-head curve includes a very significant static head component. To illustrate this effect, I have plotted in Fig. 1.74 the performance of the same two pumps operating in a system in which the static head is 110 ft. In this particular case, the combined head-capacity curve of pumps A and B intersects the system-head curve at 1900 GPM. Pump A contributes 1200 GPM to this total and pump B only 700 GPM. If the speed of pump B were reduced further to 2070 RPM, its shutoff head would become 116 ft. Under these conditions, the head-capacity curve of pump B would intersect the system head at 1280 GPM and 116 ft, and pump B would be unable to deliver any flow.

This possibility indicates that protection should be provided to avoid running one of the pumps against a closed check valve: A bypass must be taken off between the pump discharge nozzle and the check valve, leading somewhere back to the source of suction supply where the heat added to the water within the pump can be dissipated.

Returning to the problem of the two steam turbine-driven chilled-water pumps, the problem is ameliorated when a reduction in demand for chilled water reduces the pump speed by throttling the valve that controls the admission of steam to the turbine. Under the reduced load demand conditions, the difference in steam piping arrangement will have less effect on the operating speed of the pumps. Thus, if the two pumps and turbines are reasonable duplicates of each other, they will probably split the total capacity fairly evenly at flows lower than the maximum. You can readily test this in your installation by measuring the speed of the two pumps and their individual capacity at several reduced loads.

Question 1.50 Variable-Speed Operation of Multiple-Pump Installation

We have a tentative planned application in the municipal waterworks field that consists of a remote, unattended 5 million gal ground reservoir and booster pump station. It is desired that the pumping plant have the following characteristics: constant discharge pressure control at 150 ft head; variable capacity from 375 to 3750 GPM. We have selected three units for the job as follows:

1. One 375 to 750 GPM centrifugal pump with variable-speed drive to vary the speed from 1620 to 1750 RPM.

Chapter 1

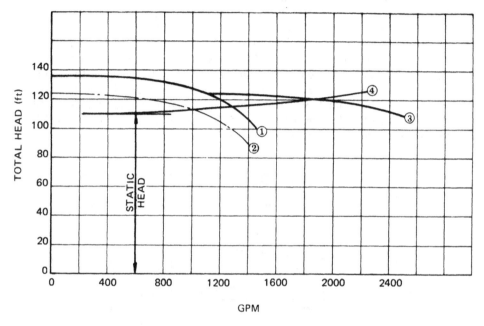

Figure 1.74 Performance of the two pumps from Fig. 1.73 operating in a system in which the static head is 110 ft.

2. Two 750 to 1500 GPM centrifugal pumps with variable-speed drive to vary the speed from 1450 to 1750 RPM.

We hope to be able to control the speed of the pumps by sensing system pressure and give the station the capacity to deliver a night winter load of 375 GPM to a peak day summer load of 3750 GPM.

In your opinion, can a centrifugal pump be designed to vary capacity from 50 to 100% at a reasonably constant discharge head?

Answer. Let me first state unequivocally that centrifugal pumps intended for waterworks service can be and are designed to vary from 50 to 100% in capacity at constant head either by throttling excess pressure or, what is more economical, by operating at variable speed. As a matter of fact, they are frequently operated over an even wider range of capacities. However, your solution of using multiple units is generally more economical, since the individual pumps are never required to operate in that capacity range in which their efficiency deteriorates excessively.

You mention a speed range for both sizes of pump. I do not know on what information you have based your selection of this speed range, but at first glance the minimum speed of 1450 RPM given for the 1500 GPM pump appears to be unnecessarily low. I have plotted a typical curve for this size pump in Fig. 1.75. Since you did not state the suction conditions that will exist in this installation, I have assumed zero suction lift (or suction head), and consequently the pump is shown to be designed for 150 ft total head. You will note that the speed required to maintain 150 ft. constant head varies from 1750 RPM at the full design capacity of 1500 GPM down to 1620 RPM at zero flow. At 50% capacity, that is, at 750 GPM, the required speed is approximately 1645 RPM.

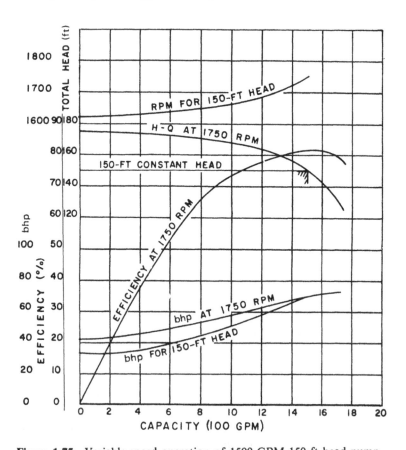

Figure 1.75 Variable-speed operation of 1500 GPM 150 ft head pump.

You do not specify the means to be employed to vary the pump speed. It is probable that you intend using one of the following three arrangements:

1. Wound-rotor motor
2. Hydraulic coupling between an induction motor and the pump
3. Magnetic drive between an induction motor and the pump

Although the first solution eliminates one piece of equipment, it has the disadvantage of being unable to provide a stepless control of the discharge pressure. Commerical speed controls cut the resistance into the rotor circuit in steps, so that the resulting speed change is likewise in steps. For a finer adjustment of speed, it becomes necessary to use a control with a large number of contact points or a liquid slip rheostat. Since the particular application under consideration requires a very narrow range of speed control, the finer adjustment obtainable from a hydraulic coupling or from a magnetic drive appears to be preferable.

The range of pressure within which the variable-speed control can operate these pumps depends strictly on the sensitivity that you will specify. Remember that for any control to operate, it is necessary for the pressure to deviate in one or another direction from the desired pressure, so that the variance can be detected by the sensing element and an impulse can be generated to adjust the pump speed. To some degree, therefore, the greater the sensitivity you will require (that is, the less the permitted variance) the more expensive will be the control.

Since this is to be a remote and unattended station, it is probably best that the control be electrical rather than hydraulic or pneumatic. The control would incorporate a transducer that would generate a signal proportional to the pressure being measured. A differential control relay would compare the value measured by the transducer with the desired pressure and send an impulse to the actual speed-varying device employed.

It will also be necessary to incorporate a multicontact timer that will switch pumps on and off as necessary and automtically in a preselected sequence. Means should preferably be incorporated for changing the sequence of operation of the pumps to equalize the wear.

Finally, consideration should be given to the conditions that will arise whenever electric power supply is restored after an interruption. For instance, if all three pumps are restarted simultaneously, an overload may occur on the electrical power supply system. In addition, an undesirable hydraulic surge may be created. It would be best, therefore, to incorporate some form of time-delay relay into the motor starter circuits so that if it becomes necessary to start more than one pump, they will be started sequentially.

If the pumps operate with a suction lift, it will be necessary to provide a central automatic priming system. There are a number of different systems of this type available on the market, and it would be preferable to have the pump manufacturer incorporate this with the equipment, as a package. As a safety precaution, the system should be capable of preventing operation of the pumps in an unprimed condition. This control depends on the type of priming system used, but in most cases it uses either some form of float switch or electrodes placed in the liquid.

Question 1.51 Constant- and Variable-Speed Pumps in Parallel

Do you recommend as a satisfactory arrangement the use of two half-capacity boiler feed pumps, of which one would be operated at constant speed and the second one at variable speed by means of a hydraulic coupling? This second pump would then take up the load fluctuations. Would the same spare rotor be suitable for both pumps, allowing for the slip of the coupling?

Answer. To my recollection, a few such installations were made some 30 or 35 years ago, and of course they can be made to work. Nevertheless, I am definitely not in favor of such a setup. The savings correspond to the initial cost of one variable-speed device. But they are reduced by the fact that the system still requires the use of a feedwater regulator and its cost and maintenance must be charged against this arrangement.

What is more important, however, is that there are a number of unfavorable factors in this type of arrangement. There are two ways of operating these pumps in parallel. The first of these is the one you mention, that is, with the constant-speed pump operating between its design and its maximum capacity and the second pump taking up load fluctuation by speed variation, as shown on Fig. 1.76. Thus, the savings that are obtainable from variable-speed operation are reduced quite appreciably by virtue of the fact that over a very wide range of operation the variable-speed pump delivers a significantly reduced flow and therefore operates at very reduced efficiencies. Examine, for instance, what takes place when the total flow is slightly above the flow that can be delivered into the system with a single pump. The constant-speed pump operates to the right of its design capacity and, therefore, at a slightly lower efficiency than design. The variable-speed pump must make up a very small additional portion of the demand flow and delivers an insignificant percentage of its rated flow. Its efficiency under these conditions will be only a fraction of its normal rated efficiency. As a matter of fact, it might even be required to deliver

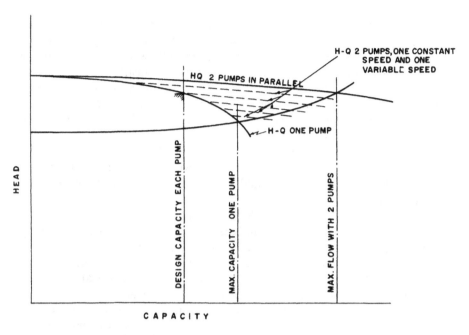

Figure 1.76 Operation of two boiler feed pumps in parallel, one at constant speed and one at variable speed.

less than its minimum permissible flow and the recirculation bypass will have to be open. It will be found in general that all of the power savings have been wiped out and that the cost of supplying one of the pumps with a hydraulic coupling drive is no longer justifiable. Thus, if the constant-speed pump is operated out on its curve and the variable-speed pump only makes up the difference between the flow from the constant-speed pump and the total demand, either both pumps should have been provided with hydraulic couplings or both should be constant speed.

Of course, there is another method of operating two such pumps in parallel and we should examine that method. Here, a master control is used so as to load up both pumps evenly. In other words, the total flow is split between the two pumps so that their individual capacities are equal. The variable-speed pump control actuates the position of the oil scoop in the hydraulic coupling, and the feedwater regulator of the constant-speed pump throttles its discharge to permit equal delivery from each pump. Now the overall savings are essentially half the savings that would be available if both pumps were operated at variable speed. The investment, however, is not cut exactly in half: it remains necessary to provide a feedwater regulator in the discharge of one of the pumps and the controls are somewhat more complex than if both units were the same. But even were we to assume that both savings and extra investment are cut in half, what sort of reasoning can be used to justify the arrangement? If we can justify the purchase of one hydraulic coupling by the value of the resulting savings, why should not doubling these savings justify double the investment? If even one coupling cannot be justified, then doubling the savings will not justify twice the investment.

I should add that there is one more overwhelming disadvantage to the arrangement you suggest and that has to do with the inner assembly, which is normally carried as a spare and may have to be used at sometime or another with either one of the two pumps.

The variable-speed pump will operate at approximately 2 to 3% lower speed than the constant-speed pump because of the slip in the hydraulic coupling. If a spare rotor is purchased and if both pumps are provided with the same impeller diameter, the variable-speed pump will have a lower head-capacity curve at its rated speed than the constant-speed pump. This would further aggravate the poor distribution of load between the two pumps. It would also introduce difficulties in starting up the variable-speed pump whenever the constant-speed pump was running alone at very low flows, since the variable-speed pump could not develop sufficient pressure to lift its check valve.

As a result, the two pumps should be designed to develop identical head-capacity curves, each at its own rated full speed. This would require the variable-speed pump to be built with impellers 2 to 3% larger than those of the constant-speed pump. The same should be true of the spare rotor, which must be capable of developing the design head and capacity regardless of which pump is to be rebuilt with it. If, then, the spare rotor has maximum diameter impellers, it will produce from 4 to 6% more head if placed into the constant-speed pump. The alternative is to cut these impellers when so using the rotor. This adds to the cost and time and renders the rotor no longer suitable for use with the variable-speed pump.

All in all, I can see no sufficient justification for this arrangement. If variable-speed drive can be justified, it will generally be justifiable for both pumps. If load conditions and evaluation methods cannot justify two hydraulic couplings, putting in only one seems to be a wasteful and complicated compromise.

Question 1.52 Rated Speeds for High-Pressure Boiler Feed Pumps

I was looking at a recent power plant equipment survey. I was surprised to notice that the highest boiler feed pump speed shown was only 6260 RPM.

The majority of the higher pump speeds was in the the 5400 to 5900 RPM range, with some pumps rated as low as 4500 to 4600 RPM. This seems to be a lot slower than the speeds cited in the equipment lists of a few years ago. As I do not recall seeing anything in the published literature relating to serious problems with the early high-speed pumps—which, I believe, ran at 8000 or 9000 RPM—I am somewhat puzzled by the industry's apparent abandonment of these speeds in favor of the much lower ones cited above. I would appreciate your comments on this matter.

Answer. I can well understand your puzzlement by what appears to be a trend toward lower speeds for so-called high-speed boiler feed pumps. Although it is true that the speeds of pumps of this type are generally decreasing, this does not represent a departure from the design concepts under which the original 8000 to 9000 RPM pumps were developed. As a matter of fact, if one were to compare pumps of the same capacity, one would find that today's high-speed pumps are actually running somewhat faster than those of a few years ago.

Let us first reexamine the reasons that led to the concept of high-speed boiler feed pumps. Conventional 3600 RPM high-pressure boiler feed pumps were generally designed to develop from 230 to 350 psi per stage, that is, from about 600 to 900 ft. As operating pressure kept increasing, these heads per stage led to the necessity of using constantly greater numbers of stages. This in turn increased the pump shaft span and the shaft deflection, and pump reliability was being unfavorably affected. It was the desire to maintain or even improve boiler feed pump reliability that led to the development of higher speed pumps that could have materially higher heads per stage and, consequently, develop the required pressure with a significantly lesser number of stages. For instance, the first high-speed boiler pump to go into commercial service in 1954 at the Kearney Generating Station of Public Service Electric & Gas Co. of New Jersey ran at 9000 RPM and developed 1650 ft per stage in a four-stage configuration. In other words, the prime purpose of the new concept was to increase the head generated per stage, rather than merely to run pumps at higher speeds.

The reductions in speed that you have noticed have been primarily caused by the increases in pump capacities required by larger turbine-generator units as well as by the fact that a greater number of units use full-capacity boiler feed pumps rather than half-capacity ones. In a number of cases, as well, the reduction in speed may have been caused by the necessity of limiting the pump speed to within the capabilities of its driver. Finally, in a few cases, the desire to eliminate a booster pump may have led the station designer to request the pump manufacturer to drop the pump speed sufficiently to reduce the required NPSH to within the available limits. (This last situation, parenthetically, is not necessarily the most economical or the most sound solution.)

Let us look at the effect capacity can have on pump speed. The geometry and hydraulic performance of a centrifugal pump impeller is determined by its specific speed type. This is expressed by the relation

$$n_s = \frac{N\sqrt{Q}}{H^{3/4}}$$

For a particular specific speed impeller, the RPM at which it must operate is a function of its design capacity and of the head it will be designed to develop. Rearranging the formula given for specific speed, we get

$$N = \frac{n_s H^{3/4}}{\sqrt{Q}}$$

Specific speeds for high head per stage boiler feed pumps range between 1000 and 1800. A typical head per stage at the best efficiency capacity would be 1600 ft. If, for instance, a specific speed of 1400 and a head per stage of 1600 are substituted in the relation given above, one may determine how speed would vary with capacity for that particular impeller configuration.

$$N = \frac{1400(1600)^{3/4}}{\sqrt{Q}} = \frac{354{,}200}{\sqrt{Q}}$$

In other words, to keep within a given specific speed and head per stage, the pump speed must vary inversely as the square root of the pump design capacity. Thus, a 2000 GPM impeller with specific speed of 1400 would operate at 7940 RPM, but a 10,000 GPM impeller designed for the same head per stage and with the same specific speed would operate at only 3540 RPM.

Note that in slowing down to 3540 RPM we have not decreased the head per stage. The same high head per stage that characterizes a 2000 GPM, 7940 RPM pump also applies to a 10,000 GPM pump operating only 45% as fast. But since the head per stage is the same, it must mean that the peripheral speed of the two impellers is also the same. The larger, and slower pump has a larger impeller diameter, reversely in proportion to the operating speed.

Figure 1.77 shows the relation between speed and capacity for several specific speed impellers designed for 1600 ft head. You can see that regardless of the specific speed we choose (with a fixed head per stage), the pump RPM must decrease with increasing capacity.

Chapter 1

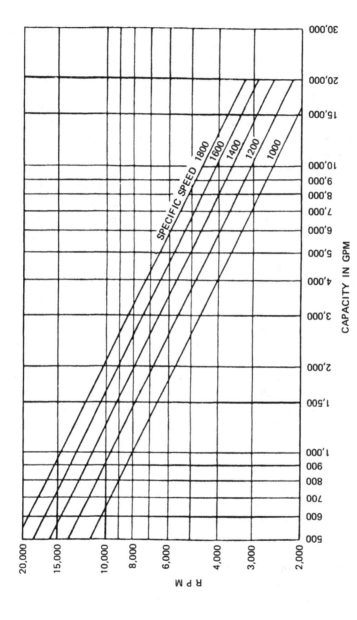

Figure 1.77 Effect of design capacity on operating speed to high-speed boiler feed pumps designed for a head of 1600 ft per stage.

Because of the desire to further reduce the number of stages of boiler feed pumps, there is a tendency to use heads per stage higher than the 1600 ft mentioned. Given a desirable specific speed range of between 1400 and 1800, this in turn will increase the operating speed over the entire capacity range, but this does not affect the downward influence of higher capacities on pump speeds. Thus, if we designed a 10,000 GPM pump for 2000 ft per stage and a specific speed of 1800, the pump speed would be 5385 RPM, which is still only about two-thirds of the 8000 RPM that is typical of a 2000 GPM pump. Even at a head per stage of 3000 ft, which is today about the highest under consideration for pumps with 13% chrome steel internals, the speed of a 10,000 GPM, 1800 specific speed pump would be only 7300 RPM. Since a specific speed of 1800 is about the maximum that will provide hydraulic characteristics suitable for boiler feed pump service, it appears that for the present, increasing capacities will keep the operating speeds of high-speed pumps below their earlier range.

The other factor governing the design speed of a boiler feed pump is the allowable speed characteristic of its driver. I am quite certain, for example, that most of the pumps selected for the survey to which you refer were driven by auxiliary steam turbines. When properly integrated into the station heat balance, the use of auxiliary steam turbines can provide significant savings in overall heat rate and for this reason there is a strong trend toward their use as boiler feed pump drives.

The subject of steam turbine speed limitations is a complex one, but it can generally be stated that most condensing steam turbines (the most popular type for the larger units) are generally limited to speeds of 4500 to 6000 RPM, depending upon their design hp and back pressure. Both increasing vacuum at the turbine exhaust and increasing design horsepower tend to reduce the turbine maximum permissible speed. Since these steam turbines are considerably more expensive than the pumps they drive, the development of special or new turbines to match existing pump designs is usually economically unsound. For this reason, the usual practice is to use an existing pump that matches the available turbine speeds. It is even not uncommon to develop a new pump in order to meet available drive turbine speed limitations.

If the pump is to be driven by a motor or a main turbine-generator, it is unlikely that its maximum speed could be affected by the design criteria governing the step-up gear between the prime mover and the pump. At the moment, the gears most commonly used with boiler feed pump drives are of the double-helical horizontally offset type. Although these are available in higher ratios, 5:1 is a safe rule of thumb to use to be sure that the gear price will not become uneconomical. Even a ratio of 5:1 means that gear design limits are not apt to restrict high-speed pump speeds, at least for the present, as capacity considerations alone will place the pump RPM somewhere under 9000 in any event. Thus, any pump speed that may be required will be obtainable as long as the drive speed is 1800 RPM or over.

Question 1.53 Main Shaft Drive of Feed Pumps

In the late 1950s and the 1960s, boiler feed pumps were often driven by the main turbogenerator. Why has this practice apparently disappeared?

Answer. The evolution of boiler feed pump drives in the period you mention is an extremely interesting example of the principle of the "survival of the fittest." Just prior to the introduction of the turbo-generator as the boiler feed pump drive, a number of factors united to make the steam turbine preferable to the electric motor for this purpose. The growth in the size of individual units, coupled to the higher operating steam pressures, led to an appreciable increase in the size of the pump drivers. At the same time, developments in boiler feed pump design led to the use of speeds well above the 3600 RPM maximum available with electric motors. Of course, these higher speeds could be reached by using step-up gears, and as a matter of fact, the combination of 1750 RPM motors and step-up gears had become the accepted boiler feed pump drive. But the many advantages of the steam turbine contributed to making it replace the electric motor.

Among these advantages were the fact that a steam turbine is inherently a variable-speed machine and did not require the use of a variable-speed device, such as the hydraulic coupling or the magnetic drive, but most important of all, the use of the steam turbine increased the net power plant output by eliminating the electric horsepower required by the boiler feed pumps.

Just as the steam turbine drive was coming into its own, the possibility of saving the cost of a separate turbine led to the idea of driving the boiler feed pump directly from the main turbine-generator unit. Incidentally, the idea was not exactly new: the invention of the steam pump by Henry R. Worthington in 1840 was prompted by his desire to abandon the same type of arrangement. Until then, the feedwater pump had been operated directly by the steam engine it served.

I must admit that at the time I was less than enamored by this approach. It had a number of defects in my opinion. Step-up gears had to be used if advantage was to be taken of the high-speed pump. A hydraulic coupling had to be used to obtain variable speed. And unless a cross-compound unit was a unit that could drive two half-capacity pumps, a tandem unit forced the power plant designer to apply a full-capacity pump, which is not always the best solution. Finally, main shaft drive always required the installation of an electric motor-driven start-up boiler feed pump that was also necessary for a hot restart even after a very short trip-out. Nevertheless, the idea of mainshaft drive caught on rather fast and my predictions that the fad would not last very long threatened for a while to remain unfulfilled.

But with time, the principle of the survival of the fittest took over. The many points of superiority of individual steam turbine drive asserted themselves, helped along by one other development in turbine-generator construction. As their size continued to grow, it became desirable to unload the low-pressure end if possible. This could be done if the boiler feed pump turbine were a condensing turbine.

And that, essentially, is what happened. The preferred boiler feed pump drive today is a condensing turbine, and to my knowlege, there has not been a main shaft-driven boiler feed pump installation for many years.

Question 1.54 Paralleling Centrifugal and Reciprocating Pumps

We have a service application in which the centrifugal pump we had installed several years ago is no longer sufficient in capacity. We have a small reciprocating pump that we could install to operate in parallel with this centrifugal pump and get just about enough capacity from the two for our needs. Is this considered good practice?

Answer. Although such arrangements have been used, I definitely do not consider it good practice. The performance of the centrifugal pump will be affected both mechanically and hydraulically by the pulsations of the reciprocating pump. Of course, the more cylinders the reciprocating pump has, the less effect it will have on the centrifugal pump. A quintuplex pump would be ideal, a triplex pump a little less satisfactory, and actual trouble would be certain to arise if a single-acting, single-cylinder pump were used.

I would use this solution strictly as a last resort. Maybe the centrifugal pump you have can be accomodated with a larger capacity impeller. It would certainly be wise to check this with the manufacturer.

But if you do install these two pumps in parallel, make sure to use separate suction lines, especially if the pumps handle a high suction lift.

Question 1.55 Effect of Water Temperature on Pump bhp

I have been told that the power consumption of a boiler feed pump increases as the feedwater temperature increases. How can this be, since the gravity of water decreases with higher temperature and it should take less power to handle a lighter fluid?

Answer. Your informants are quite correct, and the power consumption of a feed pump does increase exactly as an inverse ratio of the specfic gravity of the feedwater it handles. The apparent paradox arises from the fact that a boiler feed pump must be selected to handle a given weight of feedwater of so many pounds per hour and to develop a certain pressure in pounds per square inch rather than a total head of so many feet.

The formula for brake horsepower is

$$\text{bhp} = \frac{\text{GPM} \times \text{head in feet} \times \text{sp gr}}{3960 \times \text{efficiency}}$$

Thus, as long as a fixed volume in gallons per minute is pumped against a fixed head in feet, the bhp will decrease with specific gravity.

However, if we convert pounds per hour into gallons per minute and pounds per square inch pressure into feet of head, we find that the relationship with specific gravity changes,

$$\text{GPM} = \frac{\text{lb per hr}}{500 \times \text{sp gr}}$$

and

$$\text{Head in feet} = \frac{\text{psi} \times 2.31}{\text{sp gr}}$$

and therefore

$$\text{bph} = \frac{\text{lb per hr}/(500 \times \text{sp gr}) \times (\text{psi} \times 2.31)/\text{sp gr} \times \text{sp gr}}{3960 \times \text{efficiency}}$$

By simplifying this relation, we get

$$\text{bhp} = \frac{\text{lb/hr} \times \text{psi}}{857{,}000 \times \text{efficiency}} \times \frac{1}{\text{sp gr}}$$

and we see that as the temperature increases and the specific gravity decreases, there is an increase in power consumption.

Question 1.56 Design Capacity of Condensate Pumps

I have noticed that the rated capacity of condensate pumps is generally selected with considerably more margin over the maximum expected condensing steam flow than is the case with boiler feed pumps when their rated capacity is compared with the maximum turbine throttle flow. When I say more margin, I really mean a considerable difference. For instance, I have seen cases in which as much as 30% or more is added to the maximum condensing steam flow in selecting condensate pumps. Is this difference influenced by the thought that condensate pumps may operate under cavitating conditions and that therefore their effective capacity will be reduced much below their rated capacity?

Answer. The effect of cavitation is not the determining factor in selecting condensate pump capacity. To begin with, the use of cavitation to control the delivery of a condensate pump (this is generally referred to as *submergence control* or *self-regulation*) is restricted to open feedwater systems, in which the condensate pump delivers into a direct-contact heater. This arrangement used to be quite popular sometime ago, but it is less frequently used today, and the delivery of the condensate pumps is controlled by a throttling valve that takes its impulse from either the condenser hot-well level or the level in the storage space of the direct-contact heater.

Moreover, even when condensate pumps are operated on submergence control, their submergence requirements are so chosen that they can deliver their full rated capacity under prevailing submergence conditions, and no limitation on their effective capacity is allowed to be imposed by the static elevation conditions of the installation.

Condensate pumps must be designed for a considerably higher capacity than they normally will have to handle because of the very nature of the usual feedwater cycle. If you refer to Fig. 1.78, you will see that the typical regenerative cycle includes a series of closed heaters located between the condensate pump and the boiler feed pump. Generally, the heater drains from these heaters are cascaded down to the lowest pressure heater in the system, from which they are pumped back into the condensate discharge header by a heater drain pump. In this manner, the heater drain pump handles the difference between the condensed steam flow and the flow required from the boiler feed pump.

In the event abnormal conditions arise, as, for instance, the necessity for bypassing the closed heaters or the unavailability of the heater drain pump, a condensate pump designed for just the normal condensing steam flow would be inadequate. If the heaters are bypassed and no extraction steam is taken from the bleed stages, the condensing steam flow will be much increased over its normal value, and the condensate pump must be capable of evacuating the hot well.

If, on the other hand, the heater drain pump is unavailable for some reason, the heater drains will be diverted and dumped into the condenser

Chapter 1

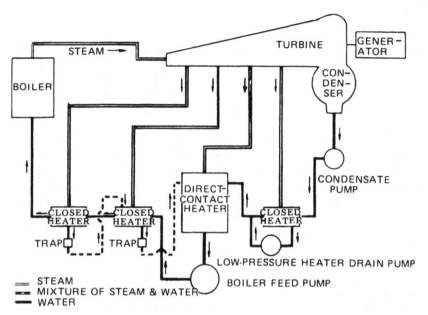

Figure 1.78 Typical present-day steam-electric cycle employs a heater drain pump to handle the difference between condensed steam flow and the flow required from the boiler feed pump.

hot well, swelling the flow from the turbine exhaust. In this case as well, it becomes necessary to have sufficient capacity built into the condensate pump to handle all the flow that is entering the condenser hot well.

Thus, whereas the boiler feed pump is designed for something like 8 to 15% excess capacity—and that only to take care of boiler swings and of pump wear—condensate pumps may easily be rated for up to 30% in excess of normal condensate steam flow.

Question 1.57 Should Circulating Pump Handle Cold or Warm Brine?

We are laying out a brine system in which the receiver for the warm brine return, which acts as the surge tank, is located about 25 ft above the shell and tube brine cooler. The arrangement in the plant is such that it would be more convenient to install the brine circulating pump, which is a centrifugal unit, beyond the cooler so that it will handle cold brine. Are there any troubles likely to develop because of this location, and why is it customary to install the brine circulating pump ahead of the cooler?

Answer. The handling of cold brine that results from locating the circulating pump beyond the brine cooler introduces both hydraulic and mechanical complications that are best avoided. When the pump takes its suction directly from the surge tank, it operates with the maximum available head on suction. If the pump is located beyond the cooler, the suction head is reduced by the amount of the friction losses in the cooler, and depending on the magnitude of these losses, the pump may have to operate with a suction lift. Any excessive fouling of the cooler tubes will increase the friction losses to an extent that the suction lift may become higher than permissible for the required capacity and the pump may be subject to cavitation. This will restrict the available capacity, cause noisy operation, and reduce the life of the pump through wear of the impeller.

Another unfavorable effect resulting from handling cold brine is the increase in power consumption. At lower temperatures, the viscosity of the brine increases very rapidly and reduces pump efficiency to an appreciable extent. On the other hand, the effect of viscosity on pump efficiency when the brine handled is warm is almost negligible. In addition, a lower efficiency means more heat dissipated in the pump and therefore more heat added to the brine handled by the pump. This reduces the effectiveness of the cooler and of the cooling system.

Handling cold brine instead of having the pump take its suction direclty from the warm-brine return tank can lead to mechanical problems centering at the bearings. Of course for either location, the temperature of the brine will be below the freezing temperature of water. The bearings of a brine pump are therefore provided with some sort of a seal that will prevent or at least reduce the tendency of the moisture in the surrounding atmosphere to enter and condense inside the bearing housings. Likewise, a lubricant is chosen that will give satisfactory service at the prevailing low temperatures. Both these problems are magnified when the pump takes its suction from the cooler. It may become necessary to circulate warm water through the jackets of the bearing housings (if such are provided) or possibly to locate electric heating coils within the bearing housings.

Although a definite decision on the pump location must perforce be made after weighing all these factors against whatever problems will arise if the pump is located directly under the warm-brine return tank, it would appear that these problems cannot be completely unsurmountable and that the advantages of handling warm brine are sufficient to provide an incentive to reexamine the installation plans.

Question 1.58 Testing Pumps at Reduced Speed

Our existing test laboratory is suitable for standard running tests of pumps requiring up to 400 hp at 1470 RPM (50 cycles). For bowl assemblies requiring more than 400 hp, it is necessary to reduce either the number of stages or the speed.

With regard to reducing speed, we would like to have your advice as to the lowest speed recommended in order to justify correct application of the affinity formulas.

Answer. The limitations introduced in your testing program by the lack of sufficient power facilities arise frequently even in the case of pump manufacturers. After all, once a test stand has been installed with whatever size driver that may be chosen because of practical and economic limitations, the natural course of progress will always result in there being a pump with a power consumption that exceeds the test stand facilities. It is then the practice to resort to one of the two methods you indicate, that is, to test at reduced speed or if a multistage pump is involved, to test at full speed with fewer stages.

I shall deal in a moment with the question of relative accuracy and of the effect on the affinity laws. Let me first state that my preference would almost always be in favor of testing a pump with all its stages and at reduced speed. After all, a test is not run only for the purpose of checking out the head-capacity curve and the power consumption but also, and to an equal degree, the mechanical performance of the pump—particularly to establish that it will run without undue vibration or without rubbing at the running joints. Dismantling the pump to remove stages and running at full speed and the reassembling it will not give as sound an indication of the ultimate mechanical performance of the pump as a test of the complete pump at reduced speed.

In addition, a broader range of power consumption reduction can be obtained by running at reduced speed. For instance, if a pump is designed for 1450 RPM, testing it at say 725 RPM will reduce the power consumption by 8:1 (the cube of the speed ratio, which would be 2). To get an equal power reduction by removing stages would require the removal of seven-eighths of the stages. What does one do then with a five-stage pump?

AWWA test specifications make specific provision for deep-well vertical turbine pumps when the horsepower exceeds 200. They permit testing only that number of stages of the unit that come within this power requirement. No efficiency correction is permitted for testing a lesser number of stages than the actual. The head and horsepower are to be increased in direct proportion to the number of stages in the final assembly, compared with the number of stages used in the laboratory test.

These same test specifications permit model tests of pumps when the size of the bowl exceeds 20 in. This, however, does not apply to your specific case.

Finally, on bowl assemblies that have an OD exceeding 20 in. or that require more than 200 hp, these specifications permit testing the actual bowl assembly at a speed slower than that at which the pump will run in the field, instead of making a model test. Here again, no efficiency correction is allowed when translating the performance of the slower speed test to full-speed performance. In other words, such reduced speed tests are to be stepped up following the classic affinity laws:

$$Q_2 = Q_1 \left(\frac{n_2}{n_1} \right)$$

$$H_2 = H_1 \left(\frac{n_2}{n_1} \right)^2$$

$$P_2 = P_1 \left(\frac{n_2}{n_1} \right)^3$$

where Q_1 = capacity, in gallons per minute at speed n_1
Q_2 = capacity, in gallons per minute at speed n_2
H_1 = head, in feet, at speed n_1, for capacity Q_1
H_2 = head, in feet, at speed n_2, for capacity Q_2
P_1 = brake horsepower, at speed n_1, at H_1 and Q_1
P_2 = brake horsepower, at speed n_2, at H_2 and Q_2

Thus, the capacity varies directly with the speed, the head varies as the square of the speed, and the bhp varies with the cube of the speed. (See Fig. 1.79.)

The Hydraulic Institute Standards also have a specific reference to conditions when laboratory power facilities do not permit a full-speed test. No reference to any specific correction for the variation in speed is made, and the relationship of capacity, head, and power consumption is implied to follow exactly the affinity laws. The sole caution given is that of testing the pump at the same sigma value. This, really, can be interpreted to mean testing at the same suction specific speed.

You will note that measurement tolerances are indicated in the Hydraulic Institute Standards for various measurement means. For instance, it is stated that capacity tolerances of 2% for cast venturi tubes, 0.75% for accurately machined venturis, 1.5% for orifices, and so on, are applicable. Obviously, similar tolerances in the measurement of power consumption may be expected.

The Hydraulic Institute Standards also codify model testing. The most accurate possible model is that provided by the very same pump

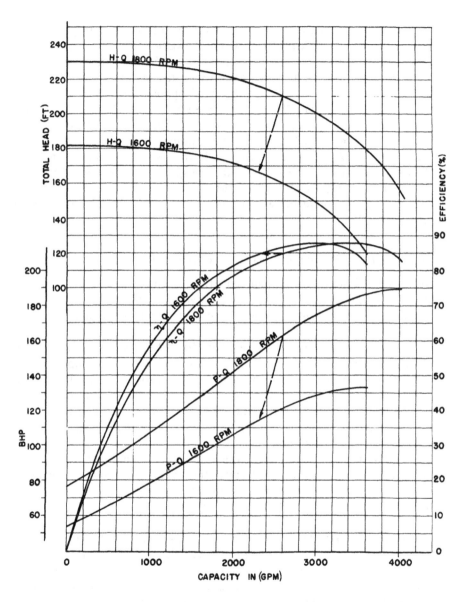

Figure 1.79 Effect of speed change on pump characteristics.

tested at a different speed. The range of the correction factors indicated is rather broad, with the losses (that is, the value of 1.0 minus the efficiency) varying as a power of the ratio of the scale factor, this power itself varying from 0 to 0.26.

I would say, therefore, that any error introduced by testing a pump at reduced speed would be probably within the value of the errors, tolerances, and so on, as long as the speed was not reduced by more than 50%. If you have 400 hp available, a pump that on your test stand required 400 hp at 725 RPM would be one that at the rated operating speed of 1450 RPM would be taking about 3200 hp. I assume that this would more than meet any pump application that you may encounter. If it did not, for instance, if you wished to test a 5000 hp, 1450 RPM pump, you would require a power reduction of 12.5:1, which corresponds to a speed reduction of about 2.33:1, or 620 RPM. This is not a 50 cycle speed and you would be forced to test at about 580 RPM, but I would still prefer to do this. The efficiency may fall off one point or two, but that would be within the commercial tolerance.

Question 1.59 Pressurizing Suction of Hydraulic Pumps

We recently installed individual centrifugal pumps to pressurize the inlet ports of high-pressure variable-volume hydraulic oil pumps. We also installed full-flow oil filters between the centrifugal pumps and the high-pressure pumps.

The centrifugal pumps are rated at 150 GPM at 50 psi, 7½ hp at 3600 RPM. The oil is 300 sec hydraulic oil. The high-pressure pumps are 0 to 135 GPM at 860 RPM with supercharging and 0 to 95 GPM without. The filters are rated at 150 GPM, 5 psi pressure drop, 20 μm filtration.

The recommendations of the high-pressure pump manufacturer include a statement that the supercharge pump, if a centrifugal type, be a "rapid-recovery" type pump.

What is a rapid-recovery type centrifugal pump, and how do its characteristics differ from a conventional centrifugal pump? We checked a number of local pump suppliers prior to purchasing the pumps we installed. None of them could interpret this term.

Answer. I am afraid that the description "rapid-recovery-type centrifugal pump" has no accepted meaning in the centrifugal pump terminology, at least none that I have ever encountered. If the manufacturer of your high-pressure variable-volume positive displacement pumps does not also manufacture centrifugal pumps, they may have coined this phrase.

Reading between the lines, I image that they may be making reference to a pump with a flat head-capacity curve (Fig. 1.80) which would be preferable to a steep curve, since the former would permit the pump to carry farther out in capacity with a reduced discharge pressure than a steep curve.

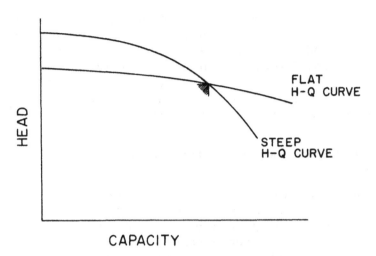

Figure 1.80 Comparison of flat and steep head-capacity curves.

It is my opinion, however, that the system in which the centrifugal pump is made to operate plays a greater part in providing satisfactory operation and supercharging than the shape of the centrifugal pump curve. If the variable-volume hydraulic pump changes flow very rapidly, the system may not permit the centrifugal booster pump to adequately "supercharge" the hydraulic pump. This would be the case, for instance, if the centrifugal pump were installed with very little margin in available NPSH over the required NPSH .

Rapid acceleration of the suction column could reduce the NPSH below the minimum required value for the rated capaicty, the centrifugal pump would start cavitating, and the pressure at the suction of the hydraulic pump could fall off very rapidly.

The same problem may exist on the discharge side of the centrifugal pump. If a very long line were involved between the centrifugal pump and the hydraulic pump it pressurizes and if the velocity in this line were high, the acceleration caused by a rapid increase in flow would be such as to reduce very appreciably the pressure at the hydraulic pump, and the effectiveness of the booster pump would have been reduced.

Frequently, the supercharging of a hydraulic pump is provided with a constant-volume positive displacement pump operating in a closed-loop system, with a relief valve. This is a certain means to maintain suction pressure at the hydraulic pump but is more complicated and more expensive than a centrifugal pump.

If the supercharging is provided by a centrifugal pump, the system must be made suitable. A reasonable margin must be provided over the minimum required NPSH. The suction line to the centrifugal pump must be as short as possible. Likewise, the piping between the centrifugal pump and the hydraulic pump must be as short as possible, and large-diameter piping must be used. This keeps the velocities low, and therefore the acceleration required by a rapid change in flow will also be low.

Question 1.60   Correct Discharge Pressure for Boiler Feed Pumps

I would like to have your comments on the correct feedwater pressure for a given system. In my experience, a great number of operators waste energy and create valve maintenance problems by overpressure in the boiler feed line in the name of conservative operation.

In our plant we have two examples of this phenomenon:

1. Three waste heat boilers operating on copper reverb furnaces, making 875 psi 905 °F steam for a 15,000 kW turbine-generator, are served by one variable-speed steam turbine-driven boiler feed pump, equipped with a constant outlet pressure control operating through a Woodward governor. Feedwater distribution lines are large, and the system has a friction loss of 12 psi between the pump and the boiler feed valve at 17.5 MW load. We currently take a 150 psi drop across our control valve at full output from all boilers.
2. Our powerhouse contains two power boilers that produce 875 psig 905 °F steam. Each unit has two pumps that are both large enough to give us a 250 psi drop across the feedwater control valve.

Maybe with some authoritative comment by you, we shall be able to stop this wearing out of pumps and valves and the waste of energy.

Answer.   Your point is very well taken, and I could not agree with you more. Too frequently the pressure margin provided in boiler feed pump installations is far more generous than need be, with the obvious result that much power is wasted and that excessive wear takes place in the regulating valves. This, of course, is particularly true for motor-driven constant-speed pumps, as the resulting power losses are much reduced if the pumps operate at variable speed. In the latter case, the pump speed is reduced to that needed to meet the prevailing conditions of service, and the power loss is only that caused by the speed reduction, as, for instance, the losses in a fluid coupling.

One must remember, incidentally, that insofar as pump selection is concerned, it is of no consequence whether excessive margin is included in the rated capacity or whether too much caution is used in setting the required discharge pressure and hence the pump total head. The only difference will appear on the nameplate rating of the pumps, but the head-capacity curve will exceed the system-head curve by too great a margin in both cases. The power consumption will have been unnecessarily increased, and too much pressure drop will take place in the feedwater regulator.

Of course, a margin in head and capacity should always be included over and above the conditions of service based on maximum flows to be encountered in operation. This margin is intended to compensate for the effect of wear in service. It also acts as a precaution against underestimating the maximum capability of the main unit or the system pressure losses. But I cannot remember a single case in my experience except one when this margin was ever fully utilized.

This exception refers to the Chalmette Plant of the Kaiser Aluminum and Chemical Company in Louisiana. The installation comprised a total of 14 boilers operated at 850 psig, and back in 1952 I was asked to make recommendations for this installation. In those days, the average discharge pressure rating for boiler feed pumps serving boilers of that pressure ran from 1150 to as high as 1250 psig. The owners and the plant designers, fortunately, were willing to follow my recommendations, and as a result the pumps serving these boilers are designed for 1024 psig discharge pressure. One full-capacity motor-driven pump serves each boiler, and four turbine-driven duplicate pumps are installed as spares for the entire installation. Although I do not have recent data on the operation of these units, I had checked approximately 12 years after the startup of the installation. At that time not one of the pumps had been opened up for overhaul, and the pumps were still capable of feeding the boilers, despite the probable fact that a certain amount of wear had taken place at the internal clearances. This is a direct example that as much as 125 to 250 psi could generally be saved in selecting the discharge pressure of boiler feed pumps serving boilers operating at this pressure.

I should add that it is not longer the practice to add overliberal margins to the maximum boiler capacity in arriving at the rated capacity of the boiler feed pump installation. As a general rule, an 8% margin in capacity is considered to be ample to compensate for possible boiler swings and for the wear that will ultimately reduce the effective capacity of the pumps. Installations in utility power plants have actually been made in which this margin has been reduced to as little as 4%.

Referring specifically now to the second example that you cite, not only does the practice of maintaining an excessive pressure drop across the feedwater regulator contribute to the wearing out of pumps and valves and constitute an unnecessary waste of energy, but operating two pumps when a single pump can carry the load further aggravates this situation.

You do not indicate the size of the powerhouse boilers, but it would appear from the fact that yours is an industrial rather than a utility installation that the size of these boilers is such as to make it uneconomical to use half-capacity boiler feed pumps. For larger units, it has frequently been the practice to use three half-capacity pumps, two of which serve the boiler under full-load conditions and the third pump acts as a standby. Lately, this standby pump has been eliminated from most half-capacity installations. The justification for using half-capacity boiler feed pumps is based on the assumption that whenever the load is reduced to the point at which a single pump can meet the demand, one of the pumps can be shut down. Only too frequently, however, operators do not shut down one of the two pumps when operating conditions are such that economical operation dictates this procedure. This practice, of course, nullifies the justification that the power plant designer would have used in reaching a decision in favor of half-capacity boiler feed pumps.

Note that the amount of wasted energy in running two pumps at half-load when a single pump could have met the conditions is a very significant one. This can best be demonstrated by reference to a so-called 100% curve as shown in Fig. 1.81. Such a curve is constructed by plotting the head-capacity, power capacity, and efficiency-capacity curves in terms of the percentage of the respective values of capacity, head, power, and efficiency at the pump capacity corresponding to maximum efficiency. To simplify our analysis, we shall assume that the pump rated conditions of service correspond to this 100% point.

If a single-half-capacity pump is operating at the half-load condition of the boiler that it serves, the power consumption will be 100% of its rated value. If, on the other hand, both pumps are kept on the line, each pump will operate at 50% capacity, taking 73% of its rated power consumption. Thus, the total power consumption of two pumps operating will be 146% of that required if only one pump were to be kept on line.

I have not frequently run into a situation in which full-capacity pumps are installed with the intention of keeping one of the pumps as a standby and both pumps are operated at all times as you describe. But here the excess power used by the two pumps as compared with single-pump operation is again of the same order of 46% excess.

An examination of the curves in Fig. 1.81 will also show that operating two pumps at 50% of their rated capacity instead of running a single pump at 100% flow makes the pumps develop approximately 116% of their rated head. This means that in addition to the pressure difference

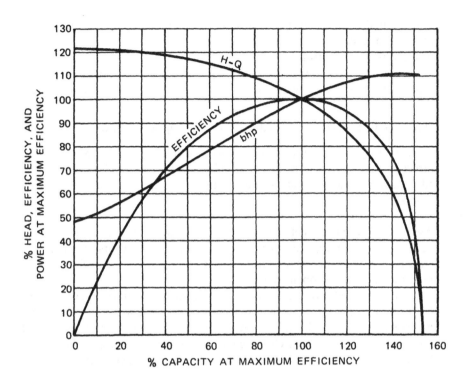

Figure 1.81 Typical performance curve of centrifugal boiler feed pump.

between the rated discharge pressure and the sum of the drum pressure plus friction losses, the feedwater regulator will also have to throttle this 16% excess in total head.

Since we are speaking of excess margins, I might bring up the controversial question whether the drum pressure at full rated load that is used to calculate the required discharge pressure should be set at 6% above the boiler safety valve setting. This would correspond to a literal interpretation of the ASME boiler code. Actually, the code makes no mention of the quantity of feedwater that should be fed to the boiler at that pressure. It is my opinion, shared by a great majority of steam power plant designers, that such an interpretation imposes an unnecessary penalty on the boiler feed pump power consumption. The average boiler feed pump head-capacity curve has a rise of 15 to 20% from its design head to shutoff, the steepest portion occurring near the rated capacity. Even if the pump had to feed water into the boiler under the adverse conditions of all the safety valves blowing, it would still be able to maintain a very respectable rate of flow into the boiler and protect it against any mishap. Thus, the drum pressure to be used in calculating rated pump conditions need not incorporate the 6% overpressure setting.

You are quite right in questioning overconservative practices. They lead to much wasted power and excessive pump and valve maintenance.

Question 1.61 Trends in Feedwater Cycles: Open or Closed?

What is the present trend as to open feedwater cycles (Fig. 1.82) with one direct-contact feedwater heater versus closed cycles (Fig. 1.83) with all closed feedwater heaters, and how does this trend influence the boiler feed pumps?

Answer. The choice between open and closed feedwater cycles affects boiler feed pump application indirectly, namely through its effect on the adequacy of the suction conditions and the possible need of a booster pump in the case of open cycles.

In fossil fuel power plants, the trend is definitely to open cycles, with 80 to 90% of the installations using direct-contact deaerators. This trend has persisted despite the fabrication problems created by the large size of the deaerators used in today's large individual units. Because of the growth in capacities and the prevalence of boiler feed pump operating speeds well

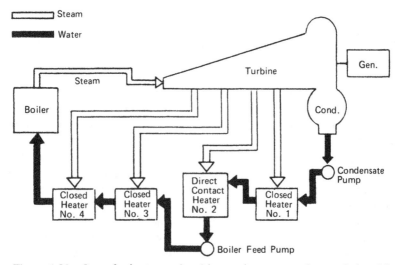

Figure 1.82 Open feedwater cycle with one deaerator and several closed heaters.

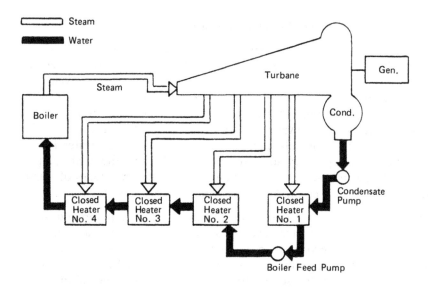

Figure 1.83 Closed feedwater cycle.

over 3600 RPM, the use of booster pumps has become almost inescapable and fully accepted by the utilities. When booster pumps are used, a few installations have located one closed feedwater heater between the booster and main feed pump. This results in some dollar savings in the cost of that closed feed water heater, but it does introduce a few complications and increases the overall power consumption so that cost savings in the feedwater heater are more than negated. Personally, I do not recommend this practice.

Two approaches are used for the booster pump drive. Some utilities prefer the "unit" concept, with both booster and main feed pump being driven by a single driver. If this driver is a steam turbine, this requires a step-down gear between the main pump and the booster. Other utilities prefer the added flexibility of separate motor drives for the booster pumps and a manifold between the two groups of pumps. This approach has the added advantage of permitting a "boot-strapping" start-up procedure, with the motor-driven booster pumps building up the boiler pressure to a level sufficient to start the main feed pump rolling.

In the case of nuclear plants, closed feedwater cycles are being used with practically no exception. In some cases the condensate pumps discharge to the main boiler feed pump suction through several closed feedwater heaters. In the majority of cases, however, the train also includes condensate booster pumps located in the cycle between the condensate and the boiler feed pumps.

Question 1.62 Split Feedwater Pumping Cycle

We are in the process of studying the feedwater cycle for our next unit, which will be 500 MW in size and will operate at 3500 psi throttle pressure. We have concluded so far that we shall use an open feedwater cycle, that is, with a direct-contact deaerating heater. We are, however, somewhat undecided over the relative merits of the direct pumping cycle with the high-pressure closed feedwater heaters located in the discharge of the boiler feed pump and of the split pumping cycle with these feedwater heaters located between a primary and a secondary feed pump, the total pumping pressure being split more or less equally between the two stages of pumping. Could you give us the benefit of your comments in this connection?

Answer. The subject of the location of the boiler feed pump and of the feedwater heaters in the feed cycle is probably one that has aroused more controversy over the years than any other subject in this general area (see Figs. 1.84 and 1.85). I might add that this problem had until recently shown all signs of being settled for good, with the last several major installations going the conventional (or direct pumping) route as opposed to split pumping. Your question and one other study I have recently carried out for another major utility appears to have reawakened sleeping dogs and reopened the argument.

Let me first review very briefly the history of this particular problem. Way back in the late 1920s and early 1930s, a few units were installed in the United States using the split pumping feed cycle, even though pressures were modest by today's standards, seldom exceeding 1500 psig at the boiler feed pump discharge. One strong reason for this sort of decision was the fact that few 3600 RPM pumps were used and that a pump designed to develop 1500 psi at 1800 RPM would have been extremely awkward. Once the decision was reached to use two pumps operating in series to develop

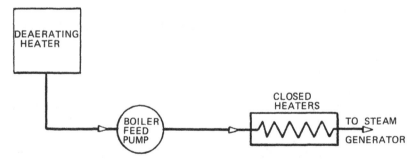

Figure 1.84 Single pump discharging through closed heaters.

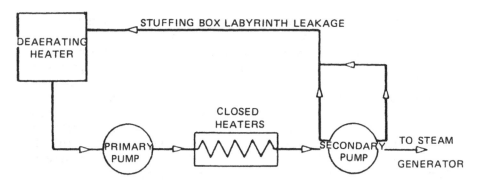

Figure 1.85 Series pump arrangement locates the closed heater between the primary and secondary pumps.

the entire required pressure, the temptation to save considerable money by buying lower pressure heaters and placing them between the two pumps was difficult to resist. Thus, for instance, two pumps in series were used in 1930 at the River Rouge plant of the Ford Motor Company. Each pump was designed to develop about 750 psi, the second pump handling feedwater at 407 °F.

But shortly thereafter, the 3600 RPM pump came into its own and the idea of split pumping cycles was abandoned entirely in the United States, even when throttle pressures climbed to 2400 psi. I shall speak of the reasons for this in a moment.

In the meantime, Great Britain held on to the split pumping cycle through thick and thin. One of the reasons for their position was that they use 50 cycle current, so that the top synchronous motor speed is 3000 RPM. With a lower operating speed than in the United States and the reluctance—until recently—to use step-up gears, it was possibly the better part of valor to split the total head to be developed between two pumps, lest the number of stages required at their lower speeds were permitted to lengthen the shaft span of the pumps beyond practical limits.

There may have been one more reason for the practice followed in Great Britain and that is the difficulty they may have encountered in designing and building feedwater heaters suitable for the pressures encountered when these heaters are located in the discharge of the boiler feed pumps.

There was a moment, it is true, when the split pumping scheme threatened to regain favor in the United States. This was brought about when the introduction of supercritical pressure steam plants imposed discharge pressures of 4500 psi and higher on the boiler feed pumps. The first supercritical installation, at the Philo plant, used two pumps in series with the closed heaters located between the two pumps. The next installation, at Eddystone, went even further, using three pumps in series with the closed heaters between the low-pressure and the intermediate-pressure pumps. (We should note, however, that in one of the papers presented by the engineers of American Power, it was stated that they would have gone the conventional route of straight-through pumping if it hadn't been that the heater manufacturers refused to build feedwater heaters for the required pressure at the time.) We shall see presently that by the time the third supercritical unit installation for the United States came along, Avon No. 8 of the Cleveland Electric Illuminating Company, the trend had reversed itself again and the feedwater heaters were located in the discharge of the boiler feed pumps.

Why this difference in the solution of the feedwater cycle arrangement? There are many methods available to compare the advantages and disadvantages of the two arrangements, and they have all been used at one time or another to calculate very precisely the overall cost of pumping and heating in each arrangement. It must be noted at this point that the methods differ widely in their approach—even among the supporters of either theory. I will avoid adding my own quantitative analysis to the literature on the subject and will restrict myself to some general qualitative observations.

The basic facts to be considered in favor of the single-pump arrangement are the following:

1. The power required by the boiler feed pump increases inversely with the specific gravity of the feedwater and, hence, increases if part of the pumping is done *after* the feedwater has gone through all the feedwater heaters.
2. This increase in power consumption is aggravated by the fact that when two pumps are used in series, the second pump must be provided with pressure-reducing breakdowns ahead of the stuffing boxes. You cannot—today—pack against 2000 psi pressure! The leakage past these breakdowns is quite significant and must be added to the capacity of the first pump in the set, increasing the power consumption quite appreciably beyond the increase caused by the higher pumping temperature of the second pump of the set.
3. Not only must this increased power consumption be added in the heat rate calculations, but it robs the turbine-generator set of a significant number of kilowatts. This kilowatt capability has a definite incremental value that must be credited in favor of the single pump scheme. (This credit does not apply if the boiler feed pumps are turbine or main-shaft driven. But the use of two pumps in series does not make these last two driving arrangements very practical.)

Chapter 1

4. The cost of the pumps and drivers and of all the valves and controls is considerably greater in the case of the split pumping scheme.
5. Reliability considerations favor the single-pump arrangement because of the lesser number of components and because the higher operating temperature and higher suction pressure introduce additional hazards. This is particularly true during transient operating conditions.

There are also factors favorable to the split pumping arrangement:

6. There is some thermodynamic advantage in putting the work of compression into the feedwater at a higher point in the feed train.
7. The cost of high-pressure heaters is appreciably higher than that of heaters subjected to a more moderate pressure, as in the case of the split pumping installation.
8. Lower pressure heaters can be considered more reliable than heaters built for the full final boiler feed pump discharge pressures.

It is obvious that when specific values are assigned to all these factors, the choice will be affected by the methods used in arriving at these values. This explains the divergence of practices between the United States and Great Britain, for instance. In Great Britain, the increased cost of the boiler feed pumps and drivers required for the split pumping scheme used to be more than compensated by the savings in heater costs. In the United States, this is not necessarily true. In addition, there are differences in the methods used to evaluate heat rate savings and the cost of incremental turbine-generator capability. I should add that the most recent large unit installations projected in Great Britain have abandoned the split pumping arrangement and are following present-day U.S. practice. Three factors seem to have united to cause this change:

1. Steam turbine drive has been introduced in Great Britain, eliminating the previous 3000 RPM pump speed limit.
2. Gear drive, where required, appears to have achieved acceptance, likewise breaking the 3000 RPM limit.
3. Greater success seems to have been achieved in the construction of high-pressure heaters.

It should be noted that the difference in power consumption in favor of the single-pump scheme is not inconsequential. In the preliminary analysis of the feedwater cycle for the supercritical Avon No. 8 unit of Cleveland Electric Illuminating Company, it was established that the difference in cost of the heaters for the two alternative schemes just about balanced the difference in cost for the pumps and their drivers. (When orders were finally placed for the equipment, the actual cost differnce swung in favor of the single-pump arrangement.) But at the rated pump flow of 2,052,000 lb/hr and with a net pressure of 4514 psi to be generated between the low head booster pumps and the steam generator, the total power consumption would have been 17,760 bhp with the split pumping cycle and only 14,720 bhp if the closed heaters were located beyond the boiler feed pumps. The penalty of 3040 bhp against the split pumping cycle could not even begin to be compensated for by the slight gain in cycle heat rate. Thus even without crediting the single pumping cycle with the intangible advantages of greater overall reliability, these power savings were sufficient to swing the decision for the boiler feed pumps serving Avon No. 8.

This decision was amply vindicated by the fact that whereas the secondary pumps of the Philo Station split pumping feed cycle encountered considerable trouble at the stuffing boxes, the boiler feed pumps at Avon No. 8 had no such problems because the stuffing boxes are designed for very nominal pressures and temperatures.

Similar studies were carried out at the time the boiler feed pumps were chosen for the supercritical units at Hudson No. 1 of Public Service Electric and Gas Company of New Jersey and Chalk Point Nos. 1 and 2 of Potomac Electric Power Company. In both cases, the final decision went to the conventional single-pump system, for good and sufficient reasons. Since those early days of supercritical steam power plants, there have been practically no installations made of the split pumping scheme.

This should not be interpreted to imply that one need not ever again carry out studies of alternative schemes. All that I am driving at is that these studies should take *all* facts into consideration, and especially that full attention should be given to the leak-off at the secondary pump stuffing boxes. This is where much of the losses occur. These studies should also take into consideration the intangibles, which involve "freedom from fear," the fear that the secondary feed pumps will have excessive maintenance and unscheduled outage. You can run a plant, if necessary, bypassing the high-pressure heaters at the expense of some Btus' increase in the heat rate. You cannot run the plant without feed pumps.

Question 1.63 Condensate Booster Pumps

In examining several heat balance diagrams of closed feedwater cycles, I have noticed that in a number of cases the condensate pump is followed by one or more closed heaters, then by a so-called condensate booster pump and several more heaters, then finally by the boiler feed pump, plus whatever high-pressure heaters remain to be accommodated.

Chapter 1

Why is it necessary to use a condensate booster pump, and can it not be eliminated, with the pressure it develops being incorporated in the condensate pump itself? It would seem to me that the addition of one more pump and one more driver to the cycle would be less economical than the use of a higher head condensate pump.

Answer. As you imply, the final choice between using a higher head condensate pump or installing a separate condensate booster pump is strictly a matter of economics. The decision to use two separate pumps to generate the total head required between the condenser hot well and the boiler feed pump suction is generally dictated by the location of the boiler feed pump in the cycle or by the fact that hydrogen coolers, demineralizers, or closed heaters located in the condensate circuit would be too expensive if they were designed for as high a pressure as would be necessary if no separate condensate booster pump were to be used. The decision, therefore, must be made with due regard to all the factors that may surround a particular case. Of course a condensate pump can be designed to develop the necessary total head, but such a solution requires that

1. All the heat exchangers located between the condensate and feed pumps can be designed for this pressure economically.
2. The total head that will be required for the condensate pump is not so high that developing it at the lower speeds required by condensate service will become uneconomical.

This last factor may become quite significant. Condensate pumps are generally designed to operate at speeds from 880 to 1750 RPM, but the condensate booster pumps—not being restricted by a low available NPSH since they can be provided with whatever NPSH it requires by the condensate pumps—will in most cases such as you describe operate at 3550 RPM, which results in considerable savings.

Although your question referred specifically to closed feedwater cycles, that is, with no deaerator in the feedwater system, it leads very logically to a consideration of a parallel problem in the case of open feedwater cycles. As a matter of fact, I would like to describe the "genesis" of what I consider to be a most elegant solution of such a problem. It arose some years ago when a certain utility company in the United States was considering its next generation of units, which were to be of 350 MW. I can recount the thinking process that led to the final solution because I was intimately involved in this process and in the discussions that took place.

The last three units installed by this utility had been 250 MW in size. Each unit was served by two 8 in. multistage horizontal condensate pumps designed for 3000 GPM and 700 ft total head, at 1150 RPM. One pump was running and the second one was a spare. Variable-speed hydraulic coupling drives had been used for these pumps, as had been the practice of this utility for all their previous units.

When it came to the 350 MW unit, we ran into a serious difficulty: there existed no horizontal condensate pump design available for the new requirement of 4140 GPM and 884 ft head. In the preceding years, as unit sizes were increasing, industry practice had shifted to the use of vertical can-type condensate pumps. On the other hand, the utility in question much preferred to avoid abandoning their established practice of using horizontal condensate pumps. This preference—which, I must add, I share wholeheartedly—was based on the following considerations:

1. For reliability and long life between overhauls, a horizontal pump with external oil- or grease-lubricated bearings is far superior to a vertical can-type pump with water-lubricated bearings.
2. A horizontal condensate pump has all the advantages from the point of view of maintenance. It can be opened up for inspection without breaking the suction and discharge connections and without moving its driver. To inspect a vertical can-type condensate pump, on the other hand, one has to remove the motor, disconnect some of the piping, and withdraw the pumping element from the can. As a matter of fact, the dismantling process is not over yet. It remains to disconnect the individual bowls and, in effect, dismantle the pump into its individual components. In the case of a single- or multistage axially split casing horizontal pump, all the running internal clearances can be measured without dismantling the rotor, once the upper half of the casing has been lifted off. (See Question 5.9.)
3. The vertical can pump could not readily accommodate the hydraulic coupling drive preferred by the utility. (I must add that at the time these deliberations were taking place, variable-frequency electric motor drives were not yet proven in the field.)

Of course, a new pattern for the necessary size of multistage horizontal condensate pump could have been developed, but such an approach may not have been too economical, considering the existing industry trend toward vertical can-type condensate pumps. But the dilemma created by the preference for horizontal pumps and the lack of the necessary patterns turned out to be a blessing in disguise. Our solution was to go to a "split pumping cycle," in other words to use two pumps in series, as illustrated on Fig. 1.86. Two half-capacity single-stage double-suction horizontal pumps were selected for the condensate portion of the service, running at

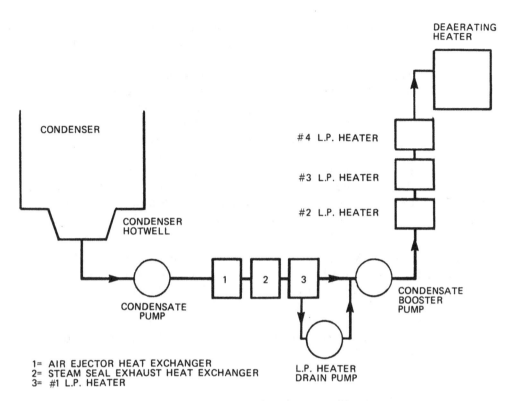

Figure 1.86 Example of the use of condensate booster pumps.

880 RPM to achieve low NPSH required values. They would pump through the air ejector, the steam seal exhaust heat exchanger, and the No. 1 low pressure (LP) heater into two-stage condensate booster pumps. The latter could run at 1750 RPM since this no longer presented an NPSH problem and pump to the deaerator through the remaining LP heaters. The condensate pumps would run at constant speed and the condensate booster pumps at variable speed. Note that the LP heater drain pump takes the cascaded drains from LP heater 1 and discharges them in parallel with the condensate pumps to the suction of the condensate booster pumps.

In summing up, the splitting of the total head requirement from the condenser hot well to the deaerator between two separate pumps provided many benefits beyond that of solving the constraint imposed by the absence of a suitable single-casing pump pattern:

1. The possibility of retaining a low speed for the first stage of the condensate pumping service, necessary to hold required NPSH values to a minimum.
2. The ability of using a higher operating speed for the major portion of the total head generation, thus reducing the cost of those pumps and of their variable-speed hydraulic couplings.
3. The imposition of the lowest possible pressure on the air ejector heat exchanger, the steam seal heat exchanger, and LP heater 1.
4. The simplification of the heater drain pumping service, by pumping forward in parallel with the condensate pumps. This made the heater drain pump a simple single-stage pump, handling low-temperature drains, and eliminated the problems usually imposed on heater drain pumps required to develop the full head up to the deaerator pressure.

Three such 350 MW units are now in successful operation in this utility's system. The solution has been incorporated in two more units for this utility, this time 500 MW in size, the first of which went into commercial operation in 1982.

It may sound trite to point out that this is an excellent illustration of the proverb "Necessity is the mother of invention." But it might be difficult to deny the claim.

Question 1.64 Simplified Feed System

I have been studying low-cost, economical packaged steam power plants in the range of 15,000 to 20,000 kW intended for use as peak-shaving plants or at the end of lines as an additional source of power. Among various problems, I have been analyzing the most economical means of transferring condensate from the condenser hot well, operating at approximately 26 in. Hg vacuum, to a steam generating unit operating at approximately 400 psig.

After studying a number of different possibilities, I came to the conclusion that a vertical can-type pump, combining the functions of the condensate and boiler feed pumps into one unit, offers several advantages. (Fig. 1.87). This solution should certainly result in a lower initial equipment cost and should reduce the installation cost. For the particular size plant I am projecting, the vertical can pump should be designed to handle approximately 300 GPM against a discharge pressure of 500 psig. No heaters would be used between the condenser and the boiler.

I would appreciate any comments you would care to make on the merits of this solution.

Chapter 1

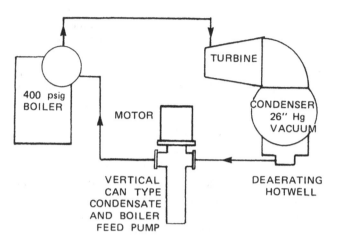

Figure 1.87 Proposed simplified feed system.

Answer. I will readily admit the merits of any solution that will reduce the initial equipment cost of a unit intended strictly for peak or end-of-line service, even if the savings are achieved at the expense of slightly increased operating costs. In this particular case, however, there are a number of considerations that tend to make the solution less desirable than it appears at first glance.

By definition, condensate pumps should operate at low speeds in order to maintain the lowest possible required NPSH. Likewise, by definition, boiler feed pumps should operate at the highest possible speeds for the high pressures that they have to develop. Higher speeds result in compact units, less costly and more reliable than low-speed pumps. The best speed for a condensate pump in the range of capacities you mention would be 1750 RPM. On the other hand, 1750 RPM is a very poor speed to develop the 1190 ft total head required here, and the boiler feed pump should operate at 3500 RPM. As a result, the idea of combining these two functions into a single pump may appear to be desirable from the point of view of first cost, but once you get into the problems of incompatibility between the two services, much of this attraction disappears.

This does not mean that such a pump could not be developed and built. It would be highly impractical to design it for operation at 1750 RPM, as this might require from 16 to 20 stages. At 3500 RPM, a seven-stage pump with 7¼ in. impellers could be built for 300 GPM and 1190 ft total head. A rather special design would be required, with heavier walls for the individual stage bowls, heavier bolting throughout, and special bearings.

Vertical can-type condensate pumps are generally built so that the motor carries the axial thrust. In this case, it would be necessary to provide a thrust bearing in the pump itself.

One very important factor to consider is that the vertical can-type pump has water-lubricated bearings. These will naturally not have the long life normally associated with horizontal shaft pumps fitted with oil- or grease-lubricated bearings. Therefore, a shorter life might be expected for this combined pump before clearances must be restored than would be the case with a separate horizontal boiler feed pump.

I do not intend to comment in detail on the question of the elimination of all heaters. This obviously reduces the cost of the unit, and the effect on its performance is apparently not a factor in your estimation. I assume that you have discussed the effect of supplying the boiler with feedwater at condenser hot-well temperature with the boiler manufacturers.

On the other hand, there will no doubt be cases in which closed heaters fed with extraction steam will be required by the ultimate customer. In that event, the effect of combining the two functions into a single pump will be unfavorable. If the heaters are located between the condensate and boiler feed pumps, they need be good for at most 100 psig. If the heater or heaters are located after the combined condensate-boiler-feed pump, they must be good for the maximum pressure developed at shutoff, which will be of the order of 600 psig. This will increase the cost of the heaters, and much of the savings contemplated will have been dissipated.

In short, it would seem to me that the more conventional arrangement with a separate condensate pump, a deaerating heater, and a boiler feed pump is a more desirable arrangement. Note that the cost of the feedwater pumping equipment represents only a few percent of the total unit cost. The savings obtainable from the scheme you have described would be of almost negligible proportion in relation to the total plant cost.

Question 1.65 Variable-Speed Condensate Booster Pumps in a Closed Feedwater Cycle

We have been analyzing the feedwater pumping sytem to serve a 350 MW, 2400 psi throttle pressure installation. We shall be using a closed feedwater system, with the pump train made up of condensate pumps, condensate booster pumps, and boiler feed pumps, with closed feedwater heaters disposed between the three groups of pumps. It is quite obvious that variable-speed operation is justified for the boiler feed pumps. On the other hand, it is equally obvious that if the condensate booster pumps were arranged for variable-speed drive instead, much smaller hydraulic couplings would be required, with a significant cost saving over that of hydraulic couplings sized for the boiler feed pumps. Since such an arrangement is so much more economical, we are surprised that it has not been used previously. Can you give us your comments?

Answer. You are quite right in that the savings that would accrue from using the much smaller hydraulic couplings needed with condensate booster pumps would be very significant. Alas, the solution is not a practical one. Let us assume that the condensate pumps generate something like 75 psi net pressure, the condensate booster pumps around 300 or 350 psi, and the boiler feed pumps the remaining pressure. If the boiler feed pump suction is, say, 325 psi and the discharge pressure is 2750 psi, the net pressure is 2425 psi. The pressure generated near pump shutoff will be of the order of 2925 psi. Adding to this the net pressure generated by the condensate and condensate booster pumps, the boiler feed pump discharge (at constant speed) would be of the order of 3400 psi. On the other hand, the total system head to be overcome between the condenser hot well and the boiler, at capacities near shutoff, will be of the order of 2500 psi at the most. The excess pressure to be eliminated by throttling or by speed variation is around 900 psi. Very obviously, even with the condensate pump running at zero speed there would still be appreciable pressure to dissipate, and this could only be done by throttling—a very uneconomical solution.

Decidedly, if variable-speed control is to be used, it must be applied to the boiler feed pump, not to the condensate booster.

Question 1.66 Variable Speed for Condenser Circulating Pumps

I would like to have you comment on the merits of installing condenser circulating pumps of the vertical type driven through variable-speed magnetic drives in order to improve part load performance and to provide for winter and summer operation at the most favorable circulating flows. In our location, we encounter water temperatures as low as 50°F in winter and up to 80°F in the summer. Our present practice is to operate one circulating pump in winter and two pumps in parallel in the summer. I believe that more efficient operation can be obtained by means of variable-speed pumps.

Answer. This is a rather broad question, and it is difficult to come up with specific recommendations without a detailed knowledge of the system, type and size of condenser, and design data on the main unit. You may carry out actual evaluation calculations if you have this information, but I believe that the probability of justifying variable speed drive for circulating pumps is rather dim.

It is quite true that operation at variable speed would result in power savings whenever the main unit is operating at part load. It is equally true that a case can be made in favor of variable-speed operation to take care of the variation in flow between summer and winter conditions. On the other hand, it should be considered that condenser circulating service is usually relatively low head service, with the result that the ratio of power or operating cost to first cost is low. This is the reason it is not the normal practice to use variable-speed drive for circulating pumps—either with magnetic drives, as you mention, or with two-speed motors, which could be applied to take care of the winter-and-summer situation. The initial cost of the added equipment cannot be evaluated from the power savings that would be obtained.

It must be pointed out that constant-speed operation with two pumps serving the condenser is not as inefficient as you may think. For instance, in the northern part of the United States, the *desired* capacity with one pump in service (in winter) is roughly two-thirds that with two pumps in service (in summer). Thus, when a single pump is operating, it handles about 133% of the capacity that it delivers when operated in parallel with the second pump. These two operating points generally fall within a very favorable pump efficiency range.

It should also be noted that with two pumps in operation during winter, there would be a definite increase in the vacuum obtained at the turbine exhaust. If the turbine and the condenser are properly selected to take full advantage of this higher vacuum, the main unit can be operated at better economy or at higher loads or a combination of both.

In order to utilize this one- and two-pump operation arrangement, a partition valve can be put in the inlet water box division plate. In this manner, one pump can be utilized to provide circulating water to both sides of the condenser.

You should note that exact matching of circulating pump flow and turbogenerator load is not required. In other words, the operation of the condenser circulating pumps does not require the fine-tuning demanded of the pumps included in the feedwater cycle, such as the boiler feed pumps. These latter must deliver to the boiler exactly as much feedwater by weight as there is steam taken from the boiler to the main turbine. On the other hand, the only possible effect of any mismatch of turbine load, condenser water temperature, and condenser circulating water flow is a slight modification of the vacuum carried in the condenser. This will cause an almost insignificant change in the heat rate of the main unit. This is why the use of a different number of pumps under different operating conditions is perfectly valid.

There is a very sound compromise solution, however, and that is the use of two-speed motors. This is less expensive than the use of a separate speed-varying device and provides more different intersection points between the combined head-capacity curve of the pumps operating in parallel and the system-head curve.

Question 1.67 Source of a Small Quantity of High-Pressure Feedwater

We have two requirements for high-pressure water:

1. We need something in excess of 100 GPM at 3000 psig to supply water to the seals of our boiler circulating pumps. This is not a 24 hr a day service but would be required only when the main unit operates at less than half-load. Normally, this seal flow is supplied from the boiler feed pump discharge, but as the boiler feed pump operates at variable speed, its discharge pressure is not sufficiently above boiler pressure at these load conditions to ensure adequate seal operation. There are four boiler circulating pumps, each requiring 25 GPM, but as the seals wear, we may expect this requirement to increase somewhat.
2. The second requirement is for boiler testing and emergency boiler feeding and would be for, say, 15 GPM at a discharge pressure of 4050 psig.

We had been thinking of using a vertical triplex power pump for the second requirement, but when the seal water requirement came up, we wondered if we could combine both services into one pump.

We understand that vertical triplex pumps are available for capacities of 115 GPM and even higher at the required pressure that could handle either condition adequately. The only problem is that of control.

What actually is required is some sort of a modulating bypass control so that the pump can operate at a constant discharge pressure. We are afraid that an accumulator and synchronized suction valve unloaders would be too complicated. The water going to the seals cannot be in contact with air, as this water is deaerated. Thus, if an accumulator is used, it would probably require the use of steam as the compressible element.

We believe that modulating bypass valves have been used quite successfully but mainly on intermittent service. The sealing water service could be required for extended periods of time, and we are afraid that the bypass valve would cut out very rapidly.

We thought that there may be two other ways of handling this sealing water problem. One would be to use a booster pump at the boiler feed pump discharge to provide the necessary discharge pressure. The second might be to use a vertical multistage pump to handle 100 to 125 GPM against the required pressure.

We would appreciate hearing your comments.

Answer. In my opinion, the use of a reciprocating pump for this service is the most logical solution, and I definitely do not recommend either of the two alternative solutions you mention.

The possible use of a booster pump taking its suction from the boiler feed pump discharge and delivering into the boiler circulator injection connection reminds me of the typical box of Quaker Oats. You will no doubt remember that these boxes carry a picture of William Penn, who holds a smaller box of Quaker Oats somewhere around his stomach. On this box there is a very clearly visible smaller picture of William Penn, who in turn is holding a still smaller box, and so on, ad infinitum. In other words, if the boiler circulating pumps require injection water, it is somewhat questionable whether a small high-suction-pressure injection pump will operate quite satisfactorily without some sort of injection as well. At any rate, I do not favor this solution.

As to a vertical multistage pump designed to develop 4050 psig pressure, assuming that you are dealing with reasonably cool water, you need something of the order of 9500 to 10,000 ft head. In the range of the small capacities required, this may need as many as 50 stages, and I doubt that this is a justifiable application for centrifugal pumps.

The use of a vertical triplex or quintuplex power pump designed for this range of pressures and capacities is not new in power plant practice. As a matter of fact, a number of such installations were made in the past to satisfy the requirement for two separate sources of feedwater supply for units served by single full-capacity turbine-driven boiler feed pumps without any motor-driven part-capacity standby pumps. A secondary use of the same power pumps was to permit restarting the units after full loss of load, as they could feed a small quantity of feedwater to the boiler to restore boiler level.

I agree with you that the control of this special-service pump does present problems. Since a power pump is a positive displacement type, its delivery cannot be controlled by a throttling valve in the discharge as with a centrifugal pump.

Volume control of a reciprocating boiler feed pump can be regulated as follows. The pump can be permitted to run at fixed speed, delivering constant capacity. A bypass line is located in the pump discharge leading back to the suction source, in this case, the direct-contact heater. A diaphragm-operated throttling bypass valve is installed in this line, being made responsive to changes in the requirements of the boiler. Essentially, this valve is similar to the normal feedwater regulator except that its action is exactly reversed; it opens wider on reduction of demand and closes on increase in load.

Unfortunately, up to the present, experience with such valves has not been too encouraging. The pressure pulsations from the discharge of a reciprocating pump have been known to cause chattering and slamming, even with a pulsation chamber on the discharge. It is possible that valve operation improvement may be obtained someday by providing them with some form of dashpot or damper on both the upward and downward stroke to minimize the effect of pump pulsation.

At the present time, however, I do not recommend using such modulating valves.

Variable capacity of reciprocating pumps can be achieved using variable-stroke pumps. Although I do not intend to present a discussion of the pros and cons of variable-stroke design, I should mention that the cost of such pumps is relatively high, their maintenance costs are higher than those of fixed-stroke designs, and other means can be found that are more economical.

I believe that the best means for varying the capacity of such auxiliary reciprocating pumps is to install a variable-speed device, such as a hydraulic coupling or a magnetic drive, between the motor and the speed-reducing gear. Such a device would permit a speed variation and so a capacity variation from 100 down to 25%. The latter is the slowest speed recommended for this type of pump for proper lubrication. Should capacities under those obtainable at 25% speed be required, a bypass would be used between the pump discharge and the suction supply, with a pressure-reducing orifice sized to handle this minimum flow of 25%.

The valve in this bypass line would be either fully closed or open, and its action would be automatic.

The efficiency of a hydraulic coupling is reduced in direct proportion to a reduction in speed, and therefore the unit will consume about the same amount of power regardless of required flow to the boiler. This is equally true of the fixed-speed variable-bypass arrangement. However, the difference in power consumption between the reciprocating pump and a centrifugal pump at the loads in question is so overwhelming that the last factor has no bearing on a comparison between the two.

Question 1.68 Bid Evaluation

I have read a number of your recent articles in which you comment on the difficulties engineers encounter in evaluating intangible factors that affect the reliability of steam power plant equipment. I am essentially in agreement with your reasoning, but I fail to see any practical suggestion to get around these difficulties. After all, it is not reasonable to say that price should not enter into the decision in selecting equipment. But if price, performance, and other factors that do lend themselves to direct comparisons are to be used in an evaluation, what means are available to introduce the effect of reliability or of other similar intangibles?

Answer. I shall plead guilty to the accusation of having stressed the need of evaluating reliability considerations. My defense rests on the fact that only too often are these considerations given no significance whatsoever and that it is highly important to call attention to this neglect. But I do not believe that reliability must be given significance to the exclusion of all other items. My thought is that an evaluation of equipment, to be valid, must be a *total* evaluation. The methods used to achieve such a total evaluation can vary widely, but if *all* factors are given some weight, an engineer's conclusion will be more logical and more valid than if certain items are neglected.

One of my friends, chief engineer of an eastern utility, once wrote me describing the particular method that he has been using in the evaluation of equipment. It is as logical an approach as I have seen advanced anywhere, and I have asked his permission to quote from his letter on the subject, a permission that he has very graciously granted me:

> For those who like numbers in the bid comparisons, I have for many years used a "point score" system wherein out of a total of 100 points, 40 are reserved for the price of the equipment. This is scored so that any equipment costing double the lowest bid scores no points on price. The remaining 60 points are divided on an arbitrary basis, such as 10 points for the manufacturer's reputation, 10 points for our experience with the equipment, 10 points for the industry's experience with the equipment, 10 or 15 points for engineering prejudice. If reliability and likelihood of failures are important, a number of points may be allocated to this feature, and, if failure may encompass heavy damage to the equipment, to other equipment, or to the public or employees, point values may be assigned to these items. The various bids are then compared and assigned a proportionate share of the point value, the equipment carrying the highest total being selected.
>
> The beauty of such a comparison is that adequate weight can be given to evaluated factors according to the judgment of the engineers. It has been our experience, when the results of such a point score are close, that we really do not care which of several alternates is selected

and that, where major differences in these point score totals occur, any amount of argument with persons of conflicting opinions will not result in enough variation of the score to change the selection of the equipment.

What could be simpler? I am sure that the point value indicated can be shifted within reason to accommodate any engineer's opinion of the relative value of the various items that are given consideration without detracting from the validity of the results.

Question 1.69   Attainable Pump Efficiencies

How do I know when the efficiency of a tendered pump represents a good, or indeed the best, value for a particular duty? Is it possible only to compare efficiencies submitted by different vendors, or is it meaningful to calculate a value of the highest efficiency attainable by a real, "as it could be constructed" pump?

I have occasionally tried the second approach, using formulas for the optimum efficiency of a first-class machine given in several textbooks on Centrifugal Pumps. However, efficiency values obtained by these methods always considerably exceed vendor figures and leave me undecided whether the quoted value may not be unnecessarily low.

Answer. I do not think that it should be too difficult to arrive at the optimum efficiency that can be obtained from commercially available centifugal pumps for any set of operating conditions. The basis for predicting pump performance is the principle of dynamic similarity, which, when applied to centrifugal pumps, expresses the fact that two pumps geometrically similar to each other will have similar performance characteristics. The term "specific speed" is the concept that links the three main parameters of these performance characteristics—capacity, head, and rotative speed—into a single term. The formula for specific speed and a discussion of this index number is given in the answer to Question 1.27.

Although one could calculate the specific speed for any given operating condition of head and capacity, the definition of specific speed assumes that the head and capacity used in the equation refer to those at the best efficiency of the pump. It should also be noted that this index number is independent of the rotative speed at which the pump is operating.

Since the middle 1940s there have been published a number of "approximate statistical averages" of the optimum attainable efficiencies of commercial centrifugal pumps plotted against specific speed. These charts include the scale effect by providing a series of curves, each for a different capacity. Such a set of curves is illustrated in Fig 1.36.* The data for these curves were accumulated some 40 years ago, and some improvements have been achieved since then, particularly in the lower range of specific speeds and for lower capacities. This is illustrated in Fig. 1.88.†

Table 1.6 illustrates the fact that for higher specific speeds and higher capacities, very little difference exists between the values given in Figs. 1.36 and 1.88, although the difference is quite marked for lower specific speeds and small pumps. Note that the efficiencies indicated on these charts are those attainable with single-stage pumps. Greater losses arising in the interstage passages and because of relatively larger diameter shafts will reduce these efficiencies from 2 to 4% for multistage pumps.

A word of caution: It should be noted that the data presented in Figs. 1.36 and 1.88 are somewhat generalized in that they recognize only two of the factors that affect pump efficiency, that is, specific speed and pump capacity. There are a number of other characteristics that affect commercially attainable efficiencies, characteristics such as surface finish, internal clearances, suction specific speed, and the effect of internal recirculation at the suction and discharge of the impeller. Up-to-date information on this subject was released in 1986 and 1987 in two papers. (The Effect of Specific Speed on the Efficiency of Single Stage Centrifugal Pumps, by E. P. Sabini and W. H. Fraser, Proceedings of the 3rd International Pump Symposium, Houston, Texas, May 1986; The Effect of Design Features on Centrifugal Pump Efficiency, by E. P. Sabini and W. H. Fraser, Proceedings of the 4th International Pump Symposium, Houston, Texas, May 1987). In order to provide a more accurate set of efficiency charts, certain specific constraints were selected for each one of these characteristics and quantitative corrections were provided for the effect of any deviation from these selected constraints. If one wishes to be more exact in establishing the attainable efficiency for any set of conditions, reference should be made to the two papers cited.

You have referred to formulas that express the optimum efficiency of centrifugal pumps of different types and sizes. Obviously, any relationship that can be presented in the shape of a set of curves can also be expressed in a formula. In turn, any formula can be presented as a curve or a set of curves. The choice between the two approaches will, of course, rest upon personal preferences. In this particular case, my preference is with curves similar to those shown in Figs. 1.36 and 1.88.

---

*From *Centrifugal Pumps*, by Karassik and Carter, McGraw-Hill, New York, 1960.

†From *Pump Handbook*, Karassik, Krutzsch, Fraser, and Messina. McGraw-Hill, New York, 1976, Fig. 5, pp. 2-10.

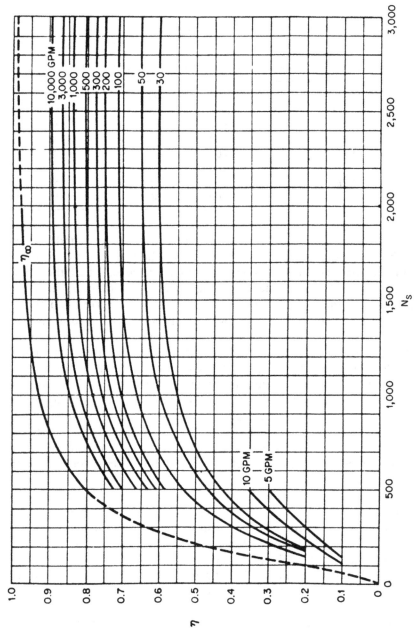

Figure 1.88 Efficiency as a function of specific speed and capacity.

Table 1.6 Chart Efficiencies

		Attainable efficiency (%)	
N_s (specific speed in U.S. units)	GPM	From Fig. 1.36	From Fig 1.88
2500	10,000	90	90
	1,000	83.5	84
2000	1,000	83	83.5
	100	69.5	71
1000	1,000	79	79
	100	62	66
500	200	55	57
	100	46	54

As you will see presently, there are certain circumstances under which the efficiencies quoted by some or even by all the manufacturers can fall below the chart efficiencies. Remember that I have said that the "type specific speed," which identifies the pump performance, must be calculated for the capacity and head conditions that correspond to the maximum efficiency for the particular pump in question. Normally, the conditions of service for which a pump is sold are relatively close to the maximum efficiency point. In such a case, the specific speed determined from the conditions of service will be a close indication of the pump type. But since no manufacturer has an infinite number of pump sizes, the conditions of service may well deviate from the best efficiency capacity of the pumps offered for any given application. In such a case, the efficiencies guaranteed by the potential vendors will fall *below* the so-called chart efficiency or the optimum efficiency calculated from the formulas you refer to. This will be particularly true of pumps with low specific speeds or with low capacities.

For example, assume that the required conditions of service are 1250 GPM and 140 ft total head. A manufacturer may offer a pump at 1760 RPM with a performance curve as illustrated in Fig. 1.89. Its best efficiency occurs at 1480 GPM and 130 ft and is 85% and compares quite well with the efficiencies predicted by Figs. 1.36 and 1.88. Yet, at the design conditions of 1250 GPM and 140 ft, the efficiency is only 83%, a full two points below what the pump can do at its b.e.p.

Another circumstance that leads to this situation involves pumps for certain services for which the need to favor some other criterion (such as very low NPSH required values) may be counterproductive to the desire of optimum efficiency performance. This is the case, for instance, with condensate pumps.

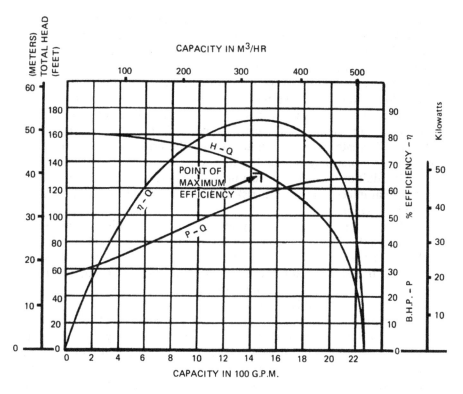

Figure 1.89 Typical performance curve of double-suction pump at 1760 RPM.

On occasion, the reverse of what you have stated may take place and guaranteed efficiencies may exceed the chart efficiency for the chosen specific speed. Whether to accept such guarantees or to question them depends on the reputation of the manufacturer in question and on the user's previous experience with this manufacturer's equipment.

In addition, it is advisable to distinguish between "attainable" and "sustainable" efficiencies. The attainable efficiency depends not only on the specific speed but also on such factors as

The design of the impeller insofar as the head coefficient is concerned
The relative dimensions of the internal clearances
The relative rigidity of the shaft design (particularly true for multistage pumps)

Higher efficiencies can generally be attained by using high head coefficients, close clearances, or smaller shaft diameters. The first of these practices will restrict the practical range of operating capacities because it will cause internal recirculation at the discharge to occur at a higher percentage of best efficiency flow and thus force the user to set a higher minimum flow limit. If the pump is not intended to operate at reduced flows, this limitation poses no problem. If it is, the energy savings at the design flow will be counterbalanced by energy losses when the pump must be operated at these reduced flows.

The net effect of the other two practices I have listed affects the "sustainability" of higher efficiencies. At best, close clearances and flimsy shafts will create conditions under which the efficiency will deteriorate rapidly. At worst, they can lead to disaster.

In addition, a user who sets up inflated evaluation bonuses on efficiency guarantees may in turn risk receiving somewhat inflated—that is, optimistic—guarantees.

Incidentally, there is a very effective means to discourage overly-optimistic efficiency guarantees. It consists of setting up what might be called "par for the course," which may correspond to the chart efficiency, corrected if necessary for the effect of multistaging. Then, the user's specification stipulates that no evaluation credit will be given for guarantees exceeding this par value.

A note of caution: It is not sufficient to compare efficiencies, and therefore power consumption, at the guaranteed point only. Lower specific speed pumps, which have lower efficiencies at the design point, at the same time have a lower power consumption at part load capacities than pumps with a somewhat higher specific speed. As a result, these lower specific speed pumps may well be more economical solutions if the pumps are intended to operate over a wide range of flows.

Question 1.70   Optimum Specific Speed Selection

Considering that pump efficiency increases as specific speed increases, would it not be sound practice to always select pumps in the range of specific speeds between 2000 and 2500?

Answer. In principle, this is correct, but only if one stays within the realm of the practical. Consider, for instance, the case of a pump selection for 2000 GPM and 250 ft. If we choose a 1750 RPM pump the specific speed will be 1266, but a 3550 RPM pump will have a specific speed of 2532 and will be more efficient—and less expensive.

Consider on the other hand the case of a pump for 100 GPM and 200 ft. If we wanted to design this pump for a specific speed of 2500, solving the specific speed equation for the pump speed would yield us 13,295 RPM, certainly an impractical solution.

In the case of multistage pumps for high heads, the desire to increase the specific speed leads to a reduction of the head per stage and hence to an increase in the number of stages required. This, in turn, will result in longer shaft spans, greater static deflections, and thus lower reliability. The desire for slightly higher efficiency under these circumstances clearly works against the user.

2
Pump Construction

Question 2.1 Packed Boxes Versus Mechanical Seals

Would you care to comment on the advantages or disadvantages of conventional packed stuffing boxes versus mechanical seals for centrifugal pumps when the pressure on the suction side is quite high, say as much as 125 psig?

Answer. My answer, of necessity, will have to be quite general in character because there are a number of factors that you do not indicate in your question that would affect the choice between packed boxes and mechanical seals. Among these factors are such items as the character of the liquid pumped, its temperature, the general style of pump in question, and the operating speed.

In general, this pressure of 125 psig is not impossible or even difficult to pack against if one is dealing with water, whether hot or cold, as long as the pump in question is designed for such service. I know of several very satisfactory installations of boiler feed pumps designed to handle feedwater at temperatures up to 350 °F and higher with as much as 250 psig pressure at the suction and a 4 in. shaft diameter, operating at 3600 RPM. On the other hand, there may be certain other factors involved that would make the use of mechanical seals preferable to the packed stuffing box. Finally, neither solution may be the best, and a condensate injection sealing arrangement as used on many modern high-pressure boiler feed pumps may be the answer.

Since both are subject to wear or failure under certain circumstances, neither packed boxes nor mechanical seals are perfect. For some applications

Chapter 2 *191*

the former is better than the latter, but for other applications the reverse is true. In some fields both give good service, and the decision as to which is used becomes a matter of personal preference or first cost.

Question 2.2 Further Comments on Mechanical Seals

Thank you for the comments you made on my question about the choice between packed boxes and mechanical seals for a centrifugal pump operating with 125 psig suction pressure. My apologies for not including complete information with the problem that I mentioned, but I was seeking some general comments rather than any specific solution. The service I had in mind dealt with the installation of chilled-water pumps handling water at 42°F and operating at 1750 RPM. The chilled-water system is a closed-circuit system. In multi-story buildings, the pumps are generally located in the basement area, and as a result, a rather high static head prevails at the suction of the pumps. The total head required by the system, on the other hand, is not high when compared with the static head of 125 psig.

In smaller installations we have experienced difficulty from time to time with mechanical seals, and when single-suction pumps are in use, this gives trouble at the bearings when the seals give way. We prefer the double-suction type of pump, which helps to eliminate this problem when this type of pump is practical in application. I have been reluctant to use mechanical seals on large pumps, as the stuffing box and glands can be readily serviced. But there is certainly a limit to the serviceability of conventional stuffing boxes when the pressure becomes high. This is the problem I had in mind, and I was wondering what the experience with mechanical seals has been on double-suction pumps and whether they prove reasonably satisfactory in comparison with conventional stuffing boxes.

Answer. Mechanical seals may be used on centrifugal pumps for two separate reasons:

1. When stuffing boxes are difficult or impractical to pack with conventional packing because of pressure or temperature conditions or because of the nature of the liquid handled by the pump
2. When the application is such that the pump location or mounting arrangement makes it difficult to have access to conventional packed stuffing boxes and mechanical seals can provide long periods of uninterrupted and unattended service

If they are properly designed, installed, and maintained, it is obvious that mechanical seals used for the first reason will give better service than conventional packed stuffing boxes. The same is essentially true in the second cited instance, except that if the service is very easy, conventional stuffing boxes can operate unattended and without replacement of the packing for very long periods of time. On the other hand, satisfactory service from mechanical seals does require, as I have stated, proper design, installation, and maintenance.

I assume that when you speak of bearing troubles when the seals give way, you are referring to excessive leakage from the seals that finds its way into the bearings housings because of inadequate sealing at these housings. Normally, a seal failure should impose neither extra radial nor extra axial thrust on the bearings. But since failure of a mechanical seal can be more sudden than the progressive deterioration of a packed stuffing box, such consequential damage to bearings is somewhat more likely to occur in a pump fitted with mechanical seals.

But the fact that a pump is of the double-suction type does not in itself guarantee that a mechanical seal failure will not result in bearing difficulties. As a matter of fact, a double-suction pump with two stuffing boxes and two mechanical seals theoretically offers a greater oppotunity for seal failure.

It should also be noted that in the case of double-suction pumps it is not only the installation of the seals that may give rise to difficulties but also the installation of the pump itself and of the suction piping and valving. If, for instance, the pump installation is such as to cause an unequal flow to the two entrances of the impeller, turbulence and partial cavitation may take place on the side to which improper flow is provided. As a result, the suction pressure on that side will fluctuate. If the condition is severe enough, premature seal failure will occur.

Someone remarked to me once that he had seen much fewer double-suction pumps fitted with mechanical seals than single-suction pumps and that he concluded that seals were less satisfactory for double-suction pumps. This conclusion is ill-founded and proceeds from the same kind of reasoning as that of the farmer who observed that black horses ate more than white horses. The farmer just had more black horses. In our case, more single-suction pumps are fitted with mechanical seals because single-suction pumps are more frequently used on services for which mechanical seals present an advantage over conventional packed stuffing boxes. Recognizing that stuffing boxes are a major source of difficulty in certain services in the chemical or petrochemical industry, pump designers developed lines of special pumps of the single-suction type for this service. Later, when mechanical seals were developed and adapted to centrifugal pump service, it was only logical that they would be most frequently applied in this same area of application in which conventional stuffing boxes had proved to be a major source of annoyance.

Chapter 2

Question 2.3 Extra Deep Stuffing Boxes

I have noticed that the centrifugal pump specifications used by our company for the purchase of new equipment include the statement "The stuffing boxes shall be extra deep." We apparently use this requirement regardless of the service for which the pumps to be purchased are intended, Is this sound practice, or are we paying an extra price for a feature that is possibly not always necessary?

Answer. This is a requirement that is found very frequently in pump specifications and that is deplorably extremely vague—one almost might say meaningless. Consider that there are no definite standards by which one can judge the relative depth of a stuffing box. In general, centrifugal pumps have stuffing boxes designed to hold from five to nine packing rings, depending on the service, pump size, and operating conditions. It must be assumed that the user will indicate the service to which a particular pump is intended, and therefore the pump manufacturer is enabled to judge the adequacy of the selection they propose to furnish. Since pump designers lay out lines of centrifugal pumps rather carefully, we must presume that barring an infrequent error—and such an error would soon be discovered and remedied—stuffing boxes in standard pump lines are as deep as need be.

Question 2.4 Sleeve Bearings Versus Antifriction Bearings

We are presenting specifications for two horizontal centrifugal pumps designed to handle 12,000 GPM each against a total head of 160 ft on cold-water service to operate at 1175 RPM. We are not certain whether we should specify sleeve bearing construction with a ball thrust bearing or whether antifriction bearings should be permitted. In looking over your articles and your book on centrifugal pumps, we find no discussion of the comparison between the two type of bearings. Can you give us an idea of the reasons for preferring one or the other type of construction?

Answer. You are quite right in noting that I have not covered the comparison between antifriction and sleeve bearing construction in any detail, and I shall try to remedy this situation here.

Although some extremely small centrifugal pumps may be fitted with internal grease-lubricated sleeve bearings, the rotor of a normal centrifugal pump is supported by two external bearings of either the sleeve (Fig. 2.1) or antifriction (Fig. 2.2) type. The choice between these two, except in certain specific cases that I shall discuss later, rests primarily upon individual

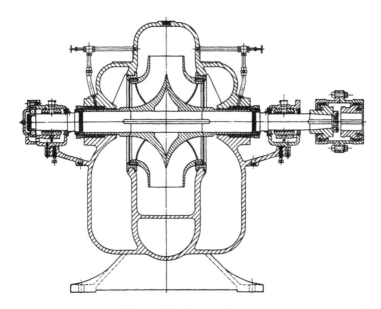

Figure 2.1 Standard lines of single-stage horizontal pumps are sometimes available with sleeve bearing construction.

preference. Such preferences are evidenced both in the design of a given line of pumps, when the choice reflects the ideas of the designer, and in the writing of specifications, when it reflects the desires and (sometimes) the past experience of the user. The controversy between the supporters of each type was a very bitter one some 40 or 50 years ago, but recent practices appear to indicate a growing trend toward the antifriction bearings in the great majority of applications. Lest I be accused of a subtle attempt to influence anyone in favor of my own preference, I should state at the very outset of this discussion that my training and my experience favor the antifriction bearing wherever it can be suitably applied.

Some of the advantages in favor of the antifriction bearings are the following:

> *Greater interchangeability of parts.* Replacement bearings are generally available throughout the world as off-the-shelf items, and bearings made by one manufacturer are interchangeable with those made by any other bearing manufacturer, size for size and type for type. They are manufactured to a high degree of precision, so that interchangeability and performance are assured.

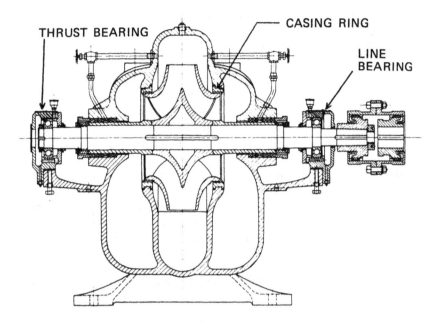

Figure 2.2 Standard lines of single-stage horizontal pumps are generally provided with antifriction bearings.

Shorter shaft span. This results in a lower shaft deflection. Although this factor may not be of significant importance in single-stage pumps such as are involved in your case, it becomes quite important for multistage designs. But even if it is not too significant, the effect is in the right direction.

Accurate alignment throughout the life of the bearings. Wear will not cause the shaft center to drop, preventing eccentricity of internal fits. This is not true of sleeve bearings, which may still operate satisfactorily after considerable wear has taken place; this wear, however, will impair alignment and may lead to other mechanical damage.

Lower starting torque. This is an important advantage, because there can be no damage at start-up in the event of a momentary absence of lubrication.

Low friction in operation. The power loss of antifriction bearings is considerably lower than that of sleeve bearings. This reduces the amount of heat generated and permits satisfactory operation with a lower volume of lubricant.

Simpler lubrication. The lubrication of antifriction bearings requires less attention on the part of the operators. Relubrication of bearings is generally required as infrequently as every 3 or 6 months. Loss or deterioration of the lubricant will shorten the life of the bearing but does not generally result in the immediate failure of the bearing.

Bearing failure is not necessarily accompanied by damage to the shaft. Maintenance procedures are available for remetallizing a shaft after it has been damaged in the journals if sleeve bearings are used, but these procedures are neither simple nor inexpensive.

Antifriction bearings are capable of carrying higher momentary overloads without failure or seizure than sleeve bearings.

I would be less than factual if I failed to mention certain advantages that sleeve bearings have over antifriction bearings:

1. They are frequently quieter in their operation.
2. Split bearings are more readily accessible for inspection or for removal without disturbing the coupling alignment.
3. They are less sensitive to small amounts of dirt or to excessive lubrication.
4. If necessary, they can be relined in the field by the maintenance personnel without the need of purchasing replacement bearings.

All these factors must be evaluated and compared when reaching a decision in favor of one or the other type. The overall evidence, however, seems to indicate that the advantages in favor of antifriction bearings outweigh those in favor of the sleeve type, because most manufacturers have standardized on antifriction bearings. It is also my impression that the majority of specifications are written today either for antifriction bearings or so as to permit either construction. Certainly I would recommend that you do so

Earlier, I had referred to specific cases in which factors other than personal preference dictate the choice of bearings to be used. There are two areas in which this is particularly true:

1. When shaft diameters are large and operating speeds are high, antifriction bearings are not applicable, and standard practice is to use sleeve bearings. This is the case, for instance, with high-pressure high-speed boiler feed pumps (Fig. 2.3).

Chapter 2 197

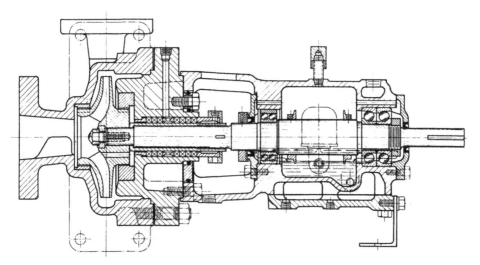

Figure 2.4 Standard end-suction, cantilever-shaft process pumps are available only with antifriction bearings.

2. When overhung cantilever construction is used, as in radially split casing-type process pumps, it is not practical to use sleeve bearings, and only antifriction bearing design are available (Fig. 2.4).

Question 2.5 Vertical Turbine Pump Bearings

We have an installation of seven vertical turbine pumps (Fig. 2.5) handling 5000 GPM of water each. The pH of the water is approximately 7.0, and the temperature ranges from 75 to 95 °F.

The water is generally free of heavy abrasives. When the cooling tower basin, from which the water is pumped, is cleaned once a year, a residue resembling mud is removed. This material is heavy enough to settle out and does not reach the pumps.

We would like to know if the bronze SAE 660 is satisfactory material for the three water-lubricated bearings on these pumps, namely, the cone discharge bearings, the bowl bearing, and the suction case bearing. If not satisfactory, please state your opinion as to the best type of material suited for this service.

Answer. As long as the water handled by these pumps is generally free from abrasives and its neutral (pH = 7.0) and cannot give rise to electrolytic action between the bronze and the ferritic materials of other pump parts, bronze should be a suitable material for the water-lubricated bearings.

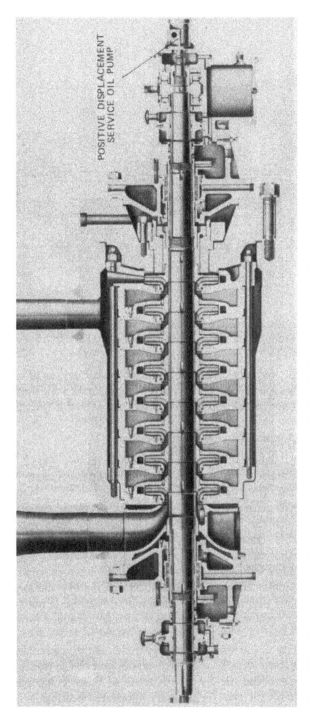

Figure 2.3 High-pressure boiler feed pumps are only available with sleeve bearings and Kingsbury-type thrust bearings.

Chapter 2 199

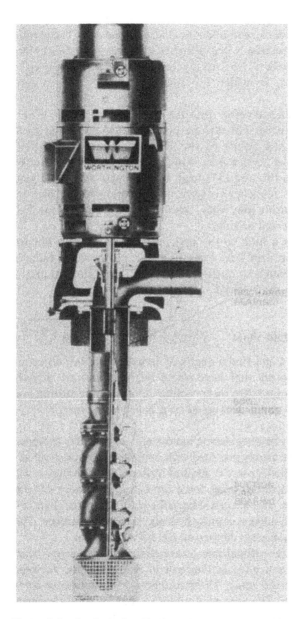

Figure 2.5 Section of vertical turbine pump illustrating water-lubricated bearings.

Various bronzes are applied to this service. Bronze SAE 660 contains 83% copper, 7% tin, 7% lead, and 3% zinc and has been very frequently employed on this service because it has a higher load-carrying ability than bronzes with greater amounts of lead. (Note that it should never be used in food industry applications, in which the zinc content will lead to contamination.)

However, you should remember that the size selection of the water-lubricated bearings used in vertical turbine pumps is seldom so marginal that the higher load-carrying ability of the 7% lead bronze is essential. Thus you may get still better service from a higher lead content bronze, for instance, 78% copper, 7% tin, and 15% lead, because with the higher lead content, there will be less tendency to seize if lubrication to the bearing fails momentarily. The bearing may wear, but the heat will cause some lead to melt out and act to prevent seizure.

Incidentally, there is a limit to the beneficial effect of lead. For instance, some 23% lead bronzes have been used but have caused failures when lubrication failed because so much lead ran out that it clogged up the radial clearances, and the shaft seized solidly at the bearings.

Question 2.6 Coupling End Float

Can you advise me where I can find a detailed discussion on the subject of *end float* of shafts of motors that have sleeve bearings and are directly coupled to centrifugal pumps? Are there similar limitations regarding end float of a shaft when the pump is coupled to a ball bearing motor?

Answer. Horizontal sleeve bearing electric motors are not generally equipped with thrust bearings but are merely provided with babbitted faces or shoulders on the line bearings. The motor rotor is allowed to float, and although it will seek the magnetic center, a rather small force can cause it to move off this center. This movement may, in some cases, be sufficient to cause the shaft collar to contact the bearing shoulders, causing heat and bearing difficulties. This is particularly true of large electric motors of 200 hp and higher.

Since all horizontal centrifugal pumps are equipped with thrust bearings, it has become the practice to use *limited end float* couplings between pumps and motors in this size range. These couplings keep the motor rotor within a restricted location. The motors are built so that the total clearance between shaft colars and bearing shoulders is not less than 1/2 in. In turn, the flexible couplings are arranged to restrict the end float of the motor rotor to less than 3/16 in. The restriction against closing the gap is provided by one of the following methods:

1. *For gear-type or grid-type couplings:* by locating a "button" at the end of the pump shaft or by inserting a predimensioned plate between the two shaft ends (see Fig. 2.6).
2. *For the flexible-disk type, such as the Thomas coupling* (Fig. 2.7): by the stiffness of the flexible disks themselves, which have inherent float-restricting characteristics. The displacement characteristics of a coupling of this type (for a 4 in. shaft diameter) are shown in Fig. 2.8.

Contact between the hubs and the coupling covers prevents excessive movement in the opposite direction for gear-type or grid-type couplings. The stiffness of the flexible disks is the restraining force in both directions in the case of the Thomas couplings.

The problem does not arise in the case of ball bearing electric motors, and therefore they do not require the use of limited end float couplings.

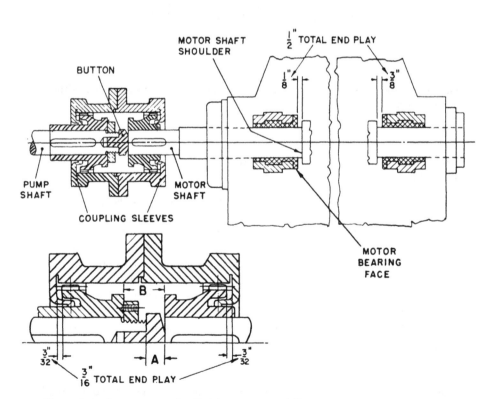

Figure 2.6 Arrangement for limiting motor end float.

Figure 2.7 Restriction against closing the clearance gap is provided by the stiffness of the flexible disks.

Question 2.7 Use of Cast Steel Casings for Boiler Feed Pumps

We have recently issued specifications for three boiler feed pumps designed for 750 psig discharge pressure. In our desire to purchase equipment with something better than cast iron for the casting material, we specified that pump casings were to be made of cast carbon steel. We were much surprised when several bidders refused to quote cast steel casings and suggested that we choose between cast iron and a 5% chrome steel. We had heard that chrome steels are used for the higher pressure range in boiler feed pumps. Can you tell us what is the reason cast steel may be unsuitable? Can you also suggest what clues we may look for in our existing installation to determine whether we need to go to more expensive materials than cast-iron casings and standard bronze fittings?

Answer. The mechanics of corrosion-erosion attacks in boiler feed pumps first became the subject of considerable attention in the early 1940s,

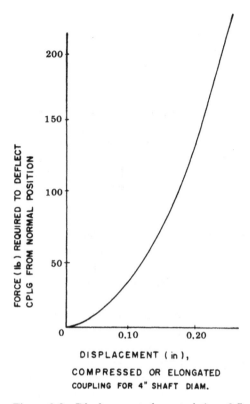

Figure 2.8 Displacement characteristics of flexible disk-type coupling.

when it became desirable to use feedwater of a scale-free character. This desire led to the use of lower pH values and to the elimination of various mineral salts that had theretofore acted as buffering agents. As you state, the practice was instituted to use chrome steels throughout the construction of high-pressure boiler feed pumps. But this does not mean that limiting this practice to high-pressure pumps only is justifiable. It is true that the feedwaters used in the lower pressure plants may not necessarily undergo the same degree of purification. Nevertheless, cases have been brought to my attention in which evidence of corrosion-erosion occured in feed pumps operating at pressures as low as 325 psi.

The exact nature of the attack and the causes leading to it are not, in my opinion, fully understood, since minor variations in the character of the feedwater or in its pH seem to produce major variations in the results. At the same time, certain very definite facts have become established:

1. If not necessarily the cause, at least it is recognized that low pH is an indicator of potential corrosion-erosion phenomena.
2. Feedwaters that coat the interior of the pump with red or brownish oxides (Fe_2O_3) generally do not lead to such trouble.
3. If interior parts are coated with black oxide (Fe_3O_4), severe corrosion-erosion may well be expected.
4. When feedwaters are corrosive, cast iron seems to withstand the corrosion infinitely better than plain carbon steel. Chrome steels, however, with a chromium content of 5% or higher, withstand the action of any feedwater condition so far encountered. Some manufacturerss have a preference for 13% chrome steels for the impellers, wearing rings, and other pump parts other than the casing.

If you install boiler feed pumps that are not fully stainless steel fitted, it is important to carry out frequent tests of the pump performance. This step will help you avoid sudden interruption of service. Corrosion-erosion attack comes on rather unexpectedly, and its deteriorating effects are very rapid once the attack has started. If protective scale formation is absent, the products of corrosion are washed away very rapidly, constantly exposing virgin metal to the attack from the feedwater.

Thus, if the original pump capacity is liberally selected, there may be no indication that anything is wrong until such time that deterioration of metal has progressed to the point that the original margin has been "eaten up." The resulting breakdown immediately assumes the proportions of an emergency, since the net available capacity is no longer sufficient to feed the boiler. Unless additional spare equipment is available, power plant operators may find themselves in an unenviable spot.

Some years ago, one of the most startling cases of this nature came to my attention. When excessive internal leakage was discovered, the pumps could no longer carry the boiler load even with the help of the standby turbine-driven pump. On dismantling, it was found that the wearing rings, which were made of cast iron, had corroded to a fantastic extent. The original clearances of 0.012 in. had increased to over 1/4 in. Other clearances had deteriorated in about the same proportion. The net capacity of the pumps (originally 300,000 lb/hr) had fallen off to less than 200,000 lb/hr. The breakdown imposed a very serious emergency condition on the power plant because no replacement parts were on hand. The reason for the unexpected nature of the trouble was that the maximum load on the boiler had seldom exceeded 650,000 lb/hr and generally hovered around 450,000 lb/hr. And yet three pumps were always kept on-line, each of these pumps carrying from 150,000 to 220,000 lb/hr load.

To avoid such an unforeseen emergency, it is recommended that complete tests of the pump performance be carried out at, say, not less than 3 month intervals if the pump is built of materials that may be subject to corrosion-erosion attack. From a more constructive point of view, it is wise to investigate the effect of the feedwater used on the materials in the pump in question. If any indication exists that these materials may be inadequate, replacement parts of stainless steel should be ordered. If it is intended to replace only the internal parts by stainless steel and to retain the original cast-iron casing, the replacement program should be carried out at the first opportunity rather than waiting to the end of the useful life of the original parts. Otherwise, the deterioration of parts that form a fit with the casing may lead to internal leakage, which, in turn, will cause the destruction of casing fits and will make ultimate repairs extremely costly.

But whatever your decision is as to the internal parts of the pumps you contemplate to purchase, I earnestly recommend that you *do not* specify cast steel casings. Too many sad experiences have been traced to their use.

Question 2.8 Taper Machining of Axially Split Casing Flanges

We have recently opened for the first time a boiler feed pump installed 11 years ago. It is designed to handle 290°F feedwater against a discharge pressure of 1125 psi. The pump is of the volute type, with an axially split casing of 5% chrome steel. Examination of the casing showed no erosion or corrosion anywhere. However, the lower half of the casing appears to have been distorted under pressure, as if the outer portion of the flange had been bent down a few thousandths. Is the pressure for which this pump was built excessive for an axially split casing? Should the lower casing half be remachined for a true flat surface?

Answer. Axially split casings are generally used for design pressures up to 1250 psi and, in such cases, tested to around 1800 psi on hydrostatic test. The appearance of the casing is not due to distortion but rather to a method of machining the two halves used to ensure full compression of the casing gasket at the internal portion inside the bolt locations. One of the two casing halves is machined on a taper as shown in Fig. 2.9, so that as the casing bolts are tightened, the gasket becomes pinched in the critical areas. The amount of the taper is generally not in excess of 0.010 in., and to simplify the machining operation, the taper is carried on one side of the casing only. This, of course, is equivalent to having both sides tapered about 5/1000 in. each. Obviously, the casing should not be remachined to remove this taper.

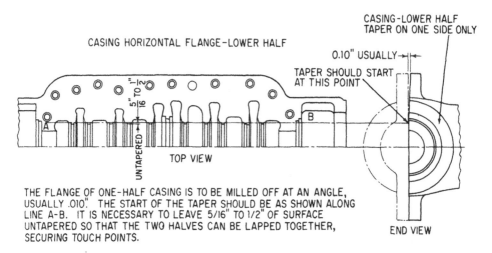

Figure 2.9 Drawing shows that only one of the two casing halves is tapered so that, as casing bolts are tightened, the gasket becomes pinched in the critical areas.

Question 2.9 Single Suction Versus Double Suction

I have noticed in the technical literature that vertical centrifugal pump installations for very large capacities, say above 10,000 GPM, and heads higher than those met with propeller pumps are always single-suction pumps. Why is it that double-suction pumps are not used for this type of application? For the same suction conditions, higher operating speeds would be permissible, and this should appreciably reduce the installed cost, since it would reduce pump size and the cost of the driving motors.

Answer. The most disputed problem with regard to the type of pump construction best suited for vertical dry-pit installations until some 40 or 50 years ago was the choice to be made between single- and double-suction pumps.

In comparing these two designs, consideration must be given to the limitations imposed by requirements of accessibility and repair. In this respect, the single-suction pump presents no problem whatsoever. Vertical single-suction pumps are built with a removable upper head, the diameter of which is larger than that of the impeller (see Fig. 2.10). Once this head has been lifted out of the way, the impeller and shaft assembly can be lifted right out.

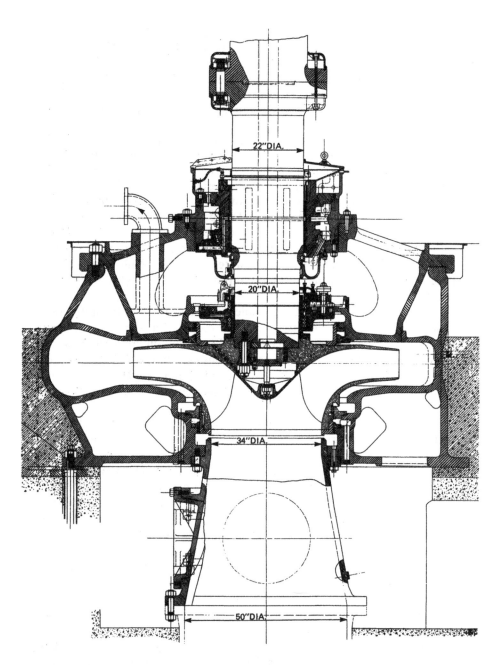

Figure 2.10 Sectional drawing of a vertical single-suction pump.

The introduction of a removable head in a double-suction pump makes it necessary to confine the suction passage within a circular ring and to depress its height so that it becomes fairly squat. It also introduces an additional circular joint between the inside of the casing and the suction passage, a joint that cannot be bolted independently and therefore cannot always be made sufficiently tight to eliminate the possibility of leakage and erosion.

The vertical double-suction pump shown in Fig. 2.11 was designed over 60 years ago. It still illustrates quite well the type of design that would have to be followed even today. The most important effect of the double-suction requirement is that the hydraulic design is seriously jeopardized by the squatness of the suction passage and by its confinement to a circular ring. It becomes impossible to provide the smooth and slow volute curvature necessary to bring a large body of water to the eye of the impeller without excessive turbulence. Figure 2.11 illustrates the abrupt changes in direction and therefore the unequal water velocities resulting at the impeller intake caused by the limitation imposed upon the design of a double-suction pump.

It would be theoretically possible to modify these unfavorable flow conditions by substituting two separate elbows for the pump suction passages. However, such a scheme would be highly impractical. It would be necessary to break the suction connections each time it is desired to gain access to the pump interior. In addition, the pump would become much higher from end to end, adding structural and mechanical disadvantages to the difficulty of access.

Another important factor in the comparison between vertical single-suction and double-suction pumps is that in the latter case unequal pressures would prevail on the upper and lower halves of the pump impeller. The danger arises that the two sides of the impeller will perform unequal work. This danger is a real one in the case of large pumps. Their operating speed is generally selected as high as possible, and therefore the pumps may be operating very near the cavitation limits. Theoretically, the maximum permissible speed for a given capacity and given suction conditions is higher for double-suction pumps than it is for a single-suction design. But if the speed of the double-suction pump is reduced to preclude the possibility of different amounts of cavitation between the upper and lower sides of the impeller, the double-suction pump automatically loses the advantage of a higher permissible speed and becomes as heavy and as expensive as the single-suction pump.

By contrast, the installation of a vertical single-suction pump has ideal suction approach conditions that do not have to compromise with accessibility requirements. And since the pressure distribution at the impeller

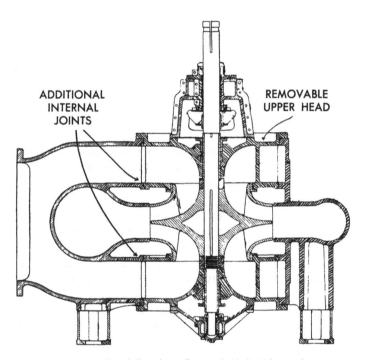

Figure 2.11 Sectional drawing of a vertical double-suction pump.

inlet is uniform, the operating speed can be carried very near to the speed at which cavitation would take place.

Another factor that favors the single-suction vertical pump is that its suction passage can be formed in its entirety out of concrete, whereas suction pipes must generally be provided for a double-suction pump, thus adding to the total cost of the installation.

Finally, the double-suction pump requires a bearing located underneath the pump proper (see Fig. 2.11). This is decidedly a poor location for a bearing. Unless it is adequately protected, there is always the danger of accidental flooding of the pump pit, damage to the bearing, and outage of the pump unit.

Thus, to sum up, the single-suction type is decidedly preferable to the double-suction type for large dry-pit vertical pump installations. The double-suction pump is an excellent solution for normal horizontal service, but when it is modified to take care of the accessibility problem in a vertical arrangement, its performance and advantages are severely handicapped.

The situation is quite different when we consider wet-pit construction. A very effective design of the double-suction type has been developed for wet-pit installation in open sumps, channels, lakes, or rivers, as shown in Fig. 2.12.

You will notice that one of the main problems caused by the double-suction arrangement disappears in this particular configuration: There is no need to bring out the two suction passages to a combined suction nozzle through two separate elbows. In this case, the pump has two separate suction bells, each leading to one of the two entrances of the double-suction impeller.

This design can also be used as a first stage for standard vertical turbine pumps in open sumps or in the can-type construction as in Fig. 2.13.

Question 2.10 Two Pumps in Series with Single Driver

Over the years I have seen a number of pumping units made up of two single-stage pumps piped up to operate in series and driven by a single electric motor or steam turbine. What is the reason for such an arrangement rather than the use of a two-stage pump in a single casing? Wouldn't a series unit be more expensive than a single two-stage pump? And if two pumps are used in series, is it better to place the motor in the middle to drive one pump at each end or to put the motor at one end and drive both pumps on the same side of the motor?

Answer. Generally, a pump manufacturer will have a line of single-stage double-suction pumps ranging from possibly 4 in. discharge all the way to

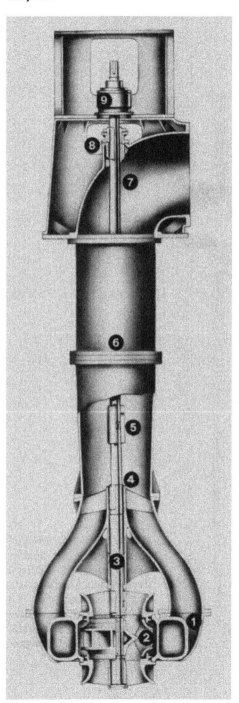

Figure 2.12 (1) Double-volute casing, (2) double-suction impeller, (3) shaft, (4) line bearings, (5) shaft coupling, (6) column pipe, (7) discharge head, (8) stuffing box, (9) thrust bearing.

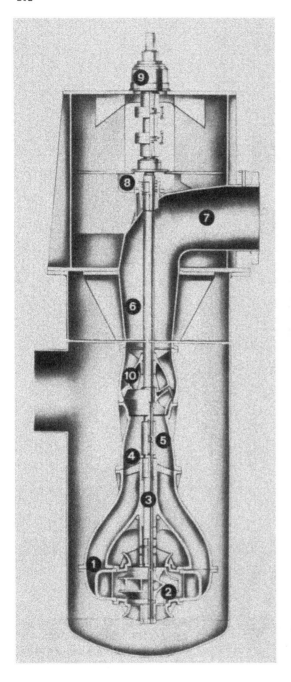

Figure 2.13 (1) Double-volute casing, (2) double-suction impeller, (3) shaft, (4) line bearings, (5) shaft coupling, (6) column pipe, (7) discharge head, (8) stuffing box, (9) thrust bearing, (10) additional mixed-flow single-suction stage.

36 and sometimes even 54 in. On the other hand, the frequency of application of two-stage pumps is somewhat restricted, and it is seldom that a standard line of pumps much larger than 8 or possibly 12 in. for two- or three stage construction is designed by any pump manufacturer. The use of single-stage pumps operated in series stems from this limitation. Whenever an application is encountered for a relatively large capacity and a total head in excess of what can be reasonably be developed by a single-stage pump and no two-stage pump is available to meet these conditions, it will always be more economical to use two single-stage pumps in series rather than develop a special pattern for a two-stage pump that may never again find another application.

Of course, in such a case, it is also more economical to use a single driver than to provide each individual pump of the series unit with its own driver. The only exception to this occurs in cases in which the total head is made up mostly of frictional losses and when a variable output may be required. Since a single pump may be used on occasion in such an installation and the second pump is needed only at the maximum flows, there is logic in providing individual drivers to permit shutting down one of the two pumps operating in series. This is quite typical, for instance, of pipeline installations.

Occasionally, a series unit made up of two single-stage pumps may be preferred to a two-stage pump even when it is available for the capacity and total head requirements. This will be the case when the power consumption of the unit is evaluated and a very high premium is placed on each point of efficiency. It is quite possible that this premium will more than offset the probable higher initial cost of the series unit over the two-stage pump.

As you state, sometimes the motor is built with a double-extended shaft and is placed between the two pumps, as in Fig. 2.14, and sometimes a standard motor is used, with both pumps driven from the same side of the motor (Fig. 2.15). In the latter case, the pump in the middle has to be provided with a double-extended shaft. This shaft must be capable of transmitting the total horsepower required by the two pumps. This may or may not be practical, even if the shaft diameter and the bearing sizes in the pump nearest to the driver have been increased. Thus, the choice between the two arrangements is sometimes dictated by pump design limitations. It will be found, generally, that it is less expensive to purchase a double-extended shaft motor than to design a special shaft and bearings for a standard pump. On the other hand, if the driver is a steam turbine, the only possible arrangement is to place both pumps on the same side of the driver, as steam turbine shafts can seldom be provided with a double extension.

214 *Pump Construction*

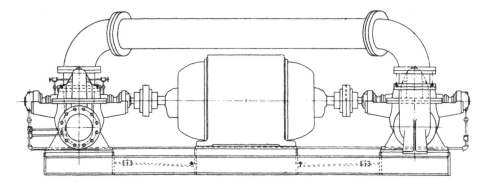

Figure 2.14 Motor with double-extended shaft between two pumps.

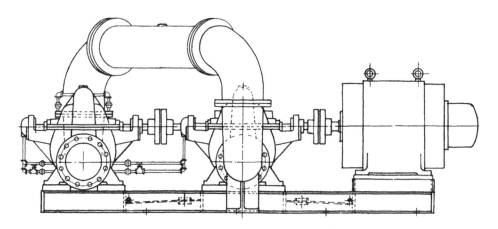

Figure 2.15 Two pumps driven from same side of motor.

Question 2.11 Balancing Holes in Impellers

We have many single-suction single-stage pumps in our plant that have back wearing rings. The impellers are also provided with holes in the back shroud, and the bulletin describing these pumps refers to these as "balancing holes." Since leakage takes place from behind the impeller into the suction passage through these balancing holes, would it not be practical to reduce this leakage by plugging up at least some of these holes?

Answer. By no means should you try to do this. Leave these holes alone! The ordinary single-suction impeller is subject to axial thrust because a portion of its front wall is exposed to a pressure fairly close to discharge pressure. To eliminate this axial thrust, one can provide "back wearing rings." As illustrated in Fig. 2.16, the pressure in the chamber created inwardly of the back wearing rings is very close to the value of the suction pressure by virtue of the balancing holes communicating between the chamber and the suction portion of the impeller. The leakage that takes place past the back wearing rings is returned into the suction area through these holes.

Although designs vary, the practice is to make the area of the balancing holes approximatley four times the area of the annular clearance at the back wearing rings. This assures that the pressure drop across the balancing holes will be almost negligible when compared with the pressure drop across the wearing rings and that the pressure in the back chamber is indeed very close to the suction pressure. In addition, it assures that as the clearance at the back wearing rings increases with time, there will not be an excessive buildup of the pressure in this chamber and no appreciable change in the axial thrust.

Let me illustrate this with an example: Let us assume that the total differential pressure across the rings *and* the balancing holes is 100 psi and that we have an area ratio of 4:1 as I mentioned earlier. The velocity through an orifice is a function of the square root of the pressure differential. Thus, with the velocity through the balancing holes 1/4 that past the wearing ring clearance, the pressure drop across the balancing holes will be 1/16 the pressure drop across the wearing ring clearance. In our case, the pressure drop across the balancing holes will be approximately 6 psi and that across the wearing rings 94 psi, for a total of 100 psi. These 6 psi in excess of the suction pressure will not affect the axial thrust significantly.

Nor will we have too much of a rise in the pressure in the back chamber after wear takes place at the wearing rings: suppose the clearance at the rings doubles while the holes remain essentially unaffected. Now the area of the holes is only double that of the clearance, the velocity there is one-half that past the rings, and the pressure drop becomes one-fourth that across the rings. The pressure in the back chamber will exceed the suction pressure by only 20 psi, with the pressure drop across the rings 80 psi.

Were you to reduce the total area of the holes until it equalled the area of the annular ring clearance when new, the pressure drop across the back wearing rings would equal that across the holes and the pressure in the back chamber would exceed the suction pressure by half the 100 psi net pressure, that is, by 50 psi. Then, after wear doubles the wearing ring clearance, one can readily calculate that the pressure in the back chamber would exceed the

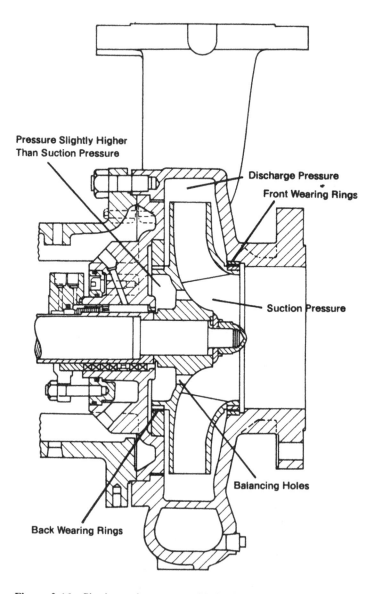

Figure 2.16 Single-suction pump with back wearing rings and balancing holes.

suction pressure by 80 psi. The effectiveness of the back wearing ring in balancing the axial thrust would be practically negated.

One final comment: the axial thrust on an impeller such as that illustrated in Fig. 2.16 is affected by the magnitude of the suction pressure. In addition to the unbalanced forces caused by different pressures acting on the various surfaces of the impeller, there is an axial force equivalent to the product of the shaft area through the stuffing box and the difference between suction and atmospheric pressure. This force acts toward the impeller suction when the suction pressure is less than atmospheric and in the opposite direction when it is higher than atmospheric. Normally, this additional force is quite negligible. Because the same pumps may be applied for many conditions of service over a wide range of suction pressures, the thrust bearing of pumps with single-suction overhung impellers is arranged to take thrust in either direction. The bearing is also selected with sufficient thrust capacity to counteract forces set up under the maximum suction pressure established as a limit for that particular pump.

Question 2.12 Impeller Pump-out Vanes

In dismantling one of our small open-impeller centrifugal pumps, we noticed that the back shroud of the impeller is provided with shallow vanes. What is the purpose of these vanes?

Answer. The vanes on the back shroud of the impeller are called pump-out vanes. Their primary function is to reduce the pressure on the stuffing box, but they also prevent much of the foreign or gritty material in suspension in the liquid being pumped from lodging in the clearance space between the back shroud and the adjacent wall of the casing. Figure 2.17 illustrates the front view of an open impeller on the left and the back view on the right.

Without pump-out vanes, the pressure at the back hub of the impeller would be only slightly less than the discharge pressure. The effect of the pump-out vanes is diagrammed in Figs. 2.18 and 2.19, which show the pressure distribution on the shrouds of impellers (closed impellers are shown in these two illustrations) without and with pump-out vanes, repectively.

Actually, the pressure acting on the shrouds of a single-suction impeller is not uniform over the entire area exposed to the discharge pressure. This is because the liquid trapped between the impeller shrouds and the casing walls is in rotation and the pressure at the periphery of the impeller is higher than that at the hub, as shown in Fig. 2.18. The effect of the pump-out vanes is to further reduce the pressure acting on the back shroud, as illustrated in Fig. 2.19.

Figure 2.17 Front (left) and rear views of semiopen impeller.

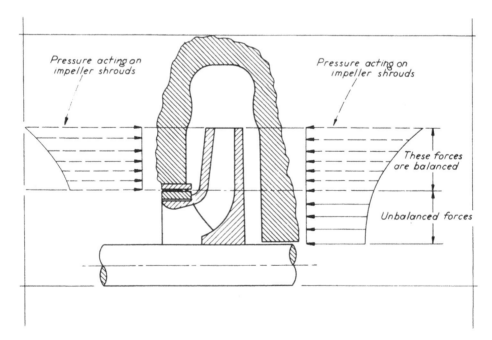

Figure 2.18 Pressure variation on shrouds as a function of impeller radius.

Chapter 2

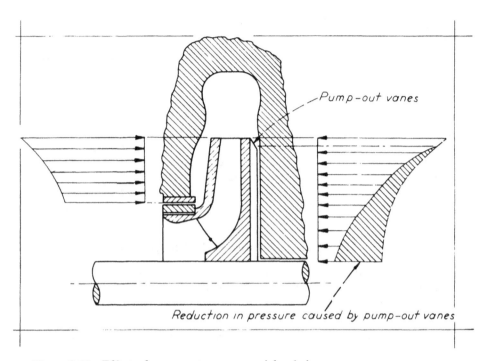

Figure 2.19 Effect of pump-out vanes on axial unbalance.

A further effect from this change in pressure distribution need be mentioned. A single-suction impeller has an unbalanced force acting toward the left in the impeller in Fig. 2.18. This creates an axial thrust. Although this thrust is not of great magnitude in small, low-head pumps, it can reach fairly high values as the total head of the pump is increased. This axial thrust would have an unfavorable effect on the ultimate life of the thrust bearing. The change in pressure distribution created by the pump-out vanes reduces the unbalanced force and hence the load on the thrust bearing.

Note that API-610 Standards that cover centrifugal pumps for refinery service do not permit the use of pump-out vanes for balancing the axial thrust of single-suction impellers. They specify that back wearing rings (as illustrated in Fig. 2.16) should be used for this purpose. The chamber located inward of the back wearing ring is connected to the suction pressure area through so-called balancing holes.

I am not familiar with the reasons that led the API committee involved to prohibit the use of pump-out vanes. One might speculate that these vanes are mostly used with open impellers, a construction seldom used in refinery service, although they are quite common in general or chemical service. Whatever the reason, one must remember that a customer has the perfect right to express a preference for this or that design feature and the manufacturer can make the choice as to whether they do or do not wish to participate in the segment of the business involved.

The prohibition against pump-out vanes is not, after all, the only change from the API-610 (fifth edition). Note, for instance, paragraph 2.1.2, which states that close-coupled, overhung two-stage and overhung single-stage double-suction pumps are not acceptable unless approved by the purchaser. Yet, I am familiar with the fact that many such pumps have operated quite successfully in refineries. Thus, one must realize that all these provisions express the consensus of the refinery people who prepared the API-610 sixth edition.

Question 2.13 Balancing Axial Thrust of Multistage Pumps

There seems to be considerable controversy regarding the construction of the inner element of multistage double-casing barrel-type boiler feed pumps. In some designs, the axial thrust of the individual single-suction impellers is compensated by a hydraulic balancing device. In other designs, commonly referred to as "opposed-impeller" construction, half the impellers face in the opposite direction from the other half and thus the axial thrust is balanced. What are the real advantages of each one of these designs, and does one of them have a superiority over the other?

Answer. Frankly speaking, and save for a minor difference that I shall discuss later, the advantages of either of the two solutions over the other one are more imaginary than factual. I have frequently said that the difference between the two designs is really a "semantic" distinction, because the pressure generated by the pump must be broken down internally. Whether this breakdown takes place across one long joint at one end of the pump or across two or three shorter joints disposed between some of the impellers is essentially immaterial. This is clearly illustrated schematically on Fig. 2.20, which shows the pressure distribution for each of the two solutions in the case of an eight-stage pump. In one case, the axial thrust is balanced by a balancing device at D, and the pressure drop across running joint D is equivalent to the pressure generated by all eight stages. In the case of an opposed-impeller construction, there is ostensibly no balancing device. There are however, three interstage running joints, A, B, and C, subject to differential pressures of four stages, three stages, and one stage, respectively. The sum of the pressure drops across these running joints is equivalent to the same eight-stage pressure drop across the balancing

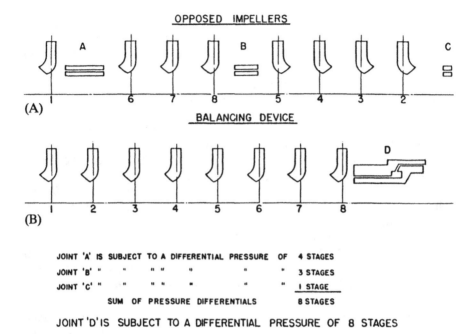

Figure 2.20 (A) Balancing axial thrust with balancing device in eight-stage pump. Joint A is subject to a differential pressure of eight stages. (B) Balancing axial thrust with opposed impellers in eight-stage pump.

device at D. This, incidentally, holds true regardless of the sequence in which the opposed-impeller stages are disposed.

Thus, no matter into how many pieces one slices it and no matter what names are given to these abbreviated pieces of balancing device, it is not possible to eliminate the function that the device performs, that is, to break down the total pressure generated by the pump. In all fairness, I should add that everything else being equal (that is, shaft span, shaft diameter, and dimensions of the running clearances), either design should give either length of service.

The one difference to which I alluded at the beginning of my answer refers to the fact that the leakage through a balancing device is returned to the pump suction via an external pipe, in which one normally locates a flow-measuring orifice. This leakage can therefore be readily measured, this measurement serving to monitor the progress of the wear experienced by the pump (see question 5.2). By inference, since the pressure drop across all the internal joints is of the same order of magnitude when expressed in terms of pressure drop across units of linear dimension, all running joints will wear at about the same rate. It is thus possible to predict when internal clearances should be renewed without dismantling the pump. On the other hand, if you separate the balancing device into several pieces and dispose them at various points within the central portion of the pump, you can no longer monitor the progress of the internal wear without dismantling the pump. This difference is not one, I should add, that all pump users hold in equal regard, and therefore personal preference usually guides the choice between the two arrangements, all other factors being equal.

There is another side to your question that I think I should discuss here. It has to do with the two different philosophies followed in the design of the inner casing itself. Some manufacturers in the United States use a radially split inner casing (Fig. 2.21) and others a casing split axially into two halves (Fig. 2.22). I say that this is another side to your question because essentially manufacturers who use opposed impellers also split their casings axially, but designs with a balancing device have radially split inner casings.

Basically, the two decisions—on the means used to balance axial thrust and on the plane or planes in which to split the inner casing—go hand in hand. Although it is possible to use an axially split inner casing and to mount the impellers sequentially in-line, following the last stage with a balancing device, this approach has, to my knowledge, never been used commercially. Thus, all axially split inner casing designs use opposed impellers. On the other hand, if the inner casing is radially split, it is rather impractical to arrange the impellers of a multistage pump in any order but sequentially, and therefore, these pumps are always provided with a balancing device.

The choice between the two solutions has remained quite controversial in the United States since the 1940s. It is interesting to analyze the reasoning behind this choice, because the difference in approach does not appear to stem so much from a basic disagreement in design principles as from a difference in the priority assigned to the importance of greater reliability on the one hand and to the ease in dismantling the pump on the other.

The proponents of a radially split inner casing consider that the symmetry of circular joints and the absence of three-way joints are preferable to the lesser symmetry of the axially split casing. Those who prefer this latter design marshall in its favor the advantage of greater accessibility (and therefore shorter dismantling time) and of the possibility of acquiring a spare assembled rotor without a spare inner casing. It must be admitted that both designs have been successful in receiving acceptance in the United

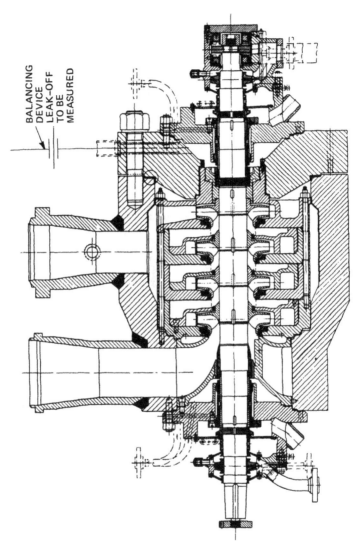

Figure 2.21 High-pressure boiler feed pump with balancing device and radially split inner casing.

224   Pump Construction

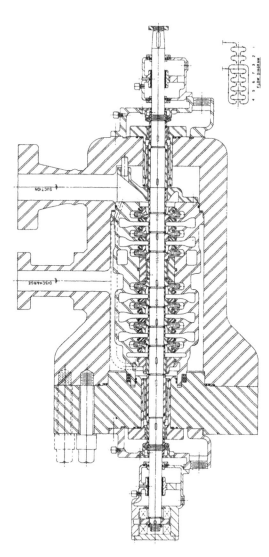

Figure 2.22 High-pressure multistage pump with opposed impellers and axially split inner casing. (Courtesy United Centrifugal Pumps)

States, proving that users have their own preferences for reliability and ease of maintenance no less than manufacturers. Thus, the choice between these two designs is somewhat subjective in nature. I cannot leave this subject without remarking that since the expected life between overhauls may range from 60,000 to as much as 140,000 hrs of operation, I find it difficult to attach much importance to a difference of a few hours in the time required to strip a pump down to individual components.

Because European pump manufacturers changed their construction of high-pressure multistage pumps to the double-casing concept much later than in the United States, they had the opportunity to weigh the respective advantages of the two concepts before proceeding with their own developments. In addition, this change took place in Europe *after* that period in the development of the high-pressure boiler feed pump during which serious troubles were besetting most installations. These early troubles were caused in the 1930s and 1940s by the inadequacy of the materials used in the construction of these pumps and—as late as the 1950s—by the lack of understanding of the phenomenon of transient operating conditions that followed sudden load rejections in steam-electric stations with open feedwater cycles. During these years, the dismantling of high-pressure boiler feed pumps was a much more frequent occasion than today. These two reasons must be why, to my knowledge, there is no European design today that employs an axially split inner casing.

Question 2.14 Single or Double Rings

I have heard frequent arguments between supporters of double wearing rings and others who think that in most applications single wearing rings are sufficient. Who is right, and are grooved wearing rings really effective against abrasion?

Answer. This is not a question to which a simple unequivocal answer can be made. Depending on the circumstances, either type of construction may be superior to the other. Originally, of course, centrifugal pumps were built without any wearing rings in either the casing or the impeller, as in Fig. 2.23. To restore original clearances after wear had taken place, it was necessary either to build up the worn surfaces by welding or brazing and then true up the parts or to buy a new impeller with an oversize hub to fit the trued-up casing fit. The idea of providing a renewable stationary ring (Fig. 2.24) that fit into the casing was an obvious improvement and, I imagine, occured very early in the history of the centrifugal pump. I suspect that the idea of providing the impeller with its own renewable wearing ring (Fig. 2.25) followed fairly quickly.

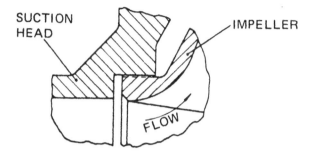

Figure 2.23 Leakage joint with no wearing rings.

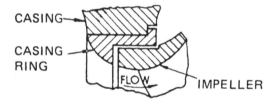

Figure 2.24 Renewable stationary casing wearing ring.

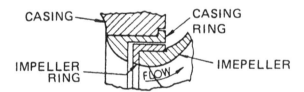

Figure 2.25 Double wearing rings.

Early centrifugal pumps were genrally applied for large capacities and were operated at relatively low speeds. Consequently, the impellers were significantly large in diameter, and the task of mounting these impellers on a lathe to true up the wearing surfaces was not a very simple one. But the reason the double-ring construction was then extended all the way down to the smallest centrifugal pumps can probably be laid at the door of the advertising department of the early centrifugal pump manufacturers. What was logical and sound for large pumps was made to appear as logical and sound for all pumps. And the enthusiasm of advertising people led them to claim that the double-ring construction permitted the user to discard both worn rings, replace them with newly bought parts, and, without any further machining or adjustment work, restore the pump to its service with clearances equal to the original dimensions.

This claim is not quite true, however. The argument that the use of double rings permits their replacement without any machine shop work is incorrect. Mounting a new wearing ring on the impeller hub can distort the concentricity of the clearance turn, regardless of whether this impeller ring is screwed on, bolted on, or shrunk on the impeller. To preserve concentricity, it is imperative that after mounting the ring on the impeller the outer turn of the wearing ring be checked by mounting the impeller on a lathe. In most cases, a slight truing-up cut will have to be taken.

Nor is the replacement of both rotating and stationary rings each time the clearances are to be restored economical. Maintenance mechanics who wish to get the greatest mileage out of their repair parts will buy casing rings with an undersize ID and impeller rings with an oversize OD. The first time clearances are restored, they will retain the old impeller rings, limiting themselves to truing up their wearing surface by mounting the impeller on a mandrel in a lathe. They then bore out the new casing rings to fit, being careful to store the worn casing rings. At the next go-round, they will true up the old casing rings taken out of storage, mount new impeller rings on the impeller, and finish machine these to match the repaired casing rings.

But generally, it is neither economical nor practical to use double wearing ring construction for small pumps—say under 4 or 6 in. It is far better to purchase replacement casing rings with an undersize bore at the impeller hub fit. A slight truing-up cut is taken on the impeller hub to restore the concentricity of the wearing surface. The repair casing wearing ring is then bored out so as to restore the initial clearance. Impeller hubs are generally provided with sufficient metal to carry out such an operation as many as three or four times.

There are exceptions to the general rules suggested. When it is desired to use expensive but more wear-resisting materials for the wearing surface, it may be best to use a double wearing ring construction regardless of pump size, because the impeller itself need not be made of any metal other than that required, let us say, for corrosion resistance. Thus, stainless steel impeller rings are frequently used on bronze or cast-iron impellers. The second exception has to do with pumps operating at relatively high speeds and fitted with high-grade impeller materials. It would be unnecessary to mount stainless steel impeller rings on a stainless steel impeller, since wear would probably be slow anyway and since the impeller hub surface can easily be restored by truing up. At the same time, the use of a wearing ring mounted on the impeller would introduce tha hazard of it loosening up under the action of centrifugal force and causing severe damage to the pump.

Thus, the prospective purchaser of a centrifugal pump will be well advised to examine the reasons a particular pump has double wearing rings or single wearing rings and to decide between the two on the basis of true merits, not that of catalog or advertisement claims.

Wearing rings with a grooved surface at the leakage joint (Fig. 2.26) are considered to afford some additional protection against the effect of foreign matter, which can be washed into the hollow space of the groove instead of jamming between the stationary and rotating parts. In a few designs, the leakage joint surface is threaded instead of just plain grooved. It is assumed that a particle of foreign matter lodging in the threads may be washed out of the leakage joint, following the threaded hollow.

Question 2.15 Reverse Threads in Wearing Rings

I am familiar with concentrically grooved wearing rings as illustrated in Fig. 2.26. I have also seen designs that use reverse threads for these wearing rings instead of concentric grooves. The intended purpose is to reduce the wearing ring leakage. Why is such a design not used more frequently?

Answer. This construction is, as you say, quite rare. The reason is that it is a most inefficient approach to the internal leakage problem. The efficiency of a reverse threaded wearing ring in pumping back the leakage would at best be in the range of 10 to 20%. It is certainly preferable to let the leakage take place and repump the leakage along with the total capacity of the pump at whatever the normal pump efficiency may be—whether 85% or even 50 or 60% as it might be for small pumps. A designer who uses such a solution is either guilty of an honest mistake—because he fails to understand the lack of logic in his solution—or of making a dishonest claim, because he assumes that the user does not understand this lack of logic.

Question 2.16 Shaft Deflection of Multistage Pumps

I have heard it said that the shaft deflection of multistage pumps varies approximately as the fourth power of the shaft span and inversely as the square of the shaft diameter. What is the origin of this relationship?

Chapter 2

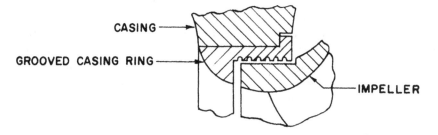

Figure 2.26 Single casing ring, grooved type.

Answer. This is a close approximation, and it can be used to compare the relative rigidity between several pump designs intended for the same service conditions. I shall explain its origin in a moment. But first, let me address myself to the possibilities of comparing the probable rate of wear between several different designs. Consider what causes wear at running joints subject to a pressure differential; wear occurs not only because of the erosive action of the leakage past the joint, but also, and to a greater measure, wear may occur by contact and through the rubbing action between two metal parts, even if the contact in momentary and too slight to cause galling and seizure. In other words, the two prime factors affecting the life of a given high-pressure boiler feed pump are the shaft deflection and the internal clearances. The greater the margin between these two, the less will be the chance of accidental contact and the longer the life of the equipment.

The deflection of a shaft supported by two external bearings is dependent upon the weight of the rotating element, the shaft span between the bearings, and the shaft diameter, the basic formula being:

$$\Delta = \frac{WL^3}{mEI}$$

where

Δ = deflection, in inches
W = weight of the rotating element, in pounds
L = shaft span between bearings, in inches
m = coefficient depending on method of support and distribution of load.
E = modulus of elasticity of shaft material
I = moment of inertia of shaft = $d^4/64$
d = shaft diameter, in inches

It is not necessary to calculate these values when a relative comparison is being made. Instead, a factor can be developed that will be representative of the relative deflections of the shafts being compared. Since the major portion of the rotor weight resides in the shaft itself and since the method of bearing support and modulus of elasticity are common to all designs being compared, the deflection can be shown to be a function of a simple factor:

$$\Delta = \text{function of } \frac{Ld^2L^3}{d^4}$$

$$= \text{function of } \frac{L^4}{d^2}$$

Therefore, the lower the (L^4/d^2) factor for a given pump, the lower will be the "unsupported" shaft deflection, essentially in proportion to this factor itself.

It is obviously of vital importance to obtain values of shaft dimensions and of all internal clearances before proceeding with the analysis of several pump designs. And although it may be interesting to compare "tabulated" values of shaft deflections given by various manufacturers, it is neither sufficient nor wise to base any comparison on such values. The reason for this is that such deflection data are not necessarily always calculated on the same basis and may reflect in some cases the support furnished to the rotor by the wearing rings, which act as water-lubricated bearings during "normal" operation. If the shaft deflection is sufficiently reduced by these means to avoid contact at the wearing rings during operation, contact will still exist at rest and wear will invariably take place during each coast-down and each start-up of the pump.

Question 2.17 Shaft Sleeves

I have noticed that whereas standard centrifugal pumps intended for general service have their shafts covered in their entirety, either by the impeller itself or by shaft sleeves, high-pressure boiler feed pumps most frequently have shafts with uncovered portions between the impellers. Is there not as much need to cover the shaft of boiler feed pumps as that of general-service pumps?

Answer. You are quite correct in noting that pump shafts are usually protected from erosion and corrosion and from wear at the stuffing boxes and leakage joints in the waterways by renewable sleeves. As in Fig. 2.2, these sleeves generally abut against the impeller, and no portion of that shaft is exposed to the liquid handled by the pump. It should be remembered, however, that general-service pumps are normally built with regular carbon steel shafts, a material highly susceptible to corrosion by almost any liquid that might be pumped, including water.

On the other hand, high-pressure boiler feed pump shafts are made of alloy steels, ranging from SAE 4340 to 13% chromium stainless steels, which are of course much less subject to corrosion attack from feedwater than would be the case with plain carbon steel. The elimination of shaft sleeves between adjacent impellers took place a number of years ago when it was found that these sleeves introduced a number of disadvantages.

In the first place, these sleeves cannot contribute to shaft stiffness. Boiler feed pump shaft diameters are generally quite substantial, so as to minimize the shaft deflection, which is aggravated by the fact that multistage pumps must be used for boiler feeding. Since there is a certain minimum shaft sleeve thickness, the diameter of shaft and shaft sleeve at the impeller eye is significantly increased by the addition of this shaft sleeve, increasing liquid pickup velocities and affecting the pump performance adversely.

But a more important reason is that it was found that when impellers abutted against each other or against intermediate sleeves, the possibility always existed that the shaft would become bowed if the matching faces were not absolutely square to the bores. This, incidentally, is true whether the pump uses *opposed impellers* to balance the axial thrust or whether the impellers are all in line and the thrust is balanced by a hydraulic balancing device as in Fig. 2.21. The axial thrust, which is quite significant in value, acts to push the impellers against each other ar against the adjacent intermediate sleeves, and the slightest deviation from a "true" face of impeller or sleeve will lead to bending forces on the shaft.

In a design such as in Fig. 2.21, each impeller is mounted individually and independently, hard up against a small shoulder, and there can be no force exerted to cause the shaft to bow. Of course, the shaft is protected against wear in the stuffing boxes by renewable shaft sleeves, regardless of whether conventional stuffing-box packing or condensate injection sealing is used.

Question 2.18 Water-Flooded Stuffing Box

What is meant by a *water-flooded stuffing box,* and under what conditions is it used?

Answer.   A water-flooded stuffing box construction refers to a modification of the pump frame that provides a water-retaining cavity surrounding the stuffing box itself. A typical design of a pump so modified is shown in Fig. 2.27, and the pump is further clarified by Figs. 2.28 and 2.29.

The walls of that portion of the frame that generally plays the part of a stuffing-box leakage drain collector are raised well above the pump centerline. The cavity thus formed can then be filled with clear, clean water. To prevent contamination of the bearing lubricant by this water, a cover is bolted to the frame adjacent to the bearings, with a double lip seal installed in this cover. A small tubing drains any condensation that may take place between the two seals.

Water-flooded stuffing boxes may be used on vacuum service, such as black liquor evaporators and hot wells. This modification provides a positive seal, ensuring the installation against loss of system vacuum. Visual observation of the water level in the cavity also provides an instant indication of packing condition.

This construction is also used on corrosive service, as it provides for the circulation of clean fluid around the shaft right outside the stuffing box. Any corrosive leakage is diluted and is carried away through the overflow connection. The overflow can be piped directly to a central drainage system. This in itself further reduces corrosion problems on bearing frames, baseplates, concrete work, and others.

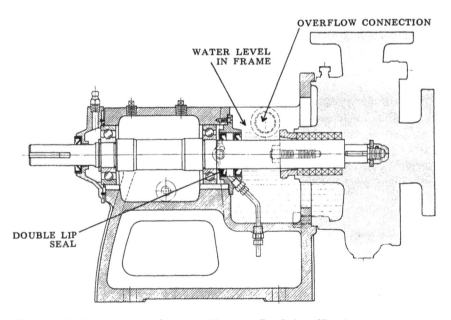

Figure 2.27 Frame-mounted pump with water-flooded stuffing box.

Figure 2.28 Pump with water-flooded stuffing box.

Figure 2.29 Pump with water-flooded stuffing box.

3
Installation

Question 3.1 Location of Discharge Nozzles on End-Suction Pumps

I have noted that the discharge nozzle of end-suction single-stage pumps is generally in a top vertical position whether the pump is close coupled or mounted on a bearing frame. We use many such end-suction pumps in our plant, but frequently a top discharge is less convenient than, say, a side or a bottom discharge when it is desired to simplify the piping for a particular installation. Is there any reason these pumps cannot be furnished with other than top discharge?

Answer. In most cases, end-suction single-stage pumps can be furnished with the discharge nozzle arranged in several alternate positions in addition to top vertical. For example, the appended elevation of an end-suction pump (fig. 3.1) indicates three additional mountings that are, obviously, obtained by simply rotating the casing with respect to the frame itself. The number of permissible locations for the discharge nozzle is, of course, dictated by the number of bolts attaching the casing to the frame or to the motor for close-coupled pumps.

There are, however, certain limitations that in some cases may interfere with some particular desired nozzle location. The pump frame, the bearing bracket, or the baseplate may sometimes interfere with the discharge flange, so that neither a bottom vertical nor bottom horizontal arrangement may be achieved. Another limitation may arise if rotating the casing makes the connection to the stuffing-box seal cage inaccessible.

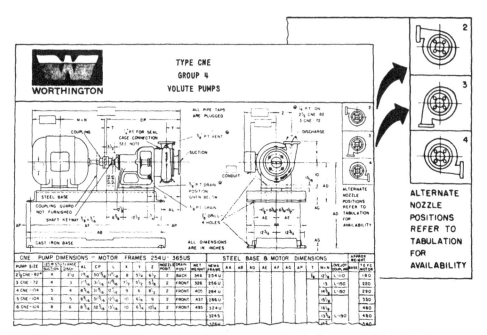

Figure 3.1 Table from end-suction pump catalog indicates available discharge nozzle positions.

Finally, there may be economic considerations involved. In some cases, such pumps are built for stock shipment, and special requirements may mean delays in delivery and extra charges for the special handling. There may also be charges for the drawings that are required unless standard drawings are available showing the various alternate nozzle arrangements.

In other words, many times these special requirements can be met, but each case has to be considered individually. Obviously, simplification of the piping may often make a small extra charge or a slight delay in delivery worthwhile.

Since writing the centrifugal pump clinic that was the source of this particular answer, horizontal, end-suction centrifugal pumps used for chemical process have become the subject of ANSI Standards (ANSI B73.1-1977). These standards prescribe dimensions for pumps from 11/2 to 8 in., as illustrated in Fig. 3.2. These standards make it impossible to provide the flexibility that was available in the past, as illustrated in Fig. 3.1. Of course, the existence of the ANSI Standards provides the user with other advantages, such as the ability to install any manufacturer's pump in

Chapter 3 237

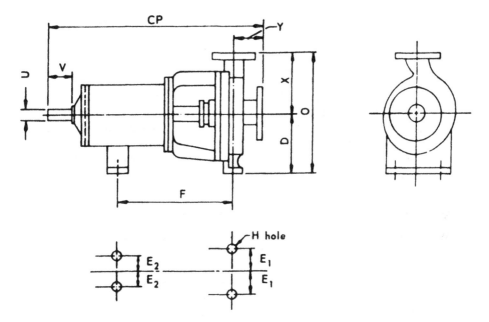

Figure 3.2 ANSI horizontal, end-suction centrifugal pumps for chemical process (see Appendix for pump dimensions).

the same physical space and piping configuration. But this trade-off illustrates very well the fact that "there is no free lunch." One cannot get the advantage of complete interchangeability without sacrificing something.

Question 3.2 Raised Faces on Pump Nozzles

We recently placed an order for a centrifugal pump designed for 600 psig discharge pressure with a cast-iron casing. When the elevation drawings were submitted for approval, we noticed that the discharge flange had a flat face, and we requested the manufacturer to change it to a raised face so as to correspond to our piping standards. The manufacturer refused to do this. Why can't a cast-iron pump casing be furnished with raised-face discharge?

Answer. Although raised-face flanges are perfectly staisfactory for cast-steel casing pumps, their use is extremely dangerous with cast-iron casings. This danger arises from the lack of elasticity in cast-iron, which frequently leads to flange breakage when the bolts are being tightened up. When a flange has a raised face, the fulcrum of the bending moment is located inward of the bolt circle, and very appreciable stresses are set up in the cast-iron flange.

As a result it is not only essential to avoid the use of raised-face flanges on a cast-iron casing, but a raised face should also be avoided on the flange of the pipe directly opposite a flat-face cast-iron flange.

If cast iron is used for the pump casing, it remains possible to use a cast steel adapter piece fashioned in the shape of a straight pipe extension, an increaser, or an elbow. This adapter piece should have a flat face on the flange that is permanently connected to the pump discharge flange. The other face, of course, can be any type of face dictated by the piping standards used in the plant in question.

Question 3.3 Eccentric Reducers at Pump Suction

We would appreciate your opinion concerning the proper application of reducers in centrifugal pump suction lines.

We have consulted many textbooks and reviewed much written material on this subject and find that there are still some aspects of this problem that are not adequately covered. For example, Fig. 3.3 is an illustration reproduced from the Hydraulic Institute Standards and is quite typical of illustrations found in a great many textbooks. Although it is the general practice to use eccentric reducers in pump suction lines, we have not found enough specific information as to what position the reducers should be placed in. Figure 3.3 indicates that with a suction line entering the pump in the horizontal plane, the eccentric reducer is so placed that the top is flat. The sketch gives no indication as to whether the source of supply is from above or below the pump. Figure 3.4 illustrates certain modifications and/or additions that we have made in an attempt to clarify this situation further.

It is our feeling that when the source of supply is from above the pump, the eccentric reducer should be so placed that its bottom is flat. The reasoning is that this would afford every opportunity for liberated vapor bubbles to pass back up the suction line instead of being trapped at the pump suction. Similarly, with the pump suction piping entering the pump from a long horizontal run or from below grade, we believe that the top of the eccentric reducer should be flat as shown in the sketch.

The proper application of suction line reducers has become quite controversial in our office. Therefore, we would be most grateful for any assistance you can offer us in arriving at an answer to our problem.

Answer. The question you have raised is an interesting one because it seems to have been given little or no specific attention in the past. Most

Chapter 3

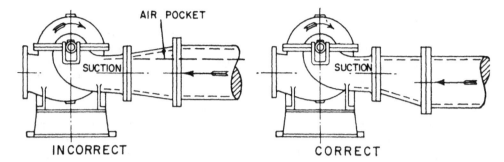

Figure 3.3 Illustration of eccentric reducer mounting from Hydraulic Institute Standards.

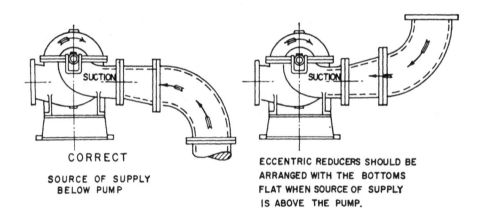

Figure 3.4 Suggested modifications for eccentric reducer mountings.

published information dealing with the subject of proper orientation of eccentric reducers has been based on the assumption that the source of the liquid being pumped is at a level lower than the pump suction flange, although as you have indicated in the case of the Hydraulic Institute Standards, this assumption has only been implied by Fig. 3.3 and has not been clearly stated.

The intent of the standards in this regard has been to implement the instruction that there should be no high point in the suction line. I quote from the Hydraulic Institute Standards:

Slope of Suction Pipe:

... Any high point in the suction pipe will become filled with air and thus prevent proper operation of the pump. *A straight taper reducer should not be used in a horizontal suction line* as an air pocket is formed in the top of the reducer and the pipe. An eccentric reducer should be used instead.

This instruction, incidentally, is applicable by proper interpretation even in cases in which the source of the liquid pumped is higher than the pump suction flange, since trapped vapors can in any event act to reduce the effective cross-sectional area of the suction line. This will result in flow velocities higher than anticipated and consequently in higher friction losses with a detrimental effect on the pump performance.

In the case of a liquid source higher than the pump suction and particularly where the suction line consists of an eccentric reducer, an elbow turned vertically upward, and a vertical length of pipe, assembled in that order from the pump suction flange upstream, I would definitely concur in your suggestion that the flat side of the eccentric reducer be placed at the bottom. The suggested modifications to Fig. 3.3 are shown in Fig. 3.4. In general, it could be stated that wherever a low point exists in a suction line, the horizontal run of piping at that point should be kept as short as possible. The preceding example then becomes a special case of this general criterion in which the horizontal run has been reduced to zero.

A further consideration which, of course, cannot be neglected in any case in which vapors must be vented against the direction of flow is that the size of the line upstream of any low point must be sufficiently ample to ensure that the velocity in the line will not exceed tha rate at which bubbles will rise through the liquid in question.*

Question 3.4 Reduction of Suction and Discharge Piping Size

We have several single-stage double-suction pumps on hand and would like to use them to provide an artificial load to test engines. The problem lies in the inlet and outlet sizes: Would we run into much trouble if we reduced the inlet size from 12 to 6 in. and the outlet from 10 to 5 in.? The suction head will be about 12 ft of 140°F water.

---

*The only exception to this recommendation of amply sizing suction lines has to do with boiler feed pump installations in which the pump takes its suction from a deaerator supplied with bled steam from the main turbine. Such installations are subject to transient conditions during sudden load reductions, and it has been established that reducing the size of the suction piping acts to attenuate the ill effects of these transients.

We assume that it may be necessary to reduce impeller size but have no way of knowing how much. Any information you could pass on would be greatly appreciated.

Answer. Normally, recommendations for a centrifugal pump installation are to use both suction and discharge piping of at least one size larger than the respective suction and discharge nozzles of the pump. On the suction side, this recommendation is directed toward providing as favorable conditions as possible and to avoid cavitation problems. On the discharge side, the intent is to minimize friction losses and therefore the power consumption of the pump. In your particular case the last problem does not exist, since you are using the pumps as power consumers and not for the primary purpose of transporting liquids from one point to another. Thus, on the discharge side at least, I can see no reason you cannot reduce the piping size to 5 in. by means of a reducer. Any valves you need to install on that side of the pump, that is, check and gate valves, can also be chosen in the 5 in. size, and this will reduce the investment of the installation.

As to the suction side, it is a little more difficult to make a decision without having some additional information. On the face of it, the installation should cause cavitation problems, even assuming that the total length of suction piping is not excessive. But to assure yourself of the adequacy of the suction conditions, you will need to obtain certain data on these pumps, including primarily the pump curve, that will show you the relationship between the capacity and the pump head, efficiency, pump brake horsepower, and required NPSH (net positive suction head). If the curves in your possession do not give you the last data, a curve of permissible suction lifts when handling cold water will do as well. To determine values of required NPSH from permissible suction lift curves, use the following equation:

Required NPSH = 33.9 − 0.6 − permissible suction lift

where

33.9 = atmospheric pressure at sea level, in feet of water

0.6 = vapor pressure of 62 °F water

Let us make a few assumptions about your installation. To begin with, I shall assume that when you say the pumps will have a suction head of 12 ft, you mean that the level of the water at the suction source will stand 12 ft above the pump centerline and that you do not refer to a gauge reading at the pump nozzle corrected to the pump centerline, that is, that you have not yet calculated and deducted the friction losses in the suction piping.

Let us further assume tha the pumps are designed for 5900 GPM, 192 ft total head at 1460 RPM and that the permissible lift at that capacity is 14 ft. The required NPSH will be 33.9 - 0.6 - 14, or 18.3 ft.

Under normal conditions, the installation might include some 50 ft of 14 in. suction piping, two elbows, a gate valve, and a 14 by 12 in. reducer. From various tables on friction losses in piping, we find that the entrance loss may be of the order of 0.25 ft, that the elbows are equivalent to 25 ft of straight piping each and the gate valve to 9 ft of straight pipe, and that the loss through the reducer is negligible, say 0.10 ft. The friction through 14 in. steel pipe at 5900 GPM is approximately 3.8 ft per 100 ft of straight pipe. Thus the total losses will be

$$0.25 + 3.8 \frac{(50 + 2 \times 25 + 9)}{100} + 0.10 = 4.5 \text{ ft}$$

We can now calculate the NPSH available when the pump is handling 5900 GPM of 140°F water with 12 ft of static suction head:

Available NPSH = atmospheric pressure − vapor pressure

+ static suction head − friction losses

= 33.9 − 6.7 + 12 − 4.5 = 34.7 ft

This available NPSH of 34.7 ft is amply in excess of the required NPSH, which we calculated to be 18.3 ft.

Now let us check what happens when we subsitute 6 in. pipe for the normal 14 in. piping that we had assumed. Friction tables do not extend far enough to show what the friction losses would be per 100 ft of piping when 5900 GPM are pumped through 6 in. pipe, but the velocity is 65.5 ft/sec, and we can estimate the losses to exceed 200 ft per 100 ft of piping. We need not carry out any calculations to see that there just would not be any NPSH left were we to use 6 in. pipe.

Let us then determine what would be the minimum size of pipe that we could use at the suction of our pump and still have sufficient available NPSH left. By setting the available NPSH as equal to the required NPSH, we can establish the permissible friction losses in the suction piping:

18.3 = 33.9 − 6.7 + 12 − friction losses

and maximum permissible suction piping friction losses are 20.9 ft.

You will find that even 8 in. pipe will not be suitable if the length of the suction piping is such as we have assumed. On the other hand, 10 in.

pipe gives us a reasonable selection: The friction losses with 5900 GPM flow are approximately 15.5 ft per 100 ft of straight pipe. Counting the elbows, the gate valve, the reducer, and the entrance losses, the total losses in the suction piping will not exceed 6 ft, well on the saft side of the 20.9 ft we have calculated as permissible.

Of course, all this has been calculated on the basis of an assumed flow and of certain assumptions of the configuration of the suction piping. You can repeat these calculations using the actual values for your projected installation for a more accurate conclusion and decision.

Insofar as the second part of your question is concerned, that is, the impeller cutdown that may be necessary, I can give you only the general rules that apply, since I have no information either on the present characteristics of the pumps at your disposal or on the range of horsepowers and speeds at which you will need to test your engines. Assuming that you have a copy of the pump characteristics, the relations that you will need to know in order to adjust the pump impeller to your needs are the following: (1) the capacity for a given point in the pump characteristic varies as the impeller diameter *and at the same time*; (2) the head varies as the square of the impeller diameter; and (3) the bhp varies as the cube of the impeller diameter. Since you may also be testing your engine at several different speeds, you need to know that the pump characteristics vary with speed exactly in the same manner as they do with impeller diameter. In other words, the pump capacity varies directly with the speed, the head varies as the square of the speed, and the power varies as the cube of the speed. Armed with these relations and with the pump characteristics at some known speed and with a given diameter, you can construct a family of curves that will permit you to select that impeller diameter that best fits the power range of your engines.

Once your pumps and engines are hooked up in the test loop, you can vary the power delivered by the engines by throttling on the pump discharge and thus changing the capacity delivered by the pump. Figure 3.5 illustrates the changes in pump characteristics obtained in a typical pump by changing impeller diameter, and Fig. 3.6 shows the characteristics at two different speeds.

Question 3.5 Are Baseplates Needed?

Centrifugal pumps are always quoted with structural steel or cast-iron baseplates under the pump and the driver. Is this really necessary, and would there not be a substantial saving in omitting the baseplates, mounting pump, and driver on soleplates embedded in concrete?

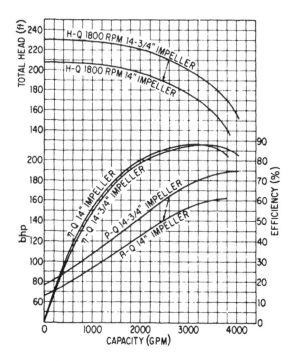

Figure 3.5 Changes in pump characteristics obtained by changing impeller diameter.

Answer. A baseplate may cost from 5 to 10% of the price of the pump, but its elimination would save only part of this cost since it would be necessary to provide machined surfaces such as soleplates, to align these soleplates with considerable care, and to fasten them to the foundation. The ultimate saving, therefore, may not be too significant.

On the other hand, baseplates provide a number of advantages that cannot be obtained with soleplates. The primary function of a baseplate is to furnish machined mounting surfaces for the pump and its driver, but it also acts to collect any drips or drains that would otherwise spill right on the floor. When a complex lubricating system is used, involving separately driven oil pumps, oil coolers, filters, and so on, these can be more readily and practically mounted on the baseplate.

Of course, as the size and weight of a unit is increased, the cost of the baseplate may exceed by a very considerable amount the cost of the field work necessary to align either individual baseplates or soleplates under pump and driver. Therefore, in the case of extremely large units, it may be

Chapter 3

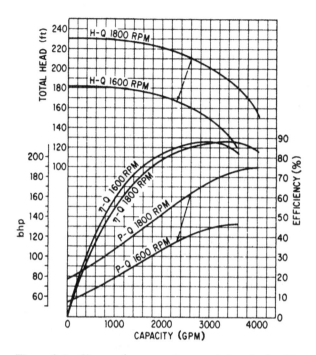

Figure 3.6 Changes in pump characteristics obtained by changing pump speed.

economical to omit the baseplate, unless the desire for improved appearance or the need to utilize the base to collect drips or to mount auxiliary equipment justifies the additional cost.

Question 3.6 Grouting of Baseplates

What is the purpose of grouting a centrifugal pump baseplate, and what is the correct procedure for placing the grout?

Answer. The baseplate of a centrifugal pump is ordinarily grouted before the piping connections are made and before final rechecking of alignment of the coupling halves. The purpose of grouting is not so much to take up irregularities in the foundations as to prevent lateral shifting of the baseplate and to reduce vibration.

The pumping unit, made up of pump and driver on the baseplate, is placed over the foundation and is supported by short strips of steel plate or shim stock close to the foundation bolts. A space of 3/4 to 2 in. should be allowed between the bottom of the baseplate and the top of the foundation for grouting. Shim stock should extend fully across the supporting edge of the baseplate.

In placing the pump on the foundation, instructions supplied by the pump manufacturer should be followed in leveling and aligning the coupling halves. When the unit is accurately leveled and aligned, the hold-down bolts should be gently and evenly tightened before grouting.

The usual mixture for pumps is composed of one part pure Portland cement and two parts building sand, with sufficient water to cause the mixture to flow freely under the base (heavy cream consistency). In order to reduce settlement, it is best to mix the grout and let it stand for a couple of hours, then remix it thoroughly before use. Do not add any more water.

The top of the rough concrete foundation on which the pump is to be installed should be well saturated with water before grouting is applied. A wooden form should be built around the outside of the baseplate to contain the grout. In some cases, this form is placed tightly against the lower edge of the base and, in other cases, is removed a slight distance from the edge. Choice is merely one of personal esthetic preference. For convenience in getting the grout under the base, one or more funnels are used at several points under the edge.

Grout is added until the entire space under the base is filled (Fig. 3.7). Grout holes provided in the baseplate serve as vents to allow air to escape. A stiff wire should be used through the grout holes to work the grout and release any air pockets.

The grout should be protected against rapid drying to prevent cracking. This is best accomplished by covering the exposed surfaces of the grout with wet burlap. When the grout is sufficiently set (about 48 hr) so that the wooden forms may be removed, the exposed surfaces of the grout and foundation should be finished to a smooth surface. When the grout is hard (72 hr or more), the hold-down bolts should be finally tightened. After this, the coupling halves should be rechecked for alignment.

Question 3.7 Shims Under Pump Baseplates

I would like to ask a question on the subject of grouting baseplates. Should the shims or wedges be removed from under the base after the grout has been set? I am curious about the concrete shrinkage, which would leave the base resting on the shims. Any comments would be appreciated.

Answer. One would think that some unanimity of opinion would exist on such a simple subject, but oddly enough this is not the case. Not only do erectors differ on this point, but apparently even experts on the use of concrete will generally hedge on this. I found an article published some years ago that contained the following two statements in two portions of the article:

Figure 3.7 Grout is added until entire space under base is filled. Holes in base (arrow) allow air to escape and permit working of grout to release air pockets.

1. Shims or wedges and screws are usually removed after the grout has hardened.
2. Wedges may be of steel or hardware, but wood wedges are suitably only for small machines. Wood wedges should be removed. Removal of steel wedges is left to the discretion of the erectors. If they are to be removed, however, care should be taken to see that *all* are removed, as a single wedge left in place may cause damage.

My opinion is that this apparent vacillation of various authorities and erectors is influenced by the type of experience they have had early in their career, in other words whether they have mainly dealt with reciprocating or rotary machinery. In the first case, removal of shims or wedges is definitely mandatory. This is true of all installations where impact, pounding action, or appreciable vibration is to be expected in normal operation. I quote the recommendations from an instruction book for vertical diesel engines:

> Before the grout has entirely set but after it will no longer flow, remove the excess grout between the engine holding down flanges and the dam and trim the grout in line with the vertical edge of the flanges. At the same time, dig around the shims and wedges sufficiently to facilitate their removal later. The grout should set at least 48 hours and longer if possible, to properly harden, after which all shims and wedges must be removed and the space they occupied regrouted. *Caution: The removal of the shims and wedges, used in the leveling of the unit, is a absolutely essential.* If not done, the unit will be suspended, with metal to metal contact, on the wedges and shims and in time, due to reciprocating forces, become loose on the foundation.

On the other hand, when dealing with rotary machinery, such as centrifugal pumps and centrifugal compressors, this danger of loosening is almost nonexistent, and to my knowledge, most erectors who specialize in this type of machinery leave the shims in place.

The danger, incidentally, is affected by the care used in choosing the grout material and in mixing it. Excess water in the grout mix is almost certain to result in shrinkage and damage to the grout. In general, a plain cement-sand mortar shrinks quite readily when it dries and is therefore a poor choice for grouting material. There are a number of grout mixes available on the market that are guaranteed not to skrink when properly used, and since they cost but little more than a plain cement-sand mixture, there is no excuse for not using them. I note in a bulletin issued by the Master Builders Co. describing their Embeco Pre-Mixed Grout the following statement: "It is not necessary to remove shims or loosen leveling screws with this grout."

Of course, no harm will result from removing the shims, and you can therefore follow your own preference in this matter as long as you are dealing with centrifugal pumps and are using suitable grouting masterial.

Whatever the practice with regard to the shims, it is imperative that pump and driver alignment be rechecked thoroughly after the grout has hardened permanently and at reasonable intervals later. In other words, the fact that a baseplate has been leveled prior to and during grouting should not be considered sufficient evidence of the permanent "evenness" of pump and driver mounting surfaces. Proper alignment and shimming between the baseplate mounting surfaces and machine feet are still necessary.

Question 3.8 Motor Doweling

We had ordered a centrifugal pump driven by a 200 hp, 3600 RPM electric motor. It has been received with the pump doweled to the baseplate but with the motor left undoweled. Instructions received with the pump indicate that the driver doweling is to be made in the field after final alignment. Why is it that the driver cannot be aligned with the pump at the factory and doweled there?

Answer. It is not practical to align drivers with their pumps at the factory prior to shipment and to have this alignment maintained with sufficient accuracy so that the unit can be started and operated satisfactorily without a final realignment in the field.

Baseplates of sufficient strength and rigidity with mounting feet and bolting on the units of sufficient size to permit operation without field realignment would be prohibitive in size and weight to the extent that they would not be acceptable to any customer.

Furthermore, the pump, the baseplate, and the driver are generally subjected to such forces during shipment, when being mounted in place and during the piping up of the unit, that serious misalignment is introduced. This misalignment is frequently sufficient to cause coupling and bearing failures and, in some cases, even shaft breakage.

Thus troubles are almost impossible to avoid unless a final field alignment is made, after which the driver can be permanently doweled to the baseplate.

If the pump is to handle high temperature liquids, the final alignment should take place with both pump and driver brought up to normal operating temperature.

This reasoning leads in many cases to havng the driver shipped directly to the installation site, without first coming to the factory where the pump has been built. This practice saves on the cost of double freighting if the driver is not required for testing the pump. In such cases, the baseplate should not be drilled for the driver feet at the factory and the drilling should take place in the field, to avoid any error and the resulting inconvenience of rewelding the holes and their redrilling.

Question 3.9 Expansion Joints

What effect does the installation of an expansion joint in the suction or discharge piping of a centrifugal pump have on the pump itself? What happens in the case of vertical pumps with bottom suction and side discharge?

Answer. The purpose of an expansion joint is to avoid transmitting any piping strains on the pump. whether these strains are caused by expansion when handling hot liquids or by misalignment of the piping and pump. On occasion, expansion joints are formed by looping the pipe as is customary in steam piping but more often they are of the corrugated diaphragm or of the slip joint type.

Expansion joints fulfill the function of eliminating piping stresses, but they often introduce an entirely new problem, that is, a reaction and a torque on the pump and its foundations. Thus, unless they are applied correctly, expansion joints can play the part of the remedy that cures the sickness and kills the patient.

The reason for the foregoing is that, in accordance with Newton's third law, action equals reaction. For example, if an expansion joint is located vertically ahead of the suction elbow (Fig. 3.8A), the reaction, or downward force, will be the product of the pressure and of an area approximately corresponding to the mean diameter of the corrugated bellows.

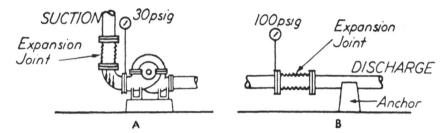

Figure 3.8 In (A), effective area of bellows-type expansion joint in 8 in. suction line would be about 93 in.². In (B), with a 6 in. discharge line, area would be approximately 53 in.².

Assuming an 8 in. pipe, the expansion joint would have 13 1/4 in. OD and 8 5/8 in. ID corrugations. The effective area would be about 93 in.², and if the internal pressure is 30 psig, the downward force would be 30 × 93, or 2700 lb. This force would apply a couple to the pump and may twist it on its foundation.

In a similar manner, when an expansion joint is located in the 6 in. discharge pipe (Fig. 3.8B) and the internal pressure is effectively 100 psig in.², the horizontal reaction will be 5300 lb tending to pull the pump off the foundation bolts. Depending on the size and number of these bolts, the stress on the foundations may become excessive.

In the case of a vertical pump with side discharge, as in Fig. 3.9, there will be a side force equal to the discharge pressure times the effective area of the expansion joint. For a 30 in. pipe, this area will be 900 in.², and if the discharge pressure is 20 psig, the side force will be 20 × 900 or 18,000 lb. Incidentally, the vertical force upward on the elbow is balanced by the axial hydraulic thrust, as far as the plate support is concerned. On the other hand, the thrust on the motor bearings is the sum of the dead weight plus this hydraulic thrust. (See Questions 3.11 and 3.12.)

Generally, expansion joints are not recommended for high-head pump installations. To illustrate the rapid rise of the reaction forces with increased sizes and pressures, consider that with a 24 in. pipe and 160 psig pressure, the reaction force is approximately 134,000 lb or 67 tons. The casing of a pump subjected to such force would have to be almost entirely sunk and anchored in concrete to hold it in place.

Question 3.10 More on Expansion Joints

One of your earlier Clinics dealt with the effect of an expansion joint on a centrifugal pump. One of the illustrations (Fig. 3.8A) showed an expansion

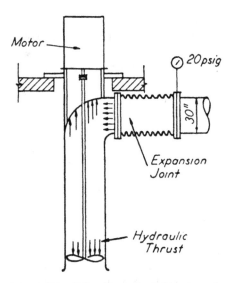

Figure 3.9 Effective area of this expansion joint is approximately 900 in.² and side force will be 18,000 lb.

joint in the suction piping of a pump operating with 30 psig pressure at the suction, and you stated that this pressure caused a moment on the pump because of the presence of the expansion joint. Is this pressure not a static pressure exerted equally in all directions, including the top of the elbow, and therefore producing an equal and opposite moment? A moment is produced on the pump by velocity pressure, the weight of the liquid column. and the weight of the pipe.

In Fig. 3.8B, the 100 psig on the discharge is a static pressure. This likewise is exerted in all directions and therefore balances out. The reaction force on the pump foundation must therefore be strictly caused by the velocity of the liquid leaving the pump. This would be the same as the air leaving a free balloon.

Answer. Your disagreement with my statements is based on three separate concepts that are not applicable in this case. You are concerned with the effect of the velocity pressure—or let us call it velocity energy. You assume that the weight of the column of water is a force that exists over and above the static pressure; finally, you do not distinguish between an expansion joint that has axial flexibility and a pipe which may be considered a completely rigid structure.

In most instances, the static pressure is the major portion of the pressure components that make up total pressure. The velocity head seldom exceeds a few percent of the total (usually from 1 to 3 psi with water as the fluid). The weight of the liquid column is a portion of the static pressure registered on the pressure gauge and need not be considered separately if you are dealing with a gauge pressure read at the point where you are interested in measuring the pressure.

For instance, if the pump took its suction from an open tank and the pressure gauge read 30 psig, this would imply that the level in the tank stood at 69.3 ft (30 psi × 2.31 ft/psi for water) above the pressure gauge, if we neglect friction losses in the suction piping.

Finally, we come to the crux of the question. An expansion joint is flexible and is designed mainly for axial movement. Although the pressure within the joint and the pipe acts in all directions, it results in a force on the anchored members at either end of the joint.

One way of visualizing this would be to imagine two pipes, one slipping within the other with a frictionless but fluid-tight slip joint and with the ends closed off. Assume further that a spring is attached internally to the two closed-off ends. Applying internal pressure to the pipes will cause the two pipes to move away from each other and will stretch the spring. An equilibrium condition will occur when the forces acting in opposite directions on the two pipes are balanced by the spring tension.

In the same manner, if the expansion joint were free to move (not being attached to anchored piping) and had closed ends, it would expand axially under internal pressure. If this pressure were high enough the joint would ultimately fail.

This applies equally to Fig. 3.8A and 3.8B. The reaction force on the pump that you believe is caused by the velocity of the liquid is negligible when compared with the force resulting from the static pressure. This static pressure will exert a force on the piping at either end of the expansion joint equal to the effective area of the joint times the pressure. This force must be corrected by the spring constant of the expansion joint, however, which will vary with the size and travel of the joint. Note that the displacement of the expansion joint is seldom of an order of magnitude where the effect of the spring constant is significant. In turn, this force is transmitted to the pipe anchors that prevent the pipe from moving.

Question 3.11 Does Axial Thrust of Vertical Turbine Pumps Act on the Foundations?

Figure 3.10 shows a typical installation of vertical turbine pumps. There seems to be a difference in opinions among pump manufacturers as to whether or not the downthrust should be considered as a factor in foundation loading in

Chapter 3 253

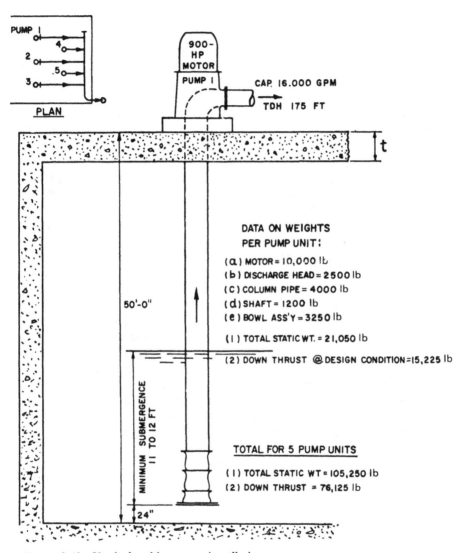

Figure 3.10 Vertical turbine pump installation.

addition to the static weight of the pump and motor combined. I shall appreciate your opinion and comment on the subject matter.

Answer. I am not aware that such a difference of opinion exists among pump manufacturers. I am sorry to have to state that if such a difference does exist, it is based on a lack of understanding of certain simple laws of mechanics. The question does arise on frequent occasions, however, and I think that it deserves being given a very thorough airing.

But first I have to admit that for some perverse reason this problem keeps reminding me of an anecdote that falls into the category of the shaggy-dog story. Some of the readers may have heard it before, but I am tempted to relate it here. A young student, hitchhiking home from his university, flagged down a small truck, which stopped in front of him. The driver motioned him to get in, at the same time swinging open the door on his own side. He grabbed a baseball bat from below his seat, jumped down, ran to the back of the truck, and beat violently on the truck panels for a full 10 sec. He then ran back to his seat, slammed his door closed, and determinedly started the truck up the road. The young hitchhiker was somewhat astonished by the peculiar sequence of these events but kept his peace for the moment. Five minutes later, the truck rolled to an abrupt stop on the side of the road, the driver again jumped down, grabbed his baseball bat, and proceeded to the rear. repeating his furious banging on the truck panels. He then again ran back to get behind the steering wheel and drove on. The hitchhiker was beginning to be more than puzzled—he started to worry over the possibility that he had accepted a ride from a maniac. When the whole crazy procedure was repeated once more after another 5 min, he hesitatingly asked the truck driver what all this meant. "Oh," replied the driver, "It's very simple. You see, this is a one-half-ton truck, and I have 1500 pounds of chickens* back there. If I don't keep one-third of them flying all the time, we'll be in real trouble."

Actually, the problem of the suspended vertical pump is quite simple when it is thought through logically. A single-suction impeller, such as is used in vertical turbine pumps, is generally unbalanced as to axial hydraulic forces and exerts an axial thrust downward. The magnitude of this thrust varies with the pressure developed by the impeller and with the unbalanced area of the impeller. The total downward thrust transmitted to the shaft of the pump is, of course, the summation of the axial thrusts of the individual impellers comprising this pump.

This thrust acts in combination with the static weight of the pump rotor and of the motor rotor. It must be counteracted by a thrust bearing, generally located within the motor. But a pump and motor mounted as in Fig. 3.10 form a self-contained entity, and all internal forces and stresses must be balanced within this entity. The foundations will react strictly to

---

*Editor's note: The way I heard it, they were canaries.

Chapter 3 255

the static weight of the pump and of its motor, plus the weight of the water contained in the pump.

Bear in mind that the weight of this water is far from being negligible. If we assume that the column pipe of the unit you are dealing with has a 24 in. diameter, the approximately 50 ft of pump will contain close to 10,000 lb of water. This weight must be added to the 21,050 lb weight of the pump and motor combined when calculating the load on the foundation.

To better illustrate the conclusion that the axial thrust is not a load on the foundation, we can develop our reasoning by analogy. The effect can be likened to the structure illustrated in Fig. 3.11. A flat plate A is resting on foundations over an opening. A second plate B is attached to it by stay bolts. A hydraulic jack is mounted on plate B and is connected to a pump by means of a hose. The load on the foundations is equal to the dead weight of plates A and B, of the stay bolts, and of the hydraulic jack. If the pump is operated and the hydraulic jack starts to push plates A and B apart, the stay bolts will be subjected to an increased tensile stress. The various stresses, however, will be entirely contained within the system made up of plates A and B, of the stay bolts, and of the hydraulic jack. The load on the foundations supporting plate A will not be affected in the slightest.

Another arrangement may consist of the same two plates and stay bolts and of a tie rod attached to the two plates and joined in the center by a turn buckle. A man stands on the lower plate, as indicated in Fig. 3.12. The load on the foundations is equal to the weight of the plates, the man, the stay bolts, and the tie rod. If the man tightens up on the turnbuckle,

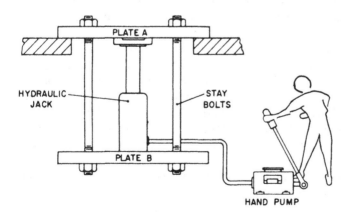

Figure 3.11 The load on the foundations does not change when the stay bolts are stretched.

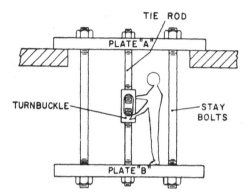

Figure 3.12 The load on the foundations does not change any more than in the case in Fig. 3.11.

shortening the tie rod, the stay bolts will be compressed, and the two plates will tend to come closer together, but the load on the foundations will not change by a single ounce. Our suspended system is self-contained.

On the other hand, if the pump and the motor are supported separately and are joined strictly by means of a rigid coupling that connects the pump and motor shafts (see Fig. 3.13), our situation will have changed radically. No longer is our system self-contained. Since the pump and motor shafts are rigidly connected, the motor thrust bearing must still counteract a downward thrust equal to the sum of the weights of the pump and motor rotors and of the axial thrust developed by the pump. But the load on the foundations when the pump is operating is no longer the same as when the unit is idle.

Let us return to our two plates, our tie rod, and the man standing on the lower plate, but let us eliminate the stay bolts that connected the two plates in Figs. 3.11 and 3.12. Let us, instead, support the lower plate on one foundation and the upper plate on another, as in Fig. 3.14. If our man tightens the turnbuckle to shorten the tie rod, the lower plate will tend to pull up on the lower foundation (reducing the load on that foundation), while the upper plate will pull down on the upper foundation (increasing the loading there).

And so it is with a pump and motor supported separately as in Fig. 3.13. The respective foundation loads will be as follows:

1. When idle:
    a. On the foundation supporting the motor: the weight of the motor plus the weight of the pump rotor

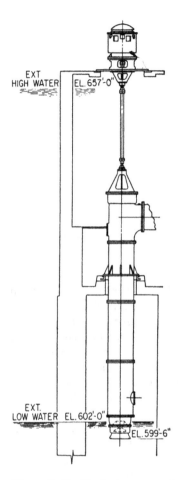

Figure 3.13 Vertical pump and driving motor supported separately.

       b. On the foundation supporting the pump: the weight of the pump stationary parts plus the weight of any water contained in the pump above the level of the water surrounding the pump
2. When the unit is running:
       a. On the foundation supporting the motor: the weight of the motor plus the weight of the pump rotor *plus the axial hydraulic thrust developed by the pump.*
       b. On the foundation supporting the pump: the weight of the pump stationary parts plus the weight of the water in the pump *less the axial hydraulic thrust developed by the pump.*

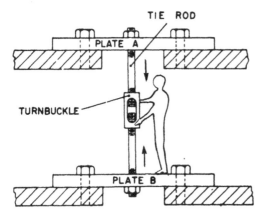

Figure 3.14 Tightening the turnbuckle increases the load exerted by plate A and decreases the load exerted by plate B.

Note that to simplify this explanation I have neglected to mention certain other forces that act on a structure like a vertical turbine pump suspended from foundations. One of these is the force caused by the acceleration of the column of water at the moment of start-up. The reaction to this accelerating force is in the form of torque and is not transferred to the foundation. This additional required torque must, however, be considered in the calculations of start-up conditions because the time required to come up to full speed depends on the difference between the available torque from the motor and the total torque required by the pump.

Another force is that caused by the change in the direction of flow at the discharge head elbow. This reaction tends to reduce slightly the vertical load on the foundation but introduces some reaction in the horizontal direction. Finally, there may be some transient reactions on the foundation if water hammer occurs in the installation caused by too rapid closure of valves.

Actually, these various forces are not of major significance in calculating the foundations. Some designers make it a practice to add an empirical 15% to the pump, motor, and water weights to cover the contingency of these forces. In my opinion this added 15% is optional and not absolutely necessary, since any foundation design will normally include an ample factor of safety over the calculated loads imposed by weight alone.

Question 3.12 Further Comments on Axial Thrust of Vertical Turbine Pumps

We have always considered a vertical pump supported as shown in Fig. 3.10 as a self-contained entity and have therefore never included the hydraulic thrust in the total foundation load. We are therefore in full agreement with you in this respect.

One phase of your discussion puzzles us, however, and we would appreciate your comments on the folowing: You state that for a pump like that shown in Fig. 3.10, "The foundation will react strictly to the static weight of the pump and of its motor plus the weight of the water contained in the pump." We are assuming here that you are speaking about the running condition. We would greatly appreciate it if you would give us your ideas as to how this water weight is transferred to the foundation while the pump is running.

After reading your answer to Question 3.11, another question was raised. In a deep-setting vertical pump mounted as shown in Fig. 3.10, should the weight of the water be included in the total load used for determining the stresses in the column pipe during the running condition? We can readily see where it should be included for a pump with check valves or a foot valve. Here, if the motor stops, the column has to take the additional dead weight of the water. But we cannot visualize how, during the running condition, the dead weight of the water could act on the column itself.

If we were to assume that the dead weight of the water is supported by the closed impellers during the running condition, then this load would be transferred through the line shaft to the motor bearing, and if this were to be the case, then this dead weight would have to be included in calculating the total thrust on the motor bearings. To the best of my knowledge, the total thrust on the motor bearing is generally considered to be the summation of the hydraulic thrust, the line and pump shaft weights, and the weight of the impellers. Your ideas on this question will also be appreciated.

Answer. With regard to your first question, it should be remembered that, unlike pressure, the weight of the water can act only in a "straight-down" direction. Thus, the projected area of the diffuser vanes must support the weight of the water in the column. This weight is therefore transmitted to the column pipe through the diffuser vanes. In turn, this load on the cloumn pipe must be transmitted to the foundation through the floor flange attached to the column pipe.

With regard to your second question, you are quite right in assuming that the weight of the water in the column is part of the foundation load and does not, *normally,* have any effect on the pump thrust. The latter consists as you say only of the weight of the rotor and of the unbalanced axial thrust. The only exception to this would be the special case in which the pump has a propeller of a diameter equal to the inside diameter of the column pipe and is not provided with a diffuser. Here, the weight of the water acting on the propeller would be transmitted to the foundation through the shaft, the motor thrust bearing, the motor frame, and the pump driving head. In such a special case, the thrust on the motor bearing would have to include the weight of the water column.

Question 3.13 Shaft Elongation of Vertical Turbine Pumps

In a deep-setting vertical turbine pump, either oil or water lubricated, when the shaft is stainless or carbon steel and the water temperature is 150°F, how do you accurately calculate the line shaft elongation due to a combination of hydraulic downthrust and 150°F temperature while the pump is running? We realize that the line shaft elongation due to hydraulic thrust can be calculated from the basic equation

$$\text{Elongation} = \frac{\text{hydraulic thrust} \times \text{length}}{\text{metal area} \times \text{modulus of elasticity}}$$

But the question remains how to calculate the additional elongation due to the heat. We thought that perhaps in your many years of experience you might have run across a similar problem.

Answer. As you have stated, the elongation of a vertical turbine pump shaft is caused by two separate and independent phenomena: first, the tensile stress due to the hydraulic thrust and, second, the thermal expansion. Strictly speaking, the tensile stress in the shaft when the pump is running arises from two components: from the weight of the shaft and of the impellers mounted thereon and from the hydraulic thrust itself. But in most cases, as you will notice presently, the initial stress caused by the weight and the elongation resulting from this stress is insignificant when compared with the stress and elongation resulting from the hydraulic thrust.

Let us look specifically at the pump described in Fig 3.10. The weight of the shaft is given as 1200 lb. We shall assume that the impellers themselves weigh an additional 400 lb. Let us further assume that the shaft diameter is 3 1/2 in. (area = 9.6 in.^{2}). The modulus of elasticity of the carbon steel shaft is 30×10^6. The length of the shaft from the thrust bearing to the bottom impeller is approximately 50 ft or 600 in. We can now calculate the elongations in question:

$$\text{Elongation caused by weight} = \frac{(1200 + 400) \times 600}{9.6 \times 30 \times 10^6} = 0.0033 \text{ in.}$$

The elongation caused by the downward hydraulic thrust is of a much higher order of magnitude:

$$\text{Elongation caused by hydraulic thrust} = \frac{15{,}225 \times 600}{9.6 \times 30 \times 10^6} = 0.0315 \text{ in.}$$

The sum of these two elongations is of the order of 35/1000 in. This is quite appreciable and must be compensated for in assembling and setting the pump so as to avoid interference contact between stationary and rotating parts. The expansion caused by temperature is completely independent of the elongation of the shaft caused by the downward hydraulic thrust. By definition,

$$\Delta L = \epsilon \, \Delta t \, L$$

where

ΔL = elongation, in inches

ϵ = coefficient of expansion, in inches per inch per degree Fahrenheit

Δt = changes in temperature, in degrees Fahrenheit

L = length of shaft, in inches

In our particular case, the elongation of the shaft caused by thermal expansion will be different depending on whether carbon steel or austenitic stainless ateel is used because the coefficients of expansion are different for the two materials. They are 0.0000065 for carbon steel and 0.0000092 for 18-8 austenitic stainless steel (type 304). Assuming then that the pump had been assembled at an ambient temperature of 70°F, the change in shaft temperature when pumping 150°F water will be 80°F. The elongation of the shaft caused by thermal expansion will therefore be

$$\Delta L = 0.0000065 \times 80 \times 600 = 0.312 \text{ in.} \quad \text{for a carbon steel shaft}$$

and

$$\Delta L = 0.0000092 \times 80 \times 600 = 0.442 \text{ in.} \quad \text{for a type 304 stainless steel shaft}$$

At first glance, this is quite a bit of expansion. But we must remember that not only will the shaft itself expand because of a change in temperature but so will the entire pump, since it is logical to assume that all parts of the pump in contact with the liquid will reach a reasonably uniform temperature fairly soon after pumping has started. Since we are more interested in the relative expansion of the rotating and stationary parts, it is obvious that should all pump parts have essentially the same coefficient of thermal expansion, no relative displacement will take place between stationary and rotating parts. Thus, if ferrous materials are used for the pump bowls, column pipe, and other parts, and carbon steel for the shaft, both stationary and rotating parts will expand by about 0.312 in., and no relative displacement will take place. If, on the other hand, the shaft is made of type 304 stainless steel, it will expand more than the rest of the pump by 0.442 − 0.312 or 0.13 in.—a little more than 1/8 in. in other words.

With this combination of pump materials, therefore, the total elongation of the shaft relative to the stationary parts would be 0.0033 in. (caused by the weight) plus 0.0315 in. (caused by the downward thrust) and plus 0.13 in. (caused by the change in temperature). The first of these three components is, as we said, negligible. But the total is about 11/64 in., and unless precautions have been taken to accomodate this displacement of the rotating element with respect to the stationary parts, difficulties will definitely be in store for this pump.

There is essentially one place within the pump where such contact could take place between rotating and stationary parts: the bottom face of the impeller hub, which is separated from the mating face in the pump bowl by a vertical clearance. This clearance, or end play as it is termed in vertical turbine pump nomenclature, is usually quite generous and can accommodate considerably more than the normal elongation encountered from the effect of hydraulic thrust (see Fig. 3.15). This end play may run from 1/8 in. for small pumps to as much as 1/2 in. for pumps of the size discussed in this question. Rating charts for different pumps prepared by the manufacturers generally list the value of this end play, and charts are also available to calculate the elongation caused by the weight of the shaft and the hydraulic thrust. Engineers making a pump selection always make a comparison between the calculated elongation and the end play to assure themselves that no interference can occur. But it will be noticed from the examples discussed here that the relative elongation between a stainless steel shaft and cast-iron bowls with steel column pipe is by far of a higher order of magnitide than the elongation caused by weight and hydraulic thrust. Thus, if the pump manufacturer is not informed of the possibility that the pump will handle liquids at considerably higher temperatures than the ambient temperature of assembly, the possibility also arises that interference will occur between rotating and stationary parts.

If the normal end play is insufficient to accommodate the extremes that may be encountered in operation, some adjustment is possible after the pump is installed by repositioning the pump shaft within the hollow shaft motor driving the pump.

Chapter 3 263

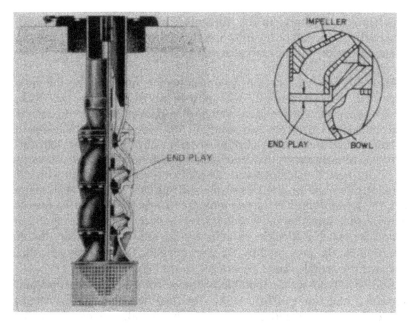

Figure 3.15 Provision for end play in vertical turbine pump.

Question 3.14   Saving Stuffing-Box Leakage

Treated makeup water for the boiler feedwater system in our high-pressure steam plant is a fairly expensive item. Exact cost is hard to figure, but we estimate that it costs us somewhere near 60 cents per 1000 gal not counting any capital charge on treating equipment.

We know accurately that we are wasting approximately 1.2 million gal of this water each year through the packing glands of our boiler feed pumps and, that represents roughly $720 per year down the drain.

Is there any practical, economical way to eliminate or greatly reduce this loss?

Answer.   A considerable amount of attention has been devoted within recent years to the problem of boiler feed pump stuffing boxes. This attention has culminated in the development of two alternative solutions to the problem, each intended to eliminate the stuffing box entirely.

One of these solutions is the use of mechanical seals; the second is based on the replacement of the packed box by a labyrinth bushing and the use of condensate injection. In the latter construction, cold condensate is brought directly from the discharge of the condensate pump into a central point of the labyrinth. A portion of the injection flow enters the pump interior, and the remainder leaks out into a collecting chamber and returns to the condenser hot well.

Each of these two solutions has its adherents, and I do not intend to examine their relative merits here. It must be noted, however, that conditions exist when the conventional design of a packed stuffing box cannot be used and either of these two solutions *must* be incorporated. On the other hand, there are other conditions of service in which the packed stuffing box can operate very satisfactorily and the major objections to its use center on the need of a maintenance schedule more frequent than with condensate injection and on the loss of the stuffing-box leakage. As a matter of fact, the last objection is the one most frequently brought out in the case of many installations that have been operating in the field for a number of years and for which difficulties of maintenance are so insignificant that no justification can be presented by the operators for the expenses of a field change to either condensate injection or to mechanical seals.

And yet the losses attributable to the stuffing-box leakage are quite appreciable, and some effort would be justified to reduce them or eliminate them almost entirely. On the other hand, attempts to reduce these losses by tightening up on the glands are generally followed by disastrous effects on the life of the packing and of the shaft sleeves. All packed stuffing boxes must have an adequate amount of leakage that acts to lubricate the packing and to carry away some of the heat of friction. Generally, this leakage drops into the bottom of the bearing bracket adjacent to the stuffing box and drains away into waste. Attempts to reclaim this leakage are not very successful because it is frequently contaminated by any of the oil leaking out into the bearing bracket from the bearings themselves, and the decontamination costs would be too expensive to be justified.

On the other hand, the cost of this leakage is quite appreciable. Since all of the feedwater is treated, the stuffing box should be charged not only with the cost of raw water drained to waste but also with the cost of treating this water. In addition, since the leakage represents a very important portion of total makeup, it might even be reasonable to charge against this leakage a portion of the initial cost of the water-treating installation. But let us make an estimate of the costs involved, even neglecting that part of the cost representing initial installation of the water-treating facilities. Let us assume that each stuffing box sustains a leakage of 2 qt/min. A boiler feed pump, therefore, will leak 1 GPM or 1440 gal per day and 525,600 gal per year. This loss takes place whether the pump is running or not because an idle pump kept on standby duty is maintained warmed up, with suction and discharge valves open, ready to be put on the line at a moment's notice.

The actual cost of treated water varies in different steam power stations. In one case, I have had it reported that water is charged at $1.50 per 1000 ft^3 of raw makeup water plus another $1.50 for the necessary chemicals to treat this water, bringing the cost to $0.47 total per 1000 gal. Another power plant charges as much as $1.15 per 1000 gal. Thus, in the case we have examined, the cost of the stuffing-box leakage may run from $250 to $600 per year per pump. An installation of three boiler feed pumps serving one main unit can sustain a loss of $750 to $1800 per year merely by allowing the stuffing-box leakage to go to waste. No wonder that serious consideration of any sound procedure to reduce this loss is justified.

The most logical and the most promising approach short of replacing the stuffing boxes by mechanical seals or by condensate injection is to prevent all possible contamination of the leakage so that it can be reclaimed easily and safely and returned to the cycle. In other words, the leakage must be collected *before* it issues into the bearing bracket and can become mixed with oil leakage from the bearings.

A test was arranged in the field on a high-pressure boiler feed pump in order to determine whether such an approach was practical and whether it would be effective. The pump was fitted with a special gland and a special shaft sleeve nut, as shown in Fig. 3.16. The nut was provided with grooves to guide the leakage into a drain in the bottom half of the gland. This drain was tapped and connected to a flexible hose, which in turn was piped to a drain return tank. In the normal construction, the shaft sleeve nut is smooth, and the leakage drips into the bearing bracket and is piped away to waste through drain opening B.

Prior to the test, the pump was repacked with the combination of packing indicated in the drawing and the packing run in and adjusted so that the stuffing-box leakage was essentially equivalent to the normal flow. The operating conditions of the pump were as follows:

   Pump capacity, 420,000 lb/hr
   Discharge pressure, 1800 psig
   Suction pressure, 140 psig
   Pumping temperature, 262 °F
   Pump speed, 3550 RPM
   Cooling-water inlet temperature, 96 °F
   Cooling-water outlet temperature, 107 °F
   Cooling-water flow, 4 GPM

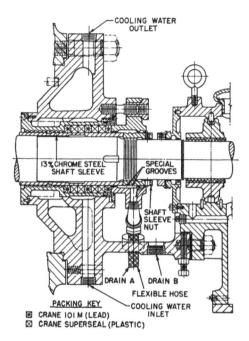

Figure 3.16 Stuffing-box arrangement with special gland and special sleeve, designed to recover stuffing-box leakage.

The test was carried on for a sufficient length of time to obtain very accurate measurements of the leakage amounts, which were found to be the following:

Leakage through drain A, 0.55 GPM
Leakage through drain B, 0.003 GPM
Temperature of leakage, 198 °F

In other words, this arrangement permitted one to reclaim all but 0.5% of the leakage. The test could certainly be considered as successful.

A few refinements should be mentioned that would improve the arrangement used for the test. In the first place, a better connection should be used at the point where the flexible hose is attached to the gland drain, because it is very difficult to use a wrench in the cramped position down in the bearing bracket cavity to tighten or loosen this connection if the gland is to be removed when repacking the stuffing box. Since the pressure in the hose is just nominal, the connection need not be more than hand-tight, so that a regular air hose coupling would be satisfactory.

In addition, a provision must be made that would permit the operator to observe tha amount of stuffing-box leakage at all times. Otherwise, there would be the danger that the gland would be tightened excessively, with resultant damage to the packing and to the shaft sleeve. An open telltale can be installed, as shown in Fig. 3.17, so that the flow would be subject to inspection.

If a water-quenched gland is used, the same arrangement can be used except that the dimensions of the drain opening in the gland and the size of the flexible hose must be adequate to carry away both the leakage through the stuffing box and the quenching flow. Of course, if raw water is used for quenching, the leakage is contaminated and must be disposed of in the same manner as the quenching water was handled prior to this modification to the stuffing box. But since the water no longer is contaminated by oil drops, it can certainly be reclaimed and returned to raw water storage.

A word of caution in connection with the use of water-quenched glands: The supply of quenching water should be stopped once in a while, and the leakage quantity should be observed at the open telltale. Otherwise, the operator will be unable to distinguish between the leakage and the quenching water, and the gland may be tightened excessively.

Question 3.15 Sealing Water for Stuffing Boxes

We are designing a plant that will have installed a large number of centrifugal pumps handling a variety of chemicals. There will also be installed several sewage-type pumps for sewage and waste disposal. The majority of these pumps are to be supplied with clean, clear water to seal the stuffing boxes. It was our intention to provide this sealing water directly from our regular water system, which is tied in with the city water supply.

Local ordinances state that no city water line may be connected to a sewage or process liquid line. Could you suggest how we should arrange the supply of sealing water without violating the local ordinances?

Answer. The simplest and most logical solution is to provide an open tank under atmospheric pressure into which city water can be admitted and from which a small pump can deliver the required quantity of sealing water. Such a water-sealing supply unit can be installed in a location from which it can serve a number of pumps. Figure 3.18 illustrates a typical arrangement. The tank is equipped with a float valve to feed and regulate the water level so that contamination of the city water supply is prevented. A small close-coupled pump is mounted directly on the tank and maintains a constant pressure of clear water at the stuffing-box seals of the battery of pumps it serves. A small recirculation line is provided from the close-coupled pump

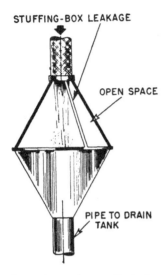

Figure 3.17 Open telltale for stuffing-box leakage.

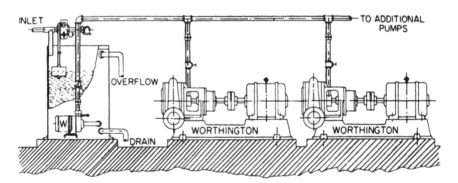

Figure 3.18 Water-sealing supply unit with tank, pump, and float valve.

discharge back to the tank to prevent operation at shutoff. Figure 3.19 is a photograph of such a water seal unit.

In selecting the unit it should be specified to have sufficient capacity plus some margin to supply all the pumps it is intended to serve. The discharge pressure of the small supply pump should be set by the maximum sealing pressure required at any of the pumps served. Supply to the individual stuffing boxes is then regulated by setting small control valves in each individual line.

Figure 3.19 Water-sealing supply unit of Fig. 3.18.

Question 3.16 Sealing Stuffing Boxes with Brine

We are planning the installation of several pumps in a leaching process at our potash plant, which involves the pumping of potassium chloride brine (KCl). The normal operating temperature is 207 °F with a possible occasional maximum of 220 °F.

We favor conventional packed boxes over mechanical seals because of the tendency of the solution to crystallize. Fresh water is at a premium in the area where our plant is located, and we therefore would like to use an NaCl brine solution for the purpose of sealing the stuffing boxes rather than fresh water.

The opinion of our engineers and operators is divided as to the suitability of the NaCl brine as sealing water, and we would like to hear your opinion on this matter.

Answer. The answer to this question will depend mainly on the concentration of the NaCl brine solution that you plan to use. Obviously, if the concentration is about equal to that of seawater, the problem is much reduced, and the brine could be used for sealing. As the concentration increases, the brine becomes less and less suitable for the purpose. It is difficult to establish an exact limit, and you may have to experiment on site to establish that rate of brine dilution that permits you to save the maximum amount of fresh water without causing an undue rate of maintenence of shaft sleeves and packing. Insofar as sleeve maintenence is concerned, you may find it most economical to use hardened alloy sleeves with a coating such as Colmonoy.

There are a number of mechanical seal designs available on the market that can be used for this application despite the tendency of KCl solutions to crystallize. They do require a supply of fresh clear water for flushing, but the quantities involved are quite insignificant. A double seal may actually be the most suitable solution for this purpose.

Finally, I may suggest a solution that will become more and more prevalent in industrial applications, especially in areas where freshwater supply is scarce and expensive, and that is the use of treated sewage effluent. Where municipal sewage treatment provides so-called complete treatment, up to 90% elimination of the suspended solids and up to 95% reduction in BOD (biological oxygen demand) are achieved in the effluent. This effluent can be used very effectively for industrial application after some rather inexpensive supplementary treatment. An outstanding example of this approach is the Nichols steam power plant of the Southwestern Public Service Co. located in Amarillo, Texas, where the effluent from the city sewage plant is used as a source of makeup water for the cooling towers

Chapter 3

serving the steam turbine condensers. At this writing, a flow of 4 million gal per day (approximately 2780 GPM) is used for this purpose. When the plant is expanded, 8 million gal per day of sewage effluent will be supplied to it.

I am certain that there are a number of additional needs for water in your plant, and I would recommend that you investigate the possibility of reducing you freshwater bill by arranging to use treated sewage effluent for all services for which this is practical.

Question 3.17 Vortex Formation in Reservoir at Pump Suction

We frequently have to take pump suction from an open shallow tank with the suction line submerged a minimum distance below the open surface of the water. A good example of this is the suction line connected into the shallow tank of a cooling tower. These pipes are connected into either the bottom or the side of the tank. Are there any data or any rules available for the minimum submergence required for the pipe to prevent whirlpool action or in the case of a side inlet lowering of the water surface? When the foregoing occurs, difficulty with air entrainment also occurs. Any answers to the above or possibly reference to technical papers or test data would be greatly appreciated.

Answer. Before answering your question, my perverse sense of humor has tempted me to insert a few lines of doggerel I wrote in May 1988 on the subject of vortices:

> For my lecture today I shall choose as my text
> The ways and the habits of a simple vortex.
> In our hemisphere, for instance, it's wise
> To expect it will not turn around clockwise.
> But if you go south far enough, I have found
> It changes its mind and goes t'other way round.
>
> I have found in utter sadness
> That it has another madness:
> Just decrease the water level
> It will start to play the devil,
> It will turn around much faster
> And contribute to disaster,
> It will fill a pump with air
> Which then quits in great despair.

In the center there's a hollow
Into which some air will follow.
It will enter into pumps
In small bubbles or big lumps.
When it does, after some time,
Pumps will groan and lose their prime.

It is remarkable that the phenomenon of whirlpool or vortex formation that you describe is well known, but the literature seems to be devoid of direct information on the means to be used in calculating minimum submergences over suction line openings. It may well be, of course, that such information does exist and that I have merely failed to locate it during my career. No doubt that a more systematic search would have uncovered much additional and interesting material. (Incidentally, should any readers have access to this type of information, I would appreciate it if they would so advise me.) The only source that I have found to contain information germane to this problem is the article by Rahm in the bibliography listed at the end of this chapter. Even here, the tests carried out do not exactly correspond to the problem you have raised. The tests that are described were made in a tank to which water was admitted over a weir and past a stilling pool. A hole was made at the center of the bottom of this tank, and a vertical pipe extending upward from the tank bottom was fitted into the hole. The water from the tank was discharged through the pipe under varying conditions of submergence. Much information is presented in this report on the rate of discharge from the pipe and on the various types of vortex formation that occur, but no direct recommendations can be developed from it as to the minimum recommended submergence over the opening.

Of course there is considerable information published on the subject of the submergence and arrangement of wet-pit pumps immersed in open channels. Even though the problem is somewhat different, one might imagine that by extension some of this information could be applied to the particular problem that we are discussing (see the bibliography at the end of this chapter). Unfortunately, even here there seems to be considerable conflicting information and difference of opinion.

Nevertheless, I shall try to give you as much information on this matter as I have and shall suggest certain simple means that are frequently used to avoid the unfavorable effect of this vortex formation.

If water is made to flow out through a hole in the bottom of a vessel or a reservoir, the mass of liquid surrounding the hole generally moves with a vortex motion. This motion is given the name *free spiral vortex* in contrast to the *forced vortex* characterized by rotating a vessel with no opening at the bottom, filled with liquid and made to rotate so that the contained liquid rotates essentially as a solid body.

Any initial disturbance in the water will determine the direction of rotation of the vortex. For instance, a tangential entrance of water *into* the reservoir will superimpose its effect on the flow *out* of the reservoir through the bottom hole and may help determine the rotation. Actually, the earth's rotation tends to establish the vortex rotation if no initial disturbance is present. This is known as the *Coriolis effect*. In the Northern Hemisphere, the vortex rotation is counterclockwise when viewed from above, and it is clockwise in the Southern Hemisphere.

When a vortex first appears, it forms as a small depression, or a dimple, on the surface of the water (as in Fig. 3.20A). It deepens gradually and forms an empty core in its center (Fig. 3.20B). Finally air is drawn down. breaking away periodically to be swept down into the opening (Fig. 3.20C). If the vortex increases in intensity, the air core lengthens to reach the opening at the bottom of the reservoir, and a continuous flow of air takes place through the central portion of the liquid outflow, as in Fig. 3.20D.

The tangential velocities of individual streamlines in the vortex are theoretically inversely proportional to the radius of the streamline, that is, to the distance between the axis of the vortex and the streamline in question. Theoretically, therefore, the streamlines adjacent to the axis of the vortex would have a velocity approaching infinity. Practically, however, surface resistances prevent the velocities near the center of rotation from reaching the high values that would follow this theoretical relationship. Thus, observed velocities on the surface profile of a vortex will always be less than calculated velocities.

This surface profile assumes the well-known general configuration of a hyperbolic funnel in order to satisfy Bernoulli's equation. (An assumption is made that no energy is added or subtracted from the liquid.) If we were to apply certain relationships developed for fluid dynamics, the approximate shape of this funnel could be predicted. Of course, it might be necessary to conduct various tests to establish some empirical correction factors to apply to our theoretical conclusions. Then, if the depth of water above the hole in the reservoir were kept to a minimum greater than the predicted depth of the funnel, the vortexing motion would be prevented from creating too unfavorable an effect on the pumping installation.

This unfavorable effect is caused primarily by the fact that air is drawn into the suction piping through the core of the vortex. As a matter of fact, in addition to any air drawn down through the core, it has been found that air may be separated out of solution from the water because of the reduced pressure in the water in the vicinity of the axis of the vortex. This air may cause the pump to lose its prime or, in general, to operate less satisfactorily.

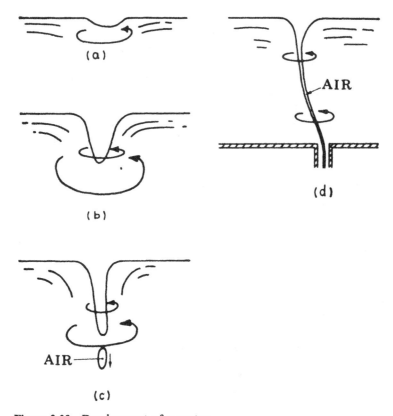

Figure 3.20 Development of a vortex.

Another effect has been observed in such cases: If the water is flowing out of a hole at the bottom of a vessel or reservoir strictly under the action of static head, the rate of flow through the hole is reduced whenever a vortex is formed. This is caused by two separate factors:

1. The effective area of the oriface is reduced by the core of the vortex.
2. The absolute direction of flow through the oriface is no longer axial but inclined at an angle of less than 90° to the plane of the orifice.

This combined effect has been observed to reduce the flow through a hole at the bottom of a vessel to as little as 25% of the flow without a vortex.

Chapter 3 275

Several other interesting observations were made during the tests described by Rahm (see the bibliography at the end of this chapter). Sometimes one and sometimes two vortices were clearly visible in the form of air-filled, rotating cores that appeared in the mass of liquid. The vortex may be centrally above the hole (as in Fig. 3.21A) or to one side (as in Fig. 3.21B). And because the rate of discharge varied considerably with the type of vortex formation, this formation itself being affected by the submergence over the pipe, a cycle consisting of different types of flow was sometimes brought about, even though the rate of supply to the tank was kept unchanged. When this occured, the water surface would fall in the tank until a flow transition took place, after which the rate of discharge was reduced and the level in the tank rose again until transition to the original flow formation took place. Then the level began to fall again.

But, as I mentioned before, I have never seen a simple formula that would relate the depth of the funnel to such variables as the diameter of the hole, the velocity of flow through the hole, the depth of water over the opening, and the size of the hole relative to the vessel or reservoir.

Of course, we could try to convert recommendations developed for wet-pit pump installations to the application we are discussing. By extension, it might be reasonable to set up the following limitations (Fig. 3.22):

1. The hole at the bottom of the sump or tank should be at least one and one-half to two diameters away from the walls.
2. The effective cross-sectional area of the sump should be at least 10 times the area of the opening. This cross-sectional area is the product of the width of the sump and of the depth over the opening.
3. If the velocity through the opening is kept to a reasonably low value (5 ft/sec maximum), the depth over the opening can be allowed to be reduced to 2 times the diameter of the opening.

In addition, it has been found beneficial to bell out the opening at the bottom of the tank. But whether these limitations are certain to prevent the formation of vortices in all cases is difficult to predict, since external disturbances frequently act very unfavorably on the flow distribution at the approach to the opening from the sump. Fortunately, means are available to reduce or eliminate the formation of vortices when they occur. Generally, these means consist of floating a raft or a flat piece of plywood directly above the opening at the bottom of the reservoir. The assistance obtained by such a piece of plywood derives from two separate effects:

1. The friction against the underside of the float retards the vortex velocities very significantly.

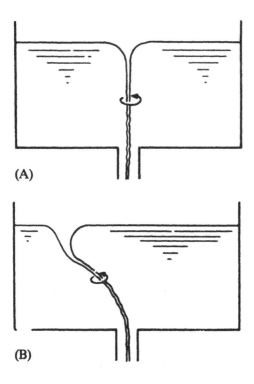

Figure 3.21 (A) Vortex center directly above opening in tank (B) Vortex may also develop off-center of opening.

2. Even if a vortex is formed, the presence of the float prevents air from being drawn into the center of the vortex and being entrained into the suction piping.

The raft should be of ample dimensions to counteract the possibility that vortices may become started at some distance away from the axis of the hole in the bottom of the tank, as in Fig. 3.21B. It is also advisable to provide means to prevent the rotation of the float by anchoring it in some manner while still permitting it to float freely on the surface of the water.

Additional improvement can be obtained by locating some form of splitter or baffle construction over the opening. This will break up the formation of vortices quite effectively.

Since first writing this, I have obtained a plot of the recommended minimum submergence above outlet levels in tanks or above the intake of submerged suction piping. Figure 3.23 represents these guidelines. The advantage of reducing the velocity at the suction piping entrance is quite evident. It

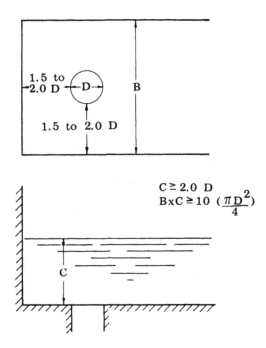

Figure 3.22 Recommended dimensional arrangement for opening in sump.

also becomes obvious that the use of bell mouths at the end of the piping is very effective in reducing the necessary submergence. Suppose, for instance, that the suction piping has a velocity of 8 ft/sec. This requires a submergence of 7 ft. If we use a bell mouth of twice the diameter of the piping, the entrance area becomes four times greater than before and the velocity is reduced to 2 ft/sec, requiring a submergence of only 1 ft.

You should also consider that the formation of vortices in a closed tank is more insidious than when a pump takes its suction from an open reservoir. In the latter case, it is generally possible to witness the vortex formation if it occurs and to quickly diagnose and correct the problem. A vortex in a closed tank is hidden and therefore can escape detection.

Question 3.18 More on Vortex Formation

Your question 3.17 treated the formation of whirlpools or vortices in open tanks from which centrifugal pumps take their suction. Your explanation of the Coriolis effect has me more than worried. If the Japan Current and the

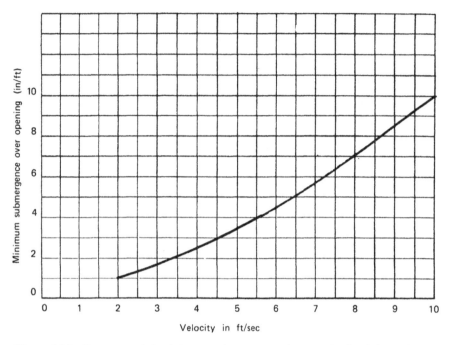

Figure 3.23 Recommended minimum submergence above outlet levels in tanks or above the intake of submerged suctional piping.

Gulf Stream have reversed direction, we are all in a lot more trouble than we realize.

Answer. I think we can all rest safely, as, from last reports, neither the Japan Current nor the Gulf Stream has chosen to change its mind.

My reference to the Coriolis effect came up in the description of the type of whirlpool that develops when water is made to flow out through a hole in the bottom of a vessel or reservoir. Generally, any initial disturbance in the water will determine the direction of rotation of the vortex. The disturbance may, for instance, be caused by a tangential flow *into* the reservoir. If no disturbance is present, the vortex rotation is influenced by the earth's rotation. This phenomenon is known as the Coriolis effect. In the Northern Hemisphere, vortex rotation is counterclockwise when viewed from above and clockwise in the Southern Hemisphere.

I am not certain whether the Coriolis effect plays a part in the formation of ocean currents. To satisfy my curiosity in that respect, I intend questioning an oceanographer at the next opportunity.

As to the validity of the Coriolis effect, I have no question right now because the theoretical explanation has just recently been verified experimentally under the most exacting conditions of accuracy. A full report on the experiments can be found in the December 15, 1962 issue of the British magazine *Nature*, in an article entitled "Bath-tub Vortex" by Ascher H. Shapiro of the Massachusetts Institute of Technology.

The experiment was conducted under perfect laboratory conditions in a perfectly symmetrical, cylindrical, flat-bottomed tank 6 ft in diameter. Precautions were taken to prevent the filling process from setting up any initial disturbance. A settling time of 24 hr was allowed in order to dissipate any residual disturbances. When water began to be drained out through a 3/8 in diameter hole, a delicate cross of two small wood slivers floating on the surface first stood still and then began to move *counterclockwise* (as predicted). Near the end of the 20 min it took to drain the tank, the cross was making one revolution every 3 to 4 sec.

This experiment was conducted in Boston. If need be, it could be repeated in Australia, but I predict that the rotation would be *clockwise* to vindicate Coriolis.

If you cannot put your hands on this particular issue of *Nature*, you might wish to write to Professor Shapiro. Or you may look up a brief description of the experiment in the March 1963 issue of the Arthur D. Little, Inc., *Industrial Bulletin*.

Question 3.19 Air Entrainment into Deep-Well Pumps

In some of our tubwells, the filter pipe is located above the pump bowl, and therefore water enters the tubwell from above as well as below the pump bowl through another piece of filter pipe. We are afraid that this will in some way produce a cavitation effect. I seem to recall that some pump manufacturer's or well drillers' association stipulates that the guarantee covering the pump and/or well will be affected by this. I have tried to locate literature covering this matter but have been unable to trace any reference to it. Would you please advise me about the possible effects of such an installation.

Answer. I do not know of any standards that make a reference to the location of filters above the pump bowl or bowls, nor is there any reference to any guarantees that may be affected thereby.

I believe that the filter pipe you refer to in your letter is the screen or the perforated section of the well casing, which admits water from the water-bearing strata to the well itself. However, the location of this screen above the pump bowl is not unusual.

In some areas, the prevailing water level has been dropping steadily, and therefore pumps may have had to be lowered many times over the past years. Thus, in many locations, the perforated portion of the well casing ends up being above the bowl assembly.

The net effect is that under some conditions water will enter the well through this perforated section of the well casing and in falling will entrain considerable amounts of air. This may affect the pump performance adversely.

Some companies make a statement on their pump curves to the effect that the well water must be free of entrained air. Other companies assume that the customer realizes this limitation and make no specific reference to it on their curves.

Anything that helps to eliminate this condition or to release the entrained air is desirable. In some cases, seals have been provided in the well to prevent air being drawn down the well.

One of the very simple solutions has been to drop several bushels of Ping-Pong balls down the wall. They help to break up the effect of the falling water and thus prevent entrainment of air into the pump (Fig. 3.24).

Question 3.20 Bypass to Drain Instead of to Suction

We are laying out a golf course sprinkler installation that will include a 2 1/2 in. single-stage close-coupled motor-driven pump designed to handle 300 GPM against a total head of 260 ft. The motor is rated 30 hp at 3550 RPM. The pump will take its suction from a reservoir, the level of which is about 20 ft above the pump centerline.

The system will be operated manually. It is intended that the attendant start up the pump with the discharge gate valve open but the sprinkler heads closed. The attendant will then walk over the course, opening such sprinkler valves as deemed necessary. However, circumstances may arise sometimes when as much as 20 min may elapse between the time the pump is started and the first sprinkler is opened.

I am concerned over the possibility that the pump will operate against dead shutoff until the attendant finally opens the first of the sprinkler heads. Under these conditions, the pump could overheat and burn up. I have decided to provide a small bypass line from the pump discharge line back to the suction, with a relief valve in this bypass line. In the event the pump is operated against shutoff, the relief valve will open and discharge a small amount of water back to the suction. Is this arrangement satisfactory? How much bypass is needed?

Chapter 3   281

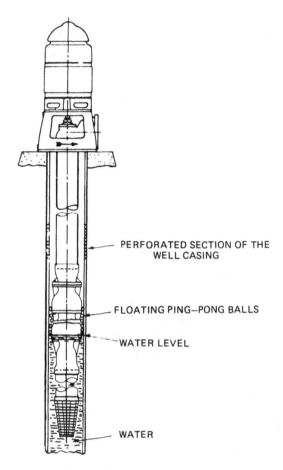

Figure 3.24 Floating Ping-Pong balls can be used to break up the effect of falling water in deep wells. This action prevents the entrained air from adversely affecting the pump performance.

Answer. Your understanding that the pump should not be operated against a closed discharge is correct. However, bypassing the water right back to the pump suction as indicated in Fig. 3.25 will not protect the pump. The heat absorbed by the water on its passage through the pump will not be dissipated, and the water returning to the pump suction will be at a higher temperature than on its first trip through the pump. Thus, on each trip through the pump the water will absorb additional heat until finally the condition you are trying to avoid will be reached.

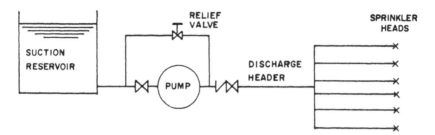

Figure 3.25 Pumping arrangement with relief valve connection to suction (incorrect).

We can even estimate very roughly how long this will take. Assuming that the pump takes approximately 15 hp at shutoff and that we want to limit the temperature rise to about 15 °F, the minimum flow should be set around 5 GPM. (The rule of thumb is that a pump should handle 30 GPM for each 100 hp at shutoff to limit the temperature rise to 15 °F.) If the total circuit from pump to bypass line to suction and back into the pump is about 15 gal in volume, pumping 5 GPM will cause a complete round trip to take 3 min. Neglecting radiation losses, this means that the water temperature will rise at the rate of 15 °F every 3 min, or 5 °F/min. In the space of 20 min, water starting at 75 °F will reach 175 °F and continue rising in temperature.

We can check our calculations in another way. The 15 gal contained in our closed loop weighs 112.5 lb. The shutoff hp is 15 or 42.4 × 15 = 636 Btu/min. Neglecting the effect of radiation losses, the temperature rise of the water will be 636 Btu/min divided by the 112.5 lb or 5.65 °F/min, which is very close to our first approximation.

If we want this heat to be dissipated, we must dump the bypass water somewhere other than into the pump suction. If you do not want to run this line all the way back to the suction reservoir (assuming that the reservoir is large enough to dissipate 636 Btu/min), you must dump it to drain as in Fig. 3.26.

The operation of the bypass can be made responsive to a relief valve as you suggest, or you can install a hand valve in the bypass line; the attendant would open it before starting the pump and close it after the sprinkler heads have been opened.

Question 3.21 Temperature-Rise Control of Recirculation Bypass

We are investigating the use of measured temperature rise across boiler feed pumps as a means of controlling boiler feed pump recirculation. Do you know

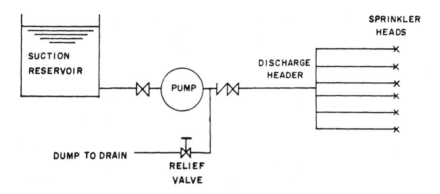

Figure 3.26 Relief valve connected back to drain (correct).

of any installations equipped with this type of recirculation control, and what is your opinion of this form of control? Incidentally, should the recirculation bypass be returned to the heater at the pump suction or to the suction line itself?

Answer. The first reaction to the use of measured temperature rise across a boiler feed pump as an impulse to operate the recirculation bypass control valve is a favorable one. One of the greatest advantages is that such a control becomes most accurate in the range of its greatest need, that is, at the greatest temperature differences between the suction and discharge. On the other hand, the flowmeter control, which is more generally used, is least sensitive at the light flows when the temperature rise requires operation of the control. Another advantage of the temperature-rise control is that it would not be necessary to calibrate it for every single pump to which it is applied, since a recommended maximum rise in temperature would be selected and would then become the impulse that would open the recirculation valve.

Unfortunately, however, temperature-rise controls would have a very appreciable time lag in registering a temperature rise such that the recirculation bypass must be opened. This time lag could lead to dangerous operation of the boiler feed pump at excessively reduced flows. This has been the major reason for the fact that flowmeter control has always been preferred to temperature-rise control as a means for actuating the recirculation bypass valves.

I have no direct knowledge of any boiler feed pump installations embodying temperature-rise controls, but some reader may possibly be familiar with such an installation. In such a case, I would be very pleased to receive any information that may be furnished to me. I do remember a descaling pump installation made some 10 or 15 years ago that included temperature-rise controls. Because the pumps were handling cold water and had a very ample suction pressure, it had been considered less dangerous to use a control with the inherent time lag involved.

The recirculation bypass flow should *always* be returned to the direct-contact heater at the pump suction (or the condenser in the case of closed feedwater cycles) rather than to the pump suction. Otherwise, a cumulative temperature rise would take place, and very shortly the permissible temperature rise would be exceeded. In other words, the bypass should return to some place in the cycle where the heat picked up by the water during its passage through the pump can be dissipated. Generally, an arrangement such as shown in Fig. 3.27 is used. Two valves are installed, one on each side of the orifice that controls the flow through the bypass. One of these valves is locked open and is intended to isolate the orifice only during inspection or renewal. The second valve is the control valve and is opened whenever the flow through the pump falls to a value such as to cause an excessive temperature rise. This valve can be operated either manually or automatically, the last arrangement being preferred at present.

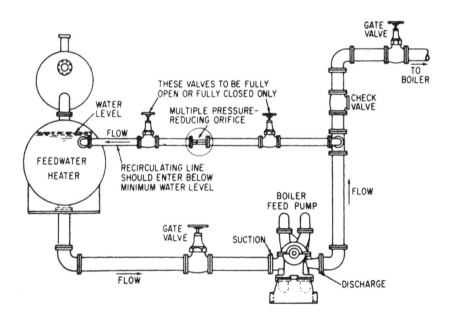

Figure 3.27 Boiler feed pump bypass piping arrangement.

Further Comments Regarding Temperature-Rise Control

The surest way always to be right is to never say anything—or at least never to state anything categorically. But this is not necessarily the best way either of learning anything new or of transmitting information that may be of value to others. The alternative is to present the latest information one has at hand and hope that any inaccuracies may be of an insignificant character and/or that you are promptly corrected by your readers.

And so it has fortunately been in the case of my answer to Question 3.21. I had stated that I had no direct knowledge of any boiler feed pump installations embodying temperature-rise controls to actuate the minimum flow recirculation. I had explained this by the fact that temperature-rise controls have an appreciable time lag in registering a temperature rise such that the recirculation bypass must be opened. This time lag could lead to dangerous operation of the boiler feed pump at extremely low flows.

After publication of this particular answer, I received a letter from Hagan Chemicals and Controls, Inc., pointing to a successful installation of recirculation control based on measured temperature rise in a large steam-electric central station. This installation, I was told, had been in operation for several years.

I am advised that the potential problem of the time lag is met by the use of high-speed thermocouples and by measuring the discharge temperature in the leak-off water from the boiler feed pump balancing device. Finally, the use of a combination of electronic and pneumatic controls further provides for rapid response.

I also received a letter from the U. S. representative of a British firm of valve and control manufacturers, Hopkinsons, Ltd., indicating that temperature-rise recirculation controls have been quite widely utilized in Great Britain. I note, however, in the last case that a dual set of controls is generally used in Great Britain, so that the impulse given by the temperature rise is backed up by a conventional flowmeter control. Whether such double protection is justified is difficult to say.

Question 3.22 Modulating Bypass Valves and Single Flow Orifice

We have an installation of two 100% capacity boiler feed pumps, one of which is on standby service. They are designed for 700,000 lb/hr. The minimum flow control valve opens when flow is reduced to 100,000 lb/hr and stays open until the flow reaches slightly over 200,000 lb/hr so as to avoid hunting. At that increased flow it will close suddenly. This imposes a severe step input to the control system and, if the pumps are on hand control, requires immediate corrective action. Is the expense of proportioning the amount of recirculation not justified?

I would also like to know why we could not utilize the feedwater flow nozzle in the discharge piping after the heaters instead of installing a separate orifice in the pump or discharge line.

Answer. Unfortunately, there is no simple means to avoid this problem if an open-and-closed recirculation valve is used. Its existence has led in a few cases to the installation of modulating control valves that maintain the sum of the flow to the boiler and of the recirculation bypass flow to the minimum specified by the pump manufacturer. The difficulty of this arrangement arises from the fact that the modulating valve is called upon to handle variable flows all the way from the full value of the minimum recirculation flow down to zero. Thus, the valve may be throttling off a pressure from a negligible 50 psi (when it is wide open) to a maximum of full shutoff net pressure of the pump (when it is almost completely closed off). This leads to a high rate of valve wear and is a possible source of high maintenence expense.

There are, however, several valve manufacturers who have developed new designs that are better able to stand up under this type of service. I shall describe and illustrate two of these designs here.

Figure 3.28 is a recirculating valve combined with a check valve and is completely self-contained. The check valve acts as the sensing and powering element for the operation of the recirculating control system. As the flow requirement of the installation is decreased and pump capacity is reduced, the spring-loaded check valve begins to descend toward its seat. The lever arm attached to the check valve stem actuates a pilot valve, which in turn permits recirculation flow to begin. The cascade design of the pressure breakdown element in the recirculation valve dissipates the high-pressure energy. The modulation control is such that the sum of the flow past the check valve and of the bypass flow is never permitted to fall below the prescribed minimum.

Another type of modulating recirculating valve is illustrated in Fig. 3.29. A disk stack in the body of the valve breaks up the flow into many labyrinth passages. The pressure breakdown takes place by virtue of the friction losses in the passages and of the many 90° turns in each passage. A movable plug located centrally of the disk stack and actuated by pneumatic or electric positioning controls regulates the number of disks exposed to fluid flow and this determines the amount of flow through the valve.

One possible solution to reduce the shock of the full change of flow is to use two orifices in parallel, each equipped with its own open-or-closed by-pass valve and to provide sequential operation for these valves as the

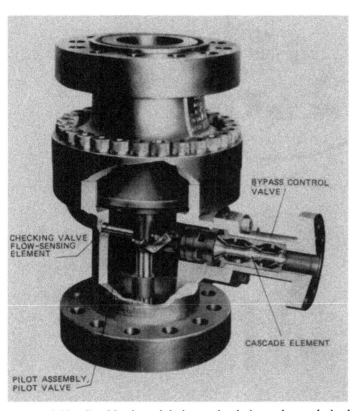

Figure 3.28 Combined modulating recirculating valve and check valve. (Courtesy Yarway Corp.).

Figure 3.29 Modulating recirculating valve with a disk stack. (Courtesy Control Components International).

need arises. The shock to the system will be less severe, but the installation is somewhat more expensive.

If, as in your case, 100% capacity pumps are involved, the feedwater flow nozzle in the discharge piping can certainly be used to actuate a minimum flow signal instead of separate flow orifices being installed in the

Chapter 3   289

pump suction or discharge lines. There is, however, a distinct risk introduced whenever the standby pump is brought on the line to replace the pump that is running. If the transfer takes place while the demand of the boiler is in the low range and pump operation is taking place in the relatively flat part of the head-capacity curve, one of the two pumps could back the other pump off the line, and the check valve in the discharge of the latter pump will close. This pump could suffer severe damage, since the main flowmeter cannot distinguish from which pump feedwater is delivered to it.

Of course, there is a means available to avoid this danger. It consists of a relay introduced into the operation of the individual bypass valves that maintains these valves *open* regardless of flow as long as both pump drivers are energized. This would not lead to any waste of power, since the period of time during which both pumps are on the line is insignificant in the case of 100% capacity pumps.

Question 3.23 Common Recirculation Line for Several Pumps

We are planning an installation of three boiler feed pumps of which two are intended to run at full load and the third is a spare. Must each pump be provided with its own bypass recirculation line, pressure-reducing orifice, and bypass control valve, or can a single common line, orifice, and valve be used for the protection of the boiler feed pumps?

Answer. Decidedly, each pump requires its own recirculation line and controls. There are several reasons for this:

1. If a common recirculation system is used (as in Fig. 3.30), the danger arises that when the flow is reduced nearly to the minimum, one of the two pumps operating may develop a slight excess of discharge pressure and shut the check valve of the other pump, allowing it to run against a fully closed discharge with no bypass.
2. Although two pumps normally operate to carry full load, there will be times when a single pump will be running at loads below 50%. At other times, when pumps are being switched, all three pumps may be running for short periods. If the orifice capacity were selected to pass a flow equal to twice the minimum flow of a single pump, the recirculation would be twice that which is necessary whenever one pump was running alone and only two-thirds of that required whenever all three pumps were on the line. The first is wasteful, and the third does not afford the necessary protection.

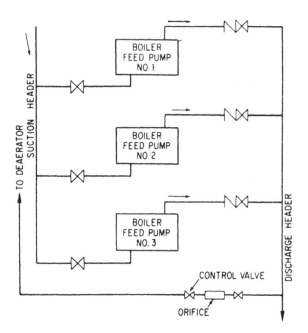

Figure 3.30 Common recirculation system for three pumps without separate bypass lines for each pump is asking for possible operating troubles.

Separate bypass recirculation lines should be provided, originating between the pumps and their check valves, as shown in Fig. 3.31. Each line must have its own orifice and its own control valve. Of course, these individual lines can be manifolded into a single return header to the deaerator on the downstream side of the pressure-reducing orifices and control valves.

Question 3.24 Location of Recirculation Control Valve

The purpose of this question is to raise a minor inquiry in connection with your answer to Question 3.23, in which you discuss boiler feed pump recirculation. You recommend individual pump recirculation rather than the use of a common bypass line and bypass orifice for all the feed pumps serving a single unit. We agree wholeheartedly with this recommendation, and we would never consider deviating from the individual pump control.

My question, with respect to this discussion, deals with the relative location of the control valve and of the orifice. It has always been our practice to locate the control valve on the upstream side of the orifice. Our

Chapter 3

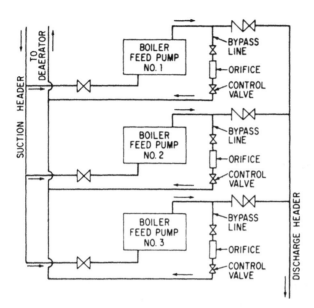

Figure 3.31 Separate bypass lines should serve each pump.

thinking in this matter has been to avoid any possibility of flashing or erosion in the control valve. Since flashing may occur in some cases, we think that it should occur in the breakdown orifice rather than in the control valve. I note that in Figs. 3.30 and 3.31 the control valve is shown to be located downstream of the orifice. This is contrary to our practice as discussed above. While this may be a minor point, I would appreciate your thoughts on this subject.

Answer. You are entirely correct with respect to the preferred location of the control valve, and I must plead guilty of inattention in preparing the sketches used to illustrate Question 3.23. Locating the valve downstream of the orifice may be conducive to erosion because flashing may take place in the valve itself under certain conditions.

Let us examine the pressures and the temperatures that might be encountered in a typical installation, as described in Fig 3.32. You will note that two valves are located in the recirculation line heading back to the deaerating heater, at A and at B, one on each side of the pressure-reducing orifice. One of these is the control valve, and the second one is used for isolation purposes. The prevailing pressures indicated on the sketch are those based on the assumption that the control valve is located at A, which,

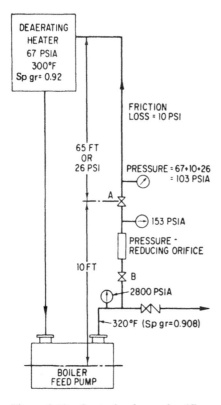

Figure 3.32 Control valve and orifice arrangement for an individual pump recirculation piping system.

as you have stated and I agree, is not the desirable arrangement.

The static pressure at point A, downstream of the control valve and of the pressure-reducing orifice, is equal to the heater pressure plus the static elevation from A to the water level in the deaerating heater plus the friction losses downstream of point A. (We shall neglect the kinetic energy in the piping or in the valve for the sake of simplicity.) In other words, the static pressure at A does not exceed the heater pressure by a very significant amount; in this case the margin is 36 psi. The heater pressure corresponds to the vapor pressure of the feedwater at the suction temperature. The temperature of the feedwater in the bypass recirculation line is considerably higher than at the pump suction by virtue of the following:

1. The temperature rise caused by losses in the pump at reduced flows

2. The temperature rise caused by the compression of the feedwater in the pump
3. The temperature rise caused by the degradation of energy through the presure-reducing orifice and the control valve

The vapor pressure corresponding to this increased temperature at point A is therefore higher than the pressure in the heater. Thus, the static pressure at point A can easily be equal to or even lower than this increased vapor pressure, and flashing will take place as soon as this happens. This flashing would lead to erosion of the control valve if it were located at point A rather than upstream of the orifice at point B.

Let us assume, for instance, that the sum of the temperature rise due to pump losses at minimum flow and of the temperature rise caused by compression is 20 °F. The water will leave the pump at a temperature of 320 °F and at a pressure of 2800 psia. The throttling process through the pressure-reducing orifice and through the control valve at A is a constant-enthalpy process. We can determine the new state downstream of the valve at A from the extended charts in *Thermodynamic Properties of Compressed Water,* by T. C. Tsu and D. T. Beecher, published by the American Society of Mechanical Engineers. From the plate VII, the original state conditions of p_1 = 2800 psia and t_1 = 320 °F give us h_1 = 295.2. At a constant enthalpy and reduced pressure, t_2 = 324.8 °F. The temperature rise caused by the throttling will have been 4.8 °F.

Consequently, the feedwater in the bypass recirculation at point A will be at a temperature of 324.8 °F. The vapor pressure corresponding to this temperature is 95.1 psia. Although this is still lower than the 103 psia at point A and flashing may be avoided, the situation is too close for comfort. Had we assumed that the temperature at the exit of the boiler feed pump were 326 instead of 320 °F, the temperature at point A would have been 330.8 °F with a correspomding vapor pressure of 104.4 psia. The static pressure at point A would have been insufficient to prevent flashing in the valve. It is obvious, therefore, that in order to ensure against flashing and erosion in the control valve, the latter should always be placed upstream of the pressure-reducing orifice.

Of course, swapping positions of control valve and orifice transfers the burden of withstanding the potential hazards of flashing and of erosion to the pressure-reducing orifice. But even if flashing does some damage to the orifice, this is a less expensive component than the control valve, and its functionings is not affected by some erosion as would be the case with the valve. At worst, it will pass somewhat more feedwater, and it can ultimately be replaced at a reasonable cost.

Finally, if it is desired to make the installation completely safe against flashing in either the control valve or the pressure-reducing orifice, it is only necessary to install another orifice as near as possible to the dearating heater so as to increase the back pressure slightly at point A. This second auxiliary orifice could be selected to have a pressure drop of some 30 to 50 psi when passing the design recirculation flow. With such an arrangement, no flashing will ever take place except at the auxiliary orifice.

Question 3.25 Location of Flowmeter for Bypass Control

Our boiler feed pumps operate in a closed feedwater cycle. What are the relative merits of a flow nozzle installed in the discharge line of a boiler feed pump versus an orifice installed in the suction line if the installation is used to meter boiler feed pump recirculation?

What is your opinion of relocating an existing installation from the discharge to the suction side of the boiler feed pumps? Would it be feasible to leave the flow nozzle in the discharge piping?

Answer. Although the unrecovered losses through a flow meter are very low, they may still have a significant value in terms of the available NPSH, and for this reason it is preferable to locate the flowmeter in the discharge piping whenever the margin between the available and the required NPSH may become critical. This is generally the case in open feedwater cycle installations but very rarely in closed feedwater cycles, as in your case. Because the cost of the flowmeter increases appreciably if it must be designed for full discharge pressure, it is more logical to install it in the suction piping when closed feedwater cycles are used.

Thus, since your installation probably has an ample margin over the required NPSH, there would be no harm from relocating the flowmeter to the suction. Nor can I see any disadvantage from leaving the present flowmeter in the discharge and using a new flowmeter in the suction. On the other hand, since you have already incurred the added expense of a flowmeter suitable for the discharge pressure, I can see no particular reason for relocating it and incurring further expense.

Should you, however, relocate the flowmeter into the suction piping, it will be necessary to reset the actuating contacts of your automatic controls so as to obtain proper operation. The reasons for resetting these contacts will become apparent from the following discussion.

The basic purpose of the bypass recirculation is to ensure that the flow through a boiler feed pump may not fall below a certain predetermined minimum. This minimum is established to hold the temperature rise across the pump to a selected maximum, regardless of the flow demand of the boiler. The bypass line branches from the discharge piping at some point between the discharge nozzle and the check and gate valves. In

Chapter 3

today's steam power plants, the recirculation bypass is controlled automatically. To this end, we must provide a means to measure the flow through the boiler feed pump either directly or indirectly.

A flowmeter located in the suction of a pump (Fig. 3.33) will give us a direct measurement. So will a flowmeter in the pump discharge as long as it is located ahead of the branching out of the recirculation bypass. However, this is seldom the case, because it is desirable to locate the flowmeter in a straight run of pipe reasonably far away from the disturbing influence of check and gate valves. This generally precludes the location of the flowmeter between the pump and the check valve.

If the flowmeter is located beyond the point at which the recirculation line leaves the discharge line (Fig. 3.34), it will give an indirect indication of the flow through the pump, since it will measure the actual flow through the pump only when the control valve in the bypass line is closed. When this valve is open, the flow through the pump will be equal to the sum of the flowmeter reading and of the flow through the bypass.

Expressing this in different words, we can say that as long as the flowmeter is located ahead of the bypass, it measures the pump flow, which is the sum of the flow to the boiler and of the recirculation bypass. If the flowmeter is placed beyond the bypass, it measures only the flow to the boiler.

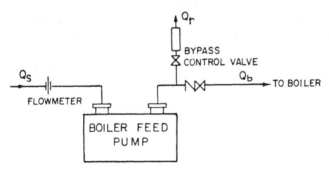

Q_s = FLOW THROUGH THE PUMP = $Q_b + Q_r$

Q_b = FLOW TO THE BOILER

Q_r = FLOW THROUGH RECIRCULATION BYPASS

Figure 3.33 If the flowmeter is located in the suction, it reads the total flow through the pump.

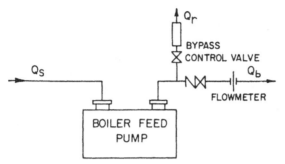

Figure 3.34 When the flowmeter is in the discharge, it measures only the flow to the boiler.

Let me add a qualification. If the pump is provided with a balancing device, the leak-off from this device is returned either to the deaerator at the suction (in an open feedwater cycle) or to the pump suction itself, quite close to the suction nozzle. In the first case, the flowmeter in the suction would measure the total flow through the pump, including the balancing device leak-off. In the second case, it registers the net flow through the pump, that is, the flow at the discharge nozzle. However, in our discussion, the matter is almost academic because the balancing device leak-off is generally but a fraction of the required minimum flow and because the pump manufacturer will have taken this matter into consideration in establishing the recommended minimum flow. Thus, in the discussion that follows, we shall neglect the effect of the balancing device leak-off.

This distinction between the measurements indicated by the flowmeter depending on whether it is located in the suction or the discharge piping has an important effect on the operation of the bypass control valve. This valve must be made to open whenever the total flow through the pump falls to a particular minimum. It can be *permitted* to close when this flow exceeds the minimum. But the opening and closing of the recirculation valve affects the flow through the pump to a major extent. Unless we arrange the control to take this fact into consideration, we shall impose a hunting condition on its operation. Consider, for instance, a boiler feed pump installation that requires a minimum flow of 300 GPM through the pump. If starting from a higher capacity, the flow to the boiler is reduced until the flow through the pump drops to 300 GPM, the bypass recirculation valve opens, and 300 GPM starts flowing through the recirculation line. If, then, the boiler demand stabilizes at just below 300 GPM—say at 250 GPM—the total flow through the pump becomes 550 GPM, which is the sum of the 300 GPM through the bypass and the 250 GPM to the boiler. Yet we cannot permit the bypass control valve to close, since the total flow would immediately fall back to 250 GPM, which is less than the permissible minimum flow.

To avoid hunting of the bypass controls, we must provide two separate contacts in the flowmeter, one of which will cause the opening of the recirculation bypass and the second its closing. The setting of the first contact will correspond to the minimum permissible flow through the pump. the choice of the second setting will depend on the location of the flowmeter. If it is located in the suction of the boiler feed pump, the flow through the pump (and therefore the flowmeter reading) doubles as soon as the bypass opens. The closing contact must therefore be set at a flow equal to twice the minimum permissible flow plus a reasonable small margin. For instance, if we consider the example cited above, the control valve should be made to open at a flowmeter reading of 300 GPM. The reading will immediately change to 600 GPM if the flow to the boiler does not change further. The valve should therefore not be permitted to close until the total flow has increased to, say, 700 GPM, of which 300 GPM would be going through the bypass and 400 GPM to the boiler.

If the flowmeter is located in the discharge, beyond the branching out of the bypass, the first contact at 300 GPM should open the bypass. The second contact should be set at 400 GPM and will close the bypass whenever the flow to the boiler rises to this value.

Note that, regardless of the location of the flowmeter and of the setting of the second contact, the flow through the pump and hence the power consumption remain the same under all operating conditions. Once the bypass has been opened, it will remain open until the flow to the boiler has increased back to 400 GPM, at which moment the flow through the pump will have increased to 700 GPM. Only the setting of the second contact will change with the flowmeter location.

All the preceding was predicated on recirculation controls utilizing a fully open or fully closed control valve. Much effort has been exerted in recent years to develop a *modulating* valve that could throttle the recirculation bypass flow and vary it so as to hold the sum of the bypass flow and of the flow to the boiler equal to the minimum permissible flow. Such a modulating valve is more economical of pump power consumption and also avoids the hydraulic shock that accompanies the sudden opening and sudden closing of control valves currently used for this service. The problem had been to find a valve that can be used on modulating service without becoming worn in an exceedingly short time. Modulating valves of improved design have been installed in steam power plants and have given a reasonably satisfactory account of themselves. If modulating valves are used, the location of the flowmeter will again affect the manner in which the impulse from the flowmeter is made to control the valve. If the flowmeter is in the suction, it reads the total flow directly, as we have seen, and the control need merely be directed at preventing the measured flow from falling below the minimum recommended. If the flowmeter is located in the discharge, it reads flow to the boiler. The impulse must be directed to position the control valve so that the bypass flow to the boiler is prevented from falling below the minimum. This requires a slightly more complex arrangement.

Question 3.26 Balancing Device Leak-off Return

In one of your earlier articles, I recall that you recommend returning the balancing device leak-off boiler feed pumps to the direct-contact heater from which they take their suction. I have heard recently that you now recommend returning this leak-off to the pump suction instead of to the deaerator. Is this true, and if it is, could you explain the reason for this change in your recommendations?

Answer. Your information is quite correct. For many years it had been the practice to recommend returning this leak-off to the direct contact heater, but more recently the practice has been changed.

Whenever multistage pumps are equipped with a balancing device to counteract the axial thrust generated by a group of single-suction impellers all facing in the same direction, it is necessary to return the balancing device into the cycle at some point where the pressure is approximately equivalent to the suction pressure. In the feedwater cycle of a steam-electric power plant, the choice of this point of return is limited to the pump suction itself and to the deaerating heater.

The earlier recommendation to return the balancing device leak-off to the deaerating heater was based on the following:

1. The balancing device leak-off undergoes a significant temperature rise as it flows through the balancing device because of the degradation of pressure energy. In a pump developing a total head of 6500 ft, this temperature rise would be in excess of 8 °F and of course, would be in addition to the temperature rise that takes place in the pump itself. Thus, if the minimum flow were set at a 15 °F rise through the pump, the total temperature rise in the leak-off could theoretically be as high as 23 to 25 °F. If this leak-off is returned to the pump suction, it will mix with the incoming feedwater and raise the temperature of the mixture. If the pump had been selected fairly close to the margin from the point of view of the NPSH conditions, this rise in the suction temperature could be dangerous.

2. On the other hand, the increase in the heat content in the balancing device leak-off could be beneficial if this leak-off is returned to the deaerator because it would mitigate to some extent the unfavorable effect of the transient operating conditions that occur following sudden load rejections or even sudden significant load reduction.

A number of reasons have led to the recent change of opinion with respect to where the balancing leak-off should be returned. To begin with, the first argument has lost much of its importance because minimum operating flows are no longer set strictly by temperature-rise considerations but are dictated by possible unfavorable hydraulic perturbations in an impeller operating in the lower range of capacities. Thus, the minimum operating capacity of a boiler feed pump today is set at a value at which the temperature rise is considerably lower than 15 °F.

In addition, most boiler feed pumps today use condensate injection-type seals. The inward leakage of cold condensate acts to reduce somewhat the temperature of the leak-off.

The second argument is likewise less significant, since power plant designers have learned how to design the feedwater system to avoid the unfavorable effects of sudden load rejection.

At the same time, some of the disadvantages of returning the balancing device leak-off to the deaerator have been better understood. To begin with, the temperature rise in the leak-off can lead to flashing in the return line unless proper precautions are taken. There is always the hazard that the isolation valve in this line may remain closed by error after a pump has been out of service for maintenence. Starting the pump with the valve closed can lead to a major failure, since the pressure in the balancing device chamber would reach full discharge pressure. Certain other potential dangers are eliminated when the balancing leak-off line is returned directly to the pump suction. (See Question 6.27 for details of this case.)

Finally, there are conditions in which a boiler feed pump may serve as a start-up or standby for two adjacent units. Returning the leak-off to the pump suction simplifies matters considerably, as it avoids the necessity of transferring the leak-off return as well as the suction and discharge from one unit to the other. The last two maneuvers are both necessary and readily remembered. The former may be overlooked.

Of course, the main advantage of returning the balancing device leak-off directly to the pump suction is that the line will be very short. As a matter of fact, it can be made an integral part of the pump piping, as in Fig. 3.35.

If in spite of this reasoning it is decided to return the leak-off to the dearator, there are certain precautions that must be taken. The manufacturer

Figure 3.35 Balancing device leak-off can be returned directly to the pump suction.

of the deaerating heater should be given full information as to the amounts and temperatures of this leak-off so that it can be provided for in the heater. The manufacturer will have to decide whether to introduce it below the waterline in the storage space or to bring it in above the water level and provide some sort of perforated pipe through which to spray the returns into the steam space. The last arrangement is frequently used when returns to the heater are at a much higher temperature than that corresponding to the heater pressure. It is important to give the heater manufacturer both the quantity of leak-off under *new pump* conditions and the estimated flow when the pump is worn.

The return line should be so arranged that the back pressure on the balancing device may not be inadvertently increased over the value of the suction pressure. Therefore, although this return piping needs to be valved off to permit the dismantling of the pump when necessary, any valve interposed between the balancing device relief chamber and the final point should be securely locked in the open position. The operators should be warned to make sure that if this valve is ever closed during inspection or overhaul, it must be opened again before the pump is allowed to be started up.

To reduce piping costs, the leak-off and recirculating lines are sometimes manifolded, as shown in Fig. 3.36. Of course, the manifolding must take place after the pressure-reducing device and after any of the valves located in the recirculation line.

Question 3.27 Check Valve in Balancing Device Leak-off Line

We would like to have your comments on the installation of a check valve in the boiler feed pump balancing device leak-off line. We are particularly interested in installations in which the leak-off from the balancing device is returned to a deaerator. Any comments you may wish to make regarding the advantages or the disadvantages of the check valve in the balancing device leak-off line between the boiler feed pump and deaerator would be appreciated.

Answer. As I have explained in my answer to Question 3.26, there are few cases in which the leak-off should be returned to the deaerator. But even in these cases, I can frankly see no basic need for a check valve in the return line. When the pump is running, the flow through the leak-off line will take place in the normal direction. When the pump is brought to a standstill, pressures are equalized, and I do not expect any flow to take place through the leak-off line in either direction.

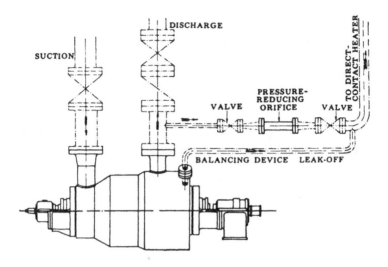

Figure 3.36 Balancing device leak-off and recirculating lines are sometimes manifolded to reduce piping costs.

There are only two possible exceptions to this statement:

1. If a booster pump is used ahead of the boiler feed pump, if the balancing device leak-off is returned to the deaerator, and if the booster pump is kept running while the feed pump is stopped, some flow will take place in the leak-off line. This flow will be in the normal direction.
2. If the feed pump is kept warmed up by means of a jumper line around the main check valve, with an orifice in this jumper line, a slight flow will again take place in the leak-off line and again in the normal direction.

The only time I can visualize that the check valve will become operative would be if the pump were dismantled for inspection and the operators had forgotten to close the isolating gate valve in the leak-off line. This is such a farfetched supposition that I frankly see no justification for the expenditure involved in installing a check valve in the leak-off line.

Question 3.28 Balancing Device Leak-off from a Lean Oil Pump

We are at present operating barrel-type pumps with a balancing device in our refinery under the following conditions:

Liquid, 0.8 sp gr lean oil

Capacity, 750 GPM
Suction pressure, 170 psig
Discharge pressure, 1550 psig
Suction temperature, 100 °F

When our present plant expansion is completed, the pumps will operate under slightly different conditions:

Liquid, 0.8 sp gr lean oil
Capacity, 800 GPM
Suction pressure, 35 psig
Discharge pressure, 1550 psig
Suction temperature, 100 °F

The slightly increased capacity and total head will be met by speeding up the turbine driver. The pumps are equipped with a 2 in. balancing return line from the balancing chamber opening in the discharge head that leads to the pump suction. A volume of oil estimated to be about 100 GPM is recirculated through this line and through each pump. We are considering the possibility of feeding the oil from the balancing chamber to a reabsorber, operating at 75 psig, instead of to the pump suction.

We realize that the correct equilibrium of forces in and around the hydraulic balancing device is required to ensure balancing of the axial thrust. However, it appears that the balancing device is not expected to take care of all the axial thrust, since the pumps are equipped with Kingsbury thrust bearings (Fig. 3.37).

We wish to know if when the suction pressure is 35 psig and the discharge pressure 1550 psig, the balancing chamber may be maintained at about 90 psig instead of about 40 psig.

In the event that it is permissible to connect the balancing chamber to a reabsorber instead of to the pump suction, we intend to maintain constant pressure on the balancing chamber by means of a back-pressure regulator and to protect the balancing line from a complete shutoff (or even just excess back pressure) by installing a relief valve that will discharge to the pump suction.

We shall appreciate it very much if you will supply an answer to our question.

Answer. In principle, there should be no major difficulty caused by reconnecting the balancing device relief line to the reabsorber and thus raising the back pressure on the balancing chamber from 40 to 90 psig. The effect of this increased back pressure is of no major significance insofar as the balancing device operation is concerned, and as you have stated, the Kingsbury thrust bearing should take care of any change in the equilibrium of forces.

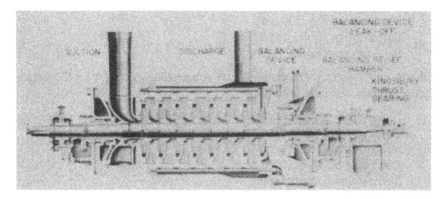

Figure 3.37 Balancing axial thrust in a barrel pump.

However, there are a number of factors that must be considered before you make the intended change. To begin with, I am not clear on what you have based your estimate of 100 GPM balancing device leak-off per pump. If the reabsorber requirements are 100 GPM per pump, you may be disappointed because such a balancing device generally passes from 35 to at most 50 GPM when in new condition. Of course, this quantity will vary with the internal clearances and is generally permitted to increase to double its initial value before renewal of parts is indicated. Thus, if 100 GPM per pump is required, you may find it necessary either to bleed additional oil from the discharge of the pump or to open up the internal clearances beyond normal recommended values.

Another factor that must be taken into consideration is the increase in oil temperature resulting from the degradation of pressure energy in the flow through the balancing device. The temperature of the oil from the balancing chamber will be from 8 to 10 °F above the oil temperature in the pump discharge. This temperature rise and its effect on the reabsorption process must be examined.

Your plan to maintain a constant pressure on the balancing chamber by means of a back-pressure regulator and to provide a relief valve in the balancing line is excellent. The relief valve will provide protection against the possibility that someone might close the valve between the pump and the reabsorber. However, this precaution should be supplemented by a hand-operated valve in a bypass line to the pump suction. The pump should be started with this bypass line open and with the line to the absorber closed.

Chapter 3305

There are two reasons for closing the line to the absorber when a pump is standing idle:
1. With 40 psig suction and 75 psig in the balancing chamber while the pump is idle, there would be a tendency to move the rotor axially toward the suction, leading to contact at the balancing device faces. If the pump starts with faces touching, these will gall, and the balancing device will fail.
2. If the reabsorber pressure is 75 psig and the pump is standing idle under 40 psig pressure, gas will blow back into the pump and vapor bind it unless the connection is valved off.

After the pump is on the line, the valve to the reabsorber can be opened and the bypass back to the suction closed. Likewise, prior to shutting down a pump, the bypass valve should be opened and the connection to the reabsorber closed.

If all these precautions are taken, and if the oil quantities obtainable from the balancing device leak-off are ample or can be supplemented, the arrangement will be satisfactory.

Question 3.29 Relief Valve in Boiler Feed Pump Suction Line

Is it common practice to install a relief valve in the suction piping of a boiler feed pump?

Some of our engineers are in favor of such a relief valve as protection against the building up of discharge pressure in the suction line when the suction line valve is closed. Others, including myself, think that our present system incorporates all the normal protective features required and that the addition of the relief valve is not justified. Building up of discharge pressure in the suction line can occur only under certain operating conditions and then only when valves have been operated incorrectly.

Figure 3.38 shows the system as now installed. Normally, one pump is on standby service with both suction and discharge valves open. Under these conditions, the discharge pressure cannot build up in the suction line. If the suction valve is closed, discharge pressure can build up in the pump and suction line only when one of the folowing conditions exists:

1. Hand valve in recirculation line closed
2. Recirculation control valve closed
3. Leakage past check valve more than 130 GPM

The hand valve is a *locked open* valve that, under normal operation, should never be closed. The recirculation control valve is open with no

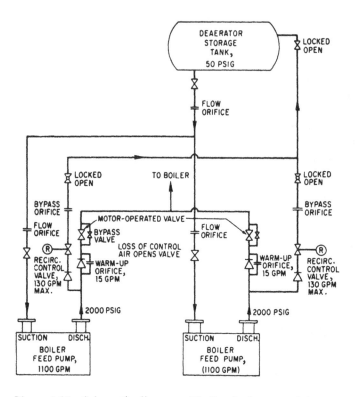

Figure 3.38 Schematic diagram of boiler feed pump piping.

flow in the suction line. This valve also fails "safe" in the open position should there be loss of control air. This leaves only excessive leakage through the check valve as a cause of excess pressure in the suction line. Since the pump operator should never close the suction valve without first closing the discharge valve, this condition is also eliminated.

To prevent reverse rotation in the pump due to excessive leakage past the check valve, it is our practice to close the discharge valve immediately after stopping a pump. The valve is then opened, and a careful check made to see that the pump is not rotating in reverse.

Your comments on this problem would be most helpful and greatly appreciated.

Answer. The practice is very far from being common, although I have seen relief valves incorporated in a few installations. It is my opinion, very frankly, that there is no justification for its use in the case you have described. As you say, it would require an unusual accumulation of circumstances and of operational errors to cause any difficulty in the absence of a relief valve. Such an accumulation would come under the heading of triple or quadruple contingency, and power plant designers cannot afford to protect a steam power plant against this. After all, there are many more areas in a plant where protection against multiple contingencies would be equally justifiable and where no such protection is employed.

In general, my reaction to a problem such as this is to start with the proposition "It is always a risk to operate a steam power plant." If we wanted to eliminate *all* risks, we could not afford to build one. Therefore, our real problem is to properly classify all the individual risks we shall encounter and establish which of these risks we can afford to eliminate, which risks we can only minimize, and which ones we must endure. My opinion is that, in the case you have described, the risk of encountering a combination of circumstances that would cause difficulty if no relief valve were provided falls into the group of risks we must endure. We have already done all that is practical to minimize it.

The only arrangement in which more serious consideration might be given to the use of a relief valve is that in which a single spare pump is used for two adjoining units (see Question 3.30 and Fig. 3.39). Here, there are two suction valves, one in each of the lines from the two deaerators. Normally, one or the other suction valve is kept open, but there is a greater chance that both valves might be closed by error. Since the recirculation lines (not shown on the illustration) will also be provided with alternative connections to the deaerating heaters so as to return the bypass flow to the proper unit, there is also a greater chance that both lines may be blocked closed. In this case, then, including a relief valve may be considered as sound protection. But even here I believe that the valve is optional and not mandatory.

Reverse rotation caused by leaking or sticking check valves is a matter that has frequently caused me concern. I am therefore very much in agreement with your procedure to close the discharge valve on a pump that is taken off the line until the operators are sure that the check valve is operating properly. I actually prefer to see the valve closed first and *then* the driver tripped out, assuming that we are dealing with a scheduled operation, not an emergency. Even with a motorized valve it may take up to 30 sec for complete closure. If the check valve is faulty and the discharge valve is *closed* after the pump is tripped out, by the time the flow is interrupted, the damage from reverse rotation may already have occured. The discharge valve can be closed while the pump is running, since the automatic bypass recirculation protects the pump against operation at shutoff.

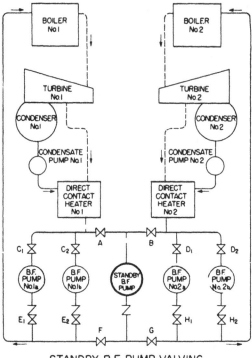

Figure 3.39 Boiler feed pump valving for standby pump serving two adjacent units.

Question 3.30 Interconnected Spare Boiler Feed Pump

Can a spare boiler feed pump be used for two adjacent and duplicate units? If the answer is yes, what are the precautions that need be taken in such an arrangement?

Answer. There have been a number of installations in which a single spare pump has been used in common for two units. This solution has been selected because of a desire to reduce first-cost investment and because the reliability of centrifugal boiler feed pumps had increased to a marked degree in the last years. It has been resorted to particularly in cases in which a 100% capacity turbine-driven pump is used to serve the boiler and a part-capacity standby or start-up motor-driven pump is desired or required. Such an arrangement is quite feasible but does present certain operating problems that must be taken into consideration in laying out the feedwater cycle.

The general arrangement of a two-unit open-cycle power plant with a common standby feed pump is illustrated in Fig. 3.39. For the purpose of simplification, only the valves that have a bearing on these problems have been indicated, and for the same reason, alternative feedwater routing lines, closed-stage heaters, and similar equipment have all been omitted.

If the starting of the standby pump were to be a prescheduled event or if the time element required to bring this pump on the line were of no specific importance, the procedure would be extremely simple. The valves connecting the standby pump suction and discharge into the unit feed system in which it is desired to operate the pump would be opened and the pump started up after the operators were assured that it will take its suction from the correct heater and discharge to the correct boiler. The valves connecting the standby pump to the other unit system would remain closed, and after the pump was started, the necessary maintenance or inspection would be carried out on the pump temporarily replaced by the standby pump.

Unfortunately, such is not always the case, and the installation of a standby pump should provide for the eventuality that its operation may be called upon under emergency conditions and that the power plant operators will find no time to manipulate any of the valves in question. In other words, the installation should be arranged in such a manner that the standby pump can be started up automatically if any one of the pumps on normal service fails to develop sufficient pressure.

Since the standby pump may have to take its suction from either of two deaerating heaters and since the two units may be operating at different loads, the suction lines to the standby pump must be valved off to prevent flow from one heater to the other. It is generally not recommended that a pump have both suction valves closed. Therefore, only one of the valves is kept closed, and the second valve is left open. Means must be made available to exchange the position of the two suction valves if it becomes necessary to include the standby pump into that unit feed system from which the suction line was closed off previously. On the other hand, both discharge valves may be kept closed or one of them remain open, connecting the standby pump to the discharge header of one feed system and isolating it from the other.

For instance, in the system illustrated in Fig. 3.39, valves A, B, F, and G may be normally closed. Whenever the standby pump is called upon to operate with unit 1, valves A and F are opened up, while valves B and G would be opened to use the pump in unit 2.

As an alternative arrangement, valves A and F may be left permanently open, so that the standby pump is always available for service with unit 1 and they are only closed and valves B and G opened whenever the pump is needed to feed boiler 2. In the latter arrangement, the pump is connected to the right unit feed system for 50% of the emergency cases that may arise, which is a statistical improvement over the first arrangement.

Normally, the switching valves A, B, F, and G are motorized and arranged to operate automatically on reduction in the discharge pressure of either one of the feed pump headers operating in units 1 and 2. The same pressure switch arrangement that is used to operate the necessary valves is used to start the standby pump, except that it is necessary to interpose a suitable time-delay mechanism into the starting circuit to assure that the pump cannot be started until the proper valves have been opened or closed (the latter to isolate the standby pump from the unit that has not called for its operation).

The various factors that may introduce operating problems in such a system are the following:

1. If both suction valves of the standby pump are kept closed until the need for putting it into operation arises, the proper suction valve opens only a fraction of a minute before the pump is put on the line.
2. If one of the suction valves is kept constantly open, it may be necessary to close it and open the corresponding valve in the adjacent unit system prior to the starting operation.
3. The same remarks apply to the condition of the discharge valves.
4. Since the standby pump may be taking its suction from either heater 1 or 2, the recirculation bypass lines must be aranged to return the bypass flow to either of the two heaters.
5. If the pumps are provided with hydraulic balancing devices and if it is preferred to return the balancing leak-off to the heater at the suction of the pump instead of directly to the pump suction, means must be provided for switching the leak-off from one heater to the other.
6. If the standby pump is kept warmed up by passing feedwater through it, since one of the units may be operating at a higher turbine load than the other and hence with a higher feedwater temperature, the standby pump may under some conditions contain feedwater at a higher temperature than the heater from which it may be called to take suction.

Let us now examine these problems in detail. If we assume that both suction valves remain normally closed, and if the standby pump is operated at very rare intervals, leakage of feedwater through the stuffing boxes may empty the pump casing to such an extent that the pump will be partially air bound. The pump will then seize immediately upon starting up, and the resultant damage may be extensive, to say nothing of the unavailability of the pump at the exact moment when it is most needed.

To prevent such an occurence, it may be preferable to keep one of the suction valves open at all times. As an alternative arrangement, a small bypass can be installed around one of the suction valves. If a line leads from a connection on the pump casing to a point where the pressure is always lower than the heater pressure, this line will also serve to keep the spare pump warmed up. It is necessary, however, to install a small orifice in this return warm-up line in order to reduce the pressure imposed upon the vessel at its termination when the pump is put into operation.

The fact that under certain conditions the sequence of operations requires the closing of some valves and the opening of some others before the standby pump is put into service should be carefully considered in the light of the time element involved in the closing and opening of the valves, since this time element will impose a definite delay in the starting of the pump and since the magnitude of this delay may impose a definite danger on the boiler. In this connection, it may be noted that the existence of such a time delay is less vital for an installation in which two half-capacity pumps are normally serving each boiler. In such a case, the standby pump would be called into service on failure of one of the pumps, and one pump would still be feeding the boiler, delaying the moment when interruption of service would result in a dangerously low boiler level condition. These considerations would indicate that the use of a common standby pump for two adjacent units is safer when half-capacity pumps are used than when a single full-capacity pump is installed for each unit.

Insofar as the recirculating bypass line is concerned, the problem is not as serious as it is frequently assumed to be. In the first place, the chance that the standby boiler feed pump may be called into service under conditions requiring the use of the recirculating system is less than that of operation at reasonably safe flows when bypass protection is not necesary. This is especially true when half-capacity pumps are used, since only one of the two pumps would be on the line at that time and the second half-capacity pump would be available for immediate starting. In such a case, the station operators would normally have sufficient time to reroute the recirculation return to the specific heater from which the pump is taking its suction. However, even if the standby pumps were to be started up under such flow conditions that the recirculating line is open and leads to the wrong deaerator, no serious damage will be done if for a few minutes the standby pump robs the heater from which it is taking suction in favor of the other heater by the amount of the bypass.

This statement can be best demonstrated by selecting a hypothetical example. If the total boiler feed pump flow were 1 million lb/hr and the minimum flow (bypass) 150,000 lb/hr, and the deaerating heater had a 5 min storage, or 83,333 lb of feedwater, the operation of the recirculating line into the wrong heater would draw on the heater storage at the rate of 2500 lb/min. In a matter of 5 min, it would have reduced the storage by about 15% and we can assume that 5 min is sufficient time in which the operators can open the recirculating line to the correct heater and close the other connection. Obviously, to avoid interconnection of the two heaters through the recirculating lines, a check valve should be located in each one of the lines after the bypass from the standby pump bifurcates to the two heaters.

The problem of the hydraulic balancing device leak-off can be solved by returning it directly to the pump suction, in which case it is always correctly piped up and no necessity arises for the manipulation of any valve.

The problem of the warm-up flow through the standby boiler feed pump is relatively complex. Under normal conditions, there are two methods in wide use for maintaining a standby pump at the temperature of the pump or pumps on the line:

1. A warm-up valve is located on the pump casing, leading to some point in the feed cycle where the pressure is appreciably lower than in the heater. Warm-up flow takes place from the heater, past the open suction valve, through the pump, and out through the warm-up valve.
2. A jumper line is installed around the pump check valve. The warm-up flow takes place from the discharge header, the pressure-reducing orifice, through the pump, and out into the suction header to the pumps on the line. (See Fig. 3.40.)

The second system is quite dangerous in that there will possibly exist times when both suction valves are closed. Unless a relief valve is interposed somewhere (and power plant operators are generally not too prone to rely on pressure relief valves), as soon as both suction valves are closed, the pressure inside the pump will rise to the discharge pressure, causing serious damage to the pump, the valving, and probably the heaters.

The first system can be employed without this danger if the arrangement is such that one of the suction valves is always open. It carries with itself another danger, however. Most modern installations have a rather wide range in feedwater temperatures between full and light load. If, then, the standby pump were to be connected into the unit that is carrying the higher load and if suddenly it were necessary to switch it to the unit operating at a rather light load, the pump would be full of water at a temperature that may actually have a vapor pressure in excess of the suction pressure, a condition that could lead to a disastrous result through

Chapter 3

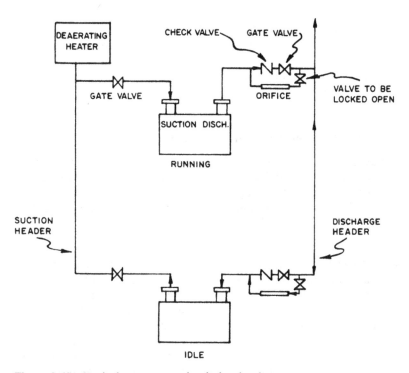

Figure 3.40 Typical warm-up recirculation hookup.

flashing and steam binding. The best solution is to manipulate the warm-up connections in such a manner that the pump is maintained at the temperature of the unit carrying the lowest load. This would, of course, mean that the standby pump must also be connected at its suction and discharge into the system of the unit carrying the lowest load.

Most of the precautions given above apply to open feed systems, that is, to systems embodying direct-contact heaters at the suction of the feed pumps. The question of suction and discharge valving, of course, does not differ in the case of closed cycles. However, in the closed cycle, the balancing device leak-off is always returned to the pump suction, and this problem disappears. Likewise, no serious problem exists insofar as warm-up connections are concerned; since the pressure at the suction actually rises instead of falls with a reduction in turbine load, and hence there is always sufficient suction pressure margin over the vapor pressure at the pump suction.

It would appear from the foregoing that it is possible, through a judicious and careful study, to reduce the number of standby boiler feed pumps in multiple-unit steam stations without sacrificing unduly reliability of operation. It is of paramount importance, however, that this study include considerations of every single factor involved and that if any automatic control is included that the fail-safe principle be incorporated for every automatic feature.

Question 3.31 Cleaning Boiler Feed Pump Suction Lines

We have experienced seizures of boiler feed pumps shortly after the initial start-up and traced these difficulties to the presence of foreign matter in the lines. This foreign matter apparently gets into the clearance spaces in the pump and damages the pump to a considerable extent. What special precautions are recommended to avoid such difficulties?

Answer. Boiler feed pumps have internal running clearances from 0.020 to as low as 0.012 in. on the diameter (that is, from 0.010 to 0.006 in. radially), and it is obvious that small particles of foreign matter such as mill scale left in the piping or brittle oxides, can cause severe damage should they get into these clearances. Incidentally, it has been the general experience that an actual seizure does not occur while the pump is running but rather as it is brought down to rest. But since the boiler feed pumps are frequently started and stopped during the initial plant start-up period, seizures are very likely to occur if foreign matter is present.

The actual method used in cleaning the condensate lines and the boiler feed pump suction piping varies considerably in different central stations. But the essential ingredient in all cases is the use of a temporary strainer located as close to the pump as possible.

Generally, the cleaning out starts with a very thorough flushing of the condenser and deaerating heater, if such is used in the feedwater cycle. It is preferable to flush all the piping to waste before finally connecting the boiler feed pumps. If possible, hot water should be used in the latter flushing operation, as additional dirt and mill scale can be loosened at higher temperatures. Some central stations use a hot phosphate and caustic solution for this purpose. Incidentally, one excellent procedure that reduces the chances of fine particles getting into the pump and remaining there consists of removing the inner element from the feed pump, closing the ends of the casing with cover plates, and flushing water—preferably hot water—through the empty casing.

The safest solution consists of using a fine-mesh strainer and flushing with the pump stationary until the strainer remains essentially clean for half a day or longer. After that, a somewhat coarser mesh can be used if it necessary to permit circulation at a higher rate. But it is very important that the pumps be turned by hand both before and after flushing to check whether any foreign matter has washed into the clearances. If the pump "drags" after flushing, it must be cleared before it is operated.

Pressure gauges *must* be installed both upstream and downstream of the screen, and the pressure drop across it watched most carefully. As soon as dirt begins to build up on the screen and the pressure drop starts to climb, the pump should be stopped and the screen cleaned out. Actually, it would probably be best to have the pressure drop actuate an alarm as a measure of safety.

Question 3.32 Suction Strainers

There seems to be a wide difference of opinion with respect to the type of suction strainers required for start-up and operation of steam power plant boiler feed pumps as well as to the recommended mesh size for these strainers. What is your opinion on this subject?

Answer. The choice of mesh has always been a compromise. Essentially, the clearance in the wearing rings should determine the mesh. But it is always a problem to get sufficient strainer area with available designs at present. Formerly, 40 mesh strainers were specified. Unfortunaltely, such strainers clogged too easily, and because quickly cleaned strainers were seldom used, this clogging necessitated frequent and tiresome shutdowns to clean the accumulation of dirt.

If 8 mesh screening is used with, let us say, 0.025 in. wire, the openings are 0.100 in., and that is too coarse to remove particles of a size that would cause difficulties at the pump clearances. If there is an appreciable quantity of finely divided solids present and the pump is stationary during flushing, some solids would be likely to pack into the clearances and cause damage when the pump is started.

As I have said, the choice of mesh has had to be a compromise, and somewhat reluctantly, I think one must accept a 20 × 20 mesh screen, backed up by a 1/4 in. mesh reinforcing screen. With 0.012 to 0.015 in. diameter wire, this 20 × 20 mesh provides openings of about 0.035 in., which, obviously, exceed the internal clearances.

This brings us to the question of the best type of strainer to use. A conical strainer such as shown on Fig. 3.41 is preferable to the conventional flat screen strainer because it has a lower pressure drop and hence permits longer periods between cleaning. It can be installed in a short spool piece that can readily be removed for cleaning. After the system has been thoroughly cleaned, the fine-mesh strainer itself can be removed and replaced with a coarser mesh.

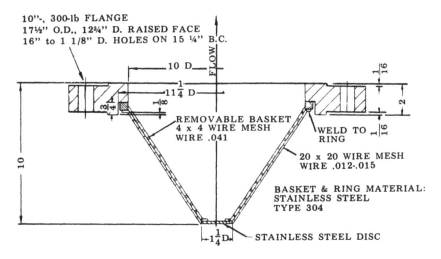

Figure 3.41 Conical strainer.

But even a conical strainer is not the most economical solution. Considerable time and expense are involved in removing, cleaning, and replacing either conventional flat screen or conical strainers, because that requires pipe fitters be on duty to break and make the pipe joints.

One of the most satisfactory types of strainers is the type illustrated in Fig. 3.42. It has a sloping screen that affords a lower pressure drop and hence longer periods between cleaning than the conventional flat screen strainer. It is provided with a draw-off cock for removing accumulated foreign matter. Thus in many cases, dirt accumulation can be dropped out by opening this draw-off cock without actually shutting down the system. The strainer is also provided with a valve opening on its downstream side, permitting backwashing the screen if necessary. Operators have reported that it is possible to clean a strainer such as this in a total of 10 min of complete outage time as against approximately 4 hr required for the conventional flat strainers.

If the size of the free straining area is properly selected to minimize the pressure drop, the start-up strainer may be left in the line for a considerable period before the internal screen is removed. Alternatively, the entire unit may be removed and replaced with a spool piece. The strainer is then available for the next initial start-up.

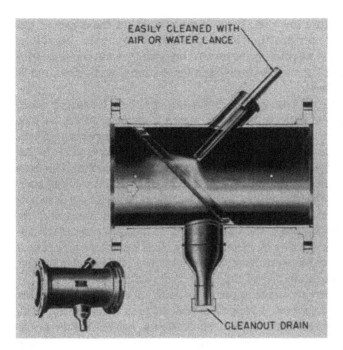

Figure 3.42 Strainer design with a draw-off cock for removing accumulated foreign matter and a valved opening downstream for backwashing the screen.

Question 3.33 Differences in Adjacent Steam Power Plant Units

I would like to ask two questions of you. The first deals with antiflash baffling in the deaerator. This baffling, which had originally been installed to reduce the unfavorable effect on the boiler feed pumps during transient operations, was removed from the deaerator. I suppose that this was done on your suggestion to our engineers during your visit here. Could you tell me what was the reason for this change from the original arrangement?

As to my second question, I am curious about a difference between the feedwater piping for our units 2 and 3 when compared with that of unit 1. This first unit has only a piston check valve in the discharge lines of each pump, but the latter units have both a tilting disk valve and a flow control valve that also acts as a check valve, by virtue of having air pressure transmitted to it by a pilot control when the pump is switched off. Why is a double-check mechanism necessary in these two later units?

Answer. To start with the question about removal of the antiflash baffling from the deaerator, I am certain that the decision to remove this baffling was considered before my visit, because the studies on the effect of this type of deaerator construction were carried on by your consulting engineers and by myself prior to the time of my visit. It is rather interesting to note that much of the work involved in these studies was carried out in Japanese power plants.

For details on the reasons for eliminating antiflash baffling I would refer you to Question 6.18. In brief, the presence of the antiflash baffling permits a significant time lag to occur in bringing the temperature of the feedwater in the storage space up to the operating temperature of the deaerator. In some cases, as much as 50 to 100°F subcooling has been observed in the storage space for as long as 1 hr or more after initial startup. If then a sudden load rejection takes place, the pressure in the deaerator may drop instantaneously until it corresponds to the vapor pressure of the feedwater in the storage space and until flashing of this feedwater at the surface can help reduce the rate of pressure decay. This instantaneous pressure drop can prove to be disastrous from the point of view of flashing in the suction header and steam binding of the boiler feed pumps.

As to the difference between unit 1 and units 2 and 3 with regard to the check valve arrangement, I suppose it was dictated by the desire to increase reliability. Experience in the United States has been that piston check valves are sometimes subject to difficulties from sticking. This in turn can cause serious damage to the boiler feed pumps. Tilting disk check valves are less liable to this trouble. Of course, the best argument that I see in this arrangement is that a gate valve would have had to be used anyway in addition to the check valve. If this gate valve is replaced by an air-assisted check valve that can also act as a stop valve, reliability will have been improved and the number of valves not increased.

Question 3.34 Location of Suction Bell of Vertical Pump

A study of standard books and articles on pumps reveals that the vertical distance between the suction bell of a vertical pump and the sump floor should be maintained at a minimum of $D/2$ to $0.75D$, where D is the suction bell diameter. Normally a strainer is fitted to the suction bell, and I am interested to know whether these recommended dimensions should be measured from the bottom of the suction bell or from the bottom of the strainer. If measured from the bottom of the suction bell, I am afraid that in some cases the strainer may come close to touching the sump bottom.

Answer. You are quite right that the literature fails to say what to do about vertical wet-pit pump installations that include a strainer, and my search failed in finding an illustration giving recommended dimensions showing a strainer.

It is my opinion that, if possible, one should measure the stipulated vertical dimension from the bottom of the sump floor to the bottom of the strainer as if it constituted the bell mouth itself. This is, of course, on the conservative side, and I imagine that if the strainer has been selected liberally as to its dimensions, one could stay within the D/2 dimension. The important thing, of course, is to prevent the buildup of foreign matter right under the strainer, as this may affect pump operation quite unfavorably.

Question 3.35 Extended Storage of Boiler Feed Pumps

It frequently happens that high-pressure boiler feed pumps are delivered to the steam power plant site a considerable period of time prior to final installation and placing into service. Could you please give us an indication of the measures recommended for proper storage of the pumps and their drivers during this period? I might add that in a few cases—particularly when the installation has been overseas—this period of time may extend to as long as 6 or 8 months.

Answer. Recommended practices for either short-term or extended storage may vary considerably among manufacturers, but they do follow a general pattern that I shall describe presently. As a result, the soundest recommendation I might make is that you follow the instructions prepared by the manufacturer of the pumps supplied to the particular installation in question. You could, therefore, consider the following suggestions as merely typical.

Boiler Feed Pumps

Today's practice in high-pressure boiler feed pump construction is to use forged steel casing barrels and 5 to 13% chrome stainless steel internals, Some of the renewable wearing parts may be made of a stainless steel with a chrome content of up to 17%. The frequent assumption that the stainless steel parts of the inner assembly are immune from corrosion in the presence of moisture and air is not borne out by experience. This is particularly true when a pump is only partially drained out after performance tests. Thus, one of the major problems in the preparation of boiler feed pumps for storage is adequate rust prevention. The degree of precaution employed to achieve rust prevention will depend somewhat on the expected duration of the storage and—obviously—on the willingness of the customer to incur the costs incidental to the various techniques employed to achieve the desired degree of precaution.

In general, the following steps should be employed prior to shipment:

1. A rust-inhibiting solution (a water-soluble oil) should be circulated through the pump prior to complete draining.
2. The bearings should be dismantled, coated with a rust preventative (such as NOX-RUST -203), and reassembled.
3. All openings should be sealed with cloth tape. This refers to such areas as between the bearing covers and the shaft sleeves of the shaft sleeve nuts and between the glands and the shaft sleeves or the holes at the breathers.
4. All nozzles should be covered with protective metal plates.

These precautions are sufficient for several months' storage. If the pumps are to be stored for an extended period, additional precautions are advisable:

5. The covers protecting the pump nozzles should be removed and a bag of silica gel suspended inside each nozzle. After the covers are replaced, seal the joint between each cover and nozzle with cloth tape.
6. Remove the hand hole cover from the oil reservoir and hang a bag of silica gel inside the reservoir. Bolt the handhole cover in place and preferably, seal it with cloth tape.
7. Once a month, remove the nozzle and handhole covers to check the silica gel bags to see whether they have become saturated with moisture and require replacement. The time between checks will depend on the atmospheric moisture and the adequacy of the sealing of the openings.
8. Do not turn the pump shaft as this will break the taped joints, permitting dirt and moisture to enter the bearings and stuffing boxes. It would also disturb the rust preventative in the bearings.

To prepare the pump for operation:

1. Remove the bags of silica gel from the nozzles and the oil tank.
2. Leave the nozzle covers in place until the adjoining piping is connected, so as to prevent dirt from getting into the pumps. This is extremely important, as the amount of dirt during power plant erection is quite significant.
3. Dismantle the bearings and clean bearing parts and shaft journals with kerosene or solvent. Reassemble the bearing housings and covers but leave out the line and thrust bearing elements during the cleaning and flushing of the oil lubrication system.
4. Fill the reservoir with turbine oil and operate the auxiliary oil pump.

5. Circulate hot water or wet steam through the oil cooler tubes to heat the oil to 175 to 200 °F. This will break free any rust or scale that may have accumulated in the lube oil piping during storage. Observe the oil pressure at the auxiliary oil pump discharge. When this pressure reaches about 10 psig, shut down the pump and clean the oil filter. Repeat this procedure until there is no further pressure buildup at the oil filter.
6. Shut down the auxiliary oil pump and drain out the lube system. Clean out the filter housing and install a new element. Clean out the oil reservoir thoroughly, including the suction screens for both service and auxiliary oil pumps.
7. Refill the reservoir with clean turbine oil and circulate for about 4 hr with the auxiliary oil pump, maintaining an oil temperature of 175 to 200 °F.
8. Shut down and remove the heating connection from the oil cooler.
9. Dismantle the bearing housings. Clean, wipe dry with lint-free rags, and reassemble the bearings. Be sure to follow the manufacturer's instructions in this reassembly.
10. Drain the lube system, wipe the reservoir dry, and fill it with oil of the type recommended by the manufacturer of the pumps.
11. If the boiler feed pump is not to be put into regular operation at this time, the auxiliary oil pump should be operated once a week to keep the oil piping coated with an oil film to prevent rusting.

Electric Motors

Storage of electric motors for either short or extended periods of time presents fewer problems than the storage of the boiler feed pumps themselves. For one thing, motors are not exposed to moisture corrosion to the same degree as pumps, since they can only be exposed to atmospheric moisture. In addition, when motor bearings are supplied with forced feed lubrication, the oil supply is generally provided from the pump lubricating system. As a result, the list of recommendations that follows is considerably simpler and shorter than the practices involved with pump storage:

1. In a heated building (50-100 °F), the motor should be covered with a plastic sheet to keep dust from entering. Under such conditions, temperature changes should be slow enough so that there would normally be no condensation in the motor. If rapid and fairly substantial temperature changes may be expected, space heaters should be placed in the motor area to keep it slightly above the ambient temperature.

2. In unheated building storage, the motor should be covered with a plastic sheet, and space heaters must be used or silica gel bags must be placed under the enclosure to absorb the moisture. All exposed machined surfaces, including the coupling, should be well slushed. If the rotor cannot be turned approximately once a month using a corrosion-inhibiting oil as lubricant, the bearings should be removed and the shaft journals slushed with rust preventative compound.
3. For outside storage, the procedure is essentially the same as in item 2 except that the covering must be more substantial and weather tight. A periodic inspection procedure must be set up to replace the silica gel bags and reslush the exposed metal parts. The seal faces on the bearing housings hould be slushed. If extended storage is foreseen, the brackets should be removed and the machine surfaces protected with a compound.
4. In all cases it may be necessary to provide protection against rats, mice, or other rodents. This will depend on local conditions, which will dictate whether traps, poison, or other means must be used.

Fluid Drives

The complexity of storing fluid drives is somewhat halfway between that for pumps and for motors. The following represent recommendations prepared by the manufacturer for this equipment:

1. *Heated building storage.* Store the fluid drive with the housing closed without any special preparation.
2. *Unheated building storage:*
    a. Apply a protective coating of oil-soluble rust preventative to all accessible external and internal unpainted surfaces. Fluid drive internal parts are accessible by removing the housing cover.
    b. Place silica gel bags to absorb moisture inside the housing.
    c. Seal all openings in the housing.
3. *Outside storage* is not recommended, but if absolutely necessary, the following procedure must be followed:
    a. Follow all steps given for unheated building storage.
    b. Seal shaft extensions and speed control extension in weather-tight enclosures that both enclose them and seal the openings where they extend from the fluid drive housings.

Chapter 3 323

   c. Seal all other housing openings weather-tight.
   d. Protect the painted external surfaces against the weather elements with a temporary cover, such as a waterproof tarpaulin.
4. *After storage:*
   a. Inspect the fluid drive to be sure it is in good condition.
   b. Remove the silica gel bags from the housing before operation.
   c. After prolonged storage, it is recommended that an order be placed for a serviceperson to inspect the fluid drive units before operation.

No mention was made of the last recommendation with respect to boiler feed pumps because it is normal practice to include the services of a start-up engineer in the price of the pumps. This practice dates back to the 1930s.

There is no mention of the precautions required in placing the lube oil system of fluid drives in operation. It is my opinion that a procedure paralleling that I outlined for boiler feed pumps would be advisable.

Question 3.36 Turning the Shaft of Pumps in Storage

I would appreciate your comments on the proper procedure of taking care of boiler feed pumps delivered at the site a considerable period of time before being installed and put into service. It seems to me that I had once heard it recommended to have someone in the erection department give the pumps a one-quarter or one-half turn by hand once a week just to keep the pumps free. This would be with the pumps in a dry condition, without there being any water in the pumps. Is this procedure acceptable or recommended today?

Answer. The procedure you have outlined used to be recommended until some 35 years ago. It was intended to prevent the pump shaft from developing a bow. The practice was similar to some extent to and was derived from that used with large steam turbines in which the turbine-generator might even be put on turning gear in the field if, after installation, it were going to be standing idle for any length of time. As a matter of fact, among the tools we used to furnish with the boiler feed pumps was a special strap wrench that had a canvas strap to be used around the shaft so as to make it easier to turn over the rotating element.

This practice was abandoned for a number of reasons. Despite the precautions taken to cover all openings, dirt may get into the pump in one manner or another. I have records of cases in which cover flanges used over

Appendix Table of Dimensions for ANSI Pumps (see Fig. 3.2)

Dimension designation	Size suction × discharge × nominal impeller diameter	CP	D	2E₁	2E₂	F	H	O	U Diameter	U Keyway	V Minimum	X	Y
AA	1½ × 1 × 6	17½	5¼	6	0	7¼	⅝	11¾	⅞	3/16 × 3/32	2	6½	4
AB	3 × 1½ × 6	17½	5¼	6	0	7¼	⅝	11¾	⅞	3/16 × 3/32	2	6½	4
A10	3 × 2 × 6	23½	8¼	9¾	7¼	12½	⅝	16½	1⅛	¼ × ⅛	2⅝	8¼	4
AA	1½ × 1 × 8	17½	5¼	6	0	7¼	⅝	11¾	⅞	3/16 × 3/32	2	6½	4
A50	3 × 1½ × 8	23½	8¼	9¾	7¼	12½	⅝	16¾	1⅛	¼ × ⅛	2⅝	8½	4
A60	3 × 2 × 8	23½	8¼	9¾	7¼	12½	⅝	17¾	1⅛	¼ × ⅛	2⅝	9½	4
A70	4 × 3 × 8	23½	8¼	9¾	7¼	12½	⅝	19¼	1⅛	¼ × ⅛	2⅝	11	4
A05	2 × 1 × 10	23½	8¼	9¾	7¼	12½	⅝	16¾	1⅛	¼ × ⅛	2⅝	8½	4
A50	3 × 1½ × 10	23½	8¼	9¾	7¼	12½	⅝	16¾	1⅛	¼ × ⅛	2⅝	8½	4
A60	3 × 2 × 10	23½	8¼	9¾	7¼	12½	⅝	17¾	1⅛	¼ × ⅛	2⅝	9½	4
A70	4 × 3 × 10	23½	8¼	9¾	7¼	12½	⅝	19¼	1⅛	¼ × ⅛	2⅝	11	4
A20	3 × 1½ × 13	23½	10	9¾	7¼	12½	⅝	20½	1⅛	¼ × ⅛	2⅝	10½	4
A30	3 × 2 × 13	23½	10	9¾	7¼	12½	⅝	21½	1⅛	¼ × ⅛	2⅝	11½	4
A40	4 × 3 × 13	23½	10	9¾	7¼	12½	⅝	22½	1⅛	¼ × ⅛	2⅝	12½	4
A80*	6 × 4 × 13	23½	10	9¾	7¼	12½	⅝	23½	1⅛	¼ × ⅛	2⅝	13½	4
A90*	8 × 6 × 13	33⅞	14½	16	9	18¾	⅞	30½	2⅜	⅝ × 5/16	4	16	6
A100*	10 × 8 × 13	33⅞	14½	16	9	18¾	⅞	32½	2⅜	⅝ × 5/16	4	18	6
A110	8 × 6 × 15	33⅞	14½	16	9	18¾	⅞	31½	2⅜	⅝ × 5/16	4	18	6
A120*	1 × 8 × 15	33⅞	14½	16	9	18¾	⅞	33½	2⅜	⅝ × 5/16	4	19	6

the pump nozzles or over the balancing device leak-off connection were removed in the field a long time before the pumps were piped up. If any dirt were to get into the pump and the pump were to be turned over by hand at frequent intervals, the danger would exist that galling might occur in the wearing rings, the balancing device, or the condensate seals.

Another source of danger to be encountered in following this practice is that dirt may get into the bearings and damage both the bearings and the shaft journals when the shaft is turned. As a matter of fact, bearings are now covered and taped up to avoid this risk.

The risk of damaging close-running clearances if foreign particles enter into the pump exists as well when the pump is filled with water. But if the pump runs at a minimum speed sufficient to develop about 50 psi per stage, the risk is somewhat minimized because the flow of water through the close clearances will tend to wash the dirt through.

Appendix

The table on page 324 lists ANSI pump dimensions (in.) (see Fig. 3.2). In those items marked with an asterisk (*), the suction connections may have tapped bolt holes.

Bibliography for Wet-Pit Pump Intakes

Brkich, A., "Rid Vertical-Pump Intake Design of Guess with Model Tests," *Power,* Feb. 1953.

Denny, D. F., "An Experimental Study of Air-Entraining Vortices in Pump Sumps," paper presented at the Nov. 4, 1955, meeting of the Institution of Mechanical Engineers in London.

Dornaus, Wilson L., *Pump Handbook,* McGraw-Hill, New York, 2nd edition, 1986, Sec. 10.1.

Fraser, W. H., "Hydraulic Problems Encountered in Intake Structures of Vertical Wet-Pit Pumps and Methods Leading to Their Solution," paper presented at the Semiannual Meeting, Toronto, Ont., Canada, June 11-14, 1951, of the American Society of Mechanical Engineers (*Transactions of the ASME,* May 1953).

Hydraulic Institute Standards for Centrifugal, Rotary and Reciprocating Pumps, 14th ed., Hydraulic Institute, 1983.

Iversen, H. W., "Studies of Submergence Requirements of High Specific Speed Pumps," *Transactions of the ASME,* May 1953, pp. 635-641.

Markland, E., and J. A. Pope, "Experiments on a Small Pump Suction Well, with Particular Reference to Vortex Formations," paper presented at the Nov. 4, 1955, meeting of the Institution of Mechanical Engineers in London.

Rahm, Lennart, "Flow Problems with Respect to Intakes and Tunnels of Swedish Hydroelectric Power Plants," *Transaction of the Royal Institute of Technology, Stockholm, Sweden,* No. 71, 1953, pp. 71-117.

Stepanoff, A. J., *Centrifugal and Axial Flow Pumps,* 2nd ed., Wiley, New York, 1957, pp. 12-16 and 357-363.

4
Operation

Question 4.1 Testing Motors for Rotation

Before 2300 V motors are returned to service following extensive electrical repairs on the distribution system, it is desirable to check rotation of equipment. We are interested to know if it is necessary to uncouple the drive from the pump before *bumping* in the case of (1) a vertical shaft propeller mixed-flow pump having a capacity of 82,500 GPM and (2) a 1600 psi, 700,000 lb/hr boiler feed pump having six stages and barrel casing construction.

Answer. Very emphatically, the answer is that the motors should be uncoupled from the pump before trying out the motor rotation, both in the case of the vertical pump and in that of the barrel-type boiler feed pump.

Your question speaks of a propeller *mixed-flow* pump, and these two terms are somewhat contradictory. A propeller pump is an almost true axial-flow pump (Fig. 4.1), but a mixed-flow pump is a lower specific speed type with both a radial- and an axial-flow component in the impeller (Fig. 4.2). Either type, however, can be damaged if the motor is started in a reverse rotation. These pumps, as well as the boiler feed pump (Fig. 4.3), have a number of shaft nuts that are generally threaded so that they tend to tighten under normal rotation. For instance, the vertical propeller circulating pump shown in Fig. 4.1 has an impeller nut, a shaft sleeve nut, and a coupling nut. The mixed-flow pump in Fig. 4.2 shows an impeller nut. The boiler feed pump in Fig. 4.3 has a coupling nut, two shaft sleeve nuts, a balancing device nut, and a thrust collar nut, all threaded to tighten from normal rotation.

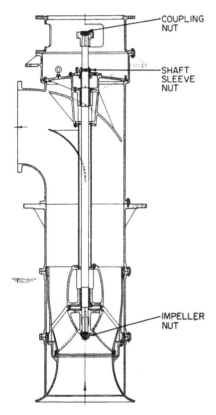

Figure 4.1 Vertical propeller circulating pump.

Of course, it is the general practice to use setscrews or some equivalent to lock these nuts in position. Nevertheless, it must be remembered that a motor will accelerate very rapidly in either direction. The danger exists of one of these setscrews shearing, and the resulting damage can be quite expensive.

This is a question that on occasion also arises in connection with small vertical turbine-type pumps that use screw-type couplings to connect individual lengths of shafting. The hand of thread used in these couplings is also such that during normal pump operation the torque exerted by the motor acts to tighten the threads. If the motor stops and reverse flow takes place through the pump, the latter acts as a turbine and runs in reverse rotation, producing torque to drive itself and the motor. Under this condition, the torque developed by the pump acts on the coupling thread exactly

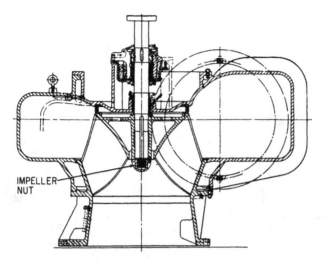

Figure 4.2 Vertical mixed-flow circulating pump.

as when the motor was driving the pump. Therefore, the coupling threads will not unscrew.

On the other hand, if the motor of this pump is wired incorrectly, the pump will start in the wrong direction, and the motor torque will act to unscrew the coupling. For this reason, motor-disengaging clutches or nonreversing ratchets are used in some of the hollow-shaft motors used to drive vertical turbine pumps. Regardless of this, it is imperative in my opinion to try out the motor rotation with the pump uncoupled. Frankly speaking, I see no particular difficulty in following this procedure, especially since the pump and driver will be uncoupled during the realignment that is necessary whenever a motor has been removed for servicing. If the motors have not been moved and only the distribution system has undergone electrical repairs, there is some possibility that the leads will have become reversed. The pumps should be uncoupled and rotation checked. Advantage can be taken of this uncoupling to check pump and driver alignment.

Further Comments about Question 4.1

During a recent visit to Columbia, S. C., my friend George L. Dibble, Vice-President of Electric Operations and Engineering of South Carolina Electric & Gas Co., and I discussed the problem of testing motors for rotation. I know that George will modestly refuse to send his comments to be printed. But since they permit me to explain the position that I have frequently had

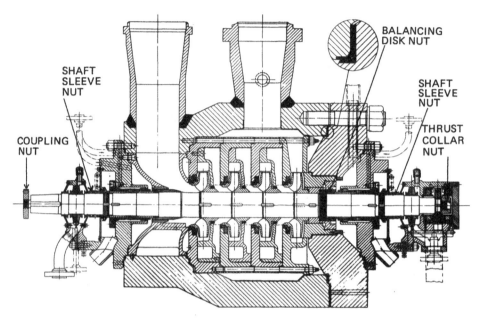

Figure 4.3 Boiler feed pump has a number of nuts on the shaft.

to take with regard to certain operating and maintenance practices, I am certain that he will forgive me if I quote him—as accurately as my memory allows me.

George Dibble mentioned that by manipulation of suitable motor starting controls, a motor can be bumped so gently and carefully that only a few degrees of rotation will take place—certainly not enough to cause the unscrewing of coupling or shaft nuts. In other words, in the hands of an experienced operator and with suitable controls, the practice of bumping a motor to determine correctness of lead connections presents little danger and will not require disconnecting the driven apparatus.

Of course, he is right. But I frequently find myself in a dilemma when discussing certain operating or maintenance practices. There are many short-cuts that, in the hands of experienced personnel, can simplify the tasks to be performed in any plant. On the other hand, these same short-cuts can lead to disastrous results when performed by inexperienced mechanics or operators. But how to distinguish between the two? And how safe is it to provide less experienced personnel with suggestions that could lead them to trouble?

This is a difficult problem. I have made it a practice, therefore, seldom to mention very risky methods and certainly never without adequate words of caution. And in this particular case, since disconnecting motor and pump is not such a time-consuming procedure, I prefer to stand on my recommendation: When testing for correct motor rotation, always disconnect the motor from the driven apparatus.

Question 4.2 Warm-up of Boiler Feed Pumps

Why should it be necessary to warm up a high-pressure boiler feed pump prior to placing it in service? I understand quite well that a pump only partially warmed up, that is, partially hot and partially cold, may be distorted sufficiently to cause major trouble from binding at the close internal clearances. But since high-pressure boiler feed pumps are generally of the double-casing barrel type, with centerline support and with suction and discharge nozzles on the vertical centerline, why can they not be started absolutely cold?

Answer. This is a rather controversial question in that it is very obvious that the answer will be affected by the design of the pump internals. However, some general considerations of this problem may be made.

In the first place, not all methods of mounting the impellers on the shaft are alike. If the impellers are mounted with a slight shrink fit, starting the pump cold will have no injurious effects. If, on the other hand, the impellers are mounted with no interference fit on a shaft the material of which will expand more rapidly than the impeller material, dependence is made on the shrink fit effected when the pump comes up to operating temperature. A cold start in such a case may lead to operation with slightly loose impellers.

The effect of cold starts on the relative position of inner assembly and casing barrel of a double-casing pump may or may not lead to difficulties, as the following analysis will show. When hot feedwater is suddenly admitted to a cold boiler feed pump, the relative expansion of the outer casing barrel and of the inner element goes through two separate and distinct phases. At first, as the inner element is much lighter than the barrel and is in more intimate contact with the hot feedwater, it expands at a considerably faster rate than the outer casing itself. To simplify our analysis, we may assume that the inner element reaches its final temperature before any appreciable temperature change has taken place in the outer casing. Then, as the pump continues in operation, the outer casing heats up and reaches its own final temperature at some later time. If the casing barrel is not lagged, the temperature on its external surface may be somewhat lower than the internal temperature, but we can neglect this difference in our analysis.

It may be interesting to go through an approximate quantitative analysis of what takes place during these two phases. Let us assume that we are dealing with a seven-stage boiler feed pump (Fig. 4.4) designed to operate at 320 °F, that the inner element is 5% chrome stainless steel, and that the outer barrel is made of forged SAE-1020 steel. The coefficients of expansion of the two metals in question can be considered to be the same and equal to 0.0000065 in./in. per °F. Let it be further assumed that the length of the inner element within the barrel (dimension A in Fig. 4.4) is 36 in.

If, when the pump is at rest on standby service, the metal temperature is permitted to fall to 120 °F, the sudden admission of 320 °F feedwater will cause an expansion of the inner element of 0.047 in. This means that by the time the inner element will have come up to its final temperature, it will have expanded 47/1000 in. with relation to the casing barrel. Ultimately, the outer casing will also come up to temperature and will expand by an equal amount, nullifying the initial expansion of the inner element.

The result of this initial relative expansion, followed by a return to the initial relative position of inner element and outer casing, will have different effects on different designs. The effects will depend on whether or not the construction of the double-casing pump permits free movement of the inner element within the barrel. If it does, the events that take place will have but little effect on the unit. If, on the other hand, the inner element is constrained within fixed limits, it becomes necessary to interpose some form of compressible gasket between the inner element and either the barrel or the discharge head. (Sometimes these gaskets are incorporated at both points.) It then falls

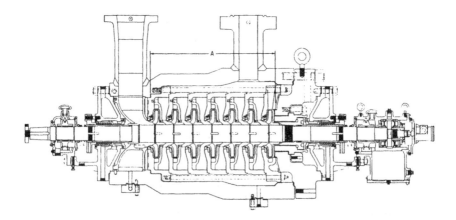

Figure 4.4 Inner element of a barrel-type boiler feed pump undergoes relatively rapid expansion with a cold start-up procedure.

to these gaskets to absorb the difference in expansion we have calculated. Whether a cold start will have an injurious effect on the unit then becomes a question of the reliability of the compressible gaskets and of their ultimate life.

Some consideration should also be given to the effect of the warm-up method selected. It is true that, in some cases, the pump will be subjected to a certain amount of distortion because heat may flow unevenly to various parts of the pump. But when warm-up has been properly completed, this distortion should disappear and, of course, will not have affected the pump while it was at rest. A careful analysis, however, is necessary to check that a pump started up cold does not undergo this type of distortion as it is coming up to temperature, since interference at the running joints or misalignment at the bearings will be a possible cause of trouble.

Thus, in conclusion, it is obvious that some pumps can be started up cold but others should not but that *all* boiler feed pumps will profit from starting up warm if the warm-up operation ensures a thorough and even distribution of heat to all parts of the pump.

Question 4.3 Should Suction or Discharge Be Throttled?

We have a 6 in. discharge single-stage centrifugal pump driven at constant speed by a squirrel cage motor operating on general service. It is at present delivering 1600 GPM at a total head of 160 ft. We find that we could do with considerably less water, say 1300 GPM. Since the pump speed cannot be changed, we can reduce the capacity by throttling either the discharge or the suction of the pump. Will this reduce the power consumption of the pump, and can we obtain a greater reduction by throttling the suction or the discharge?

Answer. A centrifugal pump operating at constant speed will deliver a capacity corresponding to the intersection of its head-capacity curve and of the system-head curve of the system in which it is operating. The system-head curve itself is made up of the sum of the static and pressure heads and of the friction losses at various capacities. In Fig. 4.5 such a system-head curve has been superimposed on a pump head-capacity curve for conditions given in the question. It has been assumed that the static and pressure head is 130 ft and that the friction losses at 1600 GPM are 30 ft.

It will be seen that the head-capacity curve at 1300 GPM is well in excess of the system head at that capacity. It is possible, however, to introduce artificial friction losses by throttling, resulting in a new system-head curve that now intersects the head-capacity curve at 1300 GPM and 176 ft total head. The power consumption will have been reduced from 77.5 to 70 hp.

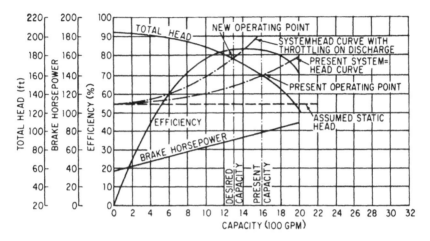

Figure 4.5 System-head curve superimposed on the pump head-capacity curve.

It is never recommended to throttle the pump suction in order to reduce pump capacity, as the effect is to change the pump head-capacity curve through cavitation and operation in the so-called *break*. The pump efficiency is seriously affected in such operation, but most important of the ill effects are the erosion and premature destruction that are caused by a cavitation that would reduce the pump capacity as drastically as desired here.

A much better solution than throttling the pump discharge is available if a permanent reduction in capacity is desired. The pump impeller can be cut down—in this case by approximately 7%—so that the pump head-capacity curve will pass through the point of 1300 GPM and 150 ft to intersect with the present system-head curve. The power consumption should be approximately 59 hp as compared with the 70 hp obtained by throttling the discharge.

Question 4.4 Can a Centrifugal Pump Be Operated Against a Closed Discharge?

Answer. Generally, this should be avoided, although there are some special cases when no significant harm can occur.

The difference between the horsepower consumed by a centrifugal pump and the useful water horsepower corresponds to the power losses within the pump. Except for a very small portion of these losses, this difference in power is converted into heat that is transferred to the liquid passing through the pump. When the pump operates with a closed discharge, no useful power is generated, and the entire brake horsepower delivered to the pump is converted into heat. As no flow takes place through the pump, this heat goes into the small quantity of liquid within the pump casing.

The temperature rise of the liquid in the casing can be easily calculated: 1 hp converted into heat corresponds to 42.4 Btu/min. If the pump takes 100 bhp at shutoff, 4240 Btu will be transferred to the liquid each minute. If the pump contains 100 lb of water (specific heat = 1 Btu/lb per °F), the temperature of this water will rise at the rate of 42.4 °F/min, and serious damage can result if the pump is allowed to operate against a closed discharge for any appreciable period of time.

On the other hand, if we have to deal with a large high-capacity and low-head pump, the volume of water contained in the casing will be very large, and the temperature of this water will increase very slowly. For instance, if we assume the same 100 bhp at shutoff but consider a pump whose casing contains 5000 lb of water, the temperature rise will be only 0.85 °F/min, and such a pump could operate for quite a long time, say 15 to 30 min, with the discharge valve closed without causing undue difficulty.

As I said, however, in general it is not recommended to operate centrifugal pumps with a closed discharge or against a closed check valve. If the conditions of installation may impose such an operation, a small bypass should be provided between the pump and its check valve so that a reasonable amount of flow will take place through the pump even when its discharge is closed off. The bypass should lead to some point in the pumping system at which the heat picked up in the pump can be readily dissipated.

Question 4.5 Starting Boiler Feed Pumps

I have been told that a centrifugal boiler feed pump should not be primed while it is running. Is that true? Is it also true that the discharge valve should not be opened in a boiler feed pump installation until the pump has reached full speed?

Answer. It is quite true that, with some very few exceptions, no centrifugal pump should ever be started until it is fully primed, that is, until it has been filled with water and all the air contained in the pump has been allowed to escape. The exceptions involve self-priming pumps and some special large-capacity, low-head, and low-speed installations where it is not practical to start with the pump primed and the priming takes place almost simultaneously with the starting.

In the case of boiler feed pumps, however, this is entirely inadmissible. Since feed pumps generally take their suction from the storage space of a direct-contact heater, priming is accomplished by opening the suction valve, holding the discharge valve closed, and leaving all the vents opened to atmosphere. Under these conditions, the water coming from the storage space under a positive head will drive out all the air contained in the pump, and the latter is ready to start. This, of course, is the practice followed for the initial start of the pump or after the pump has been opened up for examination or overhaul. Otherwise, the pump is left full of water and should need no further priming.

If the pump has been operated previously but there is reason to believe that through the closing of the suction valve or for some other reason it is no longer full of water, it is imperative to repeat this priming procedure. This will again fill the pump with water and eliminate any air pockets.

As to starting up with the discharge valve open or closed, this depends on whether reference is made to the initial starting operation or to the normal operation of a pump being brought on the line after having been on standby duty. In the first case, the pump will likely as not be discharging against a much lower pressure than normal. Under these conditions, it is preferable to start it with the discharge valve closed, opening it gradually after the pump has been brought up to speed and pressure has been built up in the boiler. It goes without saying that the recirculation bypass line provided to prevent an excessive temperature rise in the pump is held open during this operation. Starting the pump with a closed discharge valve will prevent the pump from going out on its head-capacity curve to an excessively high capacity against a very reduced head. Otherwise, there would be a serious danger to the pump both from the point of view of driver overload and from that of unsatisfactory suction conditions.

In the case of normal operation, however, the discharge header pressure is more or less equivalent to the pressure developed by the pump under rated capacity conditions. Thus, a standby pump is always held ready to start up, with its suction and discharge valves open. The discharge line check valve prevents reverse flow through the pump. When the pump is started and brought up to full speed, it will raise the check valve from its seat and will start discharging into the line in parallel with the pump or pumps already in operation.

Question 4.6 Priming Centrifugal Pumps

What are the best means to prime centrifugal pumps?

Answer. If the suction supply is above atmospheric pressure, the air, gas or vapor will be trapped in the pump and will be compressed somewhat when the suction valve is opened. But this will generally not be sufficient enough to permit the suction waterways and the eye of the impeller to be quite filled with liquid and the pump will not be properly primed. It then becomes necessary to vent the entrapped air, gas or vapor through a valve located at the highest point on the casing. Depending on the service involved, the venting may be to the atmosphere or may have to be to some enclosed vessel under a pressure lower than the suction pressure.

If the supply is below the pump itself, the air in the pump must be evacuated either by some vacuum-producing device or by providing a priming chamber in the suction line. Foot valves were commonly used in older installations, but except for certain applications, their use is much less common today, because a foot valve does not always seat tightly and the pump occasionally loses its prime.

A priming chamber is a tank with an outlet at the bottom that is level with the pump suction nozzle and directly connected to it. The size of the tank is such that the volume contained between the top of outlet and the bottom of the inlet is about three times the volume of the suction pipe. Leakage of air when the pump is shut down may cause the liquid in the suction line to leak out, but the liquid in the priming tank below the suction inlet cannot run back to the supply. When the centrifugal pump is started, it will pump this entrapped liquid out of the priming chamber, creating a vacuum in the tank. The atmospheric pressure on the supply will force liquid up the suction line into the priming chamber, replacing the liquid displaced from it.

Pumps can also be primed by using a variety of vacuum-producing devices, such as water- or steam-jet eductors as well as compressed air ejectors. However, in recent years, the desire to provide automatic priming has led to the use of motor-driven vacuum pumps. If there is more than one centrifugal pump to be primed in a common location, one priming device can serve all these pumps. Such an arrangement is called a central priming station. Such devices are readily available on the market.

Question 4.7 Priming Large Horizontal Pumps

I would like to have your opinion on the procedure for priming and starting a 30 in. large horizontal pump having a specific speed of approximately 6000, operating on a suction lift, and requiring a higher power at shutoff than at the design point. The driving motor has only a nominal margin over the bhp required at the best efficiency capacity point. I know that the procedure for vertical pumps is simpler, since the impeller can be submerged below the water level, but since our problem involves a horizontal pump with a suction lift, we need to be aware of all the necessary precautions.

Answer. The problem you have presented is difficult but not impossible to resolve. The solution depends to a great extent on a number of details of the installation, such as whether one deals with a single pump or with several pumps operating in parallel and whether the discharge line is empty or full of liquid at starting.

I presume, to begin with, that you are not dealing with an installation incorporating a foot valve, because this has become a rather uncommon arrangement. Thus, the pump cannot be primed by merely filling it and the suction piping and venting the air through a vent valve. Whatever priming method is used, it is assumed that the discharge will be closed off by a valve. Again, based on the size of the pump in question, it may well be that there will not be a check valve ahead of the discharge gate or butterfly valve.

Since the driver size is such that the pump cannot be permitted to operate against shutoff, the starting and valve opening procedures must be coordinated most carefully. It becomes necessary first to determine very exactly the valve characteristics as well as the pump and motor torque characteristics. The rate of opening the discharge valve must be such that the total head developed by the pump at some speed below full speed is sufficient to overcome or match the pressure in the discharge line, taking care simultaneously that the valve opening is not such as would permit reverse flow through the pump. This last circumstance would make the pump run backward as a turbine, and applying reverse torque on the driver may make starting hazardous to the motor.

I repeat, this is a tricky procedure. In some cases, a bypass line can be provided, so that the pump is not required to operate at shutoff. In other cases, it may be practical to use a two-speed motor, so that the power at shutoff and reduced speed does not exceed the motor rating.

Question 4.8 Protection Against Loss of Prime

We have an installation of three single-stage centrifugal pumps in our pumping station that is operated automatically without any attendants. Lately, we have been encountering a little difficulty with the suction conditions.

Do you know of any protective device that would shut the pumps down when they lose their prime and run dry? We would prefer some sort of a thermal cutout, because we may encounter difficulties on starting these units if the device were operated on pressure. The controls are already elaborate on the starting and stopping of these pumps, as they are arranged so that one, two, or three pumps are running, depending on the water demand.

Will you please inform us whether you know of some thermal device that would contact the wearing rings or other suitable parts to protect the pumps in case they run dry.

Answer. I do not know of such a device, and I rather doubt that it would be very practical. Of course, if the controls are already quite complicated, it would seem to me that the most logical approach to this problem would be to investigate why you are encountering suction difficulties. Your question implies that they did not exist initially.

A control that would be pressure operated need not interfere with the proper starting and stopping of the units if it is properly applied. Since the loss of prime means loss of discharge pressure, it would merely be necessary to locate a pressure switch with a timing element, making it inoperative during the starting cycle. The connection should be made to the pump side of the discharge check valve, as one pump may lose its prime while others remain on the line. A thermal time-delay relay should be incorporated with the pressure switch. The control should be connected to the motor side of the starter to prevent it from opening the circuit when the pump is not in use.

The sequence of operation would be the following: When the pump is being started, the pressure switch would remain closed, since there is insufficient pressure at the discharge. The delayed action of the time relay would prevent opening the circuit for a predetermined time. If, after starting, the pump operates normally, the pressure will increase sufficiently to open the circuit of the protective device. On the other hand, if the pump does not pick up suction, no pressure is developed at the discharge and, after the time delay, the protective device opens the stop circuit of the starter. Likewise, if the pump loses its suction after running normally for a period of time, the pressure switch will close, energizing the time-delay relay and ultimately cutting the pump out.

Typical circuitry for momentary and maintained contact arrangements is shown in Figs. 4.6 and 4.7, respectively.

It would probably be advantageous to attach some sort of an alarm circuit to the protective device. On the other hand, if the station is unattended, the alarm would have to be wired to some location where it can attract attention and instigate action.

There may be several other similar arrangements available on the market, and I would suggest that you contact a manufacturer of automatic

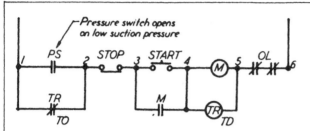

M : Line or motor controller

TR : Time-delay relay

TO : Time opening contacts, closed when relay is deenergized and for a time delay after relay TR is energized

CR : Control relay

TD : Time delay

OL : Overload relay

MOMENTARY CONTACT

Object of the circuit is to shut down the pump whenever the suction pressure is low. Initially, the pressure switch contact PS is open and will close after the pump has built up pressure. When the suction pressure is low, the contacts of the pressure switch open. To start the pump, the "start" button is pressed and the circuit to M is completed through the normally closed contact of TR 1-2, through the stop button and through the start push button. At the same time as M picks up, the coil of relay TR is energized and starts the timer timing. As the pump comes up to speed, the pressure builds up and contact 1-2 of the pressure switch closes. Therefore, after the timer has timed out and contact 1-2 on TR opens, the pressure switch contact will maintain the circuit and keep the contactor M picked up. In case the suction pressure is low, contact 1-2 of the pressure switch PS opens and drops out of the contactor in the same fashion as pressing the stop button. The timer should be adjusted for a time delay corresponding to the time it takes to build up pressure under the most adverse conditions. If the pressure switch stops the pump, it is necessary to restart the pump with the "start" button.

Figure 4.6 Circuitry of a protective device control with momentary contact buttons.

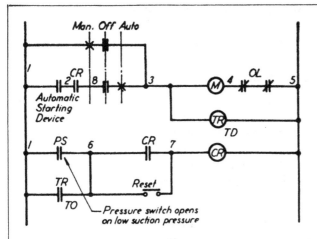

M : Line or motor controller
TR : Time-delay relay
TO : Time opening contacts, closed when relay is deenergized and for a time delay after relay TR is energized.
CR : Control relay
TD : Time delay
OL : Overload relay

MAINTAINED CONTACT CIRCUIT

Operation of this circuit is similar to the momentary contact circuit except it is necessary to add a reset button to prevent automatic restarting of the pump. For initial operation, assume the pump is operating in automatic operation and needs to be started. The "reset" push button is pressed, which causes relay CR to pick up through contact 1-6 of TR and the "reset" push button. When relay CR picks up, contact 6-7 of CR picks up and seals in the relay around the "reset" push button. When the automatic device 1-2 picks up, this picks up contactor M and relay TR. After pump is running, pressure switch contacts 1-6 close. At a time delay after starting the pump, relay TR times out and opens contact 1-6 of TR. Since by this time the PS has closed, relay CR will not drop out on starting. In case the pressure is low, contacts 1-6 of the pressure switch will open, which drops out relay CR and opens contacts 2-8. This drops out the line contactor M and stops the pump. At the same time, contact 6-7 opens, which prevents CR from picking up again until the reset button is pressed.

Figure 4.7 Circuitry of a protective device control with maintained contacts.

Question 4.9 Centrifugal Pumps in Series

In Question 2.10 you discussed the installation of centrifugal pumps in series. It is my understanding that when pumps are installed in series, some difficulty is experienced in maintaining a constant flow through both pumps when driven at constant speed and that some cavitation or overloading may be caused in one pump.

Is it necessary to design the system to prevent this possibility? Please comment on this condition.

Answer. Provided that the pumps that are to be operated in series are adequately selected, there should be no difficulty whatsoever in maintaining constant flow through the pumps whether driven at constant speed or variable speed. At least this will be the case as long as the system-head curve against which the pumps are operating does not itself vary. In other words, at any given speed, the pumps will deliver a constant flow corresponding to the intersection between their combined head-capacity curve at that speed and the system-head curve. This is true of pumps in series as well as of single pumps.

The qualification "provided that the pumps are adequately selected" refers to the exceptional case in which the first of the two pumps in series is designed for considerably less capacity than the second pump. Then, if the total head developed by the two pumps in series is relatively high compared with the system-head at the desired capacity, the two pumps will try to operate farther out on the curve until their combined head-capacity curve intersects the system-head curve. The capacity at this intersection point could be in excess of the normal range of capacities for which the first, smaller pump is designed. This could, of course, lead to cavitation of this smaller pump as well as to the overloading of the pump driver.

Normally, however, individual pumps that will be operated in series are chosen to be of the same design capacity or at least in the same range of design capacities. If because of the desire to use some existing pumps this is not possible, then it is preferable to use the larger capacity pump as the first of the series group. Such an arrangement should eliminate the possibility of cavitation. As to the overload, it can be avoided by so selecting the drivers that they are suitable for the bhp at the operating point, which, of course, is the intersection between the combined head-capacity curve and the system-head curve.

Chapter 4

As a matter of fact, to assuage you fears that pumps in series would always experience cavitation or overloading of one of the pumps, you need only consider the case of the literally hundreds of thousands (if not millions) of multistage pumps in service today. After all, a multistage pump is nothing but two or more pumps in series arranged so as to use but a single casing casting with interstage passages from one pump to the next formed therein. Essentially, water does not know whether it is passing from the first to the second stage of a multistage pump within the confines of a single casing or whether two separate casings have been used and the water has to traverse a pipe between the two casings.

Question 4.10 Frequent Starting of Electric Motors

My question deals with the frequency of starting of motor-driven boiler feed pumps. We have been told by the consulting engineers who designed and constructed our power plant that boiler feed pumps should not be switched on or off frequently, these restrictions having to do principally with the unfavorable effect on the motors. They have stated that when large induction motors are started frequently with full voltage, excessive heating of the windings and of the squirrel cage rotors takes place.

However, it is my opinion that boiler feed pumps are similar to the main stream turbines in that they are vulnerable to the transient effects of a starting cycle. These could damage internal parts running with close clearances. Thus, would the main objection to frequent starting not be based on the desire to avoid damage to the feed pumps rather than to the electric motors?

Answer. The effect of frequent starting and stopping may, as you state, be as damaging to the pumps as to the motors or it may not, depending on several factors.

If the pumps depend on support of the rotor at the wearing rings or internal bushings that act as water-lubricated bearings, there can be more damage than if the shafts are relatively stiff and the clearances are liberal. At standstill, a pump that depends on internal water-lubricated bearings has contact between rotating and stationary parts. Thus, each start and stop takes a slight toll on the ultimate pump life, even if it does not cause immediate seizure. On the other hand, if the internal clearances exceed the rotor deflection, no contact will take place between stationary and rotating parts, regardless of whether the pump is idle or running, and the starts and stops will not interfere with the integrity of the pump.

During the early operation of a steam power plant, there is more possibility of dirt in the system than after the first 3 or 6 months of operation.

In this early period, it is safer to avoid too many starts and stops, because dirt is generally more damaging during this procedure.

Frequent starting of the motors is objectionable because the starting current produces a large amount of heat in the motor windings. A typical example of this restriction is indicated by the warning tag frequently placed on boiler feed pump motors and worded as follows:

> Too frequent starting of high inertia or heavily loaded drives may result in serious injury to the motor windings. To prevent such damage, the following maximum starting duty should be observed: Two (2) successive starts, provided motor is permitted to coast to a stop between starts. A third start may be made when winding and core have cooled to within 5 °C of ambient. If motor is braked to a stop, the temperature of the windings and core should be within 5 °C of the ambient before any attempt to restart is made. Starts should not average more than four (4) per day.

I should add that these restrictions do not apply in the event the motor is operated on variable frequency. This is because a variable-frequency drive has a "soft start", that is, the motor starts with zero starting current and the problem of overheating the winding insulation does not exist.

Question 4.11 Run-out Capacity of Parallel Pumps When Run Singly

We are making an installation of two circulating pumps for cooling tower application. The pumps will be designed for 55,000 gal/min and 124 ft head and will be steam turbine driven. Because of certain circumstances specific to this installation, the pumps will have to be located at 15 ft above the normal water level. We have checked the pump installation for satisfactory conditions for the required NPSH. We are, however, concerned about what would happen if two pumps were operating in parellel and we were to lose one of the two pumps.

Obviously, it seems to us that the remaining pump would go out on its curve and intersect the system-head at a much larger flow. How can we protect ourselves against the possible further difficulties by virtue of the fact that this pump might go into serious cavitation and lose its suction?

Answer. You are quite correct in assuming that if one of the pumps is shut down, the second pump remaining on the line will go out on its curve until its head-capacity intersects with the system-head curve of the installation. Depending on the composition of the system-head curve, that is, on the percentage of static head and friction head involved, this increase in capacity will vary somewhat. The capacity at which this intersection takes place is commonly referred to as the "run-out capacity."

This intersection can take place at a somewhat higher flow than one may have foreseen, if the system-head curve has been constructed on a pessimistic or "worst case" basis. If, as illustrated in Fig. 4.8, the real system-head curve is significantly lower than the predicted curve (in this case because a higher static head and higher friction losses are anticipated than will occur in real life), the run-out flow falls to the right of the predicted intersection.

It is quite possible, as you fear, that the capacity at this intersection point is such that the resultant suction lift would be excessive for the pump, as illustrated in Fig. 4.9.

It is not, however, necessary to assume that the pump will lose its suction after it goes into cavitation. What would happen is that as the pump increases in capacity, its head-capacity curve will become more and more affected by the available NPSH until it goes into a break, that is, until further reduction in head is not accompanied by an increase in capacity.

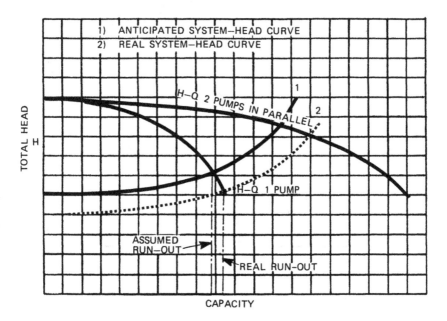

Figure 4.8 Operation of two pumps in parallel.

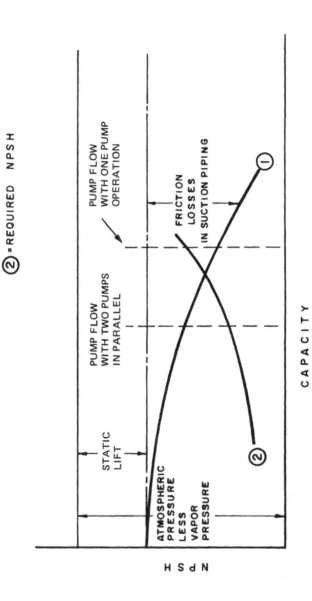

Figure 4.9 Available and required NPSH.

In this range, the head-capacity curve will assume a vertical shape. It is the intersection of this vertical "break" portion of the curve with the system-head curve that will determine the actual capacity at which the pump will operate. As a matter of fact this phenomenon has been used for many years to control the operation of condensate pumps, as illustrted in Fig. 1.33. As the load drops, the condensate pump evacuates more condensate than is returning to the hot well. Consequently the hot well level drops, the NPSH is reduced, and this in turn causes the pump to operate "in the break," reducing its capacity to that corresponding to the new reduced load.

On the other hand, running a pump that is not specifically designed for this type of operation under these circumstances for any length of time is not conducive to long life because of the resultant vibrations that might be imposed upon the pump. It would therefore be advantageous to provide some means to prevent such operation. The least expensive solution might be the installation of a pressure switch in the discharge line, this pressure switch being so tied in with the steam turbine governor control that whenever the discharge pressure falls below a certain amount the turbine speed is reduced.

Incidentally, I assume that each pump will be provided with a check valve so that when either pump stops, the check valve will prevent reverse flow and the resultant reverse rotation.

Question 4.12 Keep Flashing Pumps Running

In one of your discussions you suggested that if a boiler feed pump has reasonably ample internal clearances and a heavy enough shaft to hold the deflection to a minimum, it is safer to keep the pump running under flashing conditions rather than to shut it down. We have experienced several cases that would tend to support your recommendations. In this connection, we would like to ask you two questions:

1. Do you feel that pumps can be operated during flashing conditions for extended periods or only during the short period of time normally required to restore conditions to normal? In other words, is there any time limit?
2. In the case of seizure at full speed, would the damage to the pump not be more severe than if the pump seized while coming to rest? It seems that, especially in the case of the motor driver, this would be true.

Answer. At the time I first wrote about the wisdom of letting a pump that has flashed continue running, provided the pump did not depend on water lubrication at the wearing rings or internal bushings for shaft support, I had only three properly authenticated cases to support my theory.* Of course, I had reached this conclusion as much from a theoretical analysis of the events that accompany a seizure after shutting a pump down as from field experience. Nevertheless, it has been gratifying to collect several more documented cases that have reinforced my conclusions on this particular operating practice. In one particular case, it has been reported to me that boiler feed pumps have operated under flashing conditions for as long as 10 min during certain emergencies. Tests run on the pumps afterward indicated no noticeable damage or loss in capacity.

However, I find it difficult to answer your first question in a truly *quantitative* manner, mainly because I still feel that insufficient evidence is yet available on this score. I find it almost impossible for a set of conditions to arise that would permit us to establish time limits if such do exist. Nor can I realistically expect a utility to perform such an experiment voluntarily, considering the potential risks involved and the stakes of the game.

On the other hand, looking at it from a logical point of view, if a pump can run flashing without seizing for 10 min, why can we not expect it to run safely for 60 min? And if for 60 min, why not for 24 hr?

It probably could be done, but could it be readily proved? And if it can be done, is it an important achievement? Frankly speaking, is a time limit a necessary thing to know? I think not. I prefer to take the position that once a pump has flashed, it should be kept running (providing it meets the design criteria I have mentioned), and all efforts should be directed toward restoring adequate suction conditions in as short a time interval as possible. After order has been restored, try to find out the following three things:

1. How long has the pump run under flashing conditions?
2. Is the pump still sound, or has some damage occured?
3. What caused the flashing, so that protection against a reoccurence can be developed if at all possible?

As to the possible damage to a pump and its driver from a seizure at full speed, there is no question but that it would be much more severe than were the seizure to occur as the pump was coasting down to rest. I personally know of only one case of *full-speed seizure,* but it was of catastrophic proportions. I should add, however, that its cause was not flashing of the feedwater but rather mechanical damage to some of the impellers followed by extreme vibration and finally by overheating of the

---

*"How Do You Operate Your Boiler Feed Pumps?" by I. J. Karassik, *Power Engineering*, March 1957.

shaft at one point through rubbing contact and by breaking of the shaft. In the case in question, a hydraulic coupling was interposed between the pump and its motor driver so that the latter suffered no damage. Essentially, the potential damage of a seizure at full speed is very similar to that suffered by a steam turbine that starts throwing blades and that, if not stopped in time, will be a total loss. At the same time, I repeat that I have no direct knowledge of a pump seizure at full speed under flashing conditions. Thus, a pump that goes into severe vibration because of mechanical causes should be stopped before complete disintegration occurs, but a flashing pump that manages to operate as a centrifugal steam compressor is best left running.

This reasoning is based on the following considerations: It is practically a certainty that a pump that is brought to rest while steam bound because of inadequate suction conditions will seize and serious damage will take place. On the other hand, the chances of its surviving unscathed if kept running are very high. It is therefore sound operating practice to choose that course of action that has the greatest odds in favor of success, despite the risk of greater damage should the events depart from the law of probability.

There is a special case in which a safe procedure can be brought into play and that will ensure that the pump will not seize under flashing conditions. I refer to steam turbine-driven pumps provided with separate booster pumps driven by an electric motor that may become inoperative because of electric power failure. The steam turbine-driven pump should be kept running but brought down drastically in speed, say to 1000 or 1500 RPM. The available NPSH should be sufficient to restore normal noncavitating operation at these lower speeds. After the pump has run at this reduced speed for several minutes, it can be safely tripped out if, by that time, power has not been restored to the booster pump motors.

Question 4.13 Minimum Operating Speed

In several of your discussions on variable-speed operation of boiler feed pumps, you have mentioned that pumps should not be operated below a certain minimum speed. Could you explain the reasons for this limitation?

Answer. Actually, my recommendations on minimum speed have always referred to the initial start-up of a new steam power plant rather than to normal operation. But in general I might say that whether a minimum speed should or should not be set is a matter partly of opinion and partly of design.

If a pump is designed so as to depend on the support given by internal running joints that act as water-lubricated bearings, then certainly a water film must always be provided at these joints to act as a lubricant for the internal bearings. Therefore, some minimum pressure would have to be maintained across these internal joints—a pressure that can be available only if each stage of the pump generates some 50 to 100 psi pressure. This requirement, in turn, would dictate that the pump should not be operated below a certain minimum speed, that is, a speed sufficient to generate this pressure. This minimum speed would then apply throughout the entire lifetime of the pump.

On the other hand, if the pump design is such that no dependence is made on internal water-lubricated bearings for supporting the rotor, there is no basic mechanical requirement for any minimum operating speed. For instance, many boiler feed pumps arranged for steam turbine drive are capable of operation at speeds as low as 30 RPM on turning gear once the pumps have been adequately warmed up and reasonably uniform temperatures exist throughout the pump casing barrel.

A different problem exists during the initial operation of a steam power plant. Regardless of how carefully the piping and related equipment have been cleaned prior to the start-up, there is always the posibility that small particles of foreign material (such as mill scale or brittle oxides) will find their way into the boiler feedwater circuit. If any such particles get into the wearing ring or balancing device clearances, there is imminent danger of damage to the pump. Very little torque is available to the pump at extremely low speeds, but the torque required to overcome the dragging friction caused by particles lodged in the close clearances increases. As a result, the moment that the available torque is exceeded by the required torque, the pump will seize. In the case of boiler feed pumps that are built of stainless steel materials (and these have easily galling characteristics), the seizure is accompanied by severe damage to the metal surfaces and, potentially, by the destruction of the pump internals.

It is to avoid this possibility that I have been recommending running boiler feed pumps at a speed high enough to develop something like 100 psi per stage during the initial operation of the steam power plant. I believe that this pressure differential will cause sufficient flow to take place through the close clearances and to wash out any particles that would have a tendency to lodge in the clearances. The time element during which this minimum speed should be maintained is, of course, problematic and must be left to the operator's judgment of the relative cleanliness of the feedwater piping. Once danger of particles entering the pump clearances becomes insignificant, all minimum-speed restrictions can be removed.

Incidentally. should a pump ever suffer a seizure that may have been caused by the presence of foreign matter within the clearances, it is recommended that steps be taken to clear the obstruction before time and effort are expended in stripping the pump down. It may well be that the

Chapter 4   351

interference is not a major one and that it can be easily corrected. The pump should be uncoupled from its driver and the bearing bushings removed. If the pump is provided with conventional stuffing boxes, the packing should also be removed. Then the rotor should be moved gently both axially and up and down. If it is possible to backwash the pump at this time, the chances of removing the obstruction from the foreign matter are excellent.

Question 4.14 Basis for Minimum Flow

What dictates the minimum recommended flow for a centrifugal pump?

Answer. There are several separate and distinct unfavorable effects that can interfere with the reliable and extended operation of centrifugal pumps at reduced flows:

1. For single-volute pumps, the increased radial thrust at reduced flows may impose excessive loads on the thrust bearings. (See Question 6.31.)
2. As the pump capacity is reduced, the temperature rise of the pumped liquid increases. (See Question 4.15.)
3. At certain reduced flows, internal recirculation will occur in the suction or the discharge areas of the impeller, or in both. (See Question 1.29.)
4. High specific speed pumps have power curves that rise with reduced flows, and this may lead to driver overload. (See Fig. 4.29.)
5. When the liquid pumped contains an appreciable amount of entrained air or gas, operation below certain flows will not permit this air or gas to be swept through the pump, which can then become air or vapor bound. (See Question 1.32.)

Each one of these effects may impose a different recommended minimum flow. The final decision is then based on the greatest of these individual minimums.

Question 4.15 Elimination of Recirculation Bypass

I am trying to find a solution to a problem that confronts most users of boiler feed pumps, and I wonder if you could suggest some leads that might be pursued. This problem is that of recirculating operation of the pump during periods of light boiler load, with the attendant wasted power and erosion of the recirculating valve and associated piping. The ideal solution, of course, would be to design a pump that could operate safely at very low flows, but this does not appear too promising. If it is necessary to operate the pump at greater than required output, the problem then becomes one of reclaiming some or all of the excess energy in some other part of the cycle.

The technical literature seems to be devoid of any material on this subject. Do you know of any work that has been done, or do you have any ideas as to a solution of this problem?

Answer. As you say, it is unfortunate that the ideal solution of designing a pump that could operate safely at very low flows is not a practical one. Such a solution would require repealing several very important laws of nature. One of these laws is that energy cannot be destroyed and can only be converted into another form of energy. Thus, since it is impossible to build a pump with 100% efficiency at all flows, the power lost in frictional and hydraulic losses must perforce be transformed into heat, which is transferred to the feedwater and raises its temperature.

The temperature rise in a boiler feed pump can be calculated from the relation

$$\text{Degrees rise (°F)} = \frac{(\text{brake hp} - \text{water hp}) \times 2545}{\text{capacity in pounds per hour}}$$

This relation can be simplified into another equation that is more readily usable:

$$\text{Degrees rise (°F)} = \frac{\text{total head in feet}}{778} \left(\frac{1}{\text{efficiency}} - 1 \right)$$

Figure 4.10 shows a typical performance curve for a boiler feed pump on which there has been superimposed a curve of the temperature rise caused by frictional and hydraulic losses within the pump. This curve indicates that at or near the design conditions the temperature rise is not a significant item, because the losses are at a minimum and a large flow is available to absorb the heat generated. But the temperature rise increases as the flow through the pump is reduced because the losses which generate heat are increasing while the quantity of feedwater to which this heat is transmitted is itself reduced.*

---

*For detailed treatment of the subject of temperature rise, refer to Worthington Reprint RP-737, "How to Calculate Temperature in Centrifugal Pumps," reprinted from Industry and Power, June 1955.

Chapter 4

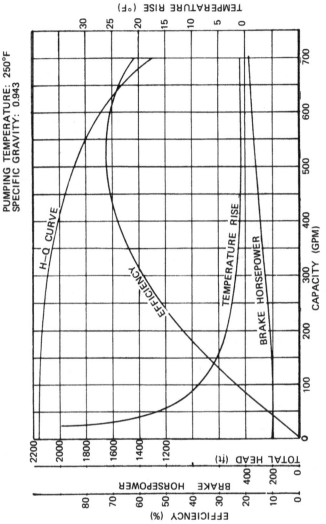

Figure 4.10 Typical characteristic curve of a multistage centrifugal pump showing, in addition, temperature rise versus capacity.

The maximum permissible temperature rise for a given pump installation is, strictly speaking, a matter of empirical recommendation. Since about 1940, it has been the practice to hold this rise to somewhere around 15 °F, although a number of installations have been made with up to 20 and even 25 °F rise. Of course, the decision on the maximum permissible rise automatically determines the minimum permissible flow through the boiler feed pump. This minimum flow is accomplished by providing a bypass recirculating line. The operation of this bypass is generally automatically controlled. The recirculated feedwater is returned to the deaerator in open cycles and to a low-pressure heater or to the condenser in closed cycles.

Greater attention has been paid to the problem of establishing a value for the minimum flow in recent years. On one hand, the larger sizes of boiler feed pumps and the higher design pressures have led to very significant values of minimum flows, requiring large and expensive valves and introducing appreciable disturbances when surges arise from the sudden increase or decrease in flow as the bypass controls open or close the recirculation. On the other hand, with the higher operating pressures, if the maximum temperature rise increased substantially to reduce minimum flows, difficulties may arise because of the effect of a decreasing specific gravity on the shape of the head-capacity curve when expressed in pounds per square inch.

To combat these problems a few installations have incorporaed a *dual bypass control*. Two separate bypass orifices and control valves are arranged in parallel. As feedwater flow decreases, the first one of the two valves opens. If the feedwater flow to the boiler decreases further, the second valve opens. The closing of these valves as feedwater demand increases follows the same pattern; first one and then the second valve closes.

To return to another portion of your question, you must remember that what you refer to as "excess energy" is generally reclaimed in the cycle. If the feedwater in the bypass is returned to the deaerator or to a low-pressure closed heater, the added enthalpy it has received in the form of pressure rise and heat in the pump is available energy; it actually reduces the amount of steam required in the heater to which it is returned. This effect can be calculated quite accurately.

On the other hand, we must consider that any loss of energy caused by recirculation generally takes place at a time when the main unit is operating at very low loads and when the overall efficiency of the plant is considerably reduced. It would hardly be practical to spend much money in order to devise some scheme whereby a greater portion of the energy in the recirculation could be reclaimed. It is for this reason that a number of systems proposed to reclaim this energy failed to find much support in the steam power industry. One such scheme, for instance, was to install a high-pressure water turbine through which the recirculating flow would take place, generating some useful horsepower in the process. The high cost of such a turbine, the small contributions that it would make to the overall plant efficiency, and the intermittent availability of power generation in that turbine have given very little incentive toward its consideration.

Question 4.16 Temperature Rise for Liquids Other than Water

In Chap. 25 of your book *Centrifugal Pumps,* on page 439 you give a formula for temperature rise in a centrifugal pump operated at reduced flows:

$$T_r = \frac{H}{778}\left(\frac{1}{e} - 1\right)$$

You also give a convenient approximation of the minimum permissible flow for general-service pumps:

$$\text{Minimum flow in GPM} = \frac{6.0 \times \text{bhp at shutoff}}{\text{permissible temperature rise in °F}}$$

Is this formula applicable to hydrocarbons? If so, are there some standard limits for the maximum permissible temperature rise when pumps are handling liquids other than water, for instance, light hydrocarbons?

Answer. These two formulas are essentially based on pumps handling water with a specific heat of 1 Btu/lb per °F. You will find a statement on that same page 439 to the effect that, when liquids other than water are handled by the pump, it becomes necessary to correct the resulting answer for the difference in the specific heats of the liquids.

I obviously should have given the relation required to make this correction, and you can be sure that I shall do so in the next edition. You may be interested to follow the derivation of the formulas given in the book, and this will help develop the necessary correction for specific heat differences.

The temperature rise is determined by multiplying the difference between the brake horsepower and the liquid horsepower by 2545 (Btu equivalent of 1 hp-hr) and thus calculating the total heat imparted to the liquid in Btu by virtue of the inefficiency of the pump. In other words,

$$\text{Total heat imparted to liquid in Btu per} = 2545\,(\text{bhp} - \text{water hp})$$

$$= 2545\,\frac{QH \times \text{sp gr}}{3960}\left(\frac{1}{e} - 1\right)$$

where Q is the pump capacity in gallons per minute.

But because specific heat is expressed in Btu per pound per degree Fahrenheit, we must restate this equation so that the pump capacity is expressed in pounds per hour instead of gallons per minute:

$$Q = \frac{W(\text{capacity in pounds per hour})}{500 \times \text{sp gr}}$$

Then,

Total heat imparted to liquid in Btu per hour

$$= 2545 \left(\frac{WH \times \text{sp gr}}{3960 \times 500 \times \text{sp gr}}\right)\left(\frac{1}{e} - 1\right)$$

$$= \frac{WH}{778}\left(\frac{1}{e} - 1\right)$$

Now, to determine the temperature rise, we need to divide this total heat by the pump capacity in pounds per hour and by the specific heat, and

$$\text{Temperature rise } T_r = \frac{H}{778}\left(\frac{1}{e} - 1\right)\frac{1}{\text{sp heat}}$$

In the formula given on page 439 for the minimum permissible flow in gallons per minute in terms of the bhp at shutoff and of the permissible rise in temperature, the factor of 6.0 was chosen as an approximation to reflect the general shape of the bhp and water hp curves of general-service pumps. It is a *very rough* approximation and considerably on the safe side. Let us see how it was derived.

if

W_p = minimum permissible flow in pounds per hour

Q_p = minimum permissible flow in gallons per minute

T_{rp} = maximum permissible temperature rise in degree Fahrenheit

then

$$W_p = \frac{\text{hp losses at } T_{rp} \times 2545}{T_{rp}}\left(\frac{1}{\text{sp heat}}\right)$$

and

$$Q_p = \frac{\text{hp losses at } T_{rp} \times 2545}{500 \times \text{sp gr} \times T_{rp}} \left(\frac{1}{\text{sp heat}}\right)$$

$$= \frac{5.09 \times C \times \text{bhp at shutoff}}{T_{rp}} \left(\frac{1}{\text{sp gr} \times \text{sp heat}}\right)$$

where

$$C = \frac{\text{hp losses at } T_{rp}}{\text{bhp at shutoff}}$$

The coefficient C will vary with the type of pump in question. Generally, for pumps on general service, it will be less than 1.0, ranging from 0.8 to 0.9, and you can well see that the factor of 6.0 used in the equation for minimum permissible flow is overconservative, having been based on a C coefficient of about 1.2.

When we come to the second half of your question, that is, what constitutes a standard or reasonable value for the maximum permissible temperature rise when pumps are handling light hydrocarbons, the answer is not too definite. It depends to a great extent on the relation between available and required NPSH values. The reason for this dependence is that some of the liquid handled by the pump will pass from the discharge of the impeller back into the suction passages of the pump through the wearing ring clearances. This short-circuiting leakage will have absorbed heat from the pump internal losses and will therefore be at a temperature in excess of the suction temperature by the amount of the temperature rise we have calculated. Mixing with the incoming liquid, the leakage will increase the temperature at the eye of the impeller and reduce the available NPSH by the amount of increase in vapor pressure.

This effect is totally negligible under normal operating conditions of capacity because the leakage is a negligible portion of the design capacity. But since the leakage remains essentially constant while capacity is being reduced, this effect can be quite important in the range of minimum flow conditions. Speaking generally, therefore, it is best to keep the temperature rise to below 10°F on hydrocarbon service and, when NPSH is critical, to about 5°F.

Question 4.17 Continuous Versus Intermittent Operation

I note that minimum flows recommended by manufacturers differentiate on occasion between "intermittent" and "continuous" operation. How does one make a decision as to what constitutes continuous and what intermittent operation?

Answer. The distinction is based on an almost subjective interpretation of the effect of the length of time a pump operates below some critical capacity on the life of that pump. I think that we are all guilty of a certain ambiguity in speaking of "continuous" and "intermittent," since we never define exactly what we mean by these two terms. I know that it would be impractical to specify exactly the number of hours that demarcate the boundaries between these two conditions, but we might try to be a little more specific. At one time I used to suggest the following approach to this problem: there are two distinct and separate periods in the operating life of a pump that should be reviewed. The first covers the period that starts once the system served by the pump is in full operation; the range of capacities and heads encountered within this period can be predicted with reasonable accuracy. For this period, we might choose some arbitrary subdivision and, for instance, define as "continuous" operation that range of conditions that will be encountered over 25 to 100% of the time. "Intermittent" operation during this period would cover that which occurs 25% or less of the time.

The second period refers to the start-up conditions of the system. Here it is difficult to predict what the duration of operation at any capacity may be. Unfortunately, this portion of the life of the pump is seldom given any consideration in the design stage of the pumping system, although the pump may operate over very wide swings in capacities for extended periods of time. The result is that frequently damage befalls the pump from operation in an undesirable range of flows even before the system has gone into full operation.

But there is danger even in such an arbitrary definition of the two modes of operation. Assume for instance that once a critical flow has been established by the manufacturer, the user finds that the pump will operate less than 25% below this critical flow. He will then assume that these conditions are intermittent. However, I submit that if they occur 25% of the time but this period is concentrated, say, in 3 consecutive months of the year, this is definitely continuous operation. Even if they occur for 6 hr every day, I hold them to be continuous.

I think, therefore, that we must choose some other approach. As I have already stated, there are several separate end results created by operation below some critical flow:

1. Excessive radial thrust, leading to shaft or bearing failure.
2. Excessive temperature rise.

Chapter 4

3. In the case of internal recirculation, hydraulic pulsations and mechanical vibrations, leading to possible mechanical failure of the pump components, including that of the impeller, of the bearings, and/or of the seals. These failures can be either progressive or catastrophically instantaneous.
4. Still in the case of internal recirculation, cavitation-type damage to the impeller. This will always be progressive, but the rate of damage will depend on a variety of factors.
5. Overload of drivers in the case of high specific speed pumps.
6. Air or vapor binding in the case of pumps handling liquids with appreciable amounts of entrained or dissolved air or gas.

All these effects must be examined separately insofar as the setting up of minimum flows is concerned. Those minimum flows dictated by effects that can lead to very rapid failure or serious deterioration of performance must be assumed to be discrete numbers, and operation below these flows should never be permitted. On the other hand, operation below the minimum flow dictated by cavitation-type damage to the impeller need not lead to an immediate destruction of the impeller but will shorten its life. Therefore, operation below this flow can be permitted to occur intermittently at the cost of a shorter life of this pump component.

To evaluate this cost we need to assume that an approximate relationship can be established between a "life" coefficient for a part like an impeller and the flow expressed as a percentage of the capacity at bep. Then, the overall life coefficient can be calculated by weighting individual coefficients by the percentage of time the pump will operate at several flows, as shown in the following table:

% Flow at bep	Assumed % of time of each flow	Life coefficient at this flow	Calculation of overall life coefficient
100.0	20	1.00	$1.0 \times 0.2 = 0.20$
75.0	50	0.80	$0.8 \times 0.5 = 0.40$
50.0	20	0.60	$0.6 \times 0.2 = 0.12$
25.0	10	0.25	$0.25 \times 0.1 = 0.025$
			Overall life coefficient = 0.745

We would then say that the overall life coefficient of the part in question is about 0.75, or that the part would last 75% of the time it would have lasted if the pump operated constantly at its best efficiency point. Of course, one would have to be able to estimate life coefficients at various percentages of flow, and this is still an arcane subject. But we'll get there one of these days.

Incidentally, all these remarks tend to give further support to my recommendation that pump selections for widely swinging loads be met by two or more pumps operating in parallel. This permits running a single pump during start-up conditions or during low-load periods, so that the minimum flow is now reduced to the minimum recommended flow of a single pump.

Question 4.18 Cooling and Sealing Water for Standby Pumps

Our plant has a great many centrifugal pumps installed on various services. Generally, a standby pump is provided for each service. Some of these pumps are started automatically from a remote location. What is the recommended practice with regard to cooling-water supply to bearings or to stuffing boxes in such cases?

Answer. Three separate solutions to this problem are available, and the choice among these three must be dictated by the specific circumstances surrounding each case:

1. A constant flow may be kept through the bearing jackets or oil coolers and through the stuffing-box lantern rings, regardless of whether the pump is running or standing still on standby service.
2. The service connections may be opened automatically whenever the pump is started up.
3. The service connections may be kept closed while the pump is idle and the operator instructed to open them within a short interval after the pump has been put on the line automatically.

The first solution is wasteful of cooling water and may be dangerous in certain cases. The necessity of regulating the amount of cooling water to the pump bearings is frequently overlooked, and generally the error is to overcool rather than to supply insufficient cooling water. A great number of ball bearing failures are due to the bearing being almost "refrigerated," so that the resulting condensation on the cold inside walls of the bearing housing mixes with the lubricating oil or grease. The rusting and pitting of the balls lead to obvious trouble. It is considered good practice not to permit the outflowing cooling water to fall much below 105 to 115 °F.

Since cooling water is frequently available at temperatures as low as 60 to 70 °F, permitting the flow of this water to take place through the bearing housing of an idle pump installed in a warm or moist atmosphere may lead to bearing troubles. It must be remembered that while the pump is not running, no heat is generated at the bearings, and the bearing housings will be maintained at exactly the temperature of the cooling water.

There are, however, certain cases in which a supply of cooling or rather sealing water to the pump stuffing boxes must be maintained whether the pump is running or not. Such, for instance, are cases where the pumps handle a liquid corrosive to the packing or a liquid that may crystallize and deposit on the shaft sleeves. The results of shutting off the sealing supply while the pump is idle are obvious in such cases. If the stuffing boxes are equipped with water-cooled jackets, leaving the connections open at all times may be wasteful but presents no particular danger.

It is possible to equip the individual water-supply lines with spring-loaded pressure control valves as illustrated in Fig. 4.11. The pressure side of the diaphragm is connected to the pump discharge by means of a pilot line, so that the valves will open as soon as the pump starts and develops pressure.

If the standby pump is motor driven, it is possible to install solenoid-operated control valves in the water-cooling supply lines. The valves remain closed under spring action as long as the solenoid is deenergized and open as soon as the motor is put on the line, energizing the solenoid.

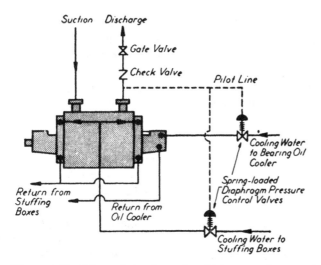

Figure 4.11 Schematic hookup of typical automatic control for cooling-water supply. Spring-loaded diaphragm valves open when pump starts and builds up pressure.

In either case, whether the valves are controlled by pressure or by solenoids, it is wise to supply them with a locking device. This will permit the operators to lock them in the open position as soon as the operators have time to attend to a pump that has been started automatically.

If operators are available near the pump location, and if the pump is of such a design and on such a service that it can be operated a few minutes without cooling- or sealing-water supply, the third solution may be the simplest.

Question 4.19 Bearing and Gland Cooling

Our plant has embarked on a drastic program of water conservation, and I have been assigned to the phase dealing with auxiliary services. One of my problems is to reduce the total amount of water used in such applications as bearing cooling and quenching at the glands for centrifugal pumps. Can you suggest any measures that would appreciably reduce the amount of water used in this service?

Answer. One of the major sources of saving in an auxiliary water supply will probably be found in the reduction of the amount of cooling water to water-cooled bearing jackets. It has been my experience that the average operator provides too much cooling water. Beside the fact that this is a wasteful practice, it can also be quite harmful to the life of the bearings. Excessive cooling has two unfavorable effects:

1. The outer race of a ball bearing is cooled excessively and contracts, while the inner race and the balls remain at a considerably higher temperature. The effect of this combination is to squeeze the bearing balls and cause damage to the bearing.
2. Excessive cooling lowers the temperature in the bearing cavities to the point that atmospheric moisture condenses and contaminates the lubricant.

A definite temperature rise should be permitted to take place in the bearing cooling-water supply. Actually, a thermostatic valve can be set in the water-cooling line, with the valve arranged to throttle the flow so as to prevent the temperature rise across the bearing jacket from falling below, say, 20°F.

Another interesting means of reducing an auxiliary water supply is to pipe the cooling services in series. In many cases, the line and thrust bearing jackets can be so arranged. The supply should come to the line bearing first (where very little heat needs to be picked up) and then to the thrust bearing. Another arrangement, illustrated in Fig. 4.12, pipes the outlet

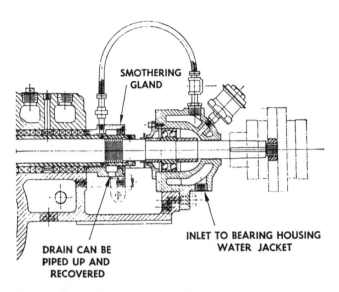

Figure 4.12 In this arrangement for conserving cooling-water supply, the outlet of the bearing water jacket is piped to the quenching gland and outgoing flow recovered from the gland drain.

from the bearing water jacket to the quenching gland connection. The drains from the water-quenching glands can be piped up and recovered.

Question 4.20 Hand-Operated Auxiliary Oil Pumps

Our boiler feed pumps have a forced-feed lubrication system. There is a small hand oil pump located within this system. What is the purpose of this pump, and does its presence indicate that the pumps cannot be started by remote control?

Answer. The purpose of a hand oil pump in such a lubricating system is to clear the air pockets out of the oil system and to ensure that the oil is ready to flow to all the bearings. Actually the hand oil pump need be used only at the time of the initial filling or after the oil system has been drained or dismantled to any extent. Of course, it is recommended that, if a pump stays idle for a long period of time, this hand oil pump be used, say, once a week. This will take care of any evaporation and will remove any sluggishness of the oil system on a new start-up. If, therefore, a standby pump is turned over once a week, as is the case in many steam power plants, the use of the hand oil pump is not required except when the lubrication system has been dismantled or the oil changed. There is sufficient oil trapped in the bearings to take care of all starting requirements.

Therefore, if standby pumps are run once a week or if the hand oil pump is used once a week on idle pumps. there is absolutely no reason why the boiler feed pumps cannot be started by remote control.

Question 4.21 Lubricating Oil Characteristics

The following material is presented not as an answer to a question but rather as an account of problems that can be encountered if one does not take into consideration all the factors that can affect the characteristics of a lubricating oil.

Operation of high-pressure boiler feed pumps requires that careful attention be paid to many details. Failure to give full consideration to these details can easily introduce serious problems and costly maintenance. Even greater costs are incurred if these problems lead, as they can, to interruption of service. It is seldom that such difficulties can be directly traced to the use of improper lubricating oil or to appreciable deviations from the required oil temperatures. Yet since such difficulties do occur and can cause considerable damage, it may be useful to relate in some detail a recent case that came to my attention and that took some time and effort to straighten out.

The installation in question consists of two full-capacity high-pressure boiler feed pumps driven by 1750 hp, 3575 RPM motors. One pump serves the unit, and the second pump is held on standby duty. Pumps incorporate a Kingsbury thrust bearing and sleeve-type line bearings with forced feed lubrication provided by an Imo-type positive displacement pump driven directly from the pump shaft as shown in Fig. 4.13. In addition, the lubricating system incorporates an auxiliary oil pump of the gear type, mounted vertically over the oil tank in the baseplate and driven by a 1 hp, 1725 RPM electric motor. The complete oil lubrication system is shown diagrammatically in Fig. 4.14.

The service oil pump delivers lubricating oil through a duplex oil filter and a vertical tubular cooler into a header from which supply connections are made to the pump thrust bearing, the pump line bearings, and the two motor bearings. Oil returns from all these bearings to a sump in the pump baseplate. The auxiliary oil pump delivers oil in parallel with the service pump, check valves being provided at the necessary points in the respective discharge lines to prevent reverse flow through an idle pump and back into the sump.

Two oil pressure switches are located in the circuit. The first of these is the switch for the main motor starter interlock, set so that the main

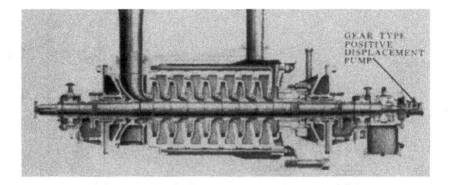

Figure 4.13 High-pressure boiler feed pump in which forced feed lubrication is provided by positive displacement pump driven by boiler feed pump shaft.

motor switch contact closes at 15 psi and opens if the pressure falls to 10 psi. It also incorporates an alarm contact that closes at 10 psi and opens at 15 psi. The second switch controls the operation of the auxiliary oil pump, stopping it whenever the oil pressure reaches 18 psi and restarting it when the oil pressure falls to 12 psi. In addition, the lubricating system incorporates the necessary oil pressure gauges and thermometers. Finally a relief valve set at 20 psi is located in the discharge from the oil cooler.

Pump specifications submitted to the pump manufacturer had stipulated that the cooling water to be used in the oil coolers would be at 100 °F.

When oil piping diagrams were submitted to the customer for approval, the required quantities of cooling water were indicated thereon and instructions were given to use oil of 250 SSU viscosity at 100 °F. Operating temperature has a great deal of effect on oil viscosity, as can be seen from Fig. 4.15. Thus, if oil rated at 250 SSU at 100 °F is used, it will have a viscosity of about 150 SSU at a temperature of 120 °F, which is approximately the temperature obtainable at the exit from an oil cooler operated with 100 °F cooling water. If the exit temperature deviates appreciable from 120 °F, the cooling-water supply must be readjusted to obtain the desired oil temperature.

This recommendation is based on the following facts:

1. Kingsbury bearings are rated for thrust load capacity at an operating viscosity of 150 SSU. Viscosities much lower than this will therefore reduce the thrust capacity of the bearing.
2. If, by virtue of oil temperatures much lower than recommended, the operating viscosity increases appreciably above 150 SSU,

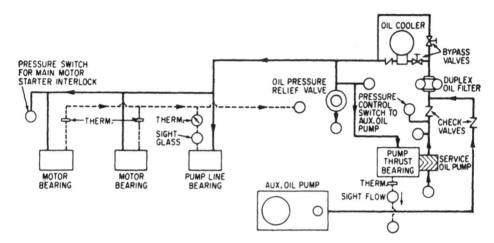

Figure 4.14 Diagram of complete lubrication system showing the arrangement of both service and auxiliary lubricating oil pumps.

power consumption of the auxiliary oil pump motor will increase to a point where serious motor overload can take place.
3. If oil temperatures are permitted to fall to 70 or 60 °F, for instance, not only will the viscosity increase excessively, but danger exists that moisture condensation will take place in the oil system and water will start accumulating in the oil tank.

Very shortly after the installation and initial operation of these pumps, I was advised that serious operation difficulties were encountered. Whenever it was desired to switch pumps and put the standby pump on the line, the operating sequence would start up the auxiliary oil pump. Immediately, the motor driving this pump would become overloaded by as much as 100%, and the thermal overload protection relays would kick the motor off the line. Thus, it was impossible to get the main motor of the standby pump started, since its starter was interlocked in such a manner with the oil pressure switch that failure of the auxiliary oil pump to provide adequate pressure prevented the starter from functioning. Conclusions reached by the operators were that the motor driving the auxiliary oil pump was inadequately sized and judging by the amount of overload involved should be replaced by a 3 hp motor.

An on-the-spot investigation disclosed several important discrepancies between the recommended oil characteristics and temperatures and the prevailing values. In the first place, an oil with 300 SSU viscosity at 100 °F

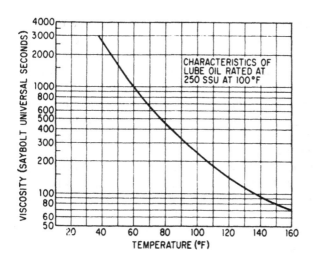

Figure 4.15 Variation of oil viscosity with temperature.

was used. The fact that contributed most heavily to the overload condition, however, was that the oil temperature leaving the oil cooler ran to as low as 75 °F during the daytime and less than this at night when the ambient temperature was lower. (This is an outdoor installation with cooling towers, and therefore the cooling-water supply is directly affected by ambient temperatures.) If reference is made to Fig. 4.15, it will be noted that viscosities of the oil may run as high as 1000 SSU under these conditions. The effect of such high viscosities on the auxiliary oil pump motor power consumption is obvious.

Discussions with the operators disclosed that 250 SSU oil at 100 °F was not readily available, and an immediate decision to permit using 225 SSU oil was reached. The question of temperatures, however, was a different matter. It developed that the specification had mentioned a 100 °F cooling-water temperature in order to be sure that the size of the oil cooler would be adequate. Cooling water was to come from the well supply and would be about 75 to 80 °F the year around.

There was another problem that we could readily see facing us: Even disregarding the cooling-water temperature, the outdoor installation would cause oil temperatures in the oil sump of the standby pump to fall to 40 to 50 °F ambient in winter nights. Should the standby pump be called upon to start under these conditions, the resulting overload on the auxiliary oil pump motor would kick this pump off the line immediately upon starting and prevent bringing the standby pump on the line. The operators, therefore, were extremely insistent on replacing the auxiliary oil pump motor by a larger unit—possibly as large as 5 hp. However, even a 5 hp motor would not do the job with 40°F oil when the viscosity would reach as high as 3000 SSU! In addition, operating the lubricating system under such conditions was absolutely unthinkable. Cold oil would in time pick up moisture from condensation, which would eventually lead to bearing failure. It was finally agreed not only that a large motor was unnecessary but that serious troubles would eventually result unless oil temperatures were kept to reasonable values.

Before a permanent solution was reached, it was necessary to provide temporary means to permit operation of the standby pump without the danger of its being shut down by the thermal overload protection of the auxiliary oil pump motor. Overloading of the auxiliary oil pump motor during start-up was due to excessive pressure drop through the filter and cooler when the oil was cold at start-up. Some means, therefore, had to be introduced for limiting the discharge pressure (and therefore power requirement) of the auxiliary oil pump.

This could best be accomplished by means of a 3/4 in. relief valve cut into the auxiliary oil pump discharge line before the check valve and discharging into the oil reservoir. Of course, this relief valve would operate only when the auxiliary oil pump was running. The valve would be set at about 40 psi, which would limit the motor horsepower to within the rated conditions. Flow through this valve would be about 5 to 6 GPM when the oil was cold. At higher oil temperatures, the relief valve would dump little or no oil. This relief valve would also acts as protection for the auxiliary oil pump in the event of complete clogging of filter or cooler. There would be no chance of damage to the bearings under these conditions, because the low-pressure cutoff switch located at the far motor bearing would trip the main breaker and take the unit out of service.

This relief valve was installed, and the oil changed to 225 SSU at 100°F rating. After proper adjustment of the relief valve, motor amperes were maintained at the rated conditions, and the problem was temporarily solved.

It remained, however, to find a permanent solution for the conditions that would prevail during winter when ambient temperatures would introduce the hazard of moisture condensation in the lube system.

Two solutions were readily available to the operators:

1. The auxiliary oil pump could be operated continuously on the standby pump. This would keep the oil in the reservoir up to reasonable temperature by virtue of the heat delivered to the oil from the auxiliary oil pump itself and the heat picked up by the

oil in its flow through the boiler feed pump bearings. The latter are heated by conduction from the pump, which is kept warm by circulating hot water through it.
2. Thermostatically controlled immersion-type electric heaters could be installed in the oil reservoir to maintain it at a desired temperature.

The first solution was finally adopted, and the installation has been operating satisfactorily since these various changes have been incorporated.

Question 4.22 Rules for Operation and Maintenance of Centrifugal Pumps

Are there some basic rules or concepts that you can suggest for the operation and maintenance of centrifugal pumps? It would seem to me that a very concise listing of such rules would help some of the younger people in the field who are still in the process of developing experience in these areas of plant operation.

Answer. As industrial processes grow in complexity, so do the tasks that face plant operators and maintenance personnel who are involved with a double problem. On one hand, this growth in complexity results in the use of more and more sophisticated equipment and in greater interaction between the different components of a plant. On the other hand, the reliability and uninterrupted service of each piece of equipment becomes more and more vital to the reliability and continued service of the entire plant. And the cost of downtime in the larger plants with which we are dealing today is growing at an alarming rate.

Little wonder, then, that so much emphasis is being placed on the importance of preventive and corrective maintenance of plant equipment. You have touched on a very important facet of the overall problem: With the growth in the number of plants of all kinds in the world, we are constantly having to add personnel in need of training to serve these plants.

You mentioned operation and maintenance of centrifugal pumps. One must also include selection, installation, and troubleshooting, because all these play a part in providing the most reliable service, the least expensive maintenance, and the longest life possible from the equipment installed in a plant.

To explain: Proper maintenance does not start with repairs or replacement of worn parts but right at the time of equipment selection. Operating demands to be placed on the equipment must be adequately anticipated over the projected life of the equipment, and both the equipment selection and

Table 4.1 Rules for Preventive Maintenance: Selection

Pumps and compressors	Turbines
1. Advise the manufacturer of the exact nature and characteristics of liquid or gas to be handled, including temperature range.	1. ←
2. Check into required capacities.	2. Check required power and speed.
3. Analyze suction or inlet conditions.	3. ←
4. Analyze discharge conditions.	4. ←
5. Advise the manufacturer whether service is continuous or intermittent.	5. ←
6. Determine what type of power is best suited for the drive. Should speed be constant or variable?	6. ←
7. Advise any space, weight, or transportation limitations involved.	7. ←
8. Advise any significant effect of location of installation (elevation above sea level, geographic location, and immediate surroundings).	8. ←
9. Be sure that sufficient spare or standby equipment is available.	9. ←
10. Keep sufficient spare parts on hand.	10. ←

the design of the system in which it must operate must be suitable for these demands (Table 4.1).

If proper selection is important, so is adequate installation. Most of us—be we manufacturers or plant operators—have seen the best possible equipment fail prematurely because some fundamental precautions were neglected at the time of installation (Table 4.2).

And finally, good maintenance depends on good operation. All the efforts on the part of a maintenance department can be wasted if there is no equal effort on the part of the production personnel to operate the equipment as it was designed to operate (Table 4.3).

When I started to answer your question, I intended to discuss just the specific piece of equipment to which you referred: the centrifugal pump. But when I proceeded to prepare the material for my answer, a strange idea

Chapter 4

Table 4.2 Rules for Preventive Maintenance: Installation

Pumps and compressors	Turbines
1. Install equipment in light, dry, and clean locations whenever possible.	1. ←
2. Foundations should be rigid	2. ←
3. Bed plates should be grouted	3. ←
4. Equipment and driver alignment must be checked under operating conditions.	4. ←
5. Piping should not impose excessive strains on equipment.	5. ←
6. Use as direct piping as possible especially at inlet.	6. ←
7. Provide vent valves at high points for pumps, drain connections for pumps and compressors.	7. ←
8. Provide warm-up and bypass connections for centrifugal pumps, relief valves for positive displacement pumps and compressors.	8. ←
9. Provide a suitable source of cooling water	9. ←
10. Install suitable gauges, flowmeters, and thermometers.	10. ←

emerged. This idea is that it is remarkable how little difference there is between the rules that should be followed for the proper maintenance of such different pieces of mechanical equipment as a centrifugal pump, a power or steam pump, a steam turbine, a compressor, or even an engine. It is obviously true that each one of these machines requires a different set of diagnostic instructions to determine why it may not be performing as intended. But when it comes to preventive maintenance rather than troubleshooting, it would seem that all these different machines are equal before the Great Engineer (Table 4.4).

As stated earlier, these fundamental rules can be broken down very readily into four separate areas: (1) selection of the equipment, (2) installation, (3) operation, and (4) maintenance. To underline the equal importance of these four areas, I have selected a total of 40 basic rules, broken down equally into 10 rules for each of the areas. These four groups of rules are

Table 4.3 Rules for Preventive Maintenance: Operation

Pumps and compressors	Turbines
1. Observe instruction book start-up and shutdown procedures.	1. ←
2. Operate equipment within range of flows, pressures, and temperatures specified by manufacturer.	2. ←
3. Do not throttle suction to reduce pump capacity.	3. Throttle inlet to vary speed and power.
4. A pump handles liquids; keep air out. A compressor handles gases; keep water out.	4. Avoid wet steam conditions.
5. Do not use excessive lubricant or excessive cooling water.	5. ←
6. Avoid shocks from sudden temperature changes.	6. ←
7. Make hourly observations.	7. ←
8. Do not run equipment if excessive noise or vibration appears.	8. ←
9. Run spare equipment occasionally to check its availability.	9. ←
10. Set up scheduled semiannual and annual inspection.	10. ←

presented in Tables 4.1 through 4.4. In each table, the rules are broken down arbitrarily into two columns, the first column covering pumps and compressors (the driven equipment) and the second covering turbines (the driver). Note how with but a couple of exceptions the second column consists mainly of arrows pointing to the left. It means that exactly as I have stated earlier, the rules are the same, regardless of the specific equipment we are considering.

It has been my experience that any equipment fares considerably better if its operator has confidence in this equipment. Contrariwise, regardless how reliable this equipment is in itself, it generally tends to develop unexpected problems if the operator lacks confidence in it. In turn, confidence can only be developed through complete understanding. Therefore, if you want to have reliable equipment that performs well and seldom causes shutdowns, learn how it is made, why it is made as it is, and

Chapter 4   373

Table 4.4 Rules for Preventive Maintenance: Repair and Maintenance

Pumps and compressors	Turbines
1. *Do not* open equipment for general inspection unless diagnosis indicates the need.	1. ←
2. Great care is required in dismantling equipment. Follow instruction book procedures.	2. ←
3. Special care is needed in examination and reconditioning of metal-to-metal fits.	3. ←
4. Clean internal surfaces thoroughly and repaint where indicated.	4. ←
5. Use new gaskets for complete overhaul.	5. ←
6. Examine parts for corrosion, erosion, and other damage.	6. ←
7. Check concentricity of parts.	7. ←
8. Restore areas subject to packing wear to proper service condition.	8. ←
9. Exercise great care in mounting anti-friction bearings or in restoring journal bearing surfaces.	9. ←
10. Keep a complete record of inspections and repairs.	10. ←

what constitutes its proper installation, proper operation, and proper maintenance.

Question 4.23   Automation of Boiler Feed Pump Operation

The current trend toward full automation of utility steam generation plants has compressed several years of system analysis and development into a period of months. This has exposed weak points in automation philosophy and forced the parallel development of basic automation concepts and final control systems.

The two major components of a steam plant, the boiler and turbine generator, are already recognized as needing fully automatic controls; they now possess the most complete control systems in the plant. But complete automation of a plant involves many auxiliary units, including the so-called *fluid-handling* group, which is comprised of the condenser, condensate pumps, deaerator, boiler feed pumps, and feedwater heaters. In these areas some questions still remain: Which items shall we automate? How shall we automate them? What equipment factors are important to operating safety or unit performance?

Would you please comment on the question of automation as it applies to boiler feed pumps. Thus, I would like to hear your opinion about what should be controlled and what degree of control the boiler feed pump requires for full automation.

Answer. The basic function of controls is to measure, evaluate measurements, and initiate sequential or corrective action in the light of these evaluations. Obviously, we would want these duties from any control system. The difference lies in the magnitude of total control applied to a plant. We may be concerned with control of a boiler feed pump, but we must analyze each control function from the point of its effect on operation of the complete system.

Of course, complete automation will not necessarily come suddenly and immediately to all new steam power plants. It will more likely be a step-by-step process in which certain functions at present filled by operators will be turned over to automated controls.

After thorough analysis you may decide to go in the direction of full plant automation, which usually means automatic start and stop plus limited operating control by computer. But you still are faced with many decisions concerning degree of automation for plant auxiliary equipment or components of the fluid-handling group. For instance, we do not absolutely need to know at all times the hydraulic performance of boiler feed pumps. We can, instead, establish certain acceptable limits of deterioration beyond which operators will be alerted to the need for remedial maintenance.

In preparing an analysis of system components, it is best to list every possible item concerning a component that may affect your decision about its operation. This includes even items remote from the equipment in question. Strictly speaking, measuring pressure differentials across check and gate valves or across a feedwater regulator and monitoring the position of these valves might be considered outside the realm of boiler feed pump automation. Nevertheless, information on these items generally will be necessary to your pump control decision making. In addition to operational controls, you will have to think in terms of supervisory controls—to protect the mechanical integrity of an unattended pump, for example.

Automatic devices are already used to either position the feedwater regulating valve or alter the pump speed so as to balance the flow of feedwater with the steam flow from the boiler. Also automated is the recirculation bypass used to protect the pump against overheating at light loads. No

Chapter 4 375

major changes are necessary in this area even if the plant operation is to be fully automated, except to integrate the operation of these devices into the overall scheme. A great many other operational or supervisory functions, however, have not yet been automated except to a very little degree.

It is rather strange that equipment like the boiler feed pump has stirred up so little interest in the line of supervisory control. The only exceptions are protection against excessive temperature rise at low flows and the use of motor-driven auxiliary oil pumps to prevent operation with insufficient oil pressure to the bearings.

Objectives of Automation

Two specific objectives are applicable to boiler feed pumps:

1. To start, vary their operating conditions, and stop, fully automatically in answer to plant demand and load conditions
2. To permit the operation of these pumps without attendance

We distinguish between two major classifications of controls: (1) operation controls and (2) supervisory controls. The first, shown in Table 4.5, are directed at such functions as starting or stopping a pump or altering its operating conditions.

Those in the second category (Table 4.6) are intended to prevent the hazards of malfunctioning of the boiler feed pump or of its operation under conditions that could lead to malfunctioning.

Subdivisions of operational and supervisory controls are somewhat arbitrary in that some control functons are borderline cases and some others belong rightly in both categories.

Automation of a piece of equipment requires the measurement of a great number of variable or constant characteristics, evaluation of these measuremants, and the weighing of them against some predetermined norms; the correction of these characteristics where possible or necessary;

Table 4.5 Operational Control Functions

1. Start up pumps for initial start or for restart of the main unit
2. Start up additional pump as station load increases.
3. Shut down one of the pumps on decrease of load.
4. Switch a spare pump for an operating pump.
5 Vary the position of the feedwater regulator or vary pump speed for station load change.

Table 4.6 Supervisory Control Functions

A. Controls for corrective action:
 1. Abnormally low flow—excessive temperature rise
 2. Insufficient NPSH
 3. Insufficient condensate injection for sealing
 4. Insufficient oil pressure or flow
 5. Excessive oil temperature
B. Controls for pump shutdown:
 1. Excessive vibration
 2. Excessive shaft runout
 3. Excessive bearing temperature
 4. Failure of oil pressure
 5. Excessive axial thrust
 6. Excessive misalignment of pump and driver
C. Controls preventing start-up:
 1. Inadequate warm-up
 2. Insufficient oil pressure
 3. Lack of condensate injection for sealing
 4. Excessive misalignment
 5. Improper axial rotor positioning

and their continuous monitoring to assure that the necessary corrective measures have been carried out. In addition, automation requires that a rigid sequence of steps be observed during start-up or shutdown of the equipment and that feedback be incorporated to assure that before any one step of the sequence is taken, all previous required steps have been satisfactorily carried out.

This evaluation and correction and this sequential operation at start-up and shutdown must be accomplished by a computer or a computer component. But a computer has no imagination. It can only "think" of whatever a human being has thought of first. We must foresee all situations and instruct the computer how to deal with each one of them.

What are the variables in and about the boiler feed pump that can affect an automatic decision? Referring to Fig. 4.16, we must first measure the pressure P_1, the temperature T_1, and the flow quantity Q_1 at the pump suction. We must also measure the corresponding quantities P_2, T_2, and Q_2 at the pump discharge. If we do not wish to measure Q_1 and Q_2 at both the suction and discharge, either of these two flows can be measured and the other calculated by appropriately adding or subtracting any in- or outflow. These measurements will, at all times, give us a complete picture of the pump hydraulic performance.

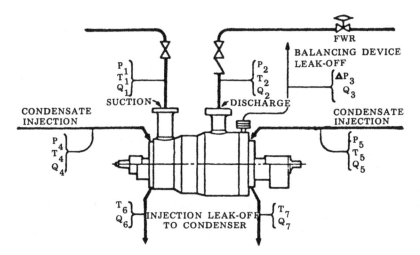

Figure 4.16 Measurement of boiler feed pump characteristics for operation control functions.

If the pump axial thrust is compensated by a balancing device, we shall want to know the flow Q_3 taking place through this balancing device. Measurement of the differential pressure ΔP_3 across the calibrated orifice in the balancing device leak-off line and comparison of this differential pressure with the orifice calibration curve will indicate this flow.

If the pump is provided with condensate injection sealing, the injection pressures, temperatures, and flows at both ends of the pump must be measured. These are P_4 and P_5, T_4 and T_5, and Q_4 and Q_5, respectively. Temperatures T_6 and T_7 and flows Q_6 and Q_7 at the condensate injection leak-offs must also be measured. At each end, the difference between the injection and leak-off ($Q_4 - Q_6$ and $Q_5 - Q_7$) will indicate the amount of condensate that flows into the pump interior.

Pump speed n_1 must be measured both for the purpose of correlating the pump performance to its initial performance and for instigating a number of supervisory control functions. If a variable-speed device like a hydraulic coupling is interposed between the pump and its driver, the input speed to the variable-speed device n_2 must also be measured.

Other measurements. Several oil pressures and temperatures must be measured. These pressure measurements will be taken at the discharge from the pump shaft-driven oil pump, at the discharge of the auxiliary motor-driven oil pump, and at the inlet to the pump bearings. Temperatures will be measured at the outlet of the oil cooler and at the outlets from the inboard line bearing and outboard line and thrust bearing. Bearing temperatures in the metal will be measured by thermocouples embedded within the bearing shells and the thrust shoes.

For the purpose of determining proper temperature distribution during warm-up and for actuating sequential warm-up valve positioning, casing surface temperatures will be measured at several points. Optionally, we may wish to monitor the differential expansion of the pump casing by means of suitably located dial indicators.

Obviously, the actual measuring devices must be located right at the point of measurement. But sensing devices can transmit information to any desired location within the power plant. The question of where this information is to proceed is a controversial one. In my opinion it is not necessary to transmit all this information to the central focus of the power plant control room. Decisions on the operation of boiler feed pumps can be made either at the local level of the pump itself or at the top level of the control room, depending on the nature of the decision.

All information as to oil pressures, bearing temperatures, shaft eccentricity or vibration, casing surface temperatures, and so on, should be fed into a local control center. The decision to shut down one of the pumps because of malfunctioning and to substitute a spare pump (if one is available) would be made there.

This decision would be transmitted to the main control room, and the sequence of steps for shutting down the pump and securing it would be initiated at the boiler feed pump control center. This does not mean that a decision to shut the pump down could not be reached in the main control room as well. But the reasons for such a shutdown would be different from a local shutdown decision: The main control room would decide strictly on the basis of the load carried by the main unit or because the entire unit is coming down.

The integration of all controls into the complete computer entity is not necessary. Only operational information need be fed to this control computer, along with any decisions reached on the local level.

Operational Controls

To meet objectives 2 and 3 of Table 4.5, if more than one pump served the boiler, the automatic control must

1. Sense the need for an increase or decrease in feedwater flow that can best be met by an increase or a decrease of operating pumps
2. Initiate an impulse that will meet this need
3. Feed back the information on this action so as to check whether the need has been met

Decisions to be reached by the operational control are the following:

1. If two pumps are operating and if the station load or total feedwater flow fall below that value that requires the operation of two pumps, one of the two pumps is shut down.
2. If one pump is operating and if the station load or total feedwater flow is increasing and approaching that value that requires two-pump operation, an additional pump must be started up.

Both station load and total feedwater flow to the boiler are monitored at the *top* control level, that is, at the main control room. Having sensed a condition requiring starting or stopping one of the boiler feed pumps, the control would send the corresponding impulse to the *local* control center, instructing it to act in accordance with the decision and proceed with the predetermined sequential operation.

Following is a typical sequence for starting a pump; it is not necessarily complete but is at least indicative of the fact that many individual actions are necessary:

1. Warm-up monitoring certifies that the pump is ready to start.
2. Suction valve monitoring certifies that the pump is ready to start.
3. Check valve monitoring certifies that the pump is ready to start.
4. The cooling-water supply to the oil cooler is opened.
5. The condensate injection valves are opened to the running position.
6. The recirculation bypass opens.
7. The warm-up valving closes for running position.
8. The auxiliary oil pump is started up.
9. Oil pressure and oil flow measurements certify that the pump is ready.
10. The motor is started up.
11. The motorized gate valve in the discharge line opens.
12. The recirculation bypass control switches to flow control.
13. The feedwater regulator or speed control device takes over and determines the operating capacity of the boiler feed pump.
14. Monitoring of operational and supervisory control continues.

An initial start-up or restart initiates similar sequence of steps, except that certain steps (such as warm-up procedures) may be omitted.

Supervisory Controls

Table 4.7 lists the five items that appear under category A in Table 4.6 and indicates the impulse that would be used to initiate the corrective action and the corresponding action itself.

Table 4.7 Supervisory Controls (Corrective)

1. Abnormally low flows or excessive temperature rise	Open recirculation bypass

2. Insufficient NPSH	1. For open feedwater cycles:
	   a. Introduce auxiliary steam to the deaerator, or
	   b. Introduce cold water to the pump suction, or
	   c. Throttle condensate supply to deaerator, or
	   d. Reduce feedwater flow to the boiler
	2. For closed feedwater cycles:
	   a. Start up an additional condensate or condensate booster pump or
	   b. Switch suction supply to auxiliary storage tank

3. Excessive condensate injection outflow temperature	Open injection control valve wide or start auxiliary injection pump
4. Insufficient oil pressure or flow	Start auxiliary oil pump or switch to auxiliary oil filter
5. Excessive oil temperature	Open cooling-water supply

Several of these controls are already widely used in modern steam-electric generating stations. However, the effect of insufficient NPSH or of improper condensate injection sealing conditions is seldom monitored and even more seldom controlled. And yet if proper protection is to be afforded the boiler feed pump, these two controls are essential even if the steam plant is not fully automated.

The condensate injection system can be controlled directly by the temperature of the leak-off. A temperature-measuring device is located in the injection leak-off at each end of the pump, sending an impulse that adjusts the setting of individual valves in the two condensate injection supply lines. Because it is generally desired to reduce the amount of cold-water injection whenever the pump is idle and on warm-up, a bias is incorporated in the control so that the injection leak-off temperature under idle conditions is permitted to exceed this temperature for running conditions.

Evaluation of Controls

It remains now to examine those supervisory controls that are intended to stop a pump or to prevent its start-up. Any abnormal condition should be reported at an annunciator panel. If the condition should be remedied but can be permitted to continue until further action is taken, nothing further is needed. But if the condition may endanger the equipment in question or related equipment, the piece of equipment must be shut down.

At first glance, this appears to be a simple statement of requirement and easy to comply with. However, many conditions are potentially dangerous to satisfactory operation. The question is how many of these dangerous conditions can we afford to monitor and use to actuate a shutdown sequence?

Protective as well as corrective controls are best evaluated if they are considered for what they really are: insurance premiums. The cost of providing these controls must be weighed against the probable expense that might be incurred should the hazards against which the controls protect the equipment be allowed to cause damage. Although it may be exceedingly difficult to place a numerical value upon the probable damage, I would suggest that of the list of controls in categories B and C in Table 4.6 the following will generally be justified: (1) adequacy of NPSH, (2) adequacy of condensate injection supply, (3) adequate supply of lube oil, (4) monitoring for vibration, and (5) monitoring warm-up operation.

The same vibration supervisory equipment normally used on steam turbine generators can be applied to monitoring boiler feed pumps. It is only necessary to monitor the shaft vibration, and this measurement can be obtained with a so-called *seismic vibration detector*. A probe with a spring-loaded bronze tip is used to produce an alternating voltage output that is proportional to the amplitude of the vibration.

The supervisory information from the vibration detectors is transmitted to a power unit. By means of the sequential programming switch, each measured point can be read every 5 to 15 sec. The indicator/alarm mechanism in the power unit can be set to give an alarm signal at one amplitude of vibration (for instance, at 0.003 in.) and to shut down the pump at some greater amplitude (for instance, 0.007 in.).

Shutdowns caused by such impulses as excessive vibration should include time-delay relays with possibly as much as a 2 or 3 min time interval. Vibration can easily be caused temporarily by a reduction in the available NPSH below the cavitation limit, and shutting down a pump under these circumstances is the most dangerous thing that can be done. The pump should be left running and all efforts directed toward restoring adequate NPSH. These efforts should generally be successful within the suggested 3 min time interval. If vibration were to continue after this, damage will probably have occured, and the pump should be brought to rest to avoid further damage.

Alternatively, an interlock with the NPSH supervisory controls can be used so that the excessive vibration impulse is prevented from shutting down the pump until after the inadequate NPSH condition has been remedied.

As to the warm-up protection, this requires the installation of an adequate number of thermocouples on the surface of the boiler feed pump casing. As soon as the pump is shut down and ready to be placed in standby position, the impulses from these thermocouples are employed to actuate the various valves used for maintaining the pump properly warmed up. At the same time, the information from the thermocouples is interlocked with the starting controls so that start-up can be prevented if the temperatures of the pump casing differ at the various monitored points by more than a prescribed maximum value.

Integration of the boiler feed pump into an automated steam power plant presents no unsurmountable problems. It requires merely a thorough understanding of the various factors that can affect the operation of the pump and a logical analysis of the operating, corrective, and protective controls that will assure trouble-free operation.

Question 4.24 Pumping Through Idle Feed Pumps to Fill the Boiler

Our steam power plant employs a closed feedwater cycle in which the condensate pumps discharge through several closed feedwater heaters en route to the suction of the boiler feed pumps. It is proposed to employ either of the two half-sized condensate pumps to fill the boiler by pumping through the idle stationary boiler feed pumps. Would you approve of this practice or would you feel that the extra cost of a valved bypass around the boiler feed pumps would be justified?

Answer. The situation that you describe is one that arises fairly frequently in closed feedwater cycles. With one exception it is perfectly practical to use the condensate pumps to fill the boiler by pumping through the idle boiler feed pumps. The exception refers to the initial start-up conditions, when it would be highly dangerous to pump water through the feed pumps if the system has not been perfectly cleaned of foreign matter, such as mill scale or particles of iron oxide. I could have also mentioned welding beads, which are not such a rare "traveler" during early plant operation.

Three methods are available to avoid getting foreign matter into the boiler feed pumps during initial start-up:

1. The boiler feed pump can be bypassed. But if such a bypass is installed, it can remain in place and serve to fill the boiler later if desired.
2. The inner assembly (assuming a double-casing pump) can be withdrawn from the pump, the casing blocked off with end plates,

and all cleaning done without the danger of getting foreign matter into the close internal clearances of the pump.
3. A suitable screen can be used in the suction line.

A screen that has given very satisfactory performance is described in Question 3.29.

Assuming that precautions are taken to avoid getting foreign matter into the boiler feed pumps, there is no objection to filling the boiler by pumping through the boiler feed pumps. As a matter of fact, a number of utilities use the condensate or condensate booster pumps to start up the main unit without starting the boiler feed pumps themselves. In several stations I have witnessed this particular procedure: At start-up, the condensate and condensate booster pumps are operated in series, pumping right through the feed pumps into the boiler. In this manner, boiler pressure is built up to somewhere near 400 to 500 psi. The main turbine is started up on this reduced pressure, brought up to speed, and synchronized. The main feed pumps, which are driven from the main turbine through a hydraulic coupling, are then started up and the boiler brought up to full pressure.

Question 4.25 Testing Pumps at Different Temperatures

We have recently run tests on one of our boiler feed pumps to check its performance. The manufacturer's curve states that this pump will handle clear water at a temperature of 370°F with a suction head of 226 psig. Performance is shown at 3500 RPM.

Our tests were run at 305°F suction temperature, and the discharge temperature under these conditions was 310°F. We calculated total head in accordance with the ASME Power Test Code for Centrifugal Pumps; that is, suction head was calculated at the suction temperature and discharge head at discharge temperature. Dead weight corrected pressures were used for both. For our purposes it was satisfactory to assume that suction and discharge areas were the same. Total head was then the difference between the discharge and suction heads.

Following the procedure described, we obtained a result of 6650 ft total head at 2600 GPM. Pump speed was held at 3500 RPM. Because of the difference between the 370°F at which the manufacturer's data are plotted and the 305 to 310°F of our test results, we are uncertain how to correctly use these two sets of data. We therefore have the following questions:

1. Do we correct the total head found at our test temperatures to 370°F, or do we enter the pump-head curve at our test capacity without corrections?

2. Do we correct the total head found at our test temperatures to 370 °F in order to determine water horsepower by using the equation

$$\text{whp} = \frac{\text{(lb/min)(head in ft-lb/lb)}}{33,000 \text{ ft-lb/hp-min}}$$

3. Do we make any correction to the 2600 GPM at 310 °F to enter the brake horsepower versus capacity in GPM curves?

Your book* states that for a given pump operating at a given speed and handling a definite volume, the energy applied to the fluids is the same for any fluid regardless of density. Therefore, we interpret this to mean that even though our test temperatures are different from the pump curve test temperature, we do not have to make corrections to enter the pump curves, and our water horsepower is found using our suction and discharge heads to find total head and is computed at our test temperature.

Answer. Essentially your interpretation of my book is correct; that is, you should enter the pump curves at the total head and capacity in GPM corresponding to your test temperatures.

In answer to your second question, you should not correct the total head found at your test temperatures to 370 °F in order to compute water horsepower. From your reference to 2600 GPM at 310 °F I assume you will use the specific gravity corresponding to that temperature to determine the pounds per minute used in the formula for water horsepower. Thus, the water horsepower for your test conditions would be calculated as follows:

$$\text{whp} = \frac{2600 \times 8.33 \times 0.914 \times 6650}{33,000}$$

$$= \frac{\text{gal/min} \times \text{lb of water/gal} \times \text{ft-lb/lb}}{\text{ft-lb/hp-min}}$$

$$= 3985$$

The answer to your third question is that you will have to make a correction for temperature to compare your 310 °F results with the brake horsepower given on the manufacturer's test curve. I assume the brake

---

*I. J. Karassik and Roy Carter, *Centrifugal Pumps, Selection, Operation and Maintenance*, McGraw Hill, New York, 1960.

horsepower on that curve is plotted for and designated as "bhp at 370 °F." To correct this to 310 °F conditions, multiply the brake horsepower at 370 °F by the ratio of the specific gravities at 310 and 370 °F.

For example, if when this pump was new its efficiency at 2600 GPM and 6650 ft total head were 78%, the brake horsepower at 370 °F would have been

$$\text{bhp} = \frac{2600 \times 6650 \times 0.88}{0.78 \times 3960} = 4915$$

where 0.88 is the specific gravity of water at 370 °F. I would therefore expect the manufacturer's curve to show 4915 bhp at these conditions. At the same capacity and head but at 310 °F, the brake horsepower required would be 4915 (0.914/0.880) or 5100. This procedure is consistent with the formula you cited for water horsepower as the correction for specific gravity takes into account the fact that 2600 GPM corresponds to 19,780 lb/min at 310 °F, whereas it corresponds to 19,780 (0.880/0.914) or 19,060 lb/min at 370 °F. More power is therefore required to pump 2600 GPM to 6650 ft at 310 °F than at 370 °F.

If you are taking only one set of head-capacity points, you can simply read the 370 °F brake horsepower from the manufacturer's test curve and correct it as we have described. If you have several head-capacity points at 310 °F, the easiest procedure would be to replot the manufacturer's brake horsepower curve for 310 °F by using the specific gravity correction procedure I have described.

Question 4.26 Minimum Flow of Condensate Pumps

We are now in the early stages of the design of a 300 MW power generating unit. My problem concerns the recirculation of the main condensate pumps.

In our previous units, we arranged the recirculation control to operate in parallel with the hot-well level control. Both the recirculation and the hot-well level control valves receive the same signal from one controller, and both operate throughout the entire band range with one valve action opposite the other. With this arrangement there will always be a recirculation flow as long as the level control valve is throttling or is not fully open. This will be the case whether the main condensate flow has dropped to the minimum requirement or not. Thus, at other than full plant load, the pump will waste a certain amount of power.

Although it is undesirable to waste pumping power, this scheme does have the advantage of always operating the condensate pump near full flow.

This is probably the reason it has been adopted in the past as our standard design practice.

Our 300 MW unit and other once-through steam generating units have a requirement of high condensate pump pressure during start-up and load shedding in order to obtain an inexpensive source of spray water for the boiler bypass system attemperator. During these conditions condensate flow is, of course, considerably less than full flow (it is only one-third during start-up).

To utilize the high pump pressure at part capacity, I am considering the installation of a solenoid valve in the air signal line of the recirculation control valve (or alternatively, the recirculation control valve can be made to operate only at a limited range of the band). The solenoid valve will connect the recirculation control valve to the controller only when the flow has dropped to the minimum requirement. Thus, the drop in pump head caused by the additional flow to the recirculation line is avoided. Under this mode of operation, however, the pump flow will not always be near the full flow.

I would like your opinion of operating condensate pumps away from their design flow. Since condensate pumps are readily subject to cavitation, will frequent or prolonged operation at part capacity hasten pump suction wear (through water shock)? What has been your experience in this matter?

Answer. The significance of operating a condensate pump away from its design flow will vary somewhat with the type of pump. Thus, a horizontal-type condensate pump with oil-lubricated bearings can be operated farther back on its curve than a vertical-type pump. This is due to the fact that the limitation as to the minimum flow of a vertical pump is apt to be affected by the compounding effect of any hydraulic roughness on the satisfactory functioning of the pump bearings.

I assume from your letter that you will probably be using vertical can-type condensate pumps on your new 300 MW unit. Under these circumstances the depth of can may be chosen to provide adequate NPSH under all load conditions. This can free you from apprehension in this regard, but you will nevertheless have to consider other effects of light load condensate pump operation.

In a vertical-type condensate pump the shaft bearings will be water lubricated. When such a pump is operated too far back on its curve, the resulting roughness of operation can seriously interfere with the successful life of the water-lubricated shaft bearings and result in less than satisfactory bearing life or even premature bearing failure. The limitations placed by manufacturers on part-load operation of vertical-type condensate pumps vary considerably, but as a rough rule of thumb, most pumps of this type may be safely operated at any capacity equal to or greater than 25% of their best efficiency point capacity.

It appears to me that the use of either the solenoid valve or the split range operated recirculation valve will aid you in obtaining the somewhat higher condensate pressure you require for part-load boiler bypass attemperation. I suggest that you review the suitability of this system and the range of operation it signifies with the manufacturer you choose for the condensate pumps for your new unit. In this manner, you may confirm its suitability for the particular pumps.

Question 4.27   Excess of Total Head in a Boiler Feed Pump

Our industrial plant generates 70,000 lb/hr of steam from a 200 psig boiler served by two motor-driven boiler feed pumps (one running, one spare), each rated at 170 GPM of 230 °F feedwater and 600 ft total head. We plan to replace this boiler by a 100,000 lb/hr unit at 400 psig but will continue operating it for several years at 70,000 lb/hr and 200 psig. The new boiler feed pumps will be rated 250 GPM and 1150 ft and will have four stages.

Can you tell us if the new boiler feed pumps can be operated without harm with an overpressure of about 200 psig? Would it be preferable to replace two stage assemblies by two dummy stages to cut the discharge pressure? Or would it be better to cut down all four impellers?

Answer. I must assume that for some reason you cannot or do not wish to keep the existing boiler feed pumps in service until such time as you are ready to step up the new boiler to its ultimate design rating.

To answer your question, I shall make some assumptions regarding the present and future operating conditions:

Deaerator pressure, 20.8 psia or 6.1 psig
Temperature, 230 °F
Specific gravity, 0.952
Available NPSH, 25 ft or 10.3 psi
Total suction pressure, 6.1 + 10.3 = 16.4 psig

	Present pump conditions	Future pump conditions
Design capacity, GPM	170	250
Total head, ft	600	1150
Net pressure, psi	247.2	474
Discharge pressure, psig	263.7	490.5

We may now construct a probable head-capacity curve for both the present and future boiler feed pumps (see Fig. 4.17).

You will note that at a capacity of 170 GPM the new pump will develop approximately 1270 ft or an excess of 670 ft (276 psi) over the pressure required to feed the boiler operating under present conditions.

You have three possible solutions available:

1. Throttle this excess pressure at the feedwater regulator. No particular harm would occur to the feed pump, but considerable horsepower would be wasted, and the regulating valve would wear just a little more rapidly.
2. Cut down all four impellers. Based on the shape of the H-Q curve that I have assumed, it would appear that as much as 30% cutdown would be required. This is much beyond the recommended limit for such a pump, even assuming that the original impellers are near to the maximum diameter. Thus, a cutdown of no more than 20% would be recommended, and some extra throttling (possibly as much as 100 psi) might still be required.

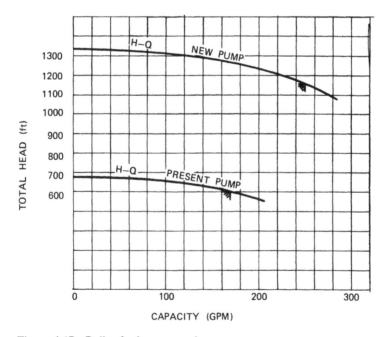

Figure 4.17 Boiler feed pump performance curves.

Thus, this solution is not very efficient. It is also not the best solution since the pump will ultimately be used at its actual design conditions, and the cutdown impellers would have to be replaced by impellers of full diameter.

3. The best solution would be to remove two of the impellers. As you assume, they should be two opposing impellers. The first stage, however, should not be removed as the result might be the introduction of NPSH problems. The exact configuration required for the two dummy stages depends on the actual pump design. One thing is certain: That portion of the shaft left uncovered by the removal of the impellers must be covered by dummy shaft sleeves. Figure 4.18 illustrates a six-stage pump that I modified as far back as 1936 to operate temporarily with only four stages with the third and fourth stages removed.

At 170 GPM, the new pump develops approximately 1260 ft total head, and when two impellers are removed, the pump would develop about 630 ft, which is barely more than the 600 ft you need for the present conditions of service.

Question 4.28 Tripping the Boiler Feed Pump During Transient Operating Conditions

We are examining alternative feedwater system pumping arrangements for future power station designs. These units will consist of modern supercritical boilers and turbine generators with turbine-driven boiler feed pumps discharging around 4300 psig full load. We are at present planning to provide motor-driven booster pumps to supply water to the boiler feed pumps. During transient operation, such as failure of a booster pump, there will be a pressure decay at the feed pump suction that may fall below the NPSH requirements for these pumps. Should this occur, we would instantly (in less than 12 sec) trip the main turbine, the boiler feed pump auxiliary turbine, and the boiler.

We would appreciate your recommendations as to design requirements for a pump for this service and if possible the answer to the following questions:

1. Would momentary cavitation during coastdown under the conditions described above be harmful to the pump?
2. If so, would a stiffer shaft or greater pump clearance enable the pump to withstand momentary cavitation? How would these changes affect the cost of the pump? Would there be a significant decrease in pump efficiency due to larger clearances?

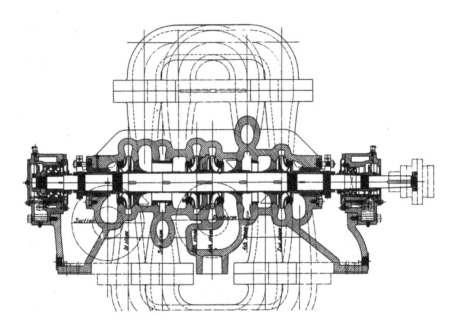

Figure 4.18 Six-stage pump modified for operation as a four-stage pump.

Chapter 4

3. If this momentary cavitation were called out in a pump specification, would you recommend we ask the pump manufacturer to guarantee that there would be no pump damage? We presume that if such a guarantee were given, the manufacturer would want to specify the maximum allowable duration of such cavitation.

Answer. Although I agree that failure of a booster pump or its motor driver would cause a pressure decay at the boiler feed pump suction of sufficient magnitude to cause the feed pumps to flash, I do not believe that instant tripping of the pump, for example, is the most advisable course of action to follow under such circumstances.

A conservatively designed boiler feed pump having a good combination of running clearances and shaft stiffness should be able to withstand momentary cavitation without seizing, provided it is allowed to continue to operate until adequate suction conditions are restored. I have advocated this approach for many years and continue to recommend its use whenever circumstances permit.

I would normally only recommend tripping a pump on low suction pressure provided the pressure at the time of the trip were still adequate to supply the pump NPSH requirement. Since the emergency condition you describe is by definition one in which adequate NPSH is almost instantly unavailable, I would suggest instead that the pump speed be reduced to perhaps 1000 or 2000 RPM rather than tripping it out. If the suction pressure can be corrected, there is much less chance of damage to the pump with this maneuver. If the suction pressure is lost completely, I would suggest reducing the speed to 1000 RPM for 5 min and, if not reestablished by then, tripping out the pump.

Whether or not this procedure can be followed without tripping out the boiler and turbine generator depends on how many feed pumps are operating when the booster is lost and whether or not the boosters and feed pumps are separately headered or arranged in separate series of booster feed pump combinations. Thus, if the boosters and feed pumps are not headered but are instead arranged in two separate booster and feed pump combinations, as has been done at some large supercritical power plants, the loss of one booster could be handled as I have suggested, and the remaining booster feed pump train would be quickly run out to one pump operation and the nominal 60% load that might be carried under these circumstances.

On the other hand, with a headered booster system both feed pumps would suffer the same loss of NPSH, the severity of which would depend on the design margin between the NPSH provided by the boosters and that required by the feed pumps. It is quite conceivable that in this case it would not be possible to maintain enough head at reduced speed to carry part unit load. Certainly under these circumstances attempting to reduce the speed of both pumps to 1000 RPM would necessitate tripping the boiler and turbine as the feed pumps could not provide enough head to maintain throttle pressure when operating at 1000 RPM.

It must be recognized that momentary cavitation under transient conditions is always fraught with some risk. Some pumps under certain circumstances will suffer virtually no damage under these conditions. The same pumps under slightly altered conditions might seize during the final moments of coastdown. Thus, if the feedwater contained an unusual amount of hard foreign matter, the presence of this material in the running clearances could cause a seizure that would not otherwise occur. The resulting damage could be negligible, and the pump could later be freed by hand. Other pumps might suffer more severe seizures, depending on their construction and clearances. This brings me to the question you asked regarding what design I would recommend for the particular conditions you describe.

This is an especially difficult question to answer for a number of reasons. I know of some manufacturers who would recommend special construction for the service you describe. Still others, with considerable experience with supercritical pumps, would vigorously defend the ability of their standard designs to withstand the conditions stipulated. Probably all manufacturers would agree that the ability of a feed pump to withstand this type of operation is in some way related to its running clearances and to the inherent stiffness given by the proportions of its shaft. I have long maintained that a stiffer pump shaft and/or greater running clearances will definitely enhance a pump's ability to endure momentary cavitation. Unfortunately, it is not at present possible to quantitatively relate a pump's ability to withstand cavitation to its specific clearance or shaft proportions, because, to my knowledge, no one has conducted the extensive tests required to establish such a relationship. Such tests would have to encompass variables in all the related operating conditions, such as suction pressure, suction temperature, capacity, speed, discharge pressure, severity of suction pressure decay, rate of suction pressure decay, and finally suction piping size, volume, and configuration. All the many permutations and combinations of these variables would have to be repeated for differing combinations of shaft stiffness and running clearances.

Although I know of a number of instances in which pumps have been severely flashed in the field without incurring damage and am also aware of many occasions of "successful" trip out under less than ideal circumstances, these experiences have almost always occurred under emergency conditions when at best only incomplete operating records were available for analysis. Furthermore, the majority of this information involves pumps of my company's design and manufacture. This is not to say

Chapter 4

that other pump manufacturers have not had the same experience but is rather to highlight the fact that each manufacturer generally only learns about the operating results of pumps with their particular shaft configurations and clearances. Since design approaches to these matters tend to remain relatively stable over fairly long periods of time, comparison between different designs is not usually feasible. This is further reinforced by the fact that new design approaches are usually resorted to only because of significant changes in operating conditions and/or requirements. Thus, the very reason for changes in design usually precludes rigorous quantitative comparisons in the reliability of these designs even when field experience with unusual operating conditions happens to be obtained for both designs. The key word in all this is *quantitative*. Although, as stated, it is almost universally agreed that larger clearances do increase pump reliability, there is little or no agreement as to either just what the right clearances and shaft stiffness are nor what precise increase in availability would be obtained by, say, doubling a particular set of clearances. Fortunately, this absence of precise quantitative correlation between pump design and reliability does not preclude the careful comparison and evaluation of pump design features in relation to a particular set of operating conditions. I would therefore recommend that you make such a comparison of the pumps bid to your specifications to determine which one offers the best combination of clearances and shaft deflection. All other things being equal, I would expect that that pump that had the greatest ratio of running clearances to its static deflection would possess the greatest ability to withstand cavitation.

The relation of clearances and shaft stiffness to pump cost is considerably easier to define than the matter of which clearance and shaft proportion to use. Changes in pump clearances should not have any significant effect on pump cost. Changes in shaft stiffness, on the other hand, could have a very substantial effect on cost as fairly large changes must be made in either shaft diameter and/or shaft length to materially affect its stiffness. Such changes may affect the impeller diameter and the number of stages. Either of these can mean major pump design modifications, which obviously are seldom inexpensive.

Pump efficiency will definitely be affected by running clearance. The magnitude of the effect varies with the pump size and initial clearance; that is, increasing clearances on a small pump will have a proportionally greater effect on efficiency than making the same absolute change on the clearances of a larger pump. In the larger sizes you are probably considering, efficiency can be expected to drop between 1 and 3% depending on the exact pump capacity and clearance in question.

As you may probably surmise from the preceding discussion, I would definitely recommend against asking any pump manufacturer to guarantee that there would be no damage to their pump in the event of the momentary cavitation you would spell out in your specifications. As I have indicated, I firmly believe that present experience simply is not adequate to statistically validate any claim to guaranteed freedom from damage even though we and others may in good faith believe that the probability of such damage is quite remote. I have to admit that faith, however strong and well founded in a general way, is not the same thing as rigorous engineering field data.

Lack of a sound engineering basis is not my only reason for recommending against such a guarantee. The matter of defining damage for the purpose of such a guarantee is really a difficult one. For example, would such a guarantee imply no contact between shaft and wearing surfaces, only minimum scratching of wearing surfaces, just a little galling but no seizure, or what? The problem of adequately and equitably defining pump internal and operating conditions prior to such an event for the purposes of adjudicating any such guarantee seem awesome to me to say the least. Obviously the defining of damage is inextricably intermixed with the matter of defining the maximum allowable time duration for such cavitation. As discussed in the earlier mentioned reference, it is quite possible that if a pump can be successfully cavitated for 10 sec, it may be just as likely that it can continue to cavitate for 10 min, or 24 hr. In essence the same lack of rigorous field data that precludes guaranteeing freedom from damage also implicitly prevents a time limit to the period of cavitation.

In summary I believe that the present state of the art of pump design is such that guarantees of no damage from flashing would of necessity have to be based mainly on commercial considerations rather than engineering ones. I would therefore recommend that you defer this requirement until such time as it can be demonstrated that the engineering content of a guarantee of this type is backed up by substantial and rigorous field data. For the present I believe that a better approach than requiring a guarantee of no damage would be to provide either three headered half-capacity boosters, that is, one spare, or to try to supply sufficient margin between booster and boiler feed pump to permit, say, 15 to 20% loss in booster head without flashing the boiler feed pump. This margin can be called for in your specifications and, furthermore, can be easily demonstrated by the NPSH testing of the individual boosters and feed pumps. An alternative solution would be to pipe up the boosters and feed pumps in two separate series arrangements, which would permit temporary reduced speed operation of one of the pumps as I have indicated.

Question 4.29 Monitoring Pump Characteristics

What boiler feed pump characteristics need be monitored to indicate the *mechanical integrity* of the pump in an automated steam power plant?

Answer. The application of monitoring instruments to a boiler feed pump so as to indicate mechanical conditions falls into five different categories. How many of these will actually be applied depends on the opinion of the power plant designer and his or her evaluation of the relative necessity of each one. These categories are (1) the adequacy of the warm-up procedures, (2) the pump speed, (3) shaft runout, (4) axial shaft position, and (5) rotor vibration.

Before arranging for automatic starting and stopping procedures, it is imperative to include instrumentation to certify that a pump is ready and capable of starting up and operating satisfactorily. One of the prerequisites to this is that a pump will be at a reasonably uniform temperature so as to avoid casing distortion. Many different methods are available to provide adequate warm-up conditions, and the choice among them frequently depends on the particularities of a specific installation. I shall not describe these methods but will restrict myself to the observation that the cost of attaching the necessary thermocouples at strategic locations of the pump casing is very modest and certainly justified. Incorporating an interlock in the pump starting controls that will prevent a start-up unless warm-up conditions have been satisfied is equally inexpensive.

The pump speed can be measured by some form of electrical tachometer. Where this information will be displayed (or recorded) will depend entirely on the arrangements used for plant automation.

Shaft runout would be monitored to determine whether the rotor is straight or whether it has been warped by uneven cooling after shutdown or otherwise damaged. To determine rotor eccentricity automatically, a spindle eccentricity meter as developed for steam turbines can be used. This eccentricity meter employs the electromagnetic principle used for measuring accurately small variations in distance as in strain gauges. Because of the relatively expensive cost of eccentricity meters, to my knowledge none have ever been installed to monitor boiler feed pump rotors.

The axial position of the shaft relative to the thrust bearing would indicate the amount of opening between the balancing device faces. It could be measured by a spindle position meter normally used on turbine generators. Here again, to my knowledge no axial position monitoring has ever been used with boiler feed pumps.

The same vibration supervisory equipment that is normally used on steam turbine-generators can be applied to monitoring boiler feed pumps. Because the boiler feed pump casing on which the bearing housings are mounted is of relatively large mass when compared with the rotor and because the pump foundations are generally sufficiently rigid, it is only necessary to monitor the shaft vibration.

Many different instruments can be used for this purpose. There are some very inexpensive vibration switches available that can be interlocked with the pump driver controls. These switches, however, cannot be too readily incorporated into the more sophisticated controls that may be required by other considerations. They may nevertheless be found satisfactory for some applications.

A more sophisticated solution would be to use a so-called *seismic vibration detector* (Fig. 4.19). A probe with a spring-loaded bronze rider tip used to produce an alternating voltage output that is proportional to the amplitude of the vibration.

Another device used to measure shaft vibration is the *proximity probe*, illustrated in Fig. 4.20. As its name implies, this device measures the amplitude of the shaft vibration without actual contact by means of a *rider*. This proximity probe indicates the amplitude of the vibration *relative* to the casing or bearing mounting, which, as already mentioned, is quite sufficient for a boiler feed pump. The same company that makes the relative vibration proximity probe illustrated in Fig. 4.20 also makes an *absolute* vibration sensor. Such a device uses a proximity probe for noncontact shaft measurement, and a seismic displacement sensor is mounted at the top of the probe stem to measure casing movement. The casing displacement is subtracted from that of the shaft by means of electronic circuits, and the shaft absolute vibration, that is, the shaft vibration relative to a point in space, is then displayed. But I do not believe that boiler feed pump monitoring requires the sophistication of measuring anything other than relative vibration.

Both horizontal and vertical readings should be made, so that four probes must be used for each pump—two for each end. Vibration power units are available for use with up to 10 vibration detectors and with means to automatically measure the output from the several vibration detectors in sequence. A single power unit is therefore capable of serving several pumps or one pump with its driver or driver components.

By means of the sequential programming switch, each measured point can be read every 5 to 15 sec. The indicator/alarm mechanism in the power unit can be set to give an alarm signal at one amplitude of vibration (for instance, 0.002 or 0.003 in.) and to shut down the pump at some greater amplitude (for instance, 0.005 or 0.007 in.).

It is necessary to note that shutdowns caused by an impulse from a vibration detector should incorporate time-delay relays with possibly as much as a 2 or 3 min time interval, because vibration can easily be caused temporarily by a reduction in available NPSH below the cavitation limit. Shutting down a pump under these circumstances is the most dangerous thing that can be done. The pump should be left running and all efforts directed toward restoring adequate NPSH conditions. These efforts should generally be successful within the suggested 2 to 3 min time interval.

Chapter 4

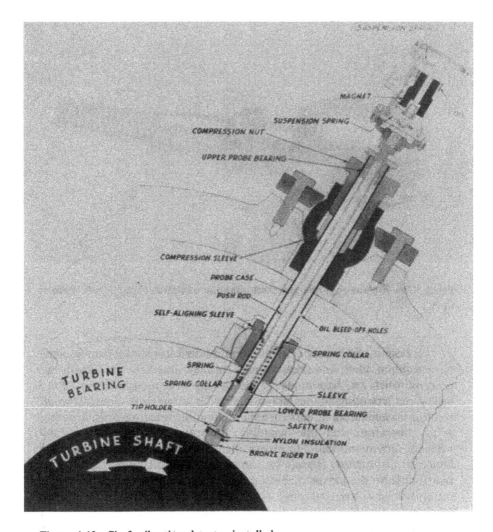

Figure 4.19 Shaft vibration detector installed.

If vibration continues after this, damage will have probably occured, and the pump should be brought to rest to avoid more serious and possibly catastrophic damage.

Alternatively, an interlock with the NPSH supervisory controls can be used, so that the excessive vibration impulse is prevented from shutting down the pump until after the inadequate NPSH condition has been remedied.

Figure 4.20 Proximity probe-type shaft vibration detector. (Courtesy the Indikon Co.)

If desired, the record of each vibration detector station can be made on a common chart in successive and periodic intervals, the length and order of which are determined by the setting of a program timer. But I believe that recording all the data monitored by supervisory controls would result in accumulating a mass of nonessential historical records. It is suggested, therefore, that the recording apparatus be triggered off only whenever any monitored data exceed their prescribed range of acceptable limits. In this manner, should the supervisory controls initiate any action that is alien to the normal operating procedure, the impulse initiating this action will have been recorded along with the time the action will have taken place. Examination of these records would permit operators to conduct any postmortem examination they may wish without necessitating the minute examination of a large mass of logged data.

Question 4.30 Checking Pump Performance Without Dismantling or Special Test Setup

Our plant has several hundred centrifugal pumps, and we need to keep track of their performance to alert us to the need of overhauls. This is a complex and time-consuming task, but we know of no easy way of carrying it out. When a new pump is delivered to us, we end up relying on what the manufacturer tells us about its performance, such as the head-capacity curve, the bhp curve, and the required NPSH, but we cannot always be sure of what the pump performance is—either when the pump is new of after it has been in service for a period of time.

On one occasion, we installed a pump on an existing service for a definite set of head and capacity conditions. The manufacturer had advised us of the dimensions that the impeller would have. On start-up, the pump appeared to be short on performance, and we spent considerable time trying to readjust the controls before we decided to open the pump to look for the cause of the trouble. We found an undersized impeller, which led to the inadequacy of the pump performance. Perhaps we could have found this immediately on start-up, or even before, but we do not want to have to dismantle every new pump or install a complex test setup.

Is there a means available to check out pumps at the time of installation and before they have to be put in regular commercial operation? What components of the normal instrumentation used in a pump installation can we rely on for clues to the performance of an operating pump? And can we judge pump conditions without opening the pump?

Answer. The performance of a pump can be determined from the results of a test that measures the following characteristics: capacity, total head, pumping temperature, speed and pump bhp. However, a field test imposes certain difficulties because it is seldom that a pump installation can meet the standards required for an acceptable degree of precision. Probably the least accurate measurement in a field test is that of the power consumption.

The question arises, How accurate does a field test need to be? The answer will be affected by the purpose of the test. If it is intended to demonstrate that the pump has met its guarantees, it is very doubtful that the degree of accuracy that can be achieved in a field test is sufficient for this purpose, unless very sophisticated means are used to measure power consumption. The user will be better served by conducting a witness test at the factory where steady-state conditions can be more readily established and where greater accuracy can be achieved. When the power consumption exceeds the test stand capability, reduced speed test readings can be projected to full-speed conditions.

If, on the other hand, the test is used to establish the degree of deterioration in performance sustained through internal wear, the problem is more difficult.

Let us first examine the effect of internal wear on the pump performance. Erosive action of the liquid flowing past wearing rings, interstage bushings, and other close clearance joints causes part of this wear. Another portion may be due to any momentary contacts between rotating and stationary parts during pump operation. As running clearances increase with wear, a greater portion of the gross capacity of the pump is short-circuited to a lower pressure area within the pump, and the net capacity of the pump at any given head is reduced by the increase in leakage. Theoretically, the leakage varies approximately as the square root of the pressure differential across the running joints and is therefore not constant at all heads. But we can disregard this effect for our purpose and assume that the increase in leakage is constant at all heads.

The effect of such an increase in leakage is illustrated in Fig. 1.58. A constant value representing this increase is deducted from the capacity at a series of total heads (H_a, H_b, H_c, H_d, and so on). The new head-capacity curve is obtained by joining the points thus obtained.

It still remains to determine this increase in leakage. Note that it is generally recommended that a pump overhaul should be carried out when the pump effective capacity has been reduced by about 4 to 5%.* But, at best, a field test will be accurate only within the same order of magnitude, that is, within 3 or 4%. It becomes obvious that such a field test cannot be very effective in guiding us with respect to the need of an overhaul unless we take certain very specific precautions.

We must assure ourselves, if possible, that whatever error creeps into our measurements remains consistent—in other words, that errors change very little in magnitude and not at all in direction. We are not so much interested in the exact values of capacity, head, or power consumption as in *changes* that take place with time in these three characteristics. It is therefore desirable to develop some additional means to determine the degree of wear that has taken place.

There are several possibilities available to the plant operator to accomplish this, assuming that a test has been run on the pump shortly after its installation and that the same instruments and methods to carry out a retest are used:

1. If the pump operates at constant speed, measure the flow for a given discharge pressure (assuming the suction pressure has not changed) at or near the rated conditions. A 4 to 5% reduction in this flow indicates need for an overhaul.
2. For variable-speed installations, measure the RPM required to meet a specific capacity and total head. Convert the data to full-speed operation, and compare with the original test data or with information obtained shortly after the initial installation.

---

*Of course, this number may be reduced further as escalation of fuel costs takes place.

Chapter 4   *401*

3. Measure the motor amps required to meet a certain capacity, and compare with readings after initial operation.
4. If the pump has an axial thrust balancing device, measure the balancing device leak-off. It will generally increase at the same rate as the leakage past other internal running joints. When the leak-off has doubled, it is probable that internal leakage has also doubled and that clearances should be restored. (See Question 5.2.)
5. If the pump has sleeve bearings, removing the bearing bushings and the packing or seals permits lifting the pump rotor up and down with a pinch bar. Comparison of total lift at initial operation and after several years of operation will give an indication of the increase in internal clearances.

Obviously, the more of these methods are used concurrently, the more reliable will be the conclusions reached from the observations.

But whatever you do, the basic approach should be to try to judge pump condition without opening the pump. As I have frequently said, with very few exceptions, the rule is *"Don't open! Let it run!"*

Question 4.31 Using Temperature Rise to Measure Pump Efficiency

Four of our electric generating units are served by steam turbine-driven boiler feed pumps. Each pump is rated to handle 2,350,000 lb/hr of feedwater at 355 °F, developing a net pressure of 3106 psi. The normal operating conditions correspond to a flow of 1,955,000 lb/hr and a net pressure of 2810 psi.

We have considered the possibility of determining the performance of these turbine-driven boiler feed pumps by using the temperature rise in the pump. This measurement would be fed into the computer to be installed on these units, and the pump performance would be calculated automatically and integrated into the station performance calculations.

We would appreciate it if you were to comment on this proposed method of determining boiler feed pump performance. You may possibly suggest some alternative procedure for monitoring pump performance.

Answer. There has been considerable agitation from certain quarters to develop a reliable method for evaluating the power consumption of boiler feed pumps by means of temperature-rise measurements. It is well known that the temperature rise across a pump is caused by two separate phenomena: the isentropic compression temperature rise and the rise generated by the inefficiency of the pump. The data necessary to calculate the first component are readily available from steam tables and charts of the thermodynamic properties of compressed water.* It would thus be necessary to measure the total temperature rise across the pump, subtract the isentropic compression temperature rise, and convert the remainder into a value representing the pump efficiency. This efficiency, in turn, would be used to calculate power consumption after the pump capacity and total head had been measured.

The problem is not as simple as it sounds. In the first place, we must account for certain losses that do not raise the feedwater temperature, such as external bearing losses. These have to be estimated or measured by conducting a heat balance around the oil cooler. Heat losses by radiation can probably be neglected since they are not of a significant order of magnitude, especially when the boiler feed pumps are properly lagged and insulated. On the other hand, the feedwater temperature may be affected significantly by the injection and admixture of cold condensate, so that the amount and temperature of this injection must be measured.

But troublesome as these corrections may be, they present the least of our difficulties. The most vexing of these is the fact that extremely small inaccuracies in temperature measurement have a most significant influence on the accuracy of the results.

Three charts are appended to illustrate this effect. Figure 4.21 shows the temperature rise caused by the isentropic compression of water for several temperatures and over a range up to 3000 psi. Figure 4.22 shows the relation between total head, pump efficiency, and temperature rise caused by losses. Finally, Fig. 4.23 indicates the required accuracy of the temperature-rise measurement to provide an accuracy of 1% in the measurement of the efficiency.

For example, in the case you describe, the isentropic temperature rise at the normal operating conditions will be of the order of 4.8°F. Assuming an efficiency of about 82%, the temperature rise caused by the pump losses will be approximately 2°F. Finally, we can determine from Fig. 4.23 that if the efficiency is to be accurate within 1%, the maximum permissible error in the measurement of the temperature rise cannot exceed 0.125°F. And not only must we use instruments capable of measuring temperature differences within such close limits, but the instruments must be so placed in the suction and discharge piping as to provide temperature readings that are truly representative of the whole stream of feedwater.

Thermodynamic Properties of Compressed Water, by T. C. Tsu and D. T. Beecher, published by the American Society of Mechanical Engineers, New York, copyright 1957.

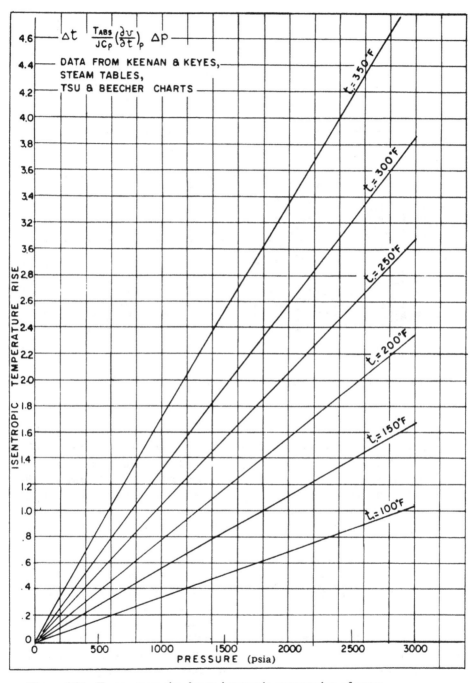

Figure 4.21 Temperature rise due to isentropic compression of water.

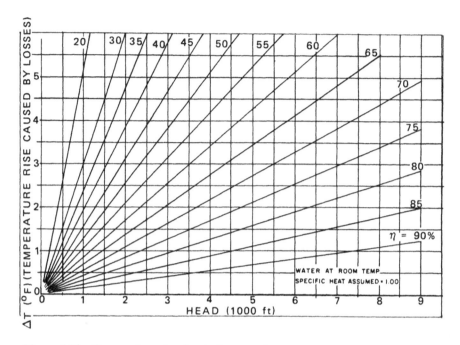

Figure 4.22 Temperature rise due to losses only.

To meet the first requirement, it is probably best to employ a bridge circuit and resistance thermometers and to measure the temperature difference directly. But it is extremely difficult to meet the second requirement because even a slight temperature stratification in the feedwater will introduce gross errors in the results. The stratification may be caused either by minute variations in the temperature of the water entering the suction piping or leaving the discharge nozzle or by virtue of some heat transfer through the piping itself.

Tests using the temperature-rise method have been conducted both on the shop test stand and in the field. So far I am not familiar with any success in developing techniques that could be considered as acceptable. Comparison of test results by temperature-rise methods and by using calibrated torque bars have shown as much as six points of efficiency discrepancy in the shop and up to nine points in the field. Further tests are continuing, but it is my considered opinion that this method cannot be made sufficiently accurate to find much acceptance in steam power plants. At best, the method can be used to obtain comparative data over a period of time.

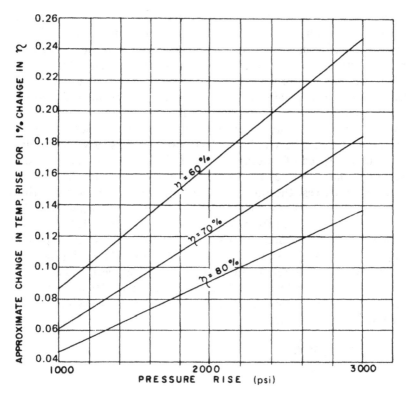

Figure 4.23 Required accuracy of temperature measurement for efficiency within 1%. (Pressure rise varies slightly with temperature. Average values for 100-300°F.)

As I have said before, knowledge of the pump performance, and thus of the increase in leakage that has taken place through wear, is needed to determine when to renew internal running clearances. In a number of my articles in the past I have said that clearances should be renewed when wear had increased them approximately to twice their original dimensions. I am afraid that this falls under the heading of a "sin of generalization." It is true that I was addressing myself principally to the case of centrifugal pumps used on boiler feed service. But even in that particular case I should have qualified my recommendations with the warning that they were based on energy costs at the particular time when I was making these recommendations. Obviously, I should have added, major changes in energy costs would affect the decisions to be made with respect to internal overhauls. I should have also added that the type of pump in question would likewise have a major effect on the benefits to be gained from an overhaul.

The effect of the wear on pump performance has been discussed in Question 1.40 and illustrated in Fig. 1.58. To make a decision on when to renew internal clearances, we must compare the cost of restoring these clearances with the value of the power savings we may obtain from operating a pump with original size clearances. This cost is relatively easy to determine: we can obtain prices on new parts and estimate the cost of labor hours required to carry out the task. But how about the savings? The fact is that these savings are not the same for every pump. Both analytical and experimental data have indicated that leakage losses vary considerably with the specific speed of a pump. This is an index number that is used to predict such characteristics as the shape of the head-capacity and the power-capacity curves, the range of efficiencies obtainable from various types of centrifugal pumps, and a number of other characteristics. The specific speed is calculated from formula (1.9) in Question 1.27.

One of the pump characteristics that is related to the specific speed of a pump is the leakage loss, and Fig. 4.24 shows the relationship between the leakage losses of double-suction pumps and their specific speed.

Let us examine a few typical cases. For instance, let us consider a pump that has its best efficiency at 3500 GPM and 160 ft when operated at 1760 RPM. Its specific speed is

$$n_s = \frac{1760 \sqrt{3500}}{160^{3/4}} = 2314$$

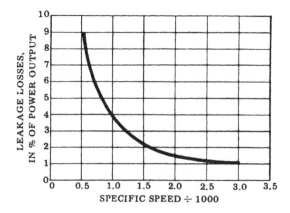

Figure 4.24 Leakage losses for double-suction pumps.

From Fig. 4.24 we can estimate that its leakage losses when the pump is new are of the order of 1.2%. Thus, when the internal clearances have increased to the point that this leakage has doubled, we can regain approximately 1.2% in power savings by restoring the pump clearances.

But if we are dealing with a pump design for 180 GPM and 250 ft head at 3550 RPM, the specific speed is

$$n_s = \frac{3550\sqrt{180}}{250^{3/4}} = 755$$

Such a pump will have leakage losses of about 5%. If the clearances are restored after the leakage losses doubled, we can count on 5% power savings.

There is yet another consideration that will bear on the decision to renew clearances. The rate of wear of internal clearances depends on many factors. To begin with, it should be obvious that it increases in some relation to the differential pressure across the clearances. It also increases if the liquid pumped is corrosive or contains abrasive foreign matter. On the other hand, the rate of wear is slowed down if hard, wear-resisting materials are used for the parts subject to wear. Finally, wear can be accelerated very rapidly if momentary contact between rotating and stationary parts occurs during the operation of the pump.

It becomes quite apparent that restoring clearances of the lower specific speed-type pumps gives us greater returns in terms of the reduction of leakage losses. It may well be, then, that economic considerations may permit us to let clearances of higher specific speed pumps increase beyond values double those of a new pump while making it attractive to renew internal clearances of low specific speed pumps before they have doubled. The final decision, at any rate, must be based on comparing the costs of an overhaul with the value of the potential power savings, considering actual energy costs.

Question 4.32 Testing Boiler Feed Pumps for Wear

The answer to Question 4.31 has been very helpful in evaluating our problem. But it has raised additional questions that we must examine. In our pursuit of a means to convert a change of speed into a useful measure of performance, when comparing a worn boiler feed pump with its performance as new, we are unsure of the proper technique to use. What we are trying to accomplish is to feed into the computer the current pump speed, the capacity pumped, and the total head developed; compare it to the speed corresponding to the new pump test data stored in the computer; convert the speed change ΔN into incremental pump power input; and then calculate the effect of pump wear expressed as Btu/kWh.

The questions we wish to have answered are the following:

1. Is there a simple or complex relationship between ΔN and kWh input? (If so, we hope it is "simple" to facilitate the computer calculations.)
2. Is it correct to assume that any change in pump efficiency results from internal leakage or recirculation, including increased balancing device leakage?
3. Is it correct to say that theoretically, with zero leakage, the pump efficiency in percent would be essentially 100 less the mechanical losses expressed in percent? If so, what is the magnitude of the mechanical losses, and can they be assumed to be constant with a varying load?
4. If there is not a calculable relationship between a change in speed and a resulting change in power input, then why is the measurement of the pump speed a better index of performance than the change in balancing device leakage? You will recall that at the time of our conversations the latter method was rejected as a single source of information because while wear is normally distributed proportionally, it can be possibly concentrated in the balancing device.

I hope that you will find it possible to guide us in applying your suggested approach of measuring the speed change to provide a determination of the pump performance.

Answer. In answer to your four questions, I would comment as follows:

1. The relationship between the change in speed and change in power input is not simple but quite complex in that it will vary for each particular pump design. Remember that although power consumption varies as the cube of the speed and the head varies as the square and capacity directly as the speed, this relationship applied for each individual point on the pump curve. Therefore, there can be a change in capacity and power without there having been a change in speed. Conversely, under some very specific conditions, there could be a change in speed without a change in power, assuming that in the meantime the system-head curve had changed and the operating capacity likewise. The last supposition is somewhat farfetched in the case of a boiler feed pump, but I am merely mentioning this to indicate the complexity of the relationship.

On the other hand, if the speed required to meet a certain capacity or at a certain load were to be compared with the speed at the same capacity or the same load when the pump was new, the change in speed could be made to indicate a change in power consumption and hence in efficiency.

2. It is basically correct to assume that any change in pump efficiency results from increased internal leakage, including that in the balancing device.

3. No, with zero leakage, the pump efficiency is not 100% less the percent mechanical losses. There are many other pump losses in addition to the volumetric losses. One large source of loss involves hydraulic losses of energy conversion in the impellers and the casing. Another source is the disk friction. Typical values of various pump losses are indicated in Fig. 10.9 on page 208 of *Centrifugal and Axial Flow Pumps*, by A. J. Stepanoff, published by Wiley, New York, 1948. The entire subject of losses is treated in detail in Chaps. 9 and 10 of this book.

4. By using the pump curve in new condition, translating it into computer languages, and feeding it into the memory of the computer, changes in speed could be made to yield changes in power consumption. This would have to be done for all pump capacities and for the varying temperatures that may be encountered in service at the different loads. The process may be complex, but it would be a more certain index of pump performance than the change in balancing device leakage for the reason cited in your question.

Of course, the degree of involvement you wish to assign to this measurement will depend on the degree of accuracy you think is justified. Frankly, I cannot get much enthusiasm for providing computer monitoring of the boiler feed pump. The deterioration in pump performance is normally such a slow process that a monthly or semiyearly test using several indices of performance should suffice amply. Monitoring in a "gross" manner should in turn be sufficient to alert the operators that a sudden and premature deterioration has taken place.

Question 4.33 Field Tests When Proper Instrumentation Is Not Available

In many of your articles you have recommended that boiler feed pump performance be monitored at regular intervals to keep track of the gradual deterioration of this performance caused by increasing internal leakage through worn clearances. How accurate should this monitoring be, and how accurate must be the instrumentation used for this purpose?

Answer. Obviously, field tests can never be as accurate as a test carried out at the manufacturer's testing laboratory with calibrated instruments. What is required is not so much accuracy of absolute measurements, but rather observation of relative changes that take place in the pump performance with the passage of time. Thus, both capacity and total head readings taken during a field test could be subject to significant error, as long as the error is always in the same direction and of the same order of magnitude.

You should consider, however, that frequent field tests not only give you an indication of the gradual deterioration of the pump performance, but can also alert you to sudden and rapid changes in this performance. The sudden changes are caused by incidents other than the gradual wear of internal clearances and can be very useful in diagnosing such incidents. Because these changes may be quite large, the accuracy of the instrumentation need not be too precise.

I might recount a specific case in which I was involved a few years ago to illustrate such a situation. This series of events took place at a steam-electric power plant of a utility that, for obvious reasons, shall remained unnamed. The plant has two 100 MW units, each served by two motor-driven half-capacity boiler feed pumps, each handling 450,000 lb of 325°F feedwater against a discharge pressure of 1860 psig. The pumps were driven by 1250 hp, 3560 RPM motors. The first unit was installed in 1959 and the second unit in 1962. All went well until sometime in 1977. Outside the normal stuffing-box and bearing maintenance, there had never arisen the need to dismantle a single one of these four pumps and to restore the internal clearances. But in 1977, one of the pumps serving unit 1—pump 11—gave evidence of the need to renew its internal clearances. Pump 12, on the other hand, appeared to be still in good condition, despite having seen some 18 years of service without ever having been opened up. A local service shop was retained to remove the inner assembly of pump 11, to replace it with the spare assembly and to rebuild the original assembly in their shop.

The service shop reported to the utility that when the inner assembly removed from pump 11 was dismantled, some odd pieces of metal were found inside, but they could not identify these pieces and had not the faintest idea of where they had come from or of how they could have gotten into the pump. They also failed to notice that the strainers initially installed in the suction lines had been removed at some time in the past.

I should add that at the time these units were installed, knowledge about transient operating conditions during load rejections was still not widely understood. Thus, the deaerators serving these two units were designed with so-called antiflash baffling in the storage space, a practice that has since been proven to aggravate rather than to alleviate the effects of load rejection.

Ultimately, the utility itself identified the foreign material as pieces of metal from some of the deaerator internal components. The deaerator was repaired, but no one could determine why it had been damaged. It certainly appears that the local service shop did not know anything about the possible unfavorable effects of transient operating conditions that accompany sudden load rejection or that so-called antiflash baffling in the deaerator can aggravate these effects. They did seem to realize that the flowmeters and pressure gauges used with these pumps were quite inaccurate but made no recommendation to recalibrate the instrumentation or to replace it.

A short 8 or 10 months after having been rebuilt, pump 11 again showed signs of distress. This was evidenced by the fact that each pump had been able to carry almost 65 MW by itself when new, but, pump 11 could barely meet a load of 50 MW and when operated alone in that load range would, after a short while, start losing discharge pressure at an alarming rate.

This time I was asked to visit the steam power plant and to help diagnose what was happening. I did determine from some of the previous reports that neither the flowmeters nor the individual pressure gauges were to be trusted for accuracy. Yet, I wanted to run some sort of a test to determine the relative short fall in the performance of pump 11.

I considered that the turbogenerator *does not know* which particular pump is running but requires a specific flow in pounds per hour for any given electrical load, as long as readings are being taken within a short period of time, so that changes in ambient temperature or in barometric pressure cannot affect the unit heat rate. Thus, any given load on the turbogenerator will correspond to the same feedwater flow, regardless of which pump is running; therefore the station MW load could be used to represent pump capacity.

I also noticed a pressure gauge located in the control room. No one seemed to be sure where specifically this gauge was connected, except that it read the pressure in the common manifold somewhere upstream of the closed feedwater heaters beyond the boiler feed pumps. Obviously, this pressure reading could well be inaccurate but would have the same amount of error, regardless of which pump was running.

I then proceeded to run a test first on the supposedly undamaged pump, taking pressure readings at various loads from about 40 to 62 MW. Then, shifting pumps, I repeated this test with the pump that was showing a loss of capacity. The comparison of the data was dramatic! For instance, at a load of 56.5 MW the pressure gauge showed 1850 psi with the undamaged pump running. But at the same pressure of 1850 psig, the other pump could only carry 52 MW. Obviously, its flow was about 8.0% lower than the first pump. When you consider that this other pump had never had its internal clearances renewed, it is obvious that a loss of well over 11% was taking place.

I made another important observation: the undamaged pump could carry well beyond a 60 MW load, and if we didn't mind overloading the motor, we could probably have gone out to 63 or 65 MW. But with the damaged pump, once the load reached about 57.5 MW, the pressure gauge started drifting down as if it were entering a zone of serious cavitation and the second pump had to be put on in a hurry if we didn't want to lose the unit.

The readings taken during this test and the resulting performance curves are shown in Fig. 4.25. You will notice that we have a curve of pressures "somewhere downstream of the pumps" plotted against unit loads in megawatts. This curve is as revealing and as helpful as a normal head-capacity performance curve. I think that this illustrates quite well the point I was trying to make with respect to the absolute accuracy required of field tests.

As to my diagnosis, I should add the relevant fact in this series of events that pump 11 was located directly underneath the deaerator, but pump 12, which after almost 20 years continued to faithfully discharge its appointed duties, was located quite some distance from pump 11 (see Fig. 4.26). What must have happened is that the deaerator must have suffered another severe upset. Pieces of metal from the deaerator probably had traveled straight down the suction piping into the eye of the first-stage impeller of pump 11 but had not made a right-angle turn to get to pump 12. I suggested that to prevent the recurrence of this event, the antiflash baffling be removed from the deaerators of both units 1 and 2 and that the suction strainers be reinstalled. It seems superfluous to add that after the inner assembly was removed and dismantled, metal pieces were found in the eye of the first-stage impeller.

Question 4.34 Testing for Shutoff Head

Are there any reasons it may be impractical or impossible to measure the shutoff head of a centrifugal pump on test?

Answer. The answer depends completely on the type of pump involved. Certain pumps can be operated against a closed discharge valve for the short period of time required to obtain head and power readings while running at shutoff, but others should not be tested under these conditions, either because of excessive temperature rise or because of excessive power consumption or excessive discharge pressures.

First consider the temperature-rise limitations. The power losses within the pump—the difference between the brake horsepower consumed and the water horsepower developed—are converted into heat and

Chapter 4

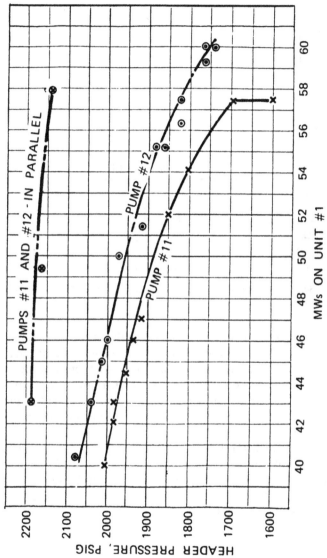

Figure 4.25 Test of boiler feed pumps 11 and 12.

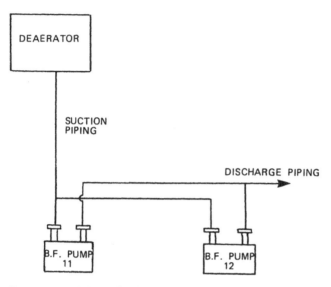

Figure 4.26 Schematic diagram of deaerator and boiler feed pump arrangement.

transferred to the liquid being pumped. When the pump is operated against a completely closed discharge valve, the power losses become equal to the brake horsepower at shutoff, and all this power goes into heating the small quantity of liquid contained in the pump casing. In turn, some of the heat in the liquid is transferred to the metal of the casing and will be dissipated to the surrounding atmosphere by radiation and convection. But this dissipation is generally negligible.

The rate at which the liquid contained in the pump casing heats up depends on the brake horsepower at shutoff and on the volume of liquid in the pump casing. If we ignore the dissipation of heat by radiation and the heat absorption by the metal in the pump casing, we can calculate the temperature rise of the liquid from the formula

$$T_r = \frac{42.4 P_{so}}{W_w C_w}$$

where
T_r = temperature rise, in degree F/min.
P_{so} = brake horsepower at shutoff
42.4 = conversion factor from brake horsepower into Btus
W_w = weight of liquid in the pump
C_w = specific heat of the liquid (1.0 if the liquid is water)

For instance, if the pump handles water and contains 100 lb of liquid and if the brake horsepower at shutoff is 100, the temperature of the water in the pump casing will increase at the rate of 42.2 °F/min. Operation at shutoff under these conditions is dangerous, and testing for shutoff conditions is not recommended.

On the other hand, if we are dealing with a low-head high-capacity pump containing 5000 lb of water that takes the same 100 hp at shutoff, the rate of temperature increase will be only 0.85 °F/min. Obviously, this is not an excessive temperature rise, and the pump could be operated at shutoff long enough for conditions to stabilize and to take head and power readings.

As mentioned, the other possible limitation to testing centrifugal pumps under shutoff conditions has to do with the shape of the pump head-capacity and power-capacity curves. Low and medium specific speed pump curves are shown in Figs. 4.27 and 4.28, and the shape of pump curves for high specific speed, axial-flow pumps is shown in Fig. 4.29. Note that in the case of low and medium specific speed pump types, power consumption falls off with a reduction in capacity, and head rise from the best efficiency flow to shutoff is not excessive. For such pumps, it is possible to test for shutoff conditions, provided no limitation exists from the point of view of temperature rise.

On the other hand, the shape of the head-capacity and power-capacity curves of axial-flow, high-specific-speed-type-pumps (as in Fig. 4.29) is such that testing for shutoff conditions may be impractical. Even if the pump casing can be subjected to the discharge pressure concomitant to a total head of 280% of the head at the best efficiency point, it is unlikely that the pump driver can accommodate the 210% of the brake horsepower taken by the pump at its design conditions.

It is therefore recommended not to test pumps of this type at shutoff conditions. If it is absolutely necessary to obtain data for shutoff operation, the pump could be tested at some reduced speed, so that total head and brake horsepower at shutoff would be appreciably reduced. Conditions at full speed are then calculated in accordance with centrifugal pump affinity laws. (Within reasonable limits of speed variation, the pump capacity varies directly with the speed, the total head as the square of the speed, and the brake horsepower as the cube of the speed.)

There is one more possible limitation imposed by the fact that radial thrust in volute pumps increases with a reduction in flow. The pump manufacturer will know whether this increase is excessive for any given centrifugal pump and therefore whether testing at shutoff is permissible or not.

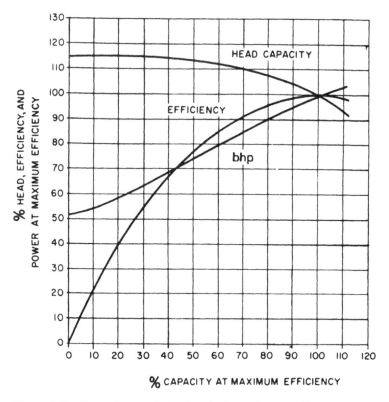

Figure 4.27 Type characteristics for single-suction centrifugal pump with specific speed of 1550.

Question 4.35 Testing Vertical Pumps Horizontally

We have ordered several vertical barrel-type condensate extraction pumps from a local pump manufacturer. They do not have a proper test stand to test these pumps in their normal vertical position, and we have been approached with the request to allow testing in a horizontal position.

We would like your opinion on whether this should be allowed and whether such a test will provide the necesary data to demonstrate that the pumps perform as guaranteed.

Answer. I presume that by "vertical barrel type" you mean a can pump such as that illustrated on Fig. 4.30. I am surprised that this pump cannot be tested in its normal position, since all that is required is a relatively small excavated pit with a diameter just slightly in excess of the diameter of the can.

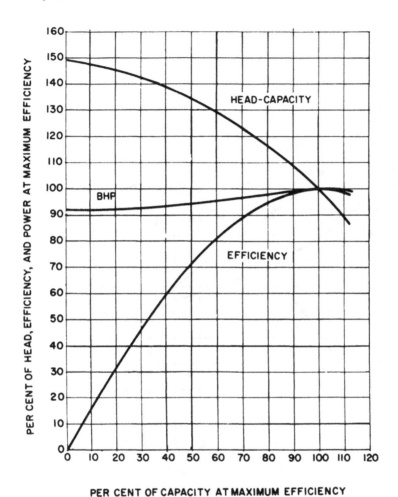

Figure 4.28 Type characteristics for single-suction centrifugal pump with specific speed of 4000.

Whatever the reasons for the inability of the manufacturer to test the pump vertically, we need to examine the effects of testing it in a horizontal position from two separate angles, namely from the points of view of the hydraulic and of the mechanical considerations.

A test in the horizontal position will give satisfactory information on the ability of the pump to perform in its normal position, provided certain precautions are taken in the test setup and during the test. Note that the

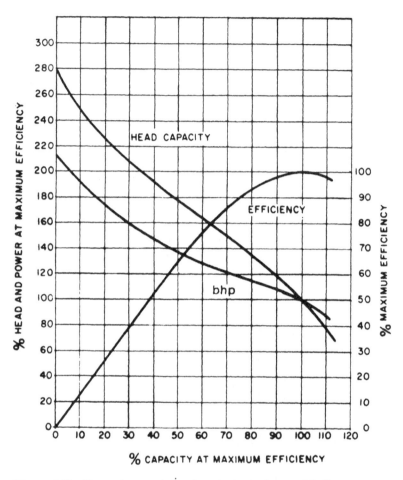

Figure 4.29 Type characteristics for single-suction, axial flow-type pump with specific speed of 10,000.

most important item to be checked that can be affected by the procedure contemplated is the cavitation characteristics of the pump. The Hydraulic Institute Standards indicate three different acceptable arrangements for cavitation tests as discussed in Question 1.30 and illustrated in Figs. 1.48 through 1.50. Although a horizontal pump is illustrated in these figures, the arrangements are equally applicable to vertical pumps.

The closed-loop test setup as illustrated in Fig. 1.48 most nearly duplicates the conditions under which a pump handling a liquid at or near boiling conditions (such as a condensate extraction pump) will operate.

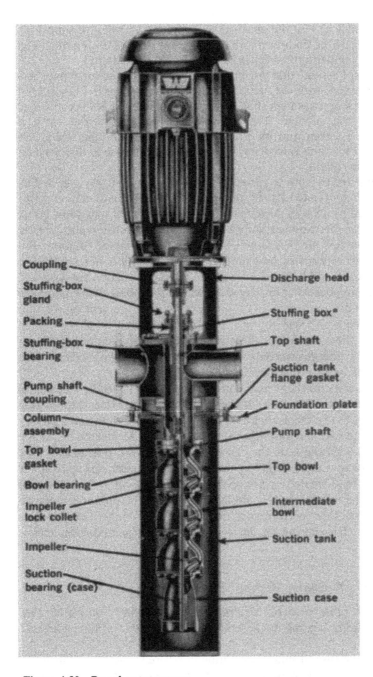

Figure 4.30 Barrel or can pump.

Since it is possible to vent out all dissolved air or gas, it will give a more accurate measure of the pump performance. It also permits testing the pump at exactly the temperature specified for the actual installation.

However, I assume that the intent is to test the condensate pumps in a horizontal position, with the suction taken from a pump as in Figs. 1.49 or 1.50. In the first arrangement (Fig. 1.49), the available NPSH is controlled by throttling the suction. The turbulence produced by the valve is dissipated by a screen and by straightening vanes. Still, the turbulence created at the throttling valve tends to accelerate the release of dissolved air or gas from the liquid.

A pump tested with a variable-level deep sump supply, as in Fig. 1.50, will be less affected by the release of air or gas, as there is no turbulence created by a valve. As a matter of fact, such a test may show better results than can be expected with a suction line of considerably greater length, because less air or gas will have time to go out of solution. But whether the test setup is such as on Fig. 1.49 or 1.50, it is necessary to be sure that there is a solid column of water in the pump suction passages and that there are no pockets of air anywhere within the piping or the pump. In other words, adequate venting is a prerequisite, and the vent tap or taps must be located at the very topmost points of the pump or of the piping.

If under these conditions the pumps demonstrate their specified performance under the guaranteed NPSH conditions, the probability is that they will do as well in their normal vertical installation.

We now come to the second consideration, that is, to the mechanical effects of testing such pumps in the horizontal position. Here, the situation is not as "rosy." The fact is that a vertical pump such as illustrated on Fig. 4.30 is provided with a multiplicity of radial bearings lubricated by the liquid pumped. Regardless of the material used for the bearings, I have serious reservations as to ability of these bearings to carry the appreciably greater radial load created by the operation of the pump in the horizontal position. At the very least, I would insist on seeing the pumps dismantled completely after their performance tests and on a complete examination not only of the bearings but of the wearing ring surfaces and clearances as well.

Question 4.36 Measuring Bearing Temperature

What is your preference for determining temperatures on boiler feed pumps: measuring bearing metal temperatures or drain oil temperatures?

Answer. In the case of pumps operating in severe services, such as boiler feed pumps, it is best to measure the bearing temperature of the metal rather than in the oil reservoir or in the oil drain flow. There is a natural time lag in transmitting the heat of the bearing metal to the oil.

The preferred method of measuring bearing temperatures of the modern "automotive"-type thin steel-backed journal bearings is shown in Fig. 4.31. The thermocouple is inserted through a drilled hole in the retainer, and its tip is made to firmly contact the back of the steel shell. The method of mounting thermocouples in the thrust shoes of a thrust bearing are also shown in Fig. 4.31. This method will detect a temperature change more quickly than if the thermocouples were measuring the temperature in the oil stream. This could be of real significance in the case of a sudden rapid rise in temperature, the cause of which might lead to severe damage or even wreck the pump before it could be shut down.

Typical recommendations of normal operating and maximum permissible temperatures are shown in Table 4.8. It will be noted that slightly lower temperatures are indicated for measurements taken in the oil flow itself compared with measurements taken in the bearing metal. The reason, of course, is that some heat dissipation does take place between the two locations.

Question 4.37 Turning Gear Operation

In one of your articles, "Trends in Boiler Feed Pumps for Large Steam-Electric Generating Plants in the USA" you state that "In the last years, the question of warm-up of boiler feed pumps before starting has been well

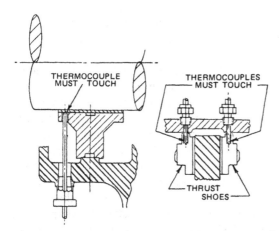

Figure 4.31 Thermocouple installation in line and thrust bearings.

Table 4.8 Recommended Bearing Temperatures

Bearing type	Temp. indicator location	Normal operating temp. range		Max. permissible temperature	
		°F	°C	°F	°C
Journal bearings	Tip in bearing retainer	130-180	54-82	200	93
	Tip in oil leaving bearing	120-160	49-71	190	88
Thrust bearing	Tip in thrust shoe	130-180	54-82	200	93
	Tip in oil leaving bearing	120-150	49-66	170	77

understood, and adequate provisions are generally provided, including the necessary monitoring of casing temperatures with thermocouples." In the same article, you refer to the fact that the warm-up of the driving steam turbines and their turning gear operation is still somewhat unresolved. Is the difference of opinion with respect to whether a boiler feed pump should or should not be operated at the steam turbine turning gear speed based on differences in pump design, or are there other reasons for the controversy?

Answer. There is only one possible difference in pump design that could affect the decision of whether to operate a pump at turning gear speeds, and that has to do with the internal clearances. If a pump is designed so as to depend on the support given by internal running joints acting as water-lubricated bearings, then a water film must always be provided at these joints to act as a lubricant for the water-lubricated bearings. Such a film can only be provided if some minimum pressure can be maintained across these internal joints. Thus, a pump with such a construction should not be operated at speeds that would not generate a pressure per stage of 50 to 100 psi. Given the fact that today's boiler feed pumps develop pressures of approximately 700 to 800 psi per stage at speeds between 5000 and 6000 RPM, minimum permissible speeds for such pumps would run between 1200 and 1500 RPM.

Fortunately, I can say that the practice of designing pumps with such thin shafts and such close clearances that the wearing rings act as internal bearings has all but disappeared.

The controversy is therefore based on considerations other than pump design. It is related instead to the relative optimism or pessimism on the part of either the pump manufacturer or of the user regarding the degree of cleanliness of the piping system. I should add that the problem created by operation at extremely low speeds arises principally during the initial start-up of the plant and during the early period of its operation.

Regardless of the care taken in flushing the system and eliminating the presence of foreign matter or mill scale, for example, in the station piping, it has been my experience that this foreign matter can never be completely eliminated. If particles of mill scale or of brittle oxides find their way into the close clearances at the wearing rings or at the balancing device and if they are not washed out of these clearances by the effect of significant pressure differentials across these clearances, they will lodge themselves in these clearances. At the turning gear speeds there is very little torque transmitted to the boiler feed pump, and the dragging friction created by the particles lodged within the clearances will exceed the available torque. At this moment, the turning gear motor will stall. In such a case, when the operator disconnects the driver from the pump to determine the cause of the problem, it will be found that the pump is mechanically bound up.

The operator may try to free the pump by removing the bearing inserts and rocking the rotating element and may or may not succeed. If not, there are two choices: withdraw the inner assembly and dismantle it to remove the foreign matter or reconnect the unit and throw the turbine on at full speed. My own tendency would be to follow the latter course, if I am reasonably certain that the problem has been created by the presence of small particles of foreign matter. At worst, one or two of the wearing ring or impeller surfaces will be grooved by the particles if it is hard enough.

On the other hand, most operators will be loath to take the chance of damaging the pump to a considerable degree if they take this course. Parenthetically, I am somewhat puzzled by such a concern: After all, motor-driven pumps are not provided with a turning gear, and an operator always starts them up by throwing the unit across the line. And yet, *he or she does not know* whether or not the pump is free.

Be it as it may, we must deal with the question of what should be considered as the best and safest practice if the operator prefers not to take a chance on starting up the pump on full speed every time the pump binds up because of the presence of foreign matter in the clearances.

To begin with, we can eliminate the use of shear sections or shear pins in the flexible coupling. These were used at the time when a number of boiler feed pumps were driven by the main turbine and when they were intended to prevent a pump seizure from damaging the main turbine. If these shear sections were adequate to prevent stalling of the turning gear motor, they would be incapable of transmitting the torque under full-load operating conditions.

The alternative is to use quick-disconnect couplings. It is true that these devices are rather expensive and that in their early period of use they were less than 100% reliable. On the other hand, their reliability has improved to the point where they are perfectly adequate for their intended service. And if they serve to avoid even a few instances of binding up during the initial stages of operation, their cost would be fully justified.

One might even consider the compromise solution of installing a quick-disconnect coupling on the first unit in a steam-electric station. By the time the second unit is to be installed, the piping system should be properly cleaned of most or all of the foreign matter. The quick-disconnect coupling can be transferred to the boiler feed pump serving the second unit and be replaced by a conventional coupling. This practice would reduce to some degree the additional expense of special couplings. And the degree of risk in operating the boiler feed pump at turning gear speeds after the system is quite clean is minimal.

I wish to emphasize the fact that if one analyzes this problem objectively, pump design can hardly be a factor in reaching a decision on whether or not to operate a boiler feed pump at turning gear speeds. Today, all pumps are essentially equal in the presence of foreign matter particles that can lodge within the internal clearances and bind up the pump. The real decision hinges on whether you are an optimist or a pessimist with respect to your ability to clean up your piping system.

Question 4.38 Changes in Shaft Deflection

Is it possible for the deflection of a boiler feed pump shaft to exceed design values under various operating conditions or after many hours of operation? If so, is it possible to monitor this deflection?

Answer. If the design deflection of a pump rotor depends on the support from the wearing rings or from some internal close clearance bushings that are designed to operate as water-lubricated bearings, wear at these running clearances will obviously permit the shaft deflection to increase beyond its design or initial value. The same would be true if the pump were permitted to become steambound since no lubrication would be available at these internal bearings.

Chapter 4 425

If, however, all internal clearances are liberally in excess of the design deflection, the latter cannot be affected by pump wear and will remain the same, regardless of the age of the pump.

It would be possible to develop some means for monitoring the shaft deflection but preferably by measuring the angularity of the shaft at some point near the bearings, that is, outside of the pump internals proper. (Measuring deflection internally would be so complex as to be impractical.) Even an outside measurement might prove to be quite expensive and difficult to justify. After all, the user must know whether the deflection and internal clearances are such that the shaft does or does not depend on internal support. If it does not, the measurment is unnecessary. If it does, measuring it under steam-bound conditions or after wear has progressed to permit contact will not prevent this contact.

Question 4.39 Impeller Cutdown

After installing a new general-service pump in our plant, we found that we had been overconservative in specifying the required total head. As a result, we are constantly wasting energy by throttling the excess head. We have decided to cut down the impeller to eliminate this waste, and calculations indicate that we should cut it down by 6%. Should we only cut down the impeller vanes, or should we also cut down the impeller shrouds?

Answer. By all means, cut down both the vanes and the shrouds. Failing to do so will not give you the maximum benefit in power reduction from the cutdown of the impeller.

One source of loss in a centrifugal pump is the disk horsepower, that is, the power consumed by the drag of the impeller shrouds in the liquid surrounding the impeller. This disk horsepower varies as the cube of the pump speed and as the fifth power of the diameter of the impeller shrouds. If you cut down only the vanes of the impeller, you reduce water horsepower (or useful work of the pump), but you will not have reduced the disk horsepower. Although I cannot estimate the reduction in disk horsepower you would obtain by cutting down the shrouds because you have not specified the pump capacity, head, and operating speed, this reduction could easily be anywhere from 1 to 2% and should not be neglected.

Question 4.40 More About Impeller Cutdown

Is it possible to cut down the impeller of a centrifugal pump in an installation that is being converted from a 50 to a 60 cycle power supply?

Answer. This depends on a number of factors. I assume that you intend either to use the existing motor (in which case you must find out if this motor is suitable for 60 cycle operation) or replace it with a new motor having the same number of poles. In this case, a pump that ran at 2950 RPM will now run at 3560 RPM, a 1450 RPM pump will now run at 1750 RPM, and a 960 RPM pump will operate at 1150 RPM.

We also must assume that the rotor dynamics of the pump will permit it to operate at six-fifths of its former speed and that the only question remaining is whether the pump hydraulics will permit you to cut the impeller diameter of the pump to approximately five-sixths its present diameter. I say "approximately" because the laws of affinity hold accurately within certain limits of impeller cutdown beyond which pump performance may deviate somewhat from that based on these laws.

For instance, in this particular case, the laws of affinity would dictate a cutdown to five-sixths of the diameter, or to 83.3% of this diameter. As a rule of thumb, I suggest cutting the impeller to not less than 85 or 86%.

Whether such a cutdown is possible or not depends on the type of impeller (that is, on its specific speed) as well as on the relation that the existing impeller bears to the maximum diameter for which the pump in question has been designed.

Low-specific-speed impellers (see Fig. 1.36) can be cut down by about 20% without adversely affecting their performance. As specific speed increases, the amount of cutdown permissible decreases.

If the pump in question was fitted with a maximum-diameter impeller, cutting this impeller to the necessary diameter may be quite feasible. But if the impeller in the pump was already less than the maximum diameter intended for the pump in question, pump performance may be affected adversely both in the shape of the head-capacity curve and in pump efficiency.

The safest approach to your problem is to contact the pump manufacturer and get advice on each pump in your installation. You may be able to convert some of the existing pumps to 60 cycle operation by cutting down or replacing impellers, but you may have to replace some pumps that are not suitable for this conversion.

Question 4.41 Use of Oversize Clearances During Early Stages of Operation

We have on order two high-pressure boiler feed pumps that will be fitted with condensate injection seals. All the necessary precautions will be taken to prevent foreign matter from getting into the injection seals, including the use of filters in the injection lines. Nevertheless, there is always the possibility that our precautions may be insufficient and that we may damage the seals during the initial operation of the unit. We understand that when mechanical seals are used on boiler feed pumps, it is recommended that conventional packing be used during the start-up period and that the mechanical seals be installed only after several months operation with packed boxes. Can such an approach be used with condensate injection seals, and if so, do you recommend it?

Answer. Your concern about the potential danger to the condensate injection seals during initial operation is a good example of preventive maintenance. It is by recognizing the possibility of "infant mortality" that we provide the best opportunity to avoid it. My answer to the first part of your question must be, of necessity, qualified, because you do not state whether the pumps are designed to operate at the conventional speed of 3500 RPM or whether they are high-speed pumps designed for 6000 to 9000 RPM. In principle, if these are conventional speed pumps, they can be packed with conventional packing. If these are high-speed pumps, the peripheral speed at the shaft sleeves is generally too high to permit the use of stuffing-box packing. There is, of course, one more problem, and that is whether the pumps are so far along in production that a change in stuffing-box design at this time is going to be very expensive or even possible. In any event, building a pump to take conventional stuffing-box packing suitable for later conversion to condensate injection seals will definitely add some cost to the price of a pump arranged for just one specific stuffing-box construction.

There is, however, one very simple provision that can be utilized when condensate injection seals are used, and it is one that I have frequently recommended. The pump is built with undersized shaft sleeves for initial operation, so that the clearances at the seals are at least twice the dimensions that the ultimate sleeves will provide. At the same time, instead of using stainless steel sleeves, these initial sleeves are made of bronze, a much softer metal and one that is not prone to galling on contact with another stainless steel part. I know of one specific installation where such precautions were taken and where considerable dirt did get into the seals without any significant damage to the pump. At worst, the bronze shaft sleeves may become grooved in the area where the foreign matter lodges between the stationary seals and the sleeves, but the danger of galling and seizure is eliminated.

Question 4.42 Minimum Start-up Time for Motor-Driven Standby Pumps

Our 325 MW, 2400 psi throttle pressure unit will be served by a full-capacity steam turbine-driven boiler feed pump. In addition, we are providing a 50%

capacity motor-driven standby pump. What is considered to be the minimum acceptable time for emergency automatic starting of the standby electric motor-driven pump? In the past we have been specifying that pumps be capable of starting in 5 sec. Is this considered to be too short? Would it be necessary to have the auxiliary lubricated oil pump running continuously while the pump is on standby duty?

Answer. It is presumed that the standby pump is held ready for immediate starting, with suction and discharge valves open and with warm-up flow through the pump maintaining it at a uniform temperature. The question then involves strictly the matter of adequate bearing lubrication at start-up.

Normally, the lubricating system is so arranged and so interlocked with the motor starter that the impulse for starting a standby pump merely energizes the auxiliary oil pump motor. As soon as this pump develops a predetermined oil pressure in the lube system, a relay energizes the main motor starter, which in turn starts the electric motor across the line. I am not certain whether the entire train of events (that is, starting of the auxiliary oil pump, operation of the pressure switch and relay and start-up, and running up to speed of the main pump motor) does or can take place within the 5 sec you specify. It would be my opinion that the time interval is more likely to be between 5 and 10 sec. This, however, should not be deemed excessive.

At the same time, if you prefer, there should be no objection to letting the auxiliary oil pump operate constantly while the pump is on standby service. The motor driving this pump ranges from 1/2 to a maximum of 2 hp, and the expenditure of power involved is not significant. I know of several steam power plants where this is standard operation. You must remember that a certain amount of heat will be picked up by the oil and that the cooling circuit must remain in operation.

Question 4.43 How to Retrofit Inadequate Minimum Flow Bypasses

We have ordered three identical so-called half-capacity boiler feed pumps to serve our 500 MW fossil fuel-fired unit. Two of these are steam turbine driven; the third is motor driven and acts as the spare pump. The design capacity is 1000 t/hr, the total head is 2400 m, and the pump speed is 5500 RPM. The available NPSH is of the order of three times the required NPSH. The suction specific speed at the design conditions is 11,300 in U.S. units.

The manufacturer recommended a minimum flow of 25% of the design flow, and the automatic bypass system is designed to provide this minimum value. We have read a number of your articles dealing with suction specific speeds and have now noted that you recommend that it does not exceed 8500 to 9000; we have also noted that when much higher suction specific speeds are used, you consider that the minimum flow should be increased to as much as 50 or 75%.

The pumps are about to be delivered and installed. We would like to know what you would recommend doing in this case. Should we have the bypass system revamped to accommodate such higher flows, or taking advantage of the wide margin between the available and required NPSH in our case, should we request the manufacturer to replace the present first-stage impellers with new ones having a suction specific speed of about 8500?

Answer. You are quite correct in concluding that I consider it more prudent to limit the value of S (suction specific speed) of the first-stage impeller to about 8500 and that minimum flows for higher S values should be increased significantly over the 25% normally recommended. To begin with, I must express my astonishment over the choice of the required NPSH in your case. If one were to instead select an impeller with an S value of 8500 instead of 11,300, the NPSH required would only increase by a factor of 1.46. This would have reduced the ratio between available and required NPSH from 3.0 to 2.05, which is still a very reasonable and most conservative value. Thousands of pump installations with S values of around 8500 have been operating for years without any problems with much lower margins and without the necessity of replacing first-stage impellers.

The considerations that tie in minimum recommended flows with the S value of the impeller are based on the effect of internal recirculation; the higher the S value, the higher the capacity at which occurs the onset of internal recirculation as a percentage of the flow at best efficiency. I need not go into too much detail in this connection, as the subject has been treated in several of my previous articles in *World Pumps*.* The formulas given in the third article were developed by W. H. Fraser and first presented in an ASME paper in 1981.† Unfortunately, since I have no access to the geometric configuration of your first-stage impeller, I cannot determine

---

*"Centrifugal Pump Suction Conditions . . . Some Reflections on the State of the Art," by I. J. Karassik, *World Pumps*, March 1982.
"Centrifugal Pump Suction Conditions . . . Further Reflections on the State of the Art," by I. J. Karassik, *World Pumps*, September 1982.
"Flow Recirculation in Centrifugal Pumps: From Theory to Practice," by I. J. Karassik, *World Pumps*, April 1983.
†"Recirculation in Centrifugal Pumps," W. H. Fraser, Texas A&M Turbomachinery Symposium, Houston, Texas, Dec. 1981.

precisely the onset of suction recirculation. But we can make a reasonable approximation, using data provided in a second paper by Mr. Fraser* and reproduced here as Fig. 4.32. You will note that for an S value of 11,300 for multistage pumps, internal recirculation at the suction occurs at about 97% of the best efficiency flow, but this onset is reduced to 76% of bep flow with an impeller having an S value of 8500.

This does not necessarily mean that minimum flows need be set at 97 and 76%, respectively. As I have mentioned, a minimum flow of 25% will be quite sufficient for the S = 8500 impeller. As to the S = 11,300 impeller, I am less sure where the minimum flow need be set, but it may well be in the 50 to 75% range. That there is a definite relationship between acceptable minimum flows and the suction specific speed can be easily

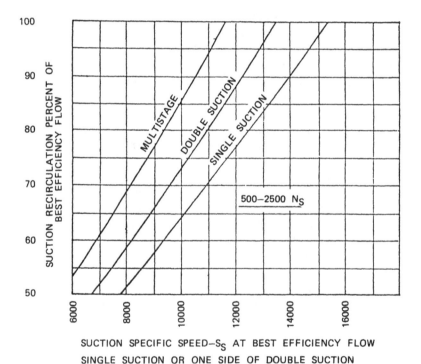

Figure 4.32 Curves developed by W. H. Fraser.

---

*"Flow Recirculation in Centrifugal Pumps," W. H. Fraser, Texas A&M Turbomachinery Symposium, Houston, Texas, Dec. 1981.

demonstrated both by actual field experience and by intuitive reasoning. Consider that for the same speed and pump capacity, the higher S value design requires a larger impeller eye diameter than does a lower S value design. This means that when the backflow starts at the eye diameter of the impeller, the peripheral velocity of this backflow will be higher with the higher S value impeller. It follows that the higher backflow velocity represents a higher energy level and that all the symptoms of operation in the recirculation zone will be intensified.

There is, of course, yet another factor that affects the recommended minimum flow and that is the degree of tolerance of the pump user to the signs of distress exhibited by the equipment. What confuses this issue is that the user is not the only one who applies such subjective judgment. It is not a secret that when comparisons are made between several pump selections, the user will find that only too frequently recommendations of minimum flow seem to bear no relation to the S value of the various offerings. Only too frequently a manufacturer with an offering that has a higher S value may quote the same or even a lower minimum flow than some other manufacturer with a more conservative design—and hence a lower recirculation flow. What this means to me is that these differences merely reflect more or less conservative estimates on the part of a designer as to what constitutes *tolerable distress*.

Let me expand on this: Starting with the proposition that the required NPSH (and hence the S value) of an impeller is determined by the conditions of service and by the geometric configuration of the impeller, it follows that two pumps with the same S value will have the same suction recirculation flow. It also follows that if an equal amount of "distress" is to be considered acceptable, the minimum flow to be recommended should be the same for equal values of S. It follows further that the higher the S value, the higher should be the value of acceptable minimum flow. I doubt that these statements can be seriously challenged.

But let us now focus on your specific case. All is not lost: there is a third alternative to the two you have cited, an alternative that will be both less expensive (to whomever has to foot the bill) and much quicker to accomplish. This alternative consists of manifolding the three bypasses, as illustrated in Fig. 4.33. In this way, without replacing the bypass valves or the existing individual piping, you could bypass either 25, 50, or 75% by manipulating the two manual control valves in the bypass manifold.

My reasoning is based on the fact that if indeed the pumps could operate satisfactorily with 25% minimum flow, this arrangement would permit you to do so and not waste energy. If, on the other hand, a higher minimum flow is (as I suspect) necessary, opening one or both the manual control valves would take care of the requirement. Consider that in normal

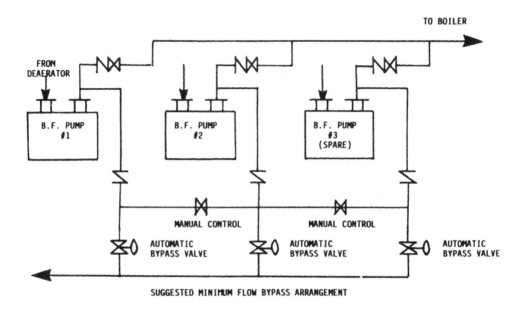

Figure 4.33 Suggested minimum flow bypass arrangement.

operation, the only time you may need to operate continuously with the bypass open is when a single pump is running. The other condition requiring the bypass to be open is during any switchover of pumps, that is, when you are bringing a second pump on the line or when you are switching on the spare pump while the two turbine-driven pumps are running, prior to shutting down one of the turbine-driven pumps. This is a condition of rather short duration, and no major problem should arise during this switchover operation.

I would start with a manual operation of the valves. But after you have had the opportunity of monitoring the operation with different values of minimum flow and have reached a decision on the proper amount of this minimum flow, the operation of the three bypass valves can be made automatic. The valves would be made to operate sequentially as follows, assuming that the minimum flow is set as 75%:

1. When the flow to the boiler falls to between 50 and 75% of a single pump capacity, one valve opens.
2. When the flow to the boiler falls to below 50% but above 25% of one pump's capacity, a second valve opens.
3. Finally, when the flow falls to below 25%, the third bypass opens.

Chapter 4 433

If, as I imagine, the bypass valves are of the "modulating" type, they can continue to be so operated. In other words, assuming that the flow to the boiler is 60%, a single one of the bypass control valves is used and is made to bypass 15%, and so on.

If the minimum flow decided upon is 50%, the only time all three bypass valves will be used is during the switching operation. Then, with two pumps momentarily on the line, each pump can be run at 37.5% of its capacity.

Question 4.44 Flashing in Boiler Feed Pump Suction Header When Booster Pumps Are Tripped

Our 120 MW unit has 3-50% feed pumps. Each pump unit consists of a booster pump and a high-pressure (HP) pump on a single motor shaft with the heaters between the pumps, as shown in Fig. 4.34.

If the standby pump is hot and the two pumps in service trip on a fault situation, how is the hot standby pump put into service when there is an obvious flash condition at the hp pump suction?

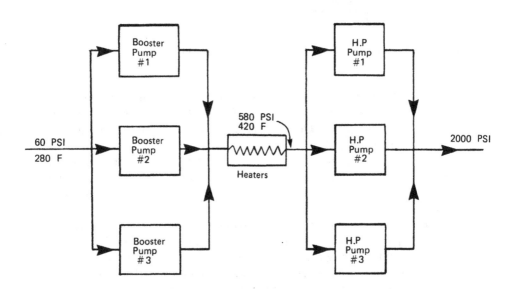

Note: Each booster and HP pump set is driven by one motor.

Figure 4.34 Note each booster and HP pump set is driven by one motor.

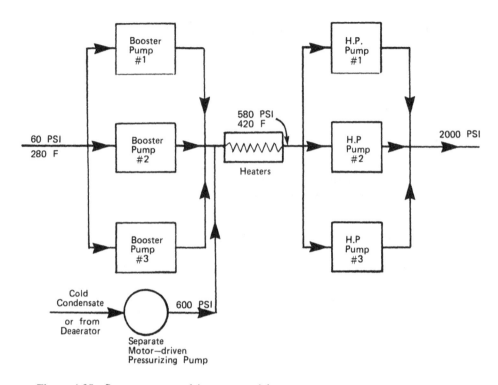

Figure 4.35 Separate motor-driven pressurizing pump.

Answer. When the two half-capacity pump units are running, the heater and the piping beyond the heater are full of 420 °F water. The standby pump is likewise full of 420 °F water since it is kept warm, presumably ready to start as needed.

If, as you say, the two pump units on the line trip out, not only is there a "flash condition" at the high-pressure pump suction, but the heater and the piping to the high-pressure pump will have flashed, since they are no longer pressurized above the vapor pressure at 420 °F.

There is a relative simple solution to this problem. It requires the installation of a separate motor-driven pressurizing pump, as shown in Fig. 4.35.

It should be designed for about 600 psig discharge pressure. As to capacity, it could be anywhere between a few hundred GPM to a duplicate of the present half-capacity booster pumps. This last may have the advantage of interchangeability of spares.

This auxiliary pressurizing pump should take its suction either directly from cold condensate (bypassing the deaerator) or from the deaerator. It would be made to start automatically whenever the boiler feed pumps on the line are tripped out. Its discharge into the suction header of the high-pressure feed pumps would ensure that these pumps coast down to rest without flashing. The pressurizing pump would continue to run until the feedwater heater and the pump suction piping are purged of water hotter than at the deaerator. It can be stopped after this, but preferably manually.

Now the boiler feed pump units can be restarted any time, without the danger of damaging the pumps.

5
Maintenance

Question 5.1 Frequency of Complete Overhauls

We are in the process of reorganizing our machine maintenance department. Because our plant has several hundred centrifugal pumps, a question that has arisen in connection with their maintenance will materially affect the reorganization plans. The question is whether centrifugal pumps should be regularly opened for inspection and complete overhaul or whether such overhauls can be carried out only when needed.

Answer. The problem of the frequency and regularity of complete overhauls of centrifugal pumps is a very controversial one. It is difficult, if not impossible, to make general rules that would apply in all cases. The type of service for which the pump is intended, the general construction of the pump, the liquid handled, the materials used, the average operating time of the pump, and the economic evaluation of overhaul costs versus possible power savings form renewed internal clearances all enter in some degree or another into the decision of the frequency of complete overhauls. Thus, some pumps on severe service may need a complete overhaul monthly, but some applications require overhauls only every 2 to 4 years or even less often. As a result, statements on the recommended frequency of conplete overhauls may be quite correct in the light of the service under consideration but could be easily challenged by an operator who has an entirely different application in mind.

Most pump designers and specialists consider that a centrifugal pump need not be opened for inspection unless evidence, of either a factual or circumstantial nature, is available to indicate that overhaul is necessary.

There are obviously some exceptions to this rule, and I shall speak of these exceptions presently.

By factual evidence I mean that the pump performance has fallen off sufficiently or that noise or driver overload indicates trouble. In this connection, it should be borne in mind that proper instrumentation is of paramount importance to the satisfactory operation and life of centrifugal pumping equipment. Means for determining pump capacities and pressures should be considered as part and parcel of the maintenance tools without which no operating engineer should consider running his plant.

A schedule should be set up for a fairly frequent complete test of the pumping unit so that the results of this test can be compared with the performance of the pump in its initial condition and so that any sudden falling off in performance can be detected without delay. This comparison, and not mere passage of a fixed number of years, months, days, and hours, should be relied upon to establish whether or not sufficient internal wear has taken place to require a complete overhaul. After all, running a complete test is a less expensive procedure than opening up a pump for inspection. In addition, it does not require taking the unit out of service.

Factual evidence in the form of noise, driver overload, excessive bearing temperatures, and similar manifestations of trouble is too obvious in character to require further explanation.

Circumstantial evidence refers to data accumulated through past experience, either with the pump in question or with similar equipment or similar service. For instance, if a group of boiler feed pumps built of chrome-stainless steel alloys has shown that 60,000 hr of continuous operation can be accumulated without the need of a complete overhaul, an additional duplicate unit need not be opened for inspection before it has in turn run up a record of 60,000 hr.

On the other hand, pumps on severe service that have required overhaul at 3 month intervals may be replaced by units of better materials or of sturdier construction. Nevertheless, until such time that the new equipment has proved itself in service and a new *experience pattern* has been established, it is best to open up the pump at the end of the same 3 month period in order to evaluate the effect of the better construction or better materials.

One of the exceptions that I mentioned earlier is related to corrosion-erosion troubles, which will not necessarily be immediately reflected in the performance characteristics of the pump obtained by means of routine tests. However, if these troubles are permitted to continue unattended, they may easily result in the total destruction of the pump, beyond any possibility of repair. But corrosion-erosion troubles are generally foreseeable. For instance, a pump handling corrosive chemicals that is built either of ordinary materials or of materials untested in that particular application may become rapidly and severely damaged. In such a case, it is advisable that the pump be opened for inspection fairly soon after initial installation and at frequent intervals thereafter until such time that the life of the pump materials under the actual operating conditions has been determined.

The other important exception refers to a condition in which operators prefer to rely on visual examination and actual measurements of clearances. They may feel insecure unless they carry out such examinations at intervals that they have established in their own minds. If operators cannot be shown proof that this procedure is unnecessary, the exception is fully warranted because without the visual evidence operators may lack confidence in their equipment, and this situation is not a healthy one. Rules were made to be broken, and the well-being and confidence of operators are much more important than the fact that a piece of equipment has been dismantled one or two extra times during its useful life.

An important warning should be added that in order to ensure rapid restoration to service in the event of an unexpected overhaul, an adequate store of spare parts should be maintained at all times. This will avoid possible delays in obtaining special repair parts from the manufacturer.

Question 5.2 Life Between Overhauls for Boiler Feed Pumps

What has experience shown with regard to the normal hours of operation between internal inspections or overhauls of modern high-pressure boiler feed pumps designed for pressures of 1800 psi and above? What is used as a criterion for normal internal inspections?

Answer. It is my opinion that, based on an analysis of economic justification, the internal clearances of a high-pressure boiler feed pump need not be renewed until they have become double their original dimensions.* The period of time that will elapse before wear will have doubled the original clearances will vary for different designs and for different installations. Since wear can take place both because of slight internal contacts during operation and because of the erosive action of flow past the clearance joints, elimination of internal contacts—or, at least, reduction in the frequency of such contacts—will act to lengthen the life of the internal clearances. Thus, pumps with stiffer shafts and with liberal internal clearances will generally have a longer life between overhauls than pumps with a greater shaft deflection and with closer clearances, since the latter type of pump will have a greater tendency toward contact between running and stationary parts during operation.

The effect of the installation—or rather of the method of operation—is also very important. A pump operated at or near its rated conditions of service

---

*See further comments about effect of pump types and of rising energy costs in Question 4.31.

a great portion of the time will wear at a much slower rate than one that operates frequently at reduced loads. This in one of the reasons I recommend operating with a single pump whenever it can carry the necessary load in the typical two half-capacity pump installations.

An exact value for the ultimate life of a high-pressure boiler feed pump between internal overhauls is therefore difficult to set. However, my estimate is that, barring accidents and assuming normal service, proper operating procedures, and an adequate design, high-pressure boiler feed pumps will not require overhaul of internal parts in less than about 50,000 to 100,000 hr of service.

Incidentally, this not quite true of most spare boiler feed pumps, which seem to have a shorter life than pumps in normal service. Possibly the reason for the difference in the ultimate life between pumps on normal service and spare pumps is that the latter are frequently operated at very reduced loads (such as during weekends) and therefore wear faster than in normal service.

Several criteria are used to determine whether a pump should be subjected to an internal inspection and, possibly, complete overhaul. One of these is the complete running test to which every pump should be subjected about once a year. Such a test gives a reasonable indication of the internal condition of the pump. Properly calibarated instruments should, of course, be used in conducting these tests. It should be noted that laboratory accuracy is not attainable in field testing. Therefore, it is not so much the absolute values determined on test that are important but rather the relative changes noted while using the same instruments that are indicative of the deterioration of the pump internal clearances.

When full tests cannot be carried out very conveniently, it is possible to get an approximate indication from the increase in power consumption at some given flow rate, from the change in pressure drop across the feedwater regulator at this rate, or from speed measurements on variable-speed units—again at some given rate of flow—with the unit in new condition and again after a certain number of operating hours.

Another criterion of internal wear is the measurement of the flow past the balancing device of boiler feed pumps that incorporate such a device. The leak-off from the balancing device is returned either to the direct-contact heater from which the pump takes its suction or to the pump itself. Generally a calibrated orifice is incorporated in the leak-off line (Fig. 5.1), and a calibration curve of the orifice (Fig. 5.2) is included in the pump instruction book. Unless damage has taken place at the balancing device, it will be found that interal pump clearances have doubled when the balancing device leakoff has increased to about double its original value. It is recommended that readings of the pressure drop across this calibrated orifice be taken daily so that any sudden and untoward change in leakage can be spotted immediately.

Finally, any odd characteristic exhibited by a boiler feed pump, such as exceessive vibration or noise, should be thoroughly investigated and analyzed

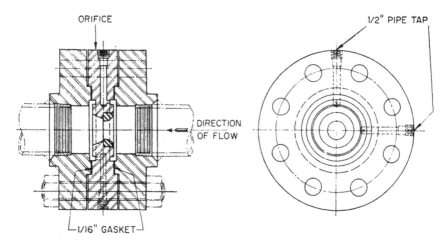

Figure 5.1 Cross section of calibrated orifice for use with boiler feed pump balancing device leak-off line.

to determine whether an inspection of the pump condition is warranted. The last, of course, is true for any pump or, as a matter of fact, for any piece of machinery.

Question 5.3 Replacement with a More Efficient Pump

We have an old centrifugal pump handling 600 GPM against a head of 200 ft. Recent performance tests indicate that its efficiency is about 71%. When the pump was new, it had an efficiency of 74%. Would it be advisable to purchase a new pump to replace the old one or to rebuild the old one?

Answer. The decision whether to renew the necessary parts to restore pump efficiency or to purchase a new pump will depend on the price you pay for electric power.

First, let us calculate the power you can save by either method. The water horsepower corresponding to your operating conditions can be calculated from the following:

$$\text{whp} = \frac{\text{gallons per minute} \times \text{total head in feet}}{3960}$$

$$= \frac{600 \times 200}{3960} = 30.1$$

Chapter 5

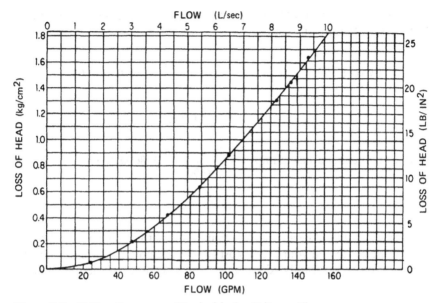

Figure 5.2 Calibration curve of typical leak-off line orifice.

$$bhp = \frac{whp}{\text{pump efficiency expressed as a decimal}}$$

Present $bhp = \dfrac{30.1}{0.71} = 42.4$

Assuming that the pump can be restored to its original performance,

$bhp = \dfrac{30.1}{0.74} = 40.7$

A modern pump for these conditions of service should be capable of meeting 81% efficiency. It will take a brake horsepower of

$bhp = \dfrac{30.2}{0.81} = 37.2$

Savings by renewing the pump = 42.4 − 40.7 = 1.7 hp

Savings by buying a new pump = 42.4 − 37.2 = 5.2 hp

Assuming 90% motor efficiency, 8000 hr of operating a year, and 3.0 cents/kWh, the savings will be

$$\text{Savings by renewing the pump} = \frac{1.7 \times 0.746}{0.90} \times 8000 \times 0.03$$

$$= \$340.90/\text{year}$$

$$\text{Savings by buying a new pump} = \frac{5.2 \times 0.746}{0.90} \times 8000 \times 0.03$$

$$= \$1042.40/\text{year}$$

These figures should be adjusted by you to reflect the actual hours of operation of the pump and your actual cost for electric power. From the adjusted figures you can readily determine which course of action is most economical.

Question 5.4   Maintenance Tools

What is considered to be an adequate set of tools to be furnished by a pump manufacturer for the maintenance of equipment?

Answer. I am just slightly afraid that this is a leading question, and I hope that the answer to it will not be used to settle some pending argument between user and supplier. Not that this answer should be so controversial, but there certainly can be some difference of opinions on this subject.

In my own opinion, the manufacturer should be expected to furnish only the special tools that might be needed to dismantle the pump. Thus, tools listed in a tool manufacturer's catalog as standard items should not be considered special even if they are not normally in a mechanic's toolbox. For example, spanner wrenches would fall into this category.

Likewise, coupling pullers, bearing pullers, slugging wrenches, and other similar convenience tools are not furnished, even if in some cases they may be somewhat special. For instance, a standard bearing puller may require extra long pull rods because of the location of the bearing on the shaft. Since the bearing puller itself would be furnished by the user, the long pull rods should also be acquired or made by the user.

On the other hand, some quite special tools frequently may be required the use of which is dictated by the particular construction of the pump in question. This is more likely to occur with special-purpose pumps or with multistage pumps. One very typical example is the cradle that is used in the dismantling of the inner asembly of a barrel-type double-casing pump (shown in use in Fig. 5.3). The category of these special tools may embrace such items as forcing-off bolts, sleeve pullers, and balancing device wrenches (Fig. 5.4).

Generally, a pump manufacturer has a reasonable good idea of the range of tools that should be expected to be found in a user's maintenance shops and provides such other tools that will complement them. The pump manufacturer cannot and should not be expected, however, to outfit these maintenance shops completely.

Question 5.5 Spare Parts for Boiler Feed Pumps

Until recently, our steam power plant had units not exceeding 15,000 kW in size and operating pressures were all under 850 psi. Our latest addition is a 60,000 kW unit operating at 1800 psi. The boiler feed pumps are of the vertically split double-casing type and are provided with a balancing device. We would like to know what stock of spare parts we should carry for these pumps.

Answer. Under normal operating conditions and based on the use of the stainless steel materials now universally used for the internal parts of high-pressure boiler feed pumps, the length of service before renewal of internal wearing parts is required should be the order of 50,000 to 100,000 hr. Boiler feed pumps operating most of the time at or near their design point will generally have a longer life before overhaul is required. Pumps operating frequently and for long periods of time at light loads may show a shorter life between overhauls. In general, replacement of wearing parts is recommended when the initial clearances have doubled.

The number of boiler feed pumps installed and the extent to which repairs of worn parts can be carried out in the field will determine to a great extent the minimum number of spare parts that should be carried in stock at the site of the installation. In the case of this 60,000 kW unit it is presumed that two pumps are used, each for the full plant capacity with one pump runnning and the second on standby service.

It is recommended that the following spares be carried:

1. A set of casing wearing rings
2. A set of stage bushings
3. A spare balancing device made up of the rotating balancing disk and of the stationary balancing disk head
4. One or two sets of shaft sleeves
5. Several sets of stuffing-box packing
6. A set of bearing bushings
7. A thrust collar and a set of thrust shoes for the thrust bearing

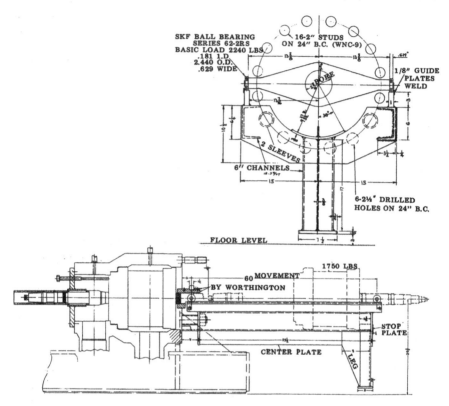

Figure 5.3 Typical arrangement for removal of rotor from barrel-type pump.

It is also frequently the practice to carry a complete inner pump assembly in stock. This element can be installed in the pump when examination or tests show that the pump has become excessively worn or if it becomes accidentally damaged. In that event, the inner assembly is withdrawn from the pump and can be immediately rebuilt in the field with the parts listed above. In the meantime, the spare inner assembly is reassembled with the pump, and no delay occurs in returning the pump into service.

Chapter 5

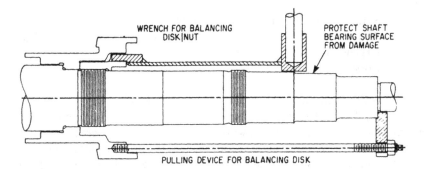

Figure 5.4 Other examples of special tools include a wrench for tightening the balancing disk nut (top half) and device for pulling the balancing disk (bottom half).

To obtain the greatest service from spare parts, certain of these parts are normally furnished with undersized bores or oversized turns. For instance, the casing rings and stage bushings carried as separate spares are furnished with undersized inside diameters. When they are applied to a rebuilt rotor, the individual impellers should be mounted on a mandrel, trued up, and miked. After these operations, the casing rings and stage bushings must be finish-machined on the inside bore in such a manner as to restore the initial clearances. The latter are generally listed in the instruction book prepared by the manufacturer.

The rotating and stationary parts of the spare balancing device are furnished with oversized turn and undersized bore, respectively. This is done in order to obtain the maximum life from the balancing device initially installed in the pump. When the first overhaul takes place, the original rotating balancing disk is mounted on a mandrel and trued up. It is then used in conjunction with the spare stationary balancing disk head, which is finish-machined in the field so as to restore the initial clearances given by the manufacturer.

At the time of the second overhaul, the original stationary balancing disk head can then be used in conjunction with the spare rotating balancing disk in the same manner as outlined above.

In many cases, the pump manufacturer outlines the procedure with respect to the use of spare parts in the instruction book and provides sketches of the spare parts, indicating the important dimensions.

Question 5.6 Repair or Replace?

What are the factors that should be considered in making a decision between replacing a part or repairing it?

Answer. I can think of four major factors that would influence such a decision:

1. Suitability of material to be repaired
2. Availability of skilled personnel and of necessary shop facilities
3. Urgency in rebuilding pump
4. Possible advantages of upgrading material

Let us consider each one of these factors in some detail.

1. Suitability of material. The repair of a part generally requires that some material be deposited on the worn part in those areas where the original material has been removed by erosion, corrosion, or intermittent contact and rubbing. The base material must be receptive to being built up by welding, brazing, or some form of coating. I have known maintenance personnel to do a creditable job in welding up cast iron, but this is a tricky procedure and I would not recommend attempting it save as an emergency measure if urgency dictates it and on the assumption that it be considered a temporary repair.

2. Availability of personnel and shop facilities. This may seem to be an obvious consideration. I list it mainly to call attention to the fact that the personnel and the facilities do not necessarily have to be available at the plant site. Many original equipment manufacturers maintain regional service shops located in the proximity of their clients' plants. These can generally provide quicker repair service than the main factory.

In addition, most plants can locate small machine shops close enough to enable one to examine and evaluate their capabilities should it be necessary to resort to their services.

3. Urgency considerations. This can become a very major factor, particularly if the user does not carry an adequate stock of spare parts on hand. Obviously, if there wil be an excessive time lag in obtaining new parts, repair, by all means! In this connection, the manner in which spare parts are ordered should be given serious scrutiny. Whenever mating parts may have to be replaced and original running clearances are to be restored, consider that parts should be carried in stock with underbored inner diameters and conversely mating parts with oversized outer diameters. For a more detailed discussion of this matter, see Questions 5.7 and 5.8, which discuss the methods used to restore wearing ring clearances.

4. Upgrading of materials. Quite frequently, the material chosen for some of the pump parts at the time the pump is purchased is not necessarily the most wear, erosion, or corrosion resistant that could have been chosen. This occurs for a variety of reasons, including sometimes the desire of the user's purchasing department to economize. In such cases, upgrading of the materials chosen for the replacement parts will often increase the time between necessary repairs by a factor of 2, 4, or even greater. In such cases, replacement is preferable to repair.

You will notice that I have not listed "cost" as a major consideration. This does not mean that I recommend that it can be ignored, but the fact is that it is very difficult to make a valid evaluation of costs because this must involve much more than just the comparison of the cost of repairs to that of the replacement parts. If the repairs are faulty, the cost of extra downtime and of the possible interruption of production will make any savings illusory and irrelevant.

Question 5.7 Wearing Ring Clearances

We have several old centrifugal pumps in our plant from which the nameplates have disappeared, so that we do not even know the name of the manufacturer. Nor do we have any data or instruction books that might give us information on some of the internal dimensions of the pump parts. We intend to overhaul these pumps if possible by producing some of the parts, such as wearing rings, in our own shop. Are there any rules that might permit us to estimate the original wearing ring clearances so that we might restore these when we overhaul the pumps?

Answer. Typical clearance and tolerance standards for wearing ring joints in nongalling metals in general-service pumps are shown in Fig. 5.5. These standards apply to the following combinations:

1. Bronze with a dissimilar bronze
2. Cast iron with bronze
3. Steel with bronze
4. Monel with bronze
5. Cast iron with cast iron

If the metals gall easily (like the chrome steels), values given in Fig. 5.5 should be increased by about 0.002 in. For multistage pumps, basic diameter clearance should be increased by 0.003 in. for large rings. The tolerance indicated is plus (+) for the casing ring and minus (−) for the impeller hub or impeller ring.

For example, let us assume that we are dealing with a single-stage pump with nongalling components and that the wearing ring joint has a diameter of 9 in. The casing ring ID should be 9.000 plus 0.003 and minus 0.000 in., and the impeller hub or impeller ring OD should be 9.000 minus 0.018 or 8.982 plus 0.000 and minus 0.003 in. The actual diametral clearance would therefore be between 0.018 and 0.024 in.

Depending on facilities available in your maintenance shops, there are several ways in which you may restore the clearances on these pumps. If the pumps have single wearing rings, you may do one of the following:

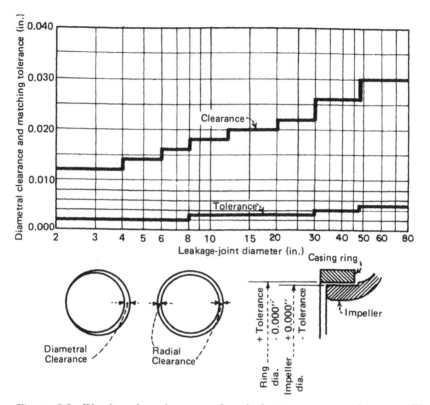

Figure 5.5 Wearing ring clearances for single-stage pumps using nongalling materials.

1. Produce a new wearing ring bored undersize and true up the impeller wearing hub by turning it down on a lathe.
2. Build up the worn surface of the wearing ring by welding, brazing, or metal spraying so that it can be bored undersize and then true up the impeller wearing ring hub.
3. True up the wearing ring by boring it oversize, build up the impeller wearing ring hub, and machine it to give correct clearance with the rebored ring.

If the pumps have double wearing rings, clearances may be renewed by one of the following methods:

1. Produce a new oversize impeller ring and use the old casing ring bored out larger.

2. Produce a new casing ring bored undersize and turn down the old impeller ring.
3. Renew both rings if necessary.
4. Build up either the casing ring or the impeller ring by welding, brazing, or metal spraying and machine the other part and finish the built-up surface to restore the original leakage joint clearance.

When truing up the impeller ring, mount it first on the impeller and carry out the truing-up operation on the assembly to avoid the possibility of distorting the impeller ring and to assure the concentricity of the running joint.

Question 5.8 More About Wearing Ring Clearances

In Question 5.7 you have discussed various methods that can be used to overhaul centrifugal pumps when the wearing ring clearances have become excessive. You gave suggestions for pumps having double wearing rings as well as single rings. For the latter case you listed three possibilities. Would it not also be possible (after wear takes place) to machine the impeller hub so that an impeller wearing ring can be mounted on it?

Answer. This is a fourth possibility in some cases but not all, as the decision will depend a great deal on the type of pump in question.

If we are dealing with a very small impeller, it is probable that wear can best be corrected by the replacement of the impeller, since any other approach would be more costly. Furthermore, you should note that not every pump impeller will have enough metal at the running clearance hub so as to accomodate an impeller wearing ring at some later date. but if sufficient metal does exist, it is logical to add this fourth solution, which I would word as follows:

4. After the impeller wearing ring hub has been trued up several consecutive times, there may not remain sufficient metal for one more truing up. In such a case, a flat impeller ring may be obtained and mounted on the ring hub. Maintenance in the future would proceed as indicated for pumps with double rings.

But this procedure would be resorted to only in the case in which economics makes it attractive when compared with the cost of purchasing a new impeller. This should never be done, however, with high-speed pumps (such as boiler feed pumps) in which impeller rings are not used to avoid loosening of the fit between impeller and ring caused by centrifugal force.

In addition, you should consider that in most cases impeller hubs are provided with sufficient metal to carry out procedure 1 several consecutive times. Ultimately, if the impeller itself is still in satisfactory condition, procedure 3 can be substituted.

Question 5.9 Measuring Internal Clearances (Axially Split Casing Pumps)

Our maintenance crew has found that the inspection of some of our multistage axially split casing centrifugal pumps can be carried out without stripping the rotor. The clearances at the wearing rings are measured by inserting a feeler gauge between the rotating and stationary parts. Some other designs, however, are such that a feeler gauge cannot be used because the stationary wearing ring has a lip that overhangs the front part of the impeller hub. Is there any short cut method that would permit us to check clearances without completely stripping the rotor?

Answer. I presume that in the first case you refer a construction similar to that shown in Fig. 5.6A, where the leakage joint is flat and therefore permits the insertion of a feeler gauge. Wearing rings of the L type, shown in Fig. 5.6B, of course, prevent the insertion of the gauge. It is still possible to obtain a reasonably accurate check of the diametral clearances in the following manner:

1. Mount a dial indicator on the impeller shroud (as in Fig. 5.7), and with the stationary ring testing on the impeller wearing ring hub, set the dial reading at zero.
2. Leaving the dial indicator in position, push up on the stationary ring from below and make a record of the maximum dial reading. This corresponds to the diametral clearance.
3. Repeat this operation for every clearance joint and make a record of all readings.

Obviously, if these readings indicate that the clearances have not changed appreciably from the initial dimensions given in the instruction book for the pump in question, no further dismantling of the rotor is necessary.

One note of warning, however: This shortcut method will disclose the extent to which clearances have increased but will give no indication as to the condition of the adjacent clearance surfaces. In other words, the presence of burrs, grooves, or indentations caused by the passage of foreign matter through the clearances and the resulting damage to the surfaces will go undetected.

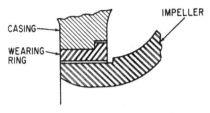

(a) - FLAT-TYPE WEARING RING

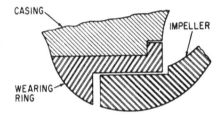

(b) - L-TYPE WEARING RING

Figure 5.6 Two different types of wearing ring construction show that the flat type permits a feeler guage to check on wear but the L type does not.

Question 5.10 Measuring Internal Clearances (Radially Split Casing Pumps)

In Question 5.9 you discussed the possibility of measuring wearing ring clearances of axially split casing boiler feed pumps without dismantling the rotor. The procedure you recommended is not, of course, applicable to radially split casing barrel-type pumps, since these have to be completely dismantled for inspection. Such an inspection, of course, also affords the opportunity to examine the wearing ring surfaces for burrs, grooves, or other irregularities.

I would like to ask you how accurately do you think wearing ring clearances of such pumps should be established on inspection? In other words, should they be measured to within 1/1000 in. or closer or less closely?

Answer. Wearing ring clearances on new high-pressure boiler feed pumps will range from 0.012 to 0.018 in. on the diameter, and the normal practice is not to renew these clearances by replacing the wearing rings until these clearances have about doubled from wear. Certainly it is not necessary to measure clearances closer than to 1/1000 in. On the other hand, it should prove to be no problem to measure these clearances to such a tolerance of measurement.

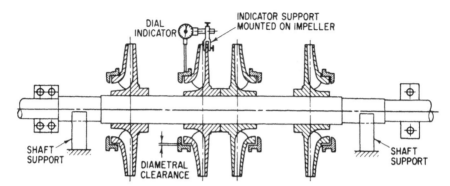

Figure 5.7 A dial indicator mounted on the impeller shroud can give a close indication of wearing ring clearances.

The normal procedure in carrying out this inspection is to measure independently the inside diameter of the wearing ring fit and the outside diameter of the impeller wearing ring hub, using inside and outside micrometers, respectively, as shown in Figs. 5.8 and 5.9. Several measurements should be taken to determine whether or not the wearing ring or the impeller has been wearing out in an egg-shaped manner. The clearance is considered to be the difference between the maximum inside diameter and the minimum outside diameter readings.

Certain maintenance mechanics measure the clearance directly by placing the impeller within the wearing ring, as in Fig. 5.10, and moving it laterally against a dial indicator to determine the diametral clearance. To discover any inequality in wear around the circumference, the impeller should be rotated, and the dial indicator should be attached to several points of the stationary part. In my opinion, the *difference* method is more reliable.

One important warning is that the impeller and wearing ring should be at the same temperature before the readings are taken. Many designs use shrunk-on impellers that have to be heated prior to removal from the shaft. The impeller may thus be heated to at least 400 °F and possibly to as much as 500 to 600 °F. Chances are that it will be allowed to cool down to something like 120 °F so that it can be handled comfortably before the measurements are taken. But if the wearing ring is at, say, 80 °F, there will be a 40 °F difference between the two parts, and this difference can be quite significant. If the coefficient of thermal expansion is taken as 0.0000065 in./in. per °F and if the wearing ring fit diameter is 8 in., the apparent clearance will be about $2/1000$ in. less than the true clearance. This error

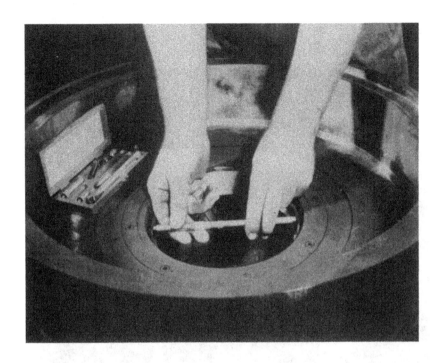

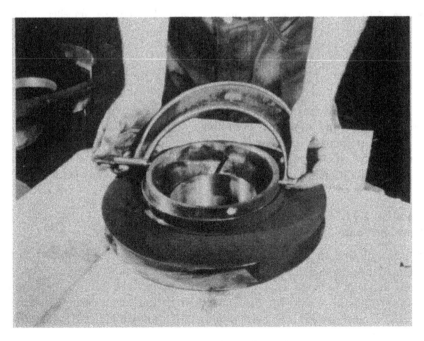

Fig. 5.9 Measuring outside diameter of impeller hub.

Figure 5.10 Measuring wearing ring clearance by moviing impeller within clearance.

will, of course, be magnified if the impeller diameter is measured when the temperature of the impeller is even higher than the 120 °F I have assumed.

This possibility of error is one that is frequently overlooked in the measurement of clearances in general, as our first reaction is to assume that such a small difference in metal temperatures cannot be of any consequence.

Question 5.11 Checking Internal Clearances Without Opening Pump

We would like to ask your assistance in the following problem: Our maintenance personnel have asked for some type of tool, device, or equipment that will indicate when overhaul of a boiler feed pump is necessary without their having to lift the upper half of the pump casing. In other words, what is needed is a device or a method that will indicate the internal

Chapter 5

condition of our boiler feed pumps without dismantling them. The purpose of this would be to prevent not only pump failures but also needless pump inspection.

A number of ideas have been discussed among our engineering and maintenance personnel, but nothing has been agreed upon that will be simple and inexpensive. These attributes are a necessity for any method or device that would be adopted. We would appreciate hearing any thoughts you may have as to the feasibility of such a project based on your experience with feed pumps.

Answer. The problem that you have raised is one that has intrigued a whole generation of centrifugal pump designers, service people, and users alike. Unfortunately, I must admit that we are not yet in the position to state that it has been solved. What renders this problem difficult is that the measurement of the running clearances of a multistage boiler feed pump would require access for probes of some sort of another at every wearing ring and every interstage bushing. Obviously such a requirement hardly complies with the criteria that you have indicated, that is, that the method or device must be simple and inexpensive.

There are, it is true, some partial solutions that I have frequently recommended to operators of boiler feed pumps and that have frequently helped reduce (at least to a partial degree) the frequency with which pumps are dismantled for inspection.

You have inferred in your question that you are dealing with axially split casing boiler feed pumps, but one of these partial solutions to which I have referred is applicable to the double-casing barrel-type pump. Thus, oddly enough, the type of pump that is more complex to dismantle is the one that allows most readily a reasonable evaluation of the wear that has taken place in the clearance joints. I should qualify this by adding that I refer to pumps with *in-line* impellers in which the axial thrust is balanced by means of a hydraulic balancing device. It can be demonstrated both theoretically and from field experience that the rate of wear in the wearing rings follows very closely the rate of wear in the balancing device. If a means is used to measure the flow past this balancing device and to compare it with the flow that was taking place when the pump was new, it is possible to estimate quite closely the increase in clearance in all the internal joints (see the answer to Question 5.2).

Another method used on occasion that requires the removal only of the bearing shells and of the packing is applicable both to the double-casing pumps and to axially split casing pumps of either the opposed impeller type or with balancing devices. After those parts that would cause interference are removed and the rotor is permitted to rest on the bottoms of its wearing rings, a dial indicator is mounted at each end of the pump over an exposed portion of the shaft. The dial indicator reading is set at zero, and the shaft is lifted first at one end and then at the other, using a pinch bar. The amount of vertical displacement of the rotor read on the indicator is a measure of the mininum internal clearance. If all clearances have worn to the same degree, this minimum will correspond to the average existing clearance.

A third method of measuring clearances requires lifting the upper casing half of the pump but not dismantling the rotor or even lifting it out of the lower casing half, as described in Question 5.9. It would seem to me, however, that if the boiler feed pump has been dismantled to the point of permitting this method of measuring internal clearances, I would like to carry the inspection one step further and lift the rotor out of the casing.

I realize that these suggestions are far from meeting the ultimate desires of your maintenance personnel, but they are all that I can suggest at this moment. Incidentally, you say that a number of ideas have been discussed among your personnel but nothing has been found that would be simple and inexpensive. If you were to describe to me some of the ideas that have been discussed, and if the readers of this book were to write me of any ideas they may have considered, it might well be that through concerted studies and discussions some one method could be improved upon to the point that it would meet the required criteria of simplicity and low cost.

Question 5.12 Replacement of Casing Gaskets

What particular precautions are necessary in replacing gaskets for the split of axially split casing pumps?

Answer. If on opening the pump it is found that the gasket is in perfect conditoin and still adheres to the lower half of the casing, it is not necessary to replace it. On the other hand, if the gasket is damaged even slightly, a new gasket must be used. The new gasket must be of the same thickness so that it will be compressed to the same extent.

To understand the importance of using the correct gasket material, it is necessary to visualize what happens to the gasket when it is applied to prevent leakage at the casing flanges. When the pump is assembled and the flanges are pulled tight by the casing bolts, these bolts are given an initial stress in tension. The gasket itself is subject to a compressive stress that is a function of the bolt stress and of the relation between the areas of the bolts and of the gasket.

Chapter 5 457

When the pump is operating, the internal pressure increases and so does the bolt stress, so that the bolts increase in length. Assuming that the initial gasket compression did not deform it permanently, that is, that the gasket stress did not exceed the yield strength of the material and that it possesses sufficient resilience, the thickness of the gasket will increase by the same amount that the bolts were increased in length. The gasket compressive stress is reduced but must remain sufficiently high to prevent leakage from taking place between the flanges.

Thus, unless the replacement gasket material is of the same quality as the original gasket and has the same physical properties, the danger exists either that it will be deformed permanently during the tightening of the casing bolts or that it will fail to "bounce back" after internal pressure is applied. In either case, leakage can take place at the casing joint.

As to the thickness of the gasket, it should be realized that too thick a gasket will generally lead to casing cutting through leakage; if the gasket is thinner than the original one, tightening the two casing halves may exert undue force on the casing wearing rings, which may become distorted. This, in turn, may lead to contact with the rotating parts and to severe damage.

When installing a new gasket, trim the inner edge squarely with sufficient overlap so that when the upper half of the casing is tightened, it will press the gasket edges against the stationary internal parts to provide sealing against leakage. No overlap should be left at the stuffing-box bores.

The trimming of the gasket is best accomplished by first cementing the gasket to the lower half of the casing, using shellac. If powdered graphite is rubbed into the gasket surface before the pump is reassembled, the gasket will not stick to the top half of the casing the next time that the pump is dismantled. It will then be possible to reuse the gasket if it has not been damaged.

Question 5.13 Further Comments on Gaskets of Axially Split Casing Pumps

In looking at your answer to Question 5.12, I find that you have overlooked a very basic concept in the design of bolted structure. You state the following: "When the pump is operating, the internal pressure increases and so does the bolt stress, so that the bolts increase in length." If the pump were designed properly in the first place, the bolt stress would not increase nor would the bolts increase in length.

Your are saying that a bolt must elongate in a bolted structure or a member before the member can take a load, and nothing is further from the truth. There is absolutely no elongation in the bolts nor increase in stress until the initial tension caused by tightening is exceeded, and this would never occur on a properly designed pump.

Answer. I regret that I have to disagree completely with the preceding comments. When a pump casing becomes subject to internal pressure, the stress in the bolting that holds the two halves together *does* increase, the bolts *do* elongate ever so slightly, and these two facts are in no way an indication of an improper design.

At the risk of touching on some quite fundamental facts of strength of materials and of machine design, I shall demonstrate the truth of these statements. Let us consider a much simplified casing such as shown in Fig. 5.11. A gasket is placed between the flanges of the two casing halves. To simplify our analysis further it is assumed that the casing is infinite in length and that we are to examine the conditions that exist in a 1 ft length of this casing. Two steel bolts are located on each side of this 1 ft casing.

The bolts are tightened until a pressure at least equal to the ultimate internal pressure is developed over the gasket surface. If the gasket were not squeezed to that extent, leakage would obviously take place once the internal pressure was applied, relieving the initial gasket stress. Therefore, the total load on the flange bolts must be at least equal to the total load of the internal pressure acting on the *active* surface of the casing and casing flanges. We may assume that this active surface is equivalent to the projected casing area between the centerlines of the bolts. (In the case assumed in Fig. 5.11, this would be the distance of 30 in. between the bolts times the 12 in. of our 1 ft long casing, or 360 in.^{2}.)

If the maximum internal pressure that is to be imposed on this casing is 250 psi, the total load on the active surface will be 90,000 lb. This is also the minimum load that must be imposed on the four flange bolts, each of which must be subjected to a load of at leasts 22,500 lb. To be on the safe side, the tightening load may be increased to 30,000 lb per bolt.

Let us assume that each bolt has a cross section of 2 in.^{2}. The initial bolt stress will be 15,000 psi.

When internal pressure is applied, a load is imposed on the casing flanges, which tends to separate them. This load, in turn, is transmitted to the bolts, and the stress in the bolts caused by tightening them is increased. The final stress in the bolts is therefore equal to the sum of initial tightening stress and that caused by the load. With each bolt subjected to a 22,500 lb load caused by the 250 psi internal pressure, the additional stress is 22,500 lb divided by 2 in.^{2}, or 11,250 psi. The final bolt stress is therefore 26,250 psi.

We can now calculate the elongation to which the bolts will be subjected by the application of the internal pressure. (Of course, the bolts have already been stretched during the tightening process, since for an elastic material any stress will carry with it a definite strain, or elongation.) As long as stresses stay within the yield strength of the material in question,

Chapter 5

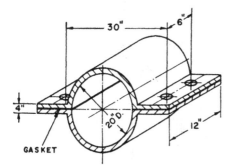

Figure 5.11 Simplified drawing of casing and bolts.

stress and strain are linked by a relation with the modulus of elasticity of the material. We can list the following relations:

$$\text{Unit stress} = \frac{\text{load}}{\text{area}} = \frac{F}{A} \tag{5.1}$$

$$\text{Unit strain} = \frac{\text{elongation}}{\text{initial length}} = \frac{e}{L} \tag{5.2}$$

$$\text{Modulus of elasticity (E)} = \frac{\text{unit stress}}{\text{unit strain}} = \frac{F}{A}\frac{L}{e} \tag{5.3}$$

Solving, we get

$$e = \frac{FL}{AE} \tag{5.4}$$

If the flanges are 4 in. thick and if we assume that the modulus of elasticity of steel is 30×10^6, we can solve Eq. (5.4) for the elongation under the additional load of the internal pressure:

$$e = \frac{22{,}500 \text{ lb} \times 4 \text{ in.}}{2 \text{ in.}^2 \times 30 \times 10^6} = 0.0015 \text{ in.}$$

In other words, our bolts will have stretched one and one-half tousandths when the internal pressure reaches 250 psi. In turn, the two flanges would have become separated by an additional one and one-half thousandths, and the gasket would have increased in thickness by the same amount over its thickness under initial tightened conditions. I should add that since strain and stress are proportional, the elongation of the bolts would have been three-thousandths had the internal pressure been 500 instead of 250 psi.

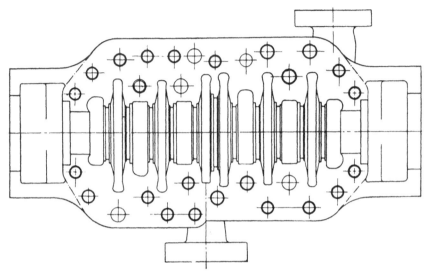

Figure 5.12 Multistage pump casing

I repeat that this design would not have been an improper or unsatisfactory design for such a flange. As a matter of fact, the final stress on the bolts of 26,250 psi is quite reasonable in comparison with the yield strength of the steel used for such bolting.

This analysis, of course, was highly simplified. A centrifugal pump casing flange is far more complex a structure. In addition, part of the casing is subjected to discharge pressure and another part to suction pressure. If we are dealing with a multistage pump casing (see Fig. 5.12), our problem is even more complex, since the internal pressure varies from suction pressure to full discharge pressure but differs for each individual stage.

As a matter of fact, bolting and flange stretch or elongation for such a pump are generally determined experimentally, as it is impossible to conduct a strictly analytical study of the "quality" of a casing flange joint under a combination of varying internal pressures at the individual stages.

Figure 5.13 shows a gauge used in determining the stretch of the two flanges of an axially split casing pump. One or more such gauges are clamped to the interstage wearing ring bores on the lower half of the casing. The gauges are so arranged that the spring pushes the soft copper wedges into the split at the adjacent portion of the casing. The plunger is retracted before installation and secured with the holding screw. To determine the gasket compression induced by the initial tightening of the bolts, small pieces of gasket are removed at various places next to the internal

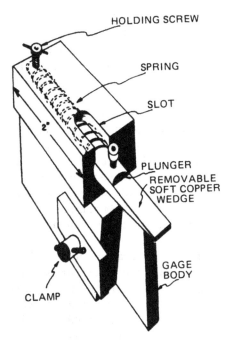

Figure 5.13 Flange deflection gauge used in experimental analysis of axially split casing design.

edge of the casing flanges. The measurement is obtained by using feeler gauges.

After the gauges have been installed, the upper half of the casing is put in place, and the casing bolts are tightened. As the shaft and impellers are not in place for this test, it is possible to reach into the stuffing-box openings and release the gauge holding screws. The soft copper wedges are thereby pressed against the closed split by the spring pressure, and they are ready to enter farther as soon as the two casing halves stretch apart under the action of the internal pressuring during the hydrostatic pressure test. It is also possible at this time to reach in with a feeler gauge and measure the the thickness of the gasket under initial tightening. Comparison of this thickness with the original gasket dimensions indicates the amount by which initial tightening has compressed the gasket material.

End plates are now secured on the stuffing-box openings and the casing subjected to the desired hydrostatic test pressure. The copper wedges penetrate into the split to the extent that it is opened up under the action of the internal pressure. After the casing has been held under test pressure for the required length of time, the pressure is released. This causes the two casing halves to return to their original position prior to their deflection under test, and an indentation is made by the edges of the casing flanges on the soft copper wedges. When the casing is opened and the gauges are removed, this indentation permits an exact determination of the casing deflection.

The comparison of this measurement with the restoration curve of the gasket material under varying stresses will indicate whether the size, number, and location of the bolts and the gasket material will assure a pump casing that will remain tight under the working pressure conditions.

Note that this test is somewhat more rigorous than the conditions that will prevail in normal operation, because the hydrostatic test pressure exceeds operating pressure by anywhere from 25 to 50% and because this pressure is applied throughout the interior of the casing, but operating pressures increase stage by stage.

I must repeat: This is standard and normal practice, and the bolts holding together the flanges of the pump casing are always stressed and stretched beyond their initial tightening stress and stretch once internal pressure is applied. This is not improper design—we must resign ourselves to the fact that things built by people must obey the immutable laws that have been imposed by an agency other than people themselves.

Question 5.14 More Comments on Casing Gaskets

I have studied your comments to Questions 5.12 and 5.13 regarding the effect of internal pressure on the bolting of axially split casing pumps. In the last of these two you have disagreed with the statement that "There is absolutely no elongation in the bolts nor an increase in stress until the initial tension caused by tightening is exceeded, and this would never occur on a properly designed pump."

I believe that this statement is true if you assume that there is no gasket between the flanges. Then if you tighten the two flanges with the bolts initially stressed so as to develop a total force equal to or larger than the pump pressure times the projected area, the two flanges will remain in contact when the pump is operating, and consequently no further stress will take place in the bolts.

The picture gets a little more confused for me when the presence of the gasket is taken into consideration. It seems to me that the elastic characteristics of the gasket will play a great role in explaining this question. As a matter of fact, the gasket has not lineal load-elongation curve, and if you have initially stressed the gasket so as to reach the "saturation" point of this curve and in an amount equal to or greater than the elongation

Chapter 5

of the bolts under pump pressure, then when the pump is operating, the stress in the bolts will be the sum of the initial force of tightening and of the force created by the pump pressure. This, however, is a very odd assumption, since it supposes that the saturation force of the gasket coincides with the initial force of tightening. Therefore I believe that the problem can be posed in a very general way with the elastic characteristics of the gasket known.

Answer. The discussion presented in Questions 5.12 and 5.13 to which you refer dealt with axially split casing pumps, which are normally built with gaskets between the two casing flanges and therefore specifically covered the case of gasketed joints. And although you state that the gasket does not have a lineal elongation curve, the materials selected for pump gaskets generally do have a curve that nearly approaches this type of relation, as shown in Fig. 5.14. Unfortunately, there is no perfect gasket material, and you will notice that a slight amount of *hysteresis* does exist. In other words, the gasket material does not possess perfect *restoration*. But the material is selected so as never to be stressed to what you call the "saturation" point, because this would be tantamount to vitiating the very purpose of the gasket.

The problem can best be visualized if we replace the gasket by coiled springs, as in Fig. 5.15. This, of course, represents an ideal gasket, but the assumption does not introduce any excessive error in our analysis. In this arrangement, let

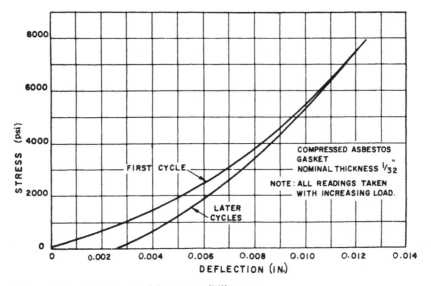

Figure 5.14 Gasket material compressibility tests.

More Comments on Casing Gaskets

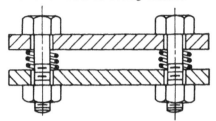

Figure 5.15 Gasket replaced by springs

F_i = initial load caused by tightening of the bolt
F_p = external load caused by internal pressure
F = final load on bolt
C = final compressive load on spring that represents our gasket
e = elongation of bolt per unit load
c = compression of spring (or packing) per unit load

When the bolt is first tightened, the initial load caused by tightening must be equal to the compressive load on the springs. Since the initial elongation of the bolt must be equal to the initial compression of the spring, we can say that

$$F_i e = F_i c \tag{5.5}$$

As the external load F_p is applied, the bolt will be further elongated by an amount $(F - F_i)e$. Since this additional elongation of the bolt must result in a final length of the bolt between the flanges equal to the final length of the spring, we can calculate this final length as

$$Cc = F_i c - (F - F_i)e \tag{5.6}$$

Solving for the final compressive load on the springs, we get

$$C = F_i \frac{e}{c}(F - F_i) \tag{5.7}$$

The final load on the bolt is made up of the sum of the compressive load on the spring and of the applied external load:

$$F = C + F_p = F_i - \frac{e}{c}(F - F_i) + F_p \tag{5.8}$$

We can simplify Eq. (5.8) to yield

$$F = F_i + \frac{c}{c + e} F_p \tag{5.9}$$

Thus, the final load that will be carried by the bolt will be equal to the initial load caused by the tightening of the bolt plus the external applied load multiplied by a factor involving the relative stiffness of the bolt and of the spring.

We have said that the use of the spring gives us a reasonable analogy to the gasket. We can now examine the effect of gaskets of varying stiffness on the value of the final load on the bolts given by Eq. (5.9). If the packing is very soft and its deformation under load c is very large compared with the elongation of the bolt under load e, the factor $c/(c + e)$ approaches unity, and the final load on the bolt is the sum of the initial load caused by tightening and of the applied load. On the other hand, if the deformation of the packing is negligible compared with the elongation of the bolt, this factor approaches zero. Finally, when we have a metal-to-metal joint, the value of c is zero, and the factor $c/(c + e)$ is zero.

Because the multiplier for the applied load is zero, the final load remains equal to the initial load of tightening. Between these two extremes, the multiplier can have various values ranging from 1.0 to zero. Empirical data are given in the book *Design of Machine Members* by Alex Vallance and Venton Lee Doughtie (2nd ed., McGraw-Hill, New York, 1943, p. 134). Table 5.1 reproduces these values and can be used to approximate the possible stretch of a gasketed joint under external load.

The case of a centrifugal pump with axially split casing approaches that for which the factor is given as 0.60. The final load on the bolts can therefore be calculated using this multiplier for the load imposed by the internal pressure. Normally, however, as stated in Question 5.13, the configuration of the pump flanges is quite complex. In addition, if we are dealing with a

Table 5.1 Values of $c/(c + e)$ for Eq. (5.9)

Type of joint	$\dfrac{c}{c + e}$
Soft packing with studs	1.0
Soft packing with through bolts	0.75
Asbestos	0.60
Soft copper gasket with long through bolts	0.50
Hard copper gasket with long through bolts	0.25
Metal-to-metal joints with long through bolts	0

multistage pump, the casing is subjected to a number of different internal pressures over different areas. It is the practice in such cases to establish the elongation of the bolts and the restoration of the gasket thickness under load experimentally.

Question 5.15 Sprayed-on Casing Gaskets

Has there ever been made an attempt to use a sprayed-on or flowed-in gasket for the horizontal joint of axially split casing single- or multistage pumps? Such a solution would obviate the necessity of carrying a stock of special gasket material and of having to cut out laboriously when a new gasket must be used.

Answer. Obviously I cannot know of all the experimental designs or maintenance procedures that may have been tried with centrifugal pumps, but I can assure you that if this procedure has ever been tried, it did not achieve a very marked success. Otherwise it would have become rather popular, and we would have heard about it.

Some years ago, in answer to a similar query our engineering department contacted five leading manufacturers of material that could be considered for this purpose, but the results were negative. Most answers indicated that no such material was available.

One reply indicated that material could be provided but that it would have a very serious shortcoming: The material would vulcanize and harden, forming a bond to the metal of such a nature that disassembly of the two mating flanges would be virtually impossible.

Question 5.16 Straightening Pump Shafts

In dismantling one of our high-pressure boiler feed pumps last year, we found that the shaft ran 14 mils out of true. We straightened the shaft by peening it with a blunt-nose chisel. Just about a year later, the pump developed considerable vibration, and it became difficult to pack the stuffing boxes satisfactorily. The pump was again dismantled, and we found the shaft cracked. Could the method used to straighten the shaft during the preceding overhaul have caused the shaft failure?

Answer. My first reaction is to ask why the shaft was found to be bent in the first place. If the causes that led to this bending were not discovered and remedied, the same causes could have accelerated the shaft failure. But certainly, additional sources of stress concentration were introduced by the method used in straightening the shaft.

In general, the process of straightening a bent shaft is a risky one. I do not say that this should never be done, but if additional life is obtained from a straightened shaft, this is a plus that must be weighed against the possibility that the pump may have to be overhauled again after a much shorter time interval than if a new shaft had been used. Of course, one also runs the even greater risk of a major failure should the shaft break during operation instead of failing progressively as in the case you described.

Any method of shaft straightening that leaves local stresses approaching the yield point of the material will almost certainly result in further warping as soon as the shaft goes through any sort of temperature cycle. This is why straightening by peening as you describe cannot be satisfactory, since the procedure merely puts small areas under compression to deform the shaft.

If a shaft is straightened, stress relieving it after any permanent deformation has taken place is essential. A suggested procedure would be the following:

1. Bend the shaft to get a permanent set of the same amount but in the opposite direction.
2. Bend it back until it is straight.
3. Stress-relieve it at 1050 to 1100 °F, holding this temperature for 4 hr.
4. If the shaft is not sufficiently straight, repeat the procedure.

When the shaft is being bent, it should be supported somewhat inside the bearing journals at each end. To avoid imposing local compressive stresses equal to or greater than the yield strength, the shaft should be supported in well-fitting half-round saddles, and the load should be applied through a similar saddle.

The stress relieving must be carried out in a muffle-type, controlled atmosphere furnace to minimize surface oxidation. The shaft should be placed in a vertical position unless it is very uniformly supported throughout its length.

As a possible means of removing a deformation of a few thousandths of an inch, it is suggested that stress relieving alone be tried. The shaft should be placed in a horizontal position, supported near the ends only and with the runout straight up. The results of such a procedure are very uncertain, but it might work in some cases.

But, I repeat, replacing the shaft or straightening it is not enough. The cause of the runout should be thoroughly investigated lest the damage recur.

Question 5.17 Eroded Casing Volute Tongue

Our plant has a battery of single-stage centrifugal pumps on general water service. We recently opened one of these pumps after it had been in operation 6 years. The pump is designed to handle 4000 GPM against a head of 250 ft at 1750 RPM. The pump casing is cast iron, and the impeller and wearing rings are bronze.

There was relatively little wear at the wearing rings, but they were replaced so as to restore the original clearances. However, we noted a certain amount of wear at the cut water of the volute casing. This wear is rather uneven, extending almost 1½ in. back, and the edge is eaten out rather sharply, almost to a knife edge. We considered filing or smoothing this cut water, but one of the operators thought that cutting it back might reduce the pump capacity. Is this true, or should we take care of the cut water at our first opportunity?

Answer. It is not infrequent that the cut water, or volute tongue as it is also called, is eroded. This happens when a pump handles water with some sand in suspension, for instance. In this particular case, this does not seem to be the cause, as otherwise the wearing rings would not be just slightly worn after 6 years of operation. Sometimes the tongue is eroded when the periphery of the impeller is located too close to it.

Another fairly common cause is galvanic action between the cast iron of the casing and the bronze fittings. The cast iron graphitizes and wears away most perceptibly in areas of high velocity, such as near the volute tongue.

The proper procedure in overhauling a centrifugal pump is to cut back the tongue so that it is straight across and file it to a smooth, rounded edge (Fig. 5.16). This will have no unfavorable effect on the pump capacity, but rather the contrary. Cutting back the volute tongue of a centrifugal pump is often resorted to if a few extra percentages of capacity have to be squeezed out of a pump without putting in a larger diameter impeller. The effect is due to the resulting increase in the casing throat area, which, with a given casing velocity, results in an increased flow.

Question 5.18 How to Pack a Pump

We have noted that as a stuffing-box packing wears, it has a tendency to be further compressed into the stuffing box and that ultimately the gland can no longer hold it under sufficient pressure. Is it good practice to add one or two rings of packing when this takes place? It seems to me that this would save considerable expenditure on new packing.

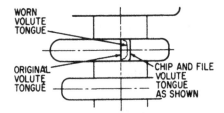

Figure 5.16 Arrow points to volute tongue. If worn, it can be cut back as shown in sketch.

Answer. Actually this practice results in a false economy, as it leads to more rapid deterioration of shaft sleeves and to more frequent shutdowns for repacking. True savings are not found in the reduction of cost of the packing itself but rather in lengthening of the life of such pump parts as shaft sleeves and in the reduction of pump shutdowns and of the labor required to attend to the stuffing boxes.

Most frequently, when the gland has to be pulled tighter and tighter all the time to hold the leakage down to a reasonable value, it is not necessarily because the packing is compressed into the box but rather because improper packing practices cause the packing to disintegrate and be washed out with the leakage as time goes on. Adding new rings of packing does not solve the problem in such cases but may even lead to its aggravation. It is best to replace the packing entirely and to follow certain sound steps in the procedure. The following represents good practice:

1. Remove old packing and clean the stuffing box thoroughly.
2. Make sure that the packing to be used is suitable for the liquid handled and for the prevailing pressure and temperature conditions.
3. Unless the packing comes preformed in sets, make sure that each ring is cut square on a mandrel of proper size to match the stuffing box.
4. Insert each packing ring separately, pushing it squarely into the stuffing box and seating it firmly. A split ring can be used to push the rings home and is very helpful.
5. Stagger successive rings so that the joints are 120° or 180° apart.
6. If the stuffing box is provided with a seal cage, make sure that it will remain located directly under the sealing liquid supply and that the insertion of the remaining rings will not displace it. Make sure that the sealing liquid supply lines are not clogged.
7. After the required number of rings has been inserted, install the gland and tighten the gland nuts firmly. Make sure that the gland

enters the stuffing box squarely and without cocking so that the packing is subjected to a uniform pressure on its entire periphery.
8 . After tightening the gland, back off the nuts until they are just fingertight. When running in new packing, it is advisable to allow quite a bit of leakage at first.
9 . Take up on the gland very slowly during the first few hours of running, say at 15 min intervals.
10. It is necessary that a certain amount of leakage always be present to carry away the heat generated by packing friction. Do not attempt to reduce the leakage excessively. If you do, the packing will burn and score the shaft sleeve. A stuffing box should be so packed that the leakage is sufficient to form a steady running stream rather than a dribble of drops. Experience will indicate the ideal amount of leakage.
11. If the pump has water-quenching glands, stop the supply of quenching water at intervals and observe whether the leakage is sufficient. Otherwise you will be unable to distinguish between the leakage and the quenching water.
12. Never let unqualified personnel adjust stuffing-box glands on the pretext that the packing leaks excessively and should be tightened. Repacking and adjusting stuffing boxes must be a task reserved for specially trained personnel.

Question 5.19 Hands-on or Hands-off?

The last observation in my answer to Question 5.18 leads me to depart from the regular question-and-answer format of this book. I am forced to do this because I cannot imagine that anyone will ask me the specific question that would permit me to vent one of my pet peeves.

Generally, my articles have dealt solely with inanimate objects, such as pumps, but this time I want to talk to you about people. The readers of this magazine are mainly interested in chemical process plants, but I want to assure you that I have seen personnel carry on in the manner that I shall describe in every kind of plant, from steam electric power stations to waterworks and sewage plants, and in refineries. Although I shall tell you what they do to centrifugal pumps, I am sure that they do it to other equipment as well. What these people do is unnecessary, wasteful, frequently dangerous and sometimes destructive. You should watch out for these people, explain to them the error of their ways and possibly reform them.

The walking thermocouple has a constant urge to place the open palm of his right hand on any bearing housing he passes by. He thinks that he has a very sensitive and very accurate built-in temperature detector in that palm and for legitimate (but mostly illegitimate) reasons, he wants to verify his suspicion that most bearings run too hot. If, indeed, he finds his suspicions justified to his satisfaction, he sets about to try and do something about it. He searches around to find cooling water piping to and from the housing of the bearing he wants to "save" and, having located the valve controlling the cooling water flow, he opens it wide open. Mostly, you can tell that he's been around by feeling the inlet and outlet piping: if you can't tell which is which, he's probably been there and done his work. The walking thermocouple is dangerous because the bearings he refrigerates will inevitably fail prematurely. The inner races will continue to expand while the outer races contract. The balls are squeezed excessively and the bearings fail.

Close relatives, the walking vibration detectors, also us their hands to estimate the amplitude of vibration at a bearing housing or, possibly, at some point on the pump casing. Most are careful to disclaim any ability to measure vibration frequency, but a few will hazard a guess even as to that. Some prefer to hold a coin between thumb and index finger and use it to supposedly sense vibration amplitudes more accurately. Whether there is a preference for nickels, dimes or quarters, I have been unable to establish.

Parenthetically, walking vibration detectors in Italy were badly thwarted in the late 1960s and the 1970s: one could get practically no coins in change there! They had disappeared from circulation. For instance, once in 1975 I bought an *International Herald Tribune* which costs 300 liras. For the change from my 500 lira note I received two 70 lira and three 20 lira stamps! Sometimes you got candy or bus tickets in change, certainly useless to measure vibrations. The walking vibration detectors are not really dangerous, as they do nothing more than shake their heads sadly when they decide that the pumps are running too roughly.

The "gauge-knocking" cousins are likewise harmless. They can't pass by a pressure gauge without stepping up to it and tapping it, generally with the fingernail of the right index finger (except for left-handed people, who use the left index finger). I am not quite certain of what it is they want to know, whether they think that the gauge needle is stuck and the tapping will dislodge it, or whether they just want to prevent it from sticking later on. They will frequently stare at the gauge after tapping it for a few seconds, then nod very knowingly and march on, only to stop at the next convenient gauge and repeat the operation. As I said, they are harmless until such time as they start using a small wrench to do the tapping. When this happens, the instrument maintenance department starts wondering why they have to replace gauge glasses so frequently. If digital readouts become very popular, the gauge-knockers will find themselves somewhat underemployed.

Believe it or not, there are still a great many centrifugal pumps equipped with stuffing box packing rather than with mechanical seals. Parenthetically, that is as it should be. The mechanical seal was a wonderful invention, but unless the liquid you are pumping is toxic, flammable, or so precious that you can't afford to lose any of it at the stuffing boxes, or unless the combination of pressures and rubbing velocities are excessively high, the mechanical seal is not necessarily the better choice. You can replace soft packing at a fraction of what a mechanical seal replacement will cost you. Of course, labor is involved for repacking. Packed stuffing boxes do, however, attract the attention of the stuffing box gland adjusters who harbor the mistaken idea that the leakage from a packed box should essentially be the same as from a mechanical seal. In their opinion, all stuffing boxes leak too much, and they are dedicated to correct this.

Of course the stuffing-box gland adjusters are wrong: one should never attempt to reduce the leakage to even a "dripping" state. It should be a steady flow, sufficient to carry away the packing friction heat. Unless enough liquid leaks through the stuffing box to remove this heat, the packing will be burnt and the shaft sleeve scored. The result is more sales for the packing companies and more shaft sleeves bought as repair parts, but that's not the way to run a railroad.

The stuffing-box gland adjuster thinks otherwise. He is a specialist. When he arrives at a pump, he generally ignores the bearing housings: he cares nought for vibrations or temperatures. He examines the leakage from the stuffing box critically, then reaches for the wrench he conveniently carries in his back pocket and starts tightening the gland bolt nuts. He is gone by the time packing smoke pours out of the stuffing boxes he has adjusted.

The grease gun is a most ingenious invention and a very helpful tool in the hands of a skilled maintenance man. When used by untrained personnel, it can also be a very effective weapon to shorten bearing life. It is probably for this last reason that the use of grease guns is diminishing and the practice of relying on grease cups is being widely revived. The "grease-gun-kid" still haunts many plants. He roams the area where pumps, motors, and other machines equipped with grease-lubricated bearings may be found and looks for his prey. He approaches it, inserts the grease gun he carries along wherever he goes, and gives the bearing as many shots of grease as he can. Of course, he doesn't carry an assortment of greases along. Grease is grease to him and he doesn't bother to find out whether different grades of greases are prescribed for the various bearings he encounters.

What he doesn't know is that a bearing fully packed with grease prevents proper grease circulation in itself and in the housing. It is generally recommended that only one-third of the void spaces in the housing be filled. An excess of grease will cause the bearing to heat up, and grease will flow out of the seals to relieve the situation. Unless the grease can escape through the seals or through the relief cock that is used on many large units, the bearing will probably fail early. More trouble is usually caused in grease-lubricated bearings by an excess of lubrication than by a lack of it—an exceptional situation in which too much attention is worse than too little.

There's one more type of person I want to describe, but he is not located in the plant. As a matter of fact, he is very far away from the plant in an office in the purchasing department. He has a very costly hobby: he collects instruction books. First, he makes sure that the specifications issued by his company call for an unconscionable number of these books to be delivered for each type of pump to be purchased. Sometimes he asks for a round dozen of them. He also specifies that *all* of them be delivered directly to the purchasing department, marked for his attention. When they arrive, he very carefully stashes them under lock and key, so that not a single copy can ever reach the plant itself.

I have not invented these people. They exist and all resemblance to living persons whom you personally know is strictly and carefully intentional. Because my sympathies have always stayed on the side of the users, of the operators and maintenance men, I have written this in hope that it will help identify and weed out their mortal enemies . . . the walking thermocouple and his cousins.

Question 5.20 Diagnosis from Worn Stuffing-Box Packing

Can the appearance of worn stuffing-box packing give a clue to the cause of premature wear?

Answer. Yes, to the eye of an experienced mechanic or operator, the appearance of the packing removed from a stuffing box will give a very clear indication of what is wrong and what corrective steps may eliminate the problem. Some of the more frequently encountered symptoms ae the following:

1. When one or two rings next to the packing gland are badly worn but the remaining rings are hardly touched, the probability is that the rings nearest to the pump interior are ineffective and that all the pressure drop takes place across the worn rings. This indicates that the packing has not been properly installed (see Question 5.18).
2. If the wear is on the OD of the rings, either the rings have been rotating with the shaft or shaft sleeve or leakage is taking place between the rings and the ID of the stuffing box. In either case, chances are that the packing size or the cutting method for the individual rings is incorrect.

3. Charring or glazing of the inner circumference of the rings is caused by excessive heating or insufficient liquid flowing past the packing to lubricate it. It will also occur when the packing is not suitable for the combination of stuffing-box pressures, temperatures, and rubbing speeds.
4. If the ID of the packing rings has been excessively increased or if the rings are heavily worn on one selective portion of the inner circumference, the probability is that the shaft rotation is eccentric because of excessively worn bearings, misalignment, or shaft *whip*.
5. If some rings are cut too short or shrink excessively, the adjacent rings will bulge and be extruded into the open space.

Question 5.21 Can Shaft Sleeves Be Reground?

We have a water-supply installation that consists of three 8 in. single-stage double-suction centrifugal pumps handling 3000 GPM each against 100 psig discharge pressure and operating under a 12 ft lift. The water handled contains a considerable quantity of fine silt, and we find that the life of the shaft sleeves is very short. We originally used bronze sleeves but later changed to a 13% chrome stainless material, and although this has lengthened the life of the sleeves, we still find it necessary to replace the sleeves about every 3 months. Is there a sleeve material available that would improve this life still further? We have a very adequate machine shop and would like to know whether it is possible to regrind shaft sleeves to reuse them. How much metal can be removed from these sleeves without jeopardizing their operation?

Answer. Although the use of material even harder than 13% chrome stainless steel, such as stellited steel, would probably require less frequent replacement, the most logical solution would be the prevention of erosion of the sleeves by the silt in the water handled. Since the pumps are operating under a suction lift, the pumps are probably provided with seal cages in the stuffing boxes to prevent in-leakage of air. Normally, water under pressure is piped to the chamber formed by the seal cage and makes an effective seal beyond which air cannot pass into the pump. This seal chamber also provides a reservoir of liquid to cool and lubricate the packing rings. In such cases, when the flow of the liquid pumped into the stuffing box and out to the atmosphere should be avoided (as when handling inflammable liquids or, as here, when the water handled erodes the shaft sleeves), the seal cages can be used to introduce a sealing liquid to prevent this outflow.

In the case at hand, clear water under pressure should be piped to the seal cages. In the event that no clear-water source is readily available, it may be possible to install a small strainer or filter in a line from the discharge of the pumps. Generally such strainers have a reasonable friction drop, and the 100 psig discharge pressure is more than sufficient to provide the required sealing water. In order that the clear water can flow freely into the pump interior and keep the packing free from the abrasives contained in the water pumped, it will be preferable to locate the seal cage as close to the pump interior as possible. For instance, if the stuffing boxes have six rings of packing in addition to the seal cage, it might be best to install two rings of packing at the bottom of the stuffing box and follow this with the seal cage and then with four more rings. Make sure that the water seal cage is located directly underneath the supply connection. It may be necessary to drill a new connection and plug the old one.

When regrinding shaft sleeves for reusage, it is extremely important to ensure concentricity of grinding, as an eccentricity in a shaft sleeve will accelerate wear. If the wear and grooving of the sleeve are of a minor nature, say under 1/32 in., the sleeve can be refinished by grinding it slightly undersize. When the sleeve is so repaired, the clearance at the bottom of the stuffing box between the sleeve and the casing or throat bushing should be checked to see that the packing will not be pushed through the enlarged space into the pump when the glands are tightened.

Question 5.22 Use of Molybdenum Disulfide on Shrink Fits

When centrifugal pumps require repairs, they are generally transported into our maintenance machine shop. The rotors are stripped there and reconditioned with new parts. We have recently run into an epidemic of damaged shafts in the process or removing ball bearings. The bearings have a press fit on the shaft, but our practice is to avoid the danger of damaging the shaft or the race, and we heat the new bearings in an oil bath and then just slip them on the shaft into position. When we want to remove them later, we use an arbor press, applying the force through a split washer as evenly as we possibly can. But in many cases we have found that the inner race drags on the shaft, damaging the latter. In several cases we have had to grind the shaft down, rough it up, metallize it, and regrind it to correct dimensions. Is there any method of removing the bearings that will avoid this difficulty?

Answer. I must assume that the bearings you use have the proper press fit dimensions and that the shaft mounting surface is free and clear of dirt before you mount the bearings. One method that I know reduces the danger of damaging the shaft surface is to use molybdenum disulfide as a lubricant before new bearings are mounted. This lubricant withstands high temperatures and remains an effective means to ease the removal of the bearings when this becomes necessary.

Molybdenum disulfide is used as a lubricant in powder form, ground to a fineness of 1½ μm. It is commercially available in a wide range of forms and in many different carriers. It is lead gray in color, rather similar to graphite. It is insoluble in water and in most organic solvents and will withstand extremely high temperatures (up to 750 °F) even in the presence of oxidizing agents.

You will find molybdenum disulfide lubricants extremely useful in a number of applications. If you shrink parts, such as impellers, onto the shaft, coating the surface of the shafts with molybdenum disulfide makes the removal of the impellers less difficult and eliminates the tendency of the impellers to stick while being mounted. It can be used before assembly to prevent galling of such parts as couplings and threaded bolts.

I know of one particular plant where it is the practice to apply a generous film of molybdenum disulfide to the rubbing surfaces of new packing before insertion into stuffing boxes. The practice, I have been told, is quite effective in preventing difficulties during the running-in period and therefore in lengthening the life of the packing and of the sleeves.

Question 5.23 Locking Wearing Rings in Place

We have encountered some difficulties in holding the casing wearing rings of our boiler feed pumps in place. These are double-casing barrel-type pumps of two different designs. In one case, the wearing rings are pressed into the stage pieces separating consecutive stages and held to these stage pieces by bolts (see Fig. 5.17), which are peened in place. Another design of pump holds the wearing ring in place by a form of tack welding to similar stage pieces. We have had some of the bolts back out after several years of operating in the case of the first design, and the tack welding has failed in the other design. Couldn't these wearing rings be screwed into the stage pieces, eliminating the need of bolts or of tack welding? As an alternative, couldn't they be mounted by means of some breech block construction?

Answer. It would be possible to screw the wearing rings into the stage pieces, but it would be quite difficult to hold reasonable concentricity between the bore of the casing wearing ring and the mating hub of the impeller. Furthermore, this mounting would still require some form of locking device, such as a setscrew or a bolt.

Chapter 5

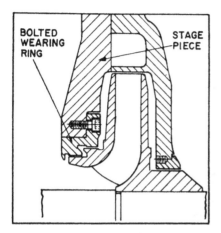

Figure 5.17 A pump design in which wearing rings are pressed into the stage piece and held in place by bolts.

A breechblock fastening is merely an interrupted thread, usually made as a heavy multiple thread to get a steeper helix angle and somewhat adjusted so that the full length of each thread segment is in contact when the closing is seated. It is useful primarily for something that must be opened and closed quickly, such as the breechblock of a firearm. But it can be only half as strong as a full thread, since half the metal is cut away. I can see no advantage and many disadvantages of this as a wearing ring fastening.

If bolts are used to fasten the wearing ring to the stage piec;e, it is important that they be adequately peened in place at the time the pump is first assembled in the manufacturer's shops and that equal care be used in peening the bolts in place should the pump be dismantled and reassembled in the field.

A very interesting alternative to peening bolts in place is illustrated on Fig. 5.18. It was developed in 1957 and has the advantage of being the least vulnerable to the human element. A small flat strip of Inconel is used, fitting into a groove counterbored in the bolted piece, immediately above the head of the bolt to be locked. The strip has a length equal to the counterbore groove diameter but is tamped into a slightly concave shape. It is punched into place during assembly and lies flat, bearing up against the bolt.

This locking device has been used very successfully for large bolts. But because the bolts holding the casing rings in place are relatively small, it is not practical to lock them in place with these strips.

Of course, someday the metallurgists will develop for us that "miracle" metal that will not wear, gall, or seize under the particular conditions

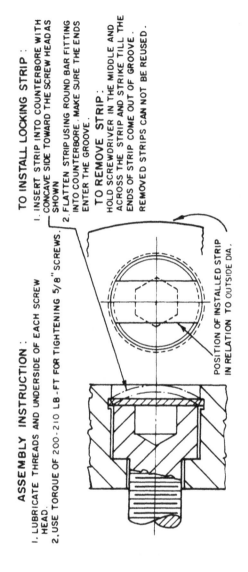

Figure 5.18 Installation of locking strips.

imposed by boiler feed service. We shall then be able to dispense with replaceable wearing rings and eliminate this fastening problem altogether. No doubt some new problem will arise to tax our ingenuities, since there is no shortcut to immortality for human-made machinery. Until the wearing ring does disappear, however, we shall just have to exert our efforts at taking proper care that it is adequately locked in place.

Question 5.24 Draining Pumps Handling Corrosive Liquids

We operate a large number of chemical pumps, some of which handle very corrosive solutions. When it is necesary to repack a stuffing box or to replace a mechanical seal, it is desirable to drain the pump before any work is done on it for safety reasons. Of course the pump can be isolated from the system by closing suction and discharge valves. But loosening the drain plug at the bottom of the casing is inconvenient and can be hazardous. What procedure would you recommend to make this easier and safer?

Answer.   Nothing could be simpler. A small 90° elbow should be used instead of the plug, followed by a nipple and a valve, as shown in Fig. 5.19. Since most chemical pumps are made of special corrosion-resistant materials, make sure that the valve and fittings are made of the same materials as the pump, or at least of a compatible material.

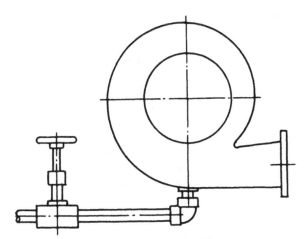

Figure 5.19 Draining corrosive liquids.

6
Field Troubles

Question 6.1 Documentation of Field Troubles

Several years ago, our engineering department instituted a program intended to improve reliability of rotating equipment installed in our plant, to reduce recurrence of breakdowns, and to extend repair intervals. Our idea was that if we were to ask maintenance personnel to document each failure and each repair operation by providing a complete history of each occurrence, we could develop certain patterns that would direct us to the real causes of trouble and help us eliminate these causes.

Our efforts seem to have produced no tangible results, and information we gathered failed to develop either recognizable patterns of conclusive evidence as to real causes of trouble encountered with our equipment. As the major portion of this equipment happens to be centrifugal pumps, I wonder whether you might comment on reasons this procedure was not successful?

Answer. The subject of equipment reliability, of diagnostic evaluation of failures, and of preventive maintenance is one that has been very close to my heart for many of the years of my career in pump engineering. I must admit that I have frequently been as baffled as you are by the inability to obtain sufficient factual evidence from the field to permit a valid diagnosis of equipment failure.

But it was not only to commiserate with you that I have chosen to answer your question. It seems to me that we must learn to distinguish between feedback of evidence and preparation of diagnostic and remedial judgment.

Field Troubles

To be useful, a failure report should include a statement describing the failure in complete detail and another statement related to diagnosis of the failure. And unless complete impartiality prevails in preparing the failure report, this information will be useless. The report must be prepared by an impartial and competent observer. This almost leads me to the conclusion that two completely separate and independent reports are required. Except for obvious cases, analytical talents required to complete the diagnostic statement require educational and experience backgrounds different from the operating and maintenance functions that are responsible for preparing the descriptive portion of a failure report.

Thus, I believe that whoever is reporting pertinent facts involved in an equipment failure should disassociate themselves from the role of diagnostician. Only after all the facts have been arrayed and analyzed does it become possible to start on failure analysis. Incidentally, the process is very similar to brainstorming, during which ideas must be listed without exercising judgment as to their validity. Later, judgmental analysis and evaluation can take place.

Assuming that the description of a failure is complete, clear, and impartial, diagnosis requires ability to distinguish between certain major categories of causes. The following classification can serve as a condensed checklist.

Causes Inherent to Equipment

1. *Faulty application.* This category covers such items as selection of a pump with inadequate suction performance and selection of a driver that will be overloaded.
2. *Faulty design.* Short of catastrophic failures of an untried and unproved new design, this category covers designs that may be marginal with respect to severity of operating conditions that may be imposed upon them. Typical examples of causes in this category could be the following:
   a. Excessive vibrations caused by improper relationship and configuration of impeller and stationary vanes
   b. Use of overly close running clearances
   c. Improper or inadequate axial thrust balance
3. *Material failures.* This category would cover both use of materials inadequate for the intended service and materials whose physical or chemical characteristics deviate from manufacturer's specifications.
4. *Manufacturing deficiencies.* The list of possible deviations from acceptable could be fairly broad, ranging from dimensional nonconformance to tolerances, through improper surface finishes,

unsatisfactory welding procedures, all the way to improper assembly.

Causes External to Equipment

1. *Faulty installation, including lack of proper protection controls.* Examples:
   a. Improper alignment
   b. Excessive forces exerted on pump by piping
2. *System-induced failures.* Example: Severe and violent cavitation induced by inadequate system conditions or operation of pump beyond normally foreseen capacities, resulting in insufficient NPSH available.
3. *Faulty operation of equipment.* This category can cover so many sins of omission or commission that it is impossible to present a complete list. Suffice it to say that this category would include such obvious errors of operation as starting up a pump without priming or without adequate warm-up provisions, shutting down a pump while it is in partial cavitation, and running a pump with a closed suction valve.
4. *Failure of auxiliary component.* Too often statistical reports of pump outages fail to differentiate between a failure of the pump itself and failure of a component, such as a hydraulic coupling or even a pump driver. In addition, routine pump maintenance may take place during an outage initiated to fix a hydraulic coupling, a leaking check valve, or other associated piece of equipment. In some cases, a report may charge the pump itself with unavailability. In other cases, it may not. Obviously, a logical distinction is necessary here.
5. *Faulty maintenance.* Here again, the list would be long and would include such diverse causes as inattention to proper lubrication and operation of pump with excessive wear.
6. *Inadequate stocking of replacement parts.* Frequently replacement parts must be obtained under "crisis" conditions. When delays are foreseen in receiving such parts from the manufacturer, emergency measures are often undertaken by obtaining parts from some other source or by manufacturing them in the plant's own shops. Although such measures are sometimes necessary and inevitable, they can readily lead to unexpected failures.

This certainly is not and cannot be an exhaustive list of failure causes, but it may prove helpful in the diagnostic process. Above all, I believe that complete familiarity with the equipment and the system in which it operates is essential for a successful diagnosis. Such a diagnosis, in turn, can lead to measures that will increase both reliability and availability of equipment.

Table 6.1 provides a list of 25 most common symptoms of troubles that beset centrifugal pumps. Each symptom is keyed in with the possible

Table 6.1 Check Chart for Centrifugal Pump Problems

Symptoms	Possible cause of trouble (each number is defined in Table 6.2)
1. Pump does not deliver liquid.	1, 2, 3, 5, 10, 12, 13, 14, 16, 21, 22, 25, 30, 32, 38, 40
2. Insufficient capacity delivered.	2, 3, 4, 5, 6, 7, 7a, 10, 11, 12, 13, 14, 15, 16, 17, 18, 21, 22, 23, 24, 25, 31, 32, 40, 41, 44, 63, 64
3. Insufficient pressure developed.	4, 6, 7, 7a, 10, 11, 12, 13, 14, 15, 16, 18, 21, 22, 23, 24, 25, 34, 39, 40, 41, 44, 63, 64
4. Pump loses prime after starting.	2, 4, 6, 7, 7a, 8, 9, 10, 11
5. Pump requires excessive power.	20, 22, 23, 24, 26, 32, 33, 34, 35, 39, 40, 41, 44, 45, 61, 69, 70, 71
6. Pump vibrates or is noisy at all flows.	2, 16, 37, 43, 44, 45, 46, 47, 48, 49, 50, 51, 52, 53, 54, 55, 56, 57, 58, 59, 60, 61, 67, 78, 79, 80, 81, 83, 84, 85
7. Pump vibrates or is noisy at low flows.	2, 3, 17, 19, 27, 28, 29, 35, 38, 77
8. Pump vibrates or is noisy at high flows.	2, 3, 10, 11, 12, 13, 14, 15, 16, 17, 18, 33, 34, 41
9. Shaft oscillates axially.	17, 18, 19, 27, 29, 35, 38
10. Impeller vanes are eroded on visible side.	3, 12, 13, 14, 15, 17, 41
11. Impeller vanes are eroded on invisible side.	12, 17, 19, 29

Table 6.1 continues

Table 6.1 *Continued*

12. Impeller vanes are eroded at discharge near center.	37
13. Impeller vanes are eroded at discharge near shrouds or at shroud or vane fillets.	27, 29
14. Impeller shrouds bowed out or fractured.	27, 29
15. Pump overheats and seizes.	1, 3, 12, 28, 29, 38, 42, 43, 45, 50, 51, 52, 53, 54, 55, 57, 58, 59, 60, 61, 62, 77, 78, 82
16. Internal parts are corroded prematurely.	66
17. Internal clearances wear too rapidly.	3, 28, 29, 45, 50, 51, 52, 53, 54, 55, 57, 59, 61, 62, 66, 77.
18. Axially-split casing is cut through wire-drawing.	63, 64, 65
19. Internal stationary joints are cut through wire-drawing.	53, 63, 64, 65
20. Packed box leaks excessively or packing has short life.	8, 9, 45, 54, 55, 57, 68, 69, 70, 71, 72, 73, 74
21. Packed box : sleeve scored.	8, 9
22. Mechanical seal leaks excessively.	45, 54, 55, 57, 58, 62, 75, 76‡
23. Mechanical seal : (a) faces are damaged. (b) sleeve damaged. (c) metal bellows fails.	45, 54, 55, 57, 58, 62, 75, 76‡
24. Bearings have short life.	3, 29, 41, 42, 45, 50, 51, 54, 55, 58, 77, 78, 79, 80, 81, 82, 83, 84, 85
25. Coupling fails.	45, 50, 51, 54, 67

‡ Also see instruction book for specific mechanical seal in question.

cause of the trouble, the numbers of these being defined in Table 6.2. Additional diagnostic assistance can be obtained from Table 6.3 (which provides a diagnosis of problems from the appearance of the stuffing-box packing) and from Table 6.4 (which shows the correlation between vibration frequencies and their causes).

Question 6.2 Suction Lift at 5200 ft Elevation

We purchased a centrifugal pump designed to handle 1000 GPM against a total head of 225 ft made up of a suction lift of 15 ft and a discharge head of 210 ft. The pump has been installed but will deliver only 700 GPM. The manufacturer advises us that we failed to mention that the pump would be installed at an elevation of 5200 ft above sea level. How does this factor affect the pump capacity?

Answer. An outside source of pressure over and above the liquid vapor pressure must be available to cause liquid to flow into a centrifugal pump impeller. When a pump takes liquid from a source below its centerline, the atmospheric pressure provides this energy. The atmospheric pressure at sea level is 33.9 ft absolute, and if we deal with cold water at 62 °F (vapor pressure = 0.6 ft), the net energy available over and above the vapor pressure is 33.9 − 0.6 = 33.3 ft. If the pump is installed so that the suction lift is 15 ft, there remains 33.3 − 15 or 18.3 ft of energy above the vapor pressure to cause flow into the impeller. This is termed the available NPSH, or net positive suction head, and must be equal to or greater than the NPSH required by the pump for the intended capacity. In this case, it would appear that the pump selected had a required NPSH of about 18 ft.

If the pump is installed at some elevation above sea level, it will still require 18 ft NPSH, but the atmospheric pressure will have been reduced and therefore so will the available NPSH, and the pump will be able to handle less suction lift. The atmospheric pressure is reduced by approximately 1 in. of mercury per 1000 ft of elevation. At 5200 ft, the atmospheric pressure will therefore be approximately 6 ft less than at sea level. The available NPSH will therefore become (33.9 − 6) − 0.6 − 15 or 12.3 ft, and the pump will be unable to handle the required 1000 GPM.

If it is desired to handle 1000 GPM with this pump, it will be necessary to reduce the suction lift by the amount of reduction in atmospheric pressure, that is, to 9 ft.

Question 6.3 Speeding up a Pump

We have an installation of several centrifugal pumps rated at 2800 GPM and 70 ft total head and running at 1160 RPM. They operate with an 18 ft

Table 6.2 Possible Causes of Trouble

A. Suction troubles
 1. Pump not primed
 2. Pump suction pipe not completely filled with liquid
 3. Insufficient available NPSH
 4. Excessive amount of air or gas in liquid
 5. Air pocket in suction line
 6. Air leaks into suction line
 7. Air leaks into pump through stuffing boxes or through mechanical seal
 7a. Source of sealing liquid has air in it
 8. Water seal pipe plugged
 9. Seal cage improperly mounted in stuffing box
 10. Inlet of suction pipe insufficiently submerged
 11. Vortex formation at suction
 12. Pump operated with closed or partially closed suction valve
 13. Clogged suction strainer
 14. Obstruction in suction line
 15. Excessive friction losses in suction line
 16. Clogged impeller
 17. Suction elbow in plane parallel to the shaft (for double-suction pumps)
 18. Two elbows in suction piping at 90° to each other, creating swirl and prerotation
 19. Selection of pump with too high a suction specific speed

B. Other hydraulic problems
 20. Speed of pump too high
 21. Speed of pump too low
 22. Wrong direction of rotation
 23. Reverse mounting of double-suction impeller
 24. Uncalibrated instruments
 25. Impeller diameter smaller than specified
 26. Impeller diameter larger than specified
 27. Impeller selection with abnormally high head coefficient
 28. Running the pump against a closed discharge valve without opening a bypass
 29. Operating pump below recommended minimum flow
 30. Static head higher than shut off head
 31. Friction losses in discharge higher than calculated
 32. Total head of system higher than design of pump
 33. Total head of system lower than design of pump
 34. Running pump at too high a flow (for low specific speed pumps)
 35. Running pump at too low a flow (for high specific speed pumps)
 36. Leaky or stuck check valve
 37. Too close a gap between impeller vanes and volute tongue or diffuser vanes

Table 6.2 continues

Table 6.2 *Continued*

B. Other hydraulic problems (continued)
38. Parallel operation of pumps unsuitable for this purpose
39. Specific gravity of liquid differs from design conditions
40. Viscosity of liquid differs from design conditions
41. Excessive wear at internal running clearances
42. Obstruction in balancing device leak-off line
43. Transients at suction source (imbalance between pressure at surface of liquid and vapor pressure at suction flange)

C. Mechanical troubles: a. General
44. Foreign matter in impellers
45. Misalignment
46. Foundations insufficiently rigid
47. Loose foundation bolts
48. Loose pump or motor bolts
49. Inadequate grouting of baseplate
50. Excessive forces and moments from piping on pump nozzles
51. Improperly mounted expansion joints
52. Starting the pump without proper warm-up
53. Mounting surfaces of internal fits (at wearing rings, impellers, shaft sleeves, shaft nuts, bearing housins, (and so on) not 100% perpendicular to shaft axis
54. Bent shaft
55. Rotor out of balance
56. Parts loose on the shaft
57. Shaft running off center because of worn bearings
58. Pump running at or near critical speed
59. Use of too long a shaft span or too small a shaft diameter
60. Resonance between operating speed and natural frequencey of foundation, baseplate, or piping
61. Rotating part rubbing on stationary part
62. Incursion of hard solid particles into running clearances
63. Use of improper casing gasket material
64. Inadequate installation of gasket
65. Inadequate tightening of casing bolts
66. Pump materials not suitable for liquid handled
67. Lack of lubrication of certain couplings

C. Mechanical troubles: b. Sealing area
68. Shaft or shaft sleeves worn or scored at packing
69. Incorrect type of packing for operating conditions
70. Packing improperly installed
71. Gland too tight, resulting in no flow of liquid to lubricate packing

Table 6.2 continues

Table 6.2 *Continued*

C. Mechanical troubles: b. Sealing area (continued)

72. Excessive clearance at bottom of stuffing box, causing packing to be forced into pump interior
73. Dirt or grit in sealing liquid
74. Failure to provide adequate cooling liquid to water-cooled stuffing boxes
75. Incorrect type of mechanical seal for prevailing conditions
76. Mechanical seal improperly installed

C. Mechanical troubles: c. At bearings

77. Excessive radial thrust in single-volute pumps
78. Excessive axial thrust caused by excessive wear at internal clearances or by failure or excessive wear of balancing device (if such is used)
79. Wrong grade of grease or oil
80. Excessive grease or oil in antifriction bearing housings
81. Lack of lubrication
82. Improper installation of antifriction bearings (damage during installation, incorrect assembly of stacked bearings, use of unmatched bearings as a pair, (and so on)
83. Dirt getting into bearings
84. Moisture contamination of lubricant
85. Excessive cooling of water-cooled bearings

Table 6.3 Diagnosis from Appearance of Stuffing-Box Packing

Symptom	Cause
Wear on one or two rings next to packing gland; other rings OK	Improper packing installation
Wear on OD of packing rings	Packing rings rotating with shaft sleeve or leakage between rings and ID of box (wrong packing size or incorrectly cut rings)
Charring or glazing or inner circumference of rings	Excessive heating; insufficient leakage to lubricate packing or unsuitable packing
ID of rings excessively increased or heavily worn on part of inner circumference	Rotation eccentric

Table 6.4 Vibration Symptoms and Causes

Vibration frequency	Cause
Several times pump RPM	Bad antifriction bearings
Twice pump RPM	Loose parts on rotor; axial misalignment of coupling; influence of twin-volute when gap is insufficient
Running speed	Imbalance of rotor; clogged impeller coupling misalignment or loose
Running speed times number of impeller vanes	Vane passing syndrome (insufficient gap between impeller vanes and collector vanes; also sometimes seen during operation with suction recirculation
One-half running speed	Oil whirl in bearings
Random low frequency	Internal recirculation in impeller or cavitation
Random high frequency	Usually resonance
Subsynchronous frequency at 70% to 90% of running speed	Hydraulic excitation of resonance

suction lift. The pumps have been on this service for the last 5 years and have given a very good account of themselves. Recently, our plant had need of a pump for a higher capacity and total head at another location. We decided to move one of these 1160 RPM pumps to this new service and to replace its motor with a larger 1760 RPM motor. We are quite familiar with the laws governing the relation of capacity, head, and pump speed. In other words, we know that the capacity should increase directly with the speed and the head as the square of the speed. Thus, at 1760 RPM, the pump should be capable of delivering 4250 GPM against a total head of 160 ft, approximately. The suction lift at the new location is 18 ft and therefore no greater than before. But after its installation, we found that the pump barely delivers 3600 GPM against a total head of 160 ft. In addition, it produces a peculiar crackling noise. Can you advise us of the possible causes for this strange behavior?

Answer. There is little strange about the behavior of this pump. The calculations for the expected head and capacity at 1760 RPM are quite correct except that they do not account for the apparent inadequacy of the suction conditions.

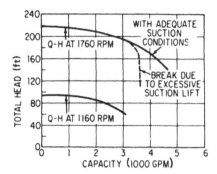

Figure 6.1 Capacity levels off and pump operates in the break when available NPSH falls below required NPSH.

Reproduced in Fig. 6.1 are the probable head-capacity curves of this pump at 1160 and 1760 RPM. The NPSH required for a given centrifugal pump varies with the operating speed according to a mathematical relation which we shall discuss presently. The available NPSH (which must be equal to or greater than the required NPSH) is the difference between the absolute suction pressure and the liquid vapor pressure at the pumping temperature. Assuming that the pump handled cold water and at an elevation corresponding to sea level, the available NPSH would be equal to 33.9 ft (barometric absolute pressure) less the 18 ft suction lift and less, say, 0.9 ft vapor pressure, or 15 ft. We know, therefore, that at 1160 and 2800 GPM the required NPSH is less than 15 ft.

The relation among speed, capacity, and required NPSH is expressed by the formula*

$$S = \frac{n\sqrt{Q}}{H_s^{3/4}}$$

where S = suction specific speed = constant
  n = speed in RPM
  Q = capacity in GPM
  H_s = required NPSH

Since S remains constant, we can say that when we change speed from 1160 to 1760 RPM and capacity from 2800 to 4250 GPM, the required NPSH would change from 15 ft or less to 34.5 ft or less.

---

*See Question 1.27.

In other words, the required NPSH changes as the square of the speed.

Incidentally, 34.5 ft NPSH under our assumed conditions corresponds to something like 1.5 ft positive suction head, which is very different from the 18 ft suction lift under which the pump was required to operate in this setup.

If the available NPSH is reduced below the required NPSH, the pump capacity is reduced, and the pump will operate *in the break*, as shown in Fig. 6.1. In other words, it will operate at the maximum capacity that it can deliver with the avaiable NPSH and at a total head below that which it can develop with a higher NPSH. The pump is operating in cavitation, and the crackling noise you report is a definite symptom of such operation.

The pump is not suitable for this installation unless you can change the suction conditions, It is possible, but not probable, that a special impeller can be furnished to enable the pump to handle 4250 GPM with 18 ft lift. More than likely, a larger pump will be required.

Question 6.4 Why Do These Pumps Lose Suction?

We experience difficulty with two 3 in. centrifugal pumps, each direct driven by a 5 hp squirrel cage three-phase motor. The pumps have a common suction and discharge; they fill a remote storage tank at 80 ft head under control of a float switch. The water spills into the top of the reservoir for aeration.

Each pump is equipped with stop valves, and the units are intended to operate separately, one acting as a standby for the other. The suctions are connected and go straight down through the concrete pump house floor in a 3 in. bore pipe that had been used for a smaller installation. The suction pipe had to be reduced to 2 1/2 in. and fitted with a vertical check valve that just slides down the bore pipe. The suction lift is about 18 to 21 ft.

Each motor has a magnetic starter with a single-phase solenoid operated by the float switch. In addition, each circuit is equipped with a phase-failure, phase-reversal cutout. The wiring is such that a two-way manual switch can put either set into operation.

Trouble has been encountered through the pump losing its suction in periods of dry weather. These pumps are intended to run unattended except for a weekly inspection visit. It is not practical to dig up the floor and put down a large bore or well with float control, which, of course, would be the simplest solution. We have been considering various alterations for protecting these units and would like some suggestions on this problem.

Answer. Although some of the important facts connected with this installation are missing, it is possible to reconstruct them in an approximate manner. Neither the capacity nor the actual total head of these two 3 in. pumps is given; however, it is probable that the 80 ft referred to is the discharge head.

Adding to this the average suction lift gives a total head of approximately 100 ft. Thus, to load a 5 hp motor, the capacity is probably between 90 and 100 GPM.

The suction lift of 18 to 21 ft mentioned is most likely the static elevation between the level of the water in the well and the pump centerline. Thus, to get the actual suction lift, it is necessary to add the frictional losses in the suction line. Assuming a capacity of 100 GPM, this loss is 12 ft per 100 ft of 2 1/2 in. pipe, or about 2.5 ft plus the loss of the check valve, which acts as a foot valve. The latter is probably extremely high. Under normal conditions, it is generally recommended that the size of a foot valve be such that its port area is at least twice the area of the suction piping. Here the physical limitations make such a relationship impossible, and the foot valve losses will probably range between 2 and 4 ft. The total suction lift may therefore vary from 22 to 27 ft.

Finally, the source of the water supply is not exactly defined. However, the reference to dry-weather difficulties implies that the water comes from a natural well of varying level.

The foregoing analysis indicates without any shadow of doubt that all the difficulties experienced in this installation are traceable to an excessive suction lift. In this particular case, this excessive lift is aggravated by the fact that the high suction inlet velocities result in the formation of eddies at the pump suction and in the entrainment of large quantities of air. Natural well water is generally well saturated with dissolved gases that, under the vacuum conditions prevailing because of the high lift, are liberated at the pump entrance and help in the creation of air and gas pockets in the suction passageways. Taking all these facts into consideration, it is obvious that the pumps will lose suction at such times when the level in the well goes to its minimum.

Depending on various factors, which can be established only from a more complete knowledge of the installation itself, of the water supply requirements, and of the relative economy of the alterations involved, there are several possible arrangements that would correct the existing difficulties. These arrangements fall into two separate categories:

1. Decrease of the unfavorable effects of high-suction-lift operation
2. Elimination of a condition whereby the pumps handle the water under a high suction lift

Inasmuch as a 3 in. bore pipe is already installed, there is no reason the friction losses and the suction velocity could not be appreciably reduced by utilizing this pipe instead of the 2 1/2 in. pipe. This might possibly be accomplished by lowering the foot valve to the bottom of the 3 in. bore pipe and sealing it at that point. The suction lift would be reduced and the amount of entrained air decreased.

In addition, it might well be worthwhile to check the installation for air leakage into the suction line from the pump that is standing idle. Since the pumps have a common suction, there appears to be nothing to prevent the leakage of air through the stuffing boxes of the idle pump and then through its suction passages into the common suction pipe. If this leakage of air were excessive, it might in itself account for the loss of suction described. To eliminate this leakage, it would be necessary to provide the stuffing boxes of both pumps (in case they are not already so fitted) with lantern seal cages and to supply these with water under pressure. This water could be diverted from the common discharge header. Since the pumps are operated intermittently by means of a float switch, the source of the sealing water should come from a point upstream of the discharge gate and check valves.

These changes are quite inexpensive and may well eliminate the problem entirely. If they do not, it may become necessary to install supplemental priming and air-handling equipment in the shape of a wet-vacuum pump or to replace the existing centrifugal pumps with regular contractor-type self-priming pumps. The latter have the advantage that they can generally handle a higher suction lift than conventional centrifugal pumps, that they have a definite air-handling capacity, and that they can operate without a foot valve, thus further reducing the friction losses in the suction piping.

Of course, the last two alterations might prove to be more expensive than digging up the floor and lowering the level of the centrifugal pump installation into a small pit. In that event, there is still one possibility, which belongs to the second category mentioned earlier in this discussion.

The last alternative solution contemplates the use of a hydraulic ejector. There are a number of these ejectors on the market that might be used to improve this installation. The ejector is installed at the lowest level to which the water supply may fall and is connected to the pump by means of two pipes. One of these is the suction pipe proper; the second is a pipe leading from the pump discharge line to the *drive* side of the ejector.

The foot valve is retained at the bottom of the suction pipe. When the system is once primed, part of the discharge capacity is diverted down the drive pipe into the ejector. At the latter, the pressure energy is converted into velocity energy, creating a suction effect at the throat of the ejector diffuser. This lifts the water through the foot valve into the ejector. Upon leaving the ejector throat, the velocity energy is reconverted into pressure energy, lifting the water into the pump suction. Such an arrangement can be used even when the static lift exceeds the theoretical maximum of 34 ft that can be handled with a pump.

It may be, of course, that the size of a commercial ejector suitable for handling the required capacity of 100 GPM would exceed the diameter limitations imposed by the existing conditions. However, although the efficiency of a homemade ejector may suffer in comparison with that of commercial models, it is entirely possible to design such a unit to fit within the 2 1/2 in. pipe. The suggested arrangement is illustrated in Fig. 6.2. The drive pipe, the ejector, the diffuser, and a sealing plate are all nestled inside the suction pipe. Since the pump would no longer handle the water under a suction lift, all the troubles that occur at present would be eliminated.

Finally, in the event that the economics of the installation were ultimately to dictate the removal of the present bore pipe, it might prove that the most satisfactory solution would be to replace the existing centrifugal pumps with vertical turbine pumps. Their control would, of course, remain exactly the same as that for the present installation.

Question 6.5 Loss of Capacity After Starting

We have a centrifugal pump installation designed to deliver 700 GPM of water into an elevated storage tank. The pump operates with a suction lift of about 17 ft. It delivers the required 700 GPM when it is started, but soon after that the capacity starts decreasing gradually, until the pump handles only 300 GPM or even less. How can we trace the source of this difficulty?

Answer. The behavior of this pump indicates that an air leak is taking place, and air is probably accumulating at the top of the casing, reducing the effective capacity of the pump. It is difficult to find the source of this air leak without examining the installation or at least looking over the piping diagram. You can, however, do this quite readily on the spot.

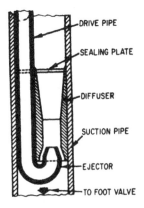

Figure 6.2 Suggested arrangement of hydraulic ejector to fit the 2½ in. pipe.

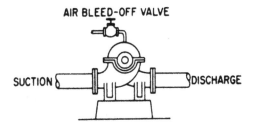

Figure 6.3 Venting accumulated air from the top of the volute casing.

When a pump operates with a suction lift, air will sometimes get into the pump through the stuffing boxes unless lantern rings are installed in these boxes and liquid under positive pressure is piped to the lantern rings. Another possibility, of course, is air leakage under the shaft sleeves. Examine the mounting of these sleeves to assure yourself that a proper seal exists between the sleeves and the impeller hub to prevent this.

Sometimes, the suction piping itself is not quite airtight and air enters into it, collecting at the top of the casing. At other times, the water handled by the pump is saturated with air, and this air may go out of solution in the pump. If trouble is due to an accumulation of air in the top of the pump volute, you can check very easily by opening the vent to atmosphere and finding whether the pump capacity is restored after this operation (Fig. 6.3).

A few additional warnings: Check the suction piping layout for air pockets. I have seen long suction lines installed with an improper pitch or with humps or high spots that permit the accumulation of air and reduce the effectiveness of the pump installation. Another possible source of trouble is the inadequacy of the suction pipe submergence at the source of supply. If the suction bell is not submerged deeply enough, vortex formation at the surface of the water will draw an appreciable quantity of air into the pump. If it is impractical to submerge the suction bell sufficiently to protect the installation against this difficulty, arrange to eliminate the formation of vortices. A very simple solution is to float a wooden screen in the shape of an egg crate on the water surface surrounding the suction pipe. This will break up the vortices very satisfactorily.

The fact that the pump will handle its rated capacity when started would indicate that there is no trouble with the pump itself. All you have to investigate is the pumping system.

Question 6.6 Further Comments on Loss of Capacity After Starting

In your solution of the problem discussed in Question 6.5 you left the impression that the air entering a centrifugal pump that is running accumulates at the top of the casing and can be bled out the vent while the pump is running. Is this correct?

Answer. I must admit that the answer I gave implied that the venting can be accomplished while the pump is running, but this is not correct.

A centrifugal pump can handle a certain amount of air in the liquid that it is pumping. The bubbles of air pass right through the pump along with the liquid. In general, the amount of air that can be handled in this manner without a significant effect is of the order of 2 to 3% by volume. When this amount is exceeded, the air will separate from the liquid, but instead of the air accumulating at the top of the casing, the centrifugal action throws the heavier liquid to the periphery, and the air is trapped in the central portion of the impeller. Under these conditions, opening the vent while the pump is running will do no good, as the liquid will be discharged but the air will stay trapped in the central core of the impeller.

As the amount of air that is released within the pump and that accumulated in the central core of the impeller increases, the head developed by the impeller vanes decreases, and the head-capacity curve of the pump falls off very appreciably. The turbulence of the water entering the impeller increases at the same time. This tends to entrain some of the air from the central core into the active portion of the impeller vanes. A balance may thus be achieved, the air leak being taken care of at the cost of reduced head and capacity. This is the cause of the trouble—the remedy is to get rid of the excess air.

If the air leak is really excessive, the air practically fills the impeller, reducing the pump head to that of the total static head and the capacity to zero. Even under these circumstances air cannot be bled out of the vent as long as the pump is running, as the air is surrounded by a ring of water thrown off by the impeller.

One of my readers also wrote me in this connection and cited two experiences that are illustrative of this problem. A horizontal open impeller pump with water-sealed glands was used to pump a measured amount (0.2 to 1 GPM) of lime slurry to a water softener. The flow was about 2 to 10% of the pump rated capacity. Slurry was measured into a small tank 6 ft above the pump and 15 ft away. Best results were obtained when the measured stream entered the hole in the bottom of the tank and flowed through the sloping 1 1/2 in. pipe to the pump. Essentially, the flow entered the pump at atmospheric pressure. The pump discharged against a 40 ft head through a check valve.

Invariably, whenever the pump was shut down and restarted, it refused to pump, and the suction line and tank filled and overflowed. When the pump was left running and the vent on top of the casing opened, the pump evacuated all the liquid out of the tank and suction line and continued to

Chapter 6 497

evacuate the inflow but did not get rid of the air. When the vent was closed, the suction line and tank refilled, but no slurry was pumped into the discharge line. When the pump was shut down and the vent opened until liquid appeared and then restarted, it continued to pump the small stream indefinitely.

The second instance that this reader called to my attention involved a case in which air was run into the suction of a vertical submerged pump to get the water saturated with air at 35 lb pressure. The air was measured with a flowscope. A certain amount of air in excess of that dissolved could be handled by the pump without a noticeable drop in discharge pressure. But as soon as this was exceeded, the pressure dropped radically, and the pump became noisy. When the air flow was reduced, the excess air was quickly flushed out of the pump and the pressure returned to normal. This can be explained only by assuming that the air separated and accumulated in the center of the impeller.

It is very apparent, therefore, that venting a pump that has become partially air bound will not be effective as long as the pump is running. The pump must be stopped so that the air can rise to the top of the casing and of the suction volutes. If the pump operates under a suction head, opening the casing and suction vents to the atmosphere will clear the air out, and the pump head-capacity curve will be restored to normal after the pump is restarted.

Of course, if the pump operates under a suction lift, opening the vents will not reprime the pump, which must be reprimed by means of an ejector or whatever other means are normally used for this purpose.

Question 6.7 Failure of a Pump to Deliver Any Capacity

The preceding two questions discussed the reasons that a pump may begin to lose capacity or even fail to deliver any flow some time after being put on the line and operating satisfactorily. What sort of circumstances may cause a pump to fail to deliver *any* flow immediately after being started?

Answer. There are many such circumstances, and I can list those that are the most common:

1. If the pump handles a suction lift, it will not deliver any liquid if it has not been primed.
2. An air pocket in the suction line can prevent a pump from delivering any flow.
3. If the total head at shutoff is less than the system pressure, the pump will be unable to lift the check valve in the discharge line and

to deliver any flow. This can occur under a variety of circumstances:
- a). If the driver fails to come up to its rated speed.
- b). When two pumps operating in parallel give a drooping head-capacity curve and the pump on the line is operating at a flow where the total head exceeds the shutoff head, the idle pump will not be able to lift the check valve on start-up (see Question 1.48).
- c). If the pump is operated in reverse rotation or if the impeller of a double-suction pump is mounted incorrectly (see Question 6.15).
- d). If the impeller is machined to a smaller diameter than specified or required.

4. If the discharge or suction valve is inadvertently closed.
5. A strainer in the suction line may be completely clogged.
6. There may be a major obstruction in the impeller eye.
7. If the level in the suction sump is reduced below the elevation required to submerge the inlet pipe.

I am sure that there are a few more possibilities, but the ones I have listed are the most probable causes. I must tell you of one specific case that was so far-fetched that I would not have imagined that it could happen. One of my colleagues told me this story, which happened some 40 years ago at a paper mill in Wisconsin. We had sold them several paper stock pumps. After they had been installed, one of the pumps was started up and came up to speed quite normally. But to the surprise of the operators, nothing came out at the end of the discharge piping. The plant engineer was summoned and after examining the installation, decided to consult the instruction book. He located the section dealing with the diagnosis and correction of a variety of pump troubles. Under the list of symptoms he found one labeled:

Pump fails to deliver rated head and capacity a short time after start-up.

He also found a miscellaneous collection of possible reasons for this dereliction of duty on the part of the pump, including wrong direction of rotation, air in-leakage, and inadequate suction conditions. But none of these causes seemed to apply. As a matter of fact, even the symptoms did not seem to cover his case: the pump delivered nothing right from the start. Of course, he could assume that his case was the extreme limit of insufficient delivery. And the pump did this right away, not *after* a period of time. The plant engineer decided that he better summon help from our factory.

Because he happened to be in the vicinity, my colleague was directed to visit the paper mill. On his arrival, the pump was started up again and again delivered no flow. But my colleague noticed that the discharge pressure gauge reading corresponded to what it should indicate under shutoff conditions. He checked the discharge valve and it was fully open. He then advised the plant engineer that there must be a solid obstruction in the discharge piping. The plant engineer refused to accept the explanation since, he said, the piping had just been installed and never used.

My colleague said "Tear down the piping." "And who's going to pay for this?" he was asked. "Tear down the piping," repeated my colleague. Tear down the piping is what they finally did. And what do you think they found? The mechanic installing the piping had cut all the flange gaskets himself and he had forgotten to cut out the hole of one of these gaskets. The shutoff pressure was not so high as to bust the gasket and he had therefore managed to provide a beautiful blank flange in his piping! So, you see, anything can happen.

Question 6.8 Unwatering a Basin

Figure 6.4 shows our centrifugal pump installation for circulating water from an outside basin through a filter and dewatering the basin when necessary. The pump is designed to handle 200 GPM against a total head of 35 ft. When we start to evacuate the water from the basin by closing valve A and opening valve B, we can pump only down to a level of 11 ft below the pump centerline. In other words, we have 4 ft of water left in the basin and we cannot get this water out. My opinion is that the existing pump was not designed for the required suction lift and will have to be replaced by one that can handle more lift, say up to 25 ft or more. What worries me is that even such a pump may not remove the last drop of water from the basin. Please let me have your opinion.

Answer. There may be two separate reasons the pump cannot lower the level beyond 11 ft from the centerline of the pump. In the first place, as you say, the pump may not be able to handle the required suction lift at the capacity at which it is required to operate. In the second place, you must realize that a very appreciable whirlpool action develops whenever the water level above the entrance of the pipe in the basin drops below a certain minimum. You have probably seen this action in a bathtub, and you may have noticed that whenever the depth of the water has been reduced below three times the diameter of the outlet, the whirlpool action is extremely pronounced. At this moment an appreciable amount of air may be drawn into the pipe. In the case of your installation, this will air-bind the centrifugal pump, which will no longer be able to handle the required capacity and may even completely lose its prime.

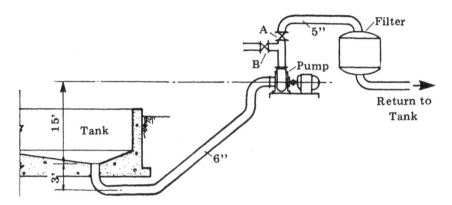

Figure 6.4 Sketch of pump installation that was incapable of pumping below 11 ft from pump centerline.

On the other hand, I am pretty sure that the present pump could be used to reduce the level below that which it has done in the past if instead of valve B being opened wide, the pump is throttled back so that the rate at which the water is withdrawn from the basin is reduced appreciably. In other words, by proper manual control, the operator can reduce the flow so that the pump can handle a higher suction lift. In addition, he or she can see whether or not a whirlpool action is developing and should throttle the pump sufficiently to prevent such formation.

In many cases, it is possible to install some baffling over the opening that will break up the action of the whirlpool and therefore obtain a better dewatering effect. Of course, these baffles need be used only during the unwatering operation. A wooden structure in the form of an egg crate, floated on the surface, has frequently been used with very satisfactory results to break up the whirlpool action.

The same precaution should be used even if a pump capable of handling a higher suction lift is used, since formation of a whirlpool will again air-bind this pump, and it will not operate properly.

It is questionable whether any pump will completely unwater the basin under the present suction arrangement. The pump will suck in air as soon as the pipe opening is uncovered. At this moment, the pump will no longer be able to discharge water, and it will be shut down. Immediately, all the water contained in the 6 in. pipe between the pump and the bottom of the basin will flow back. The use of a foot valve is questionable, as it will increase the friction losses in the suction piping and reduce the maximum permissible suction lift. One possible solution may be to provide a cap for the suction

Chapter 6 501

pipe outlet so that this cap can be screwed on it before very much water has been released back to the basin. If practical, a small deep sump could be provided at the inlet of the pipe in the basin, and very little water would then remain after the basin is unwatered.

Question 6.9 NPSH Problem When Pumping from a Reservoir or How Best to Drain a Tank

I have a problem involving the transfer of a mild acid from a tank located in the basement into another reservoir located about 25 ft above ground level. The pump that is installed at ground level cannot quite pull the suction lift required when the level in the tank at the suction falls to within 24 or even 30 in. from the bottom. The tank is cylindrical, 6 ft in diameter and 12 ft long. The suction opening is at the bottom of the tank as indicated in the attached sketch (Fig. 6.5).

This is a batch process, with liquid filling up the tank about every 2 hr. The pump is operated, therefore, about 12 times a day, for a short time. However, it would appear that the pump is not capable of pulling the full

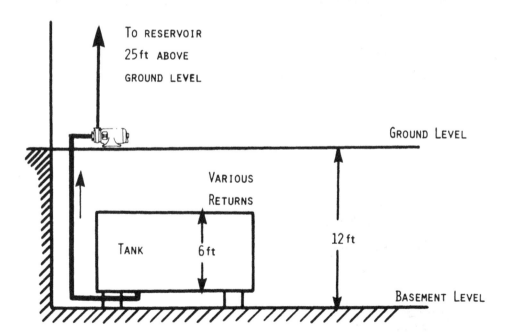

Figure 6.5 Pumping installation

lift, and as the liquid recedes to within 2 or 2 1/2 ft of the bottom, the pump loses capacity and apparently cavitates. I do not want to use a pump submerged in the tank itself, and I would prefer not to replace this pump if possible with a higher head pump or with a pump that can pull a greater lift. Is there any manner in which I can operate it without encountering this difficulty?

Answer. Before proceeding with any further steps, you should find out if the problem is created by the formation of a vortex in the bottom of the tank when the liquid level gets down to where you are experiencing trouble. This is particularly plausible if the velocity at the point of withdrawal is on the high side. After the vortex forms and increases in intensity, the air core in the center lengthens and reaches the opening at the bottom of the tank (see Fig. 6.6). After this, a continuous flow of air takes place along with the liquid that goes into the pump. This, more likely than the inability of the pump to handle the required suction lift, is probably at the root of your problem.

One possible cure against the vortex formation would be to float a form of egg crate structure over the opening. I do not know, of course, whether it might be practical to introduce this into the tank.

If vortex formation is not responsible for your dificulties, there are three different methods that can be used to permit the tank to be drained more thoroughly.

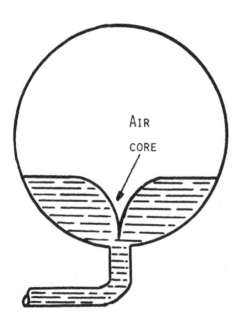

Figure 6.6 Vortex formation causes air to be drawn into the pump suction.

The first of these is the simplest, provided there is no objection to locating the pump on the basement floor. With such infrequent operation, I assume that the pump is started and stopped manually rather than automatically with float switches. The objection might arise, therefore, from the fact that you may not wish personnel to go down to the basement every time the pump must start and stop. On the other hand, it may be quite possible to install two float switches on the tank, one to start the pump and the other one to stop it. As a matter of fact, it would seem to be a simpler way to operate this installation.

Your question implies that you assume that a higher head pump must be used if it is located at a lower elevation. This is not the case. The total head that the pump will have to develop will not change after the pump is moved, since the total static elevation difference will not have changed. But the suction will have been reduced by the 12 ft difference in level between the basement and the ground floor.

The second method would consist of reducing the pump capacity as the liquid in the tank is lowered near the present limiting level. This is illustrated in Fig. 6.7. The capacity at which the pump operates normally is determined by the intersection of its head-capacity curve and of the system-head curve. As the tank is emptied and the available NPSH is decreased (the suction lift increases), a point may be reached when the available NPSH is no longer sufficient to permit normal operation. The pump starts cavitating and operates in the break.

If, however, you were to throttle the valve in the discharge line, you would create additional artificial friction, and the pump would operate against a new, steeper system-head curve. The pump head-capacity curve would intersect this system-head curve at a reduced capacity, at which the required NPSH is lower than it was at the initial capacity. This will permit the pump to operate without cavitating and to drain the tank more completely.

Incidentally, if vortexing were creating part of the problem, it should be alleviated somewhat since the velocity at the tank outlet would be reduced.

If you prefer, this throttling can be made automatically responsive to the lowering in the tank level by using a float valve that would progressively reduce the flow as the level dropped. The same control could operate a limit switch to cut the pump off completely, once a predetermined minimum level was reached.

Finally, if circumstances permit this, it would be possible to increase the available NPSH artificially without moving the pump from its present location by pressurizing the tank slightly. The available NPSH is the algebraic sum of the static pressure over the liquid and of the static level difference between the liquid and the pump centerline, less the friction losses in the suction piping and less the vapor pressure of the liquid at the pumping temperature.

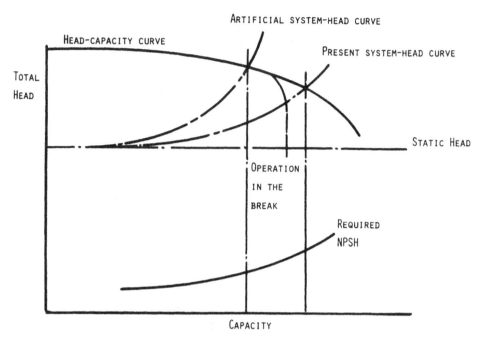

Figure 6.7 Throttling reduces capacity and required NPSH.

Thus, if we were to increase the static pressure over the liquid, we would increase the available NPSH by an equal amount without having to relocate the pump.

The amount by which we would have to pressurize the tank appears to be nominal. You state that the pump operates satisfactorily to within 24 or 30 in. of the bottom, so that an increase of 36 in. of liquid should be sufficient. If the gravity of the liquid is assumed to be 1.00, this corresponds to a pressure of 1.3 psi. To be on the safe side, we shall assume that an increase in pressure of 2 psi should be used. If the tank is structurally suitable to be pressurized to 2 psig and if no further objection exists, it will be possible to provide a small line from a compressed air supply with a regulating valve that will maintain this 2 psig pressurization while the pump is emptying the tank.

Question 6.10 Oxygen in Feedwater

At this point, instead of answering specific questions from operators of centrifugal pumps, I wish to relate two incidents that to some extent exhibit the characteristics of a whodunit. In both cases there was a "crime," in both cases the "culprit" was eventually discovered, and in both cases justice triumphed in the end, even though it was by a very narrow squeak in the second story.

I was directly involved in the first case, a good many years ago. While on a trip to the Middle West, I had been asked to stop in a small industrial plant to help settle a trouble job that centered around one of our small boiler feed pumps. Of course, the boiler feed pump gets blamed in some plants for almost everything. Beyond the common foibles of bearing or packing troubles, I have found that such things as sticky feedwater regulators, insufficient water at the suction, pressure gauges that stick, in other words anything unpleasant that may occur within a 300 ft radius of the pump can lead to a long and uncomplimentary list of epithets directed at the boiler feed pump and at anyone who had a hand with its design, construction, or application.

But this was the first time that I had heard the boiler feed pump accused of putting oxygen into the boiler. This particular installation was not operated at a very high pressure, serving, as I recall, a 200 psi boiler. The pump took its suction from a deaerating heater operating at about 3 psig and delivered the feedwater directly into the boiler. There were no closed heaters between the pump and the boiler.

Soon after the installation was completed and the plant started, rather serious troubles developed in the boiler. These troubles were thoroughly analyzed and traced to the presence of excessive oxygen in the feedwater. Obviously, the first suspicion rested on the deaerating heater. However, a modified Winkler test on the feedwater coming out of the deaerator showed less then 0.003 ml of oxygen per liter, entirely too low a concentration to have caused the damage experienced in the boiler. It was therefore at the conclusion of this test that the rather surprising theory had been advanced to the effect that the boiler feed pump was responsible and that I had been asked to make my comments and suggested solution known to the authors of the theory.

Our discussion was held in an office far removed from the power plant itself, introducing into my analysis of the facts an uncertainty with which I would have gladly dispensed. To my question as to the exact means by which it was thought the pump inroduced all this oxygen into the feedwater, I was given the answer that no doubt the oxygen must be entering through the stuffing boxes.

This statement had me stumped for a moment, as I realized that I would have to go right down to fundamentals in my defense of the innocent boiler feed pump. Apparently my audience did not seem to appreciate the fact that the stuffing-box pressures were higher than the atmospheric pressure and that therefore no in-leakage of air was possible at the boxes. I started my exposition right at the beginning, saying that all stuffing boxes, regardless of the service on which the pump is installed, have for their main function that of packing the shaft against leakage where it passes through the pump casing. But there are several variations of this function. In one case, that of the stuffing box on the suction side of a pump which takes water under a suction lift, the function of the box is *to keep the air out*, not the water in. This is obvious, since the pressure in the water chamber at the impeller suction is less than atmospheric.

On the other hand, on the discharge side of a two-stage pump such as was used in this particular plant (see Fig. 6.8), the entrance to the second-stage impeller is at a pressure in excess of atmospheric. Therefore, the function of the stuffing box on that side is *to keep the water in*.

Returning to the first function that I mentioned, that of preventing air leakage into the pump, some form of seal must be provided. The pressure of the packing against the rotating shaft or shaft sleeve is insufficient to ensure the elimination of air leakage. As a result, a water seal cage (or lantern ring) is located in the central portion of the stuffing box, as can be seen on the suction side of the pump in Fig. 6.8. Water is piped under pressure to the little chamber formed by this seal cage and makes an effective seal beyond which air cannot pass into the pump. A portion of the water piped to the seal cage enters the pump proper, and the rest leaks out to atmosphere.

But when conditions are such that the suction pressure is above atmospheric (as when a pump takes its suction from a heater operating at 3 psig located some 15 ft above the pump), the function of the seal cage is unnecessary, since air cannot leak into the pump anyway.

When the pump is handling clean, cool water, the stuffing-box seal is usually connected to the pump discharge, as shown in Fig.6.8. On the other hand, when the suction lift is very high, when the discharge pressure is low, when the water handled by the pump is dirty or gritty, or when the water temperature is high, an independent water-sealing supply must be provided.

Of course, I was going to say, a certain amount of caution is necessary in the selection of the sealing-water supply, since an appreciable amount of the sealing water will flow into the pump, as I have explained, and will mix intimately with the liquid taken in at the suction. I was also planning to add that boiler feed pumps do not, as a rule, require any sealing water. The best course, therefore, is to plug the sealing-water connection provided on small standard pumps used on boiler feed service. Such pumps always incorporate these connections because they are frequently applied to services other than boiler feeding. But before I had the time to finish my dissertation, I was interrupted by the statement than an independent seal was exactly what had been provided for the suction side stuffing box, and, apparently, even this seal was ineffective in preventing the leakage of air into the pump.

Chapter 6

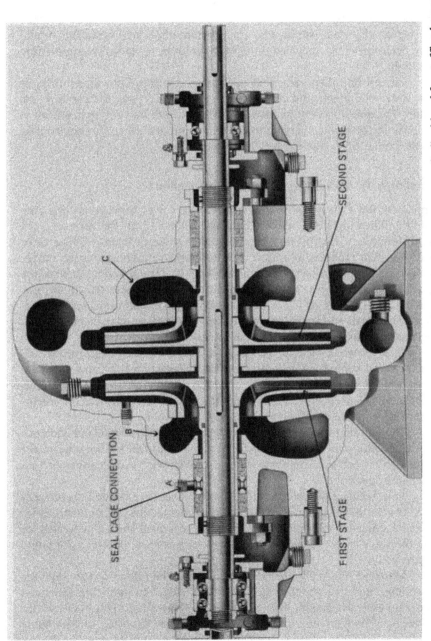

Figure 6.8 Sectional drawing of two-stage centrifugal pump showing typical arrangement of a liquid seal for stuffing boxes. A indicates connnection for sealing liquid supply; B denotes space under suction pressure; C is space under first-stage pressure.

Where did the sealing-water supply come from, I asked? The answer, almost expected by this time, was that a small line had been piped up from the nearest city water supply line at a pressure of about 45 psi. This should have been sufficient, they had thought, to hold the so-and-so pump stuffing box tight.

There I had them! My next question was if they knew how much air city water contained and how much sealing water was leaking *into* the pump. The question went unanswered. My audience had got up in unison to get back to the plant, hurrying to disconnect the sealing supply and to plug up the connection. The mystery had been solved.

Question 6.11   Admitting Air into Pump Suction

I must confess that not all pump troubles are solved as easily as this one. And the following story, told me long ago by one of our field erectors, is a good example of the mystery that surrounds some problems. This erector had been sent troubleshooting to a waterworks installation where a large centrifugal pump was behaving in a peculiar way. The superintendent of the waterworks station was complaining bitterly of the noise this pump was making and, as soon as the erector arrived, told him in no uncertain terms that he expected an immediate solution and correction of the trouble. Failing that, he was going to get the pump removed and replaced.

A day's work around the pump disclosed nothing, and the baffled erector wired home for further instructions. The wire which came in answer from headquarters read as follows: "Check suction line for air leaks. Correct if any." The check was completed the same afternoon, and the erector wired home: "Suction line absolutely tight."

The fact that the erector was no young college graduate and had been around is probably the only reason he did not lift an eyebrow when the answer to his second wire came. It read as follows: "Admit some air into the pump suction."

This, incidentally, did the trick. I have, since then, visited the installation. On the suction line, fairly near the pump suction nozzle, there is installed a small 1/8 in. pipe nipple, pointing upward. The nipple is valved, and the valve is just cracked open a bit to act as an air snifter. The pump runs quietly.

Mystery? Not really. What the serviceman had done was to mute the symptoms, not cure the disease. Very apparently, the pump was operated under an excessive suction lift, and the noise was caused by cavitation. Introducing a small amount of air into the suction had the effect of cushioning the severe collapse of water vapor bubbles that was taking place in the impeller. Had he introduced still more air, the pump would have lost its prime.

I repeat, this happened long, long ago, before much information had been compiled on the permissible suction lifts for a given capacity and pump speed condition. Obviously, the application had been incorrect. Since then, this subject has become quite common knowledge and the Hydraulic Institute Standards include an excellent series of curves that can guide us away from dangerous or even marginal suction conditions. Obviously, the real cure for this dificulty would have been the use of a slower speed pump or the relocation of the pump so as to reduce the suction lift.

Question 6.12 Float-Controlled Valve in Pump Discharge

We operate a horizontal split casing, double-suction centrifugal pump that takes the overflow from a brine settling basin and delivers it to a 25 ft high tank about 150 ft away. The settling basin overflows into a launder that feeds the pump suction, and since there is no reserve capacity in the launder, the pump suction continually pulls dry. The pump has an intermittent discharge to the tank as the suction goes dry and floods. The pump discharge cannot be throttled or the impeller cut down to overcome this condition, as the supply to the launder is variable, and the pump must have sufficient capacity to handle the peak flow. You will realize, I am sure, that this type of operation is extremely hard on the pump and that maintenance of impeller, wearing rings, bearings, and other parts is high.

We have thought of installing a small float box on the suction side of the pump, connected to a valve in the pump discharge. This valve would throttle the flow when the pump got ahead of its supply. However, there are many reasons for avoiding this installation if possible.

Can you advise me of anything that can be done to help this pump? We have noted Question 6.11 describing the installation of a small air intake on a pump suction. Would this or something similar be applicable here?

Answer. There are several means available to prevent your transfer pump from running dry. Unfortunately, all these means will require a certain amount of rearrangement and installation of valves and of new piping. To provide you with specific information and to make concrete suggestions, it would be necessary to have more information about your present installation. For instance, it would be necessary to know the rated conditions of service of your pump, to have a sketch of the settling basin arrangement and of the launder, and to have some idea of the rate at which the brine is delivered to the settling tank. Nevertheless, I can give you some general comments about this problem and some general suggestions as to the possible solutions.

A centrifugal pump operating at constant speed in a given system will deliver that capacity into the discharge piping that corresponds to the intersection of its head-capacity curve with the system-head curve. The latter is made up of the sum of the static and pressure heads and of the friction losses at various capacities. The determination of the operating conditions is illustrated in Fig. 4.5. The system-head curve is superimposed on the pump head-capacity curve, and the intersection of the two determines the operating capacity. Therefore, if the pump has been selected to be capable to handle the peak flow, it must have a head-capacity curve such that it will intersect the system-head curve at a capacity at least equal to the peak flow expected.

Unless the head-capacity curve of the pump is altered by operating the pump at variable speed or unless the shape of the system-head curve is altered by artificially throttling in the discharge, the pump will always attempt to handle a fixed capacity corresponding to the intersection of the two curves. When this capacity exceeds the overflow into the launder, the pump will rapidly lower the level in the launder.

If this lowering of the level reduces the net positive suction head available below that required by the pump, the capacity of the latter will reduce, as it will operate in the break, as shown in Fig. 6.9. However, your question implies that this does not take place, as you say that the pump runs dry. In other words, the available net positive suction head is apparently more than sufficient under all circumstances.

It is therefore probable that the pump lowers the liquid level in the launder, uncovers its suction, and loses its prime. As a matter of fact, the pump probably loses its prime even before all the brine is pumped out of the launder because a whirlpool action develops whenever the liquid level above the entrance to the suction pipe drops to a certain minimum. An appreciable amount of air can be drawn into the pump through this whirlpool, and the pump becomes air bound.

I agree with your statement that the impeller should not be cut down, as the pump could no longer handle the peak flow from the brine settling basin. Nor would it probably be economical to install means to vary the pump speed. However, I cannot understand your reluctance to throttle the pump discharge unless you imply that the throttling cannot be done manually, since the variation in supply is cyclic and frequent. Obviously, in that case, the throttling must be made automatic and responsive to the supply at the suction.

One solution would be to install a float on the suction side that would throttle the pump discharge so as to maintain a certain minimum level in the launder. Various positions of this throttling valve would change the friction component of the system-head curve, so that the latter would assume various shapes intersecting with the head-capacity curve at capacities corresponding to the inflow to the launder, as shown in Fig. 4.5.

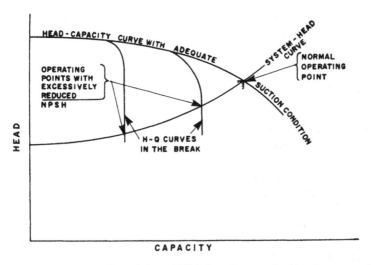

Figure 6.9 Reduction of the NPSH below that required by the pump will cause it to operate in the break.

Thus, a float control at the suction is about the only logical or practical means to accomplish what you wish to do. The only portion of the solution that is really open to choice is what to do with the impulse provided by the float control. You can either throttle the pump discharge or install a recirculation line that will bypass some of the pump output back to the suction. The float control would then actuate a valve in this bypass line, opening it more when the suction level is drooping and closing it entirely when the level is adequate. The one objection to the latter solution is that the power consumption of the pump will exceed the consumption that would occur if the throttling were done in the discharge, since the pump will always be operating at higher flows than in the latter case.

One bit of precaution: If the pump discharge is throttled and if the supply to the launder stops, the pump could be operated against a fully closed discharge. This is not a suitable manner of operating a centrifugal pump, and you should install a small recirculating line that will always permit some flow to take place through the pump. I imagine that the most practical method of operating this bypass line would be to leave the bypass flow on all the time.

The admission of a small quantity of air into the pump suction is strictly a makeshift method of reducing undesirable pump noise but will accomplish nothing in your case.

I fully realize the unfavorable effect of the present operation on the pump maintenance, and I am sure that a rearrangement of the system can reduce or eliminate this unnecessary expense.

Question 6.13 Float Switches Versus Float-Controlled Valves

Your question 6.12 dealt with the problem encountered with a transfer pump that frequently empties out all the liquid from the reservoir at its suction and runs dry. You suggested that a float control be installed in this suction reservoir, this control to operate a throttling valve in the pump discharge so as to maintain a minimum level in the reservoir.

We have about 20 tanks in our plant that collect cooling water after use, and pumps return it to the cooling towers. All these tanks are equipped with float switches that start and stop the pumps. In this way, the pumps run a minimum of time at maximum efficiency.

This seems to be a better solution than the use of a float-controlled valve in the pump discharge line, which allows the pump to run all the time, most of it at low efficiency. The float switch is also much simpler and less expensive than the float valve.

Answer. You are quite correct in suggesting that float switches can be used very advantageously to control pump operation. Whenever they can be applied so that the start-and-stop operation is not overly frequent, they may be superior to the use of float-controlled valves in that they do permit the pumps to operate nearer to their best efficiency at almost all times and that they are simpler and less expensive. On the other hand, this cannot be made into a general rule, as there are many exceptions to it. Such factors as the variation in supply to the reservoir, the size of the reservoir in relation to the normal and minimum flows, and the need or lack of need of a continuous supply from the pump all contribute to the making of the decision between float switches and float valves.

Obviously, if conditions are such that very frequent start-and-stop operations would result, the float switch is not too practical. Let us assume, for instance, that the maximum flow that we may deal with is 275 GPM and that the pump is designed for 300 GPM so as to provide a margin. Let us further assume that frictional losses are small compared with static pressure so as to simplify our problem and neglect the effect of a variable system-head curve. Whenever the incoming flow to the reservoir at the pump suction falls to, say, 150 GPM, there will be a deficit of 150 GPM between the amount that the pump can withdraw from the reservoir and the supply into it. If the volume in the reservoir between the start-and-stop levels is of the order of 1500 gal, the pump will stop 10 min after it is started and restart in another 10 min. But if that volume is only 300 gal, the period between each start and stop will have been reduced to 2 min.

Although the return to the reservoir at the suction and hence the flow required from the pump may vary, it is quite possible that the process served by the pump cannot permit a complete interruption of flow. For instance, with the exception of some household steam boilers where gravity return of condensate maintains the water level and where only makeup to compensate for small losses is required, it is not practical to use a start-and-stop operation of a boiler feed pump. The same is generally true of condensate pumps taking their suction from the condenser hot well. Such installations are always arranged for continuous operation, with either throttling valves or variable-speed operation controlling the rate of delivery.

Obviously, as I have said, it is impossible to make a general statement as to the preference between the use of a float switch and a float control for throttling the discharge. There will always be some situations in which the former is the best solution and other situations in which the reverse is true.

Question 6.14 Is Engineering an Exact Science?

In your answers to several questions you make reference to the fact that when a pump handles a liquid with an appreciable amount of entrained or dissolved air or gas, the pump can become air or vapor bound under certain conditions. You explain this be saying that at reduced flows, the velocities within the pump are no longer sufficient to sweep the air out; the air accumulates within the pump casing and air binding ensues. You also indicated (Fig. 1.51) that the overall pump performance deteriorates rapidly as the percentage of air or vapor increases.

At the same time, in Question 6.11, you state that under certain circumstances, the introduction of small quantities of air at the suction may be used to quiet noisy cavitating pumps. You have not indicated how much air would have to be introduced, nor have you said whether this would cause a significant deterioration in pump performance.

There seem to be two contradictory recommendations here:

1. Try to avoid having any entrained air or gas in the liquid pumped.
2. Admit some air at the suction to quiet a noisy pump.

Could you please expand on this matter?

Answer. First, let me address myself to the question of performance deterioration. One can only make some general references to this phenomenon. I suspect that an evaluation of the exact extent of this deterioration is too complex to permit expressing it in any mathematical formula. It is only possible to give—as I have done—a typical example of pump performance with a range of air content percentages. As a matter of fact, even such examples are no more or less accurate than the estimates of these percentages, since I know of no simple means of establishing the exact air or gas content of the liquid at the suction conditions.

To appreciate the fact that test data provide a variety of different effects of the presence of air, I refer you to Fig. 6.10, which illustrates still another typical test of the effect of air content on the performance of a pump. You will note that the deterioration appears to be much less severe in this case than in the case of the pump illustrated in Fig. 1.51. In the latter, even though the maximum amount of air content tested was 6%, one can imagine that 10% air content would completely destroy the ability of the pump to deliver any flow. This leads me to conclude that a number of variables in the design of a pump and in the geometric configuration of the impeller and of the casing play a major role in the extent of the deterioration. But the exact relationship between design and performance remains—in my opinion and to my knowledge—an unknown.

Thus, although the pump in Fig. 6.10 is better adapted to handle entrained air than the pump in Fig 1.51, I am not in a position to tell you why this is so.

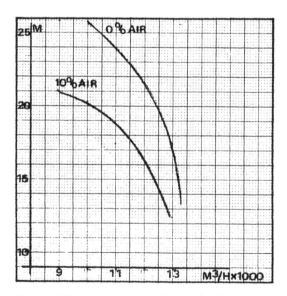

Figure 6.10 Another example of the effect of entrained air.

Chapter 6 515

I should add that although 6 or 10% air by volume does considerable harm to the pump performance, there are few reasons that such a large percentage should be tolerated. The only case with which I am familiar has to do with the production of protein from hydrocarbons. The process in question requires the use of rather large pumps that must be capable of handling as much as 15% air by volume. In essence, the pump handles something like an emulsion.

Now, let us consider the effect of admitting air at the suction for the purpose of quieting a cavitating pump. I have to admit that I do not know exactly how much air must be introduced for this purpose, but I can assure you that it is generally less than 1%; as a matter of fact I imagine that it need not exceed 1/4 or 1/2%. In all such instances that I have come across, the amount was less than would result in a deterioration of the head-capacity curve to an extent that would be observable within the accuracy of the test.

As to the contradiction you refer to, it is more apparent than real. You will certainly agree that 1/4 or 1/2% air admission is almost negligible in its effect on pump performance. You should note, however, that I cited the use of bled-in air at the suction strictly as a post factum correction and did not recommend it as the best solution for unfavorable conditions. Instead, I suggested that the pumping system should be redesigned or that a more suitable impeller be chosen for the existing conditions. Admission of air should only be a last resort if no other solution is practical.

Question 6.15 Reversed Impeller

We have a 6 in. single-stage double-suction pump that was installed about 12 years ago. It is designed to handle 2250 GPM against a total head of 175 ft at 1750 RPM. The head is all made up of friction, and there is no static head component. Recently, we had noticed that the pump capacity had fallen off somewhat. We inspected the clearances at the wearing rings and found that appreciable wear had taken place. We purchased replacement casing rings and restored the initial clearances.

Unfortunately, after the pump was rebuilt and returned to service, it failed to deliver even as much as it was capable of pumping before it was rebuilt. As near as we can make out, it can handle only 1950 GPM. The head measurement is 130 ft, indicating that the pump is operating on the system-head curve but intersecting it at this lower capacity. At the same time, electrical readings on the motor indicate a greatly increased power consumption.

We have checked every single angle we could think of. The pump has been dismantled once more, but we found neither mechanical difficulties nor any foreign matter that might have obstructed the impeller. The suction piping likewise has been examined and was found to be free and clear. The pump takes its suction from a large storage tank under several feet of positive head, and therefore there can be no air leakage into the suction. We opened all the vents before starting, and therefore the pump was fully primed, with no air pockets. We even disconnected the motor and checked it for correct rotation. Can you suggest anything that we may have overlooked?

Answer. The source of the difficulty has apparently been searched for quite thoroughly, and all possible causes that could contribute to the manufacturing have been examined save one: The test data you have obtained point to the real culprit—a reversed impeller.

Centrifugal pump impellers have backward-curved vanes; that is, they revolve *away* from the curvature of their vanes. If a pump is dismantled completely and then reassembled, an error in mounting the impeller is easily overlooked in the case of a double-suction impeller. The latter is symmetrical about its central plane and will fit on the shaft even if it is turned in the wrong direction. The effect is that shown in Fig. 6.11(B), and the hydraulic performance is that which would be obtained in a normal casing with correct rotation but with *foward-curved vanes*.

I have illustrated in Fig. 6.12 the probable performance of a pump such as you have described, rotating in the proper direction both with a properly mounted impeller and with a reversed impeller. When this performance is super imposed over a system-head curve made up strictly of friction, it becomes readily apparent why the delivery into the system is curtailed to the extent you describe. You will also note that the efficiency of a pump with reversed mounting of the impeller is seriously reduced and consequently the power consumption increased over that with normal operation.

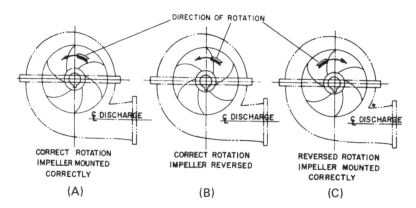

Figure 6.11 Mounting of impeller in a volute casing.

Chapter 6

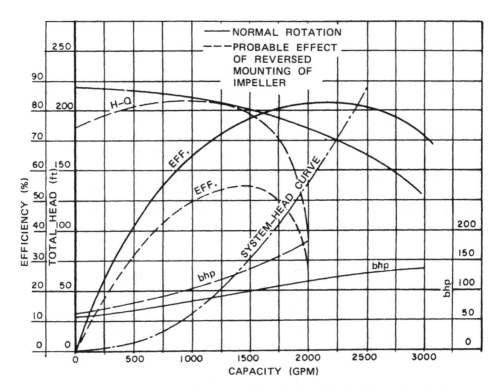

Figure 6.12 Probable effect of a reversed impeller in 6 in. double-suction single-stage pump operating at 1750 RPM.

Reversing an impeller on the shaft does not produce the same effect as running the pump in reverse rotation Fig. 6.11(C). The loss in head and capacity and the increase in power consumption in the latter case are much more severe. Figure 6.13 indicates the approximate results to be expected if the pump is operated in reverse rotation.

Question 6.16 Transient NPSH Conditions

Our plant has an excess of low-pressure (40 psig) steam. To keep from venting this steam, we tried using it in our deaerator to increase the temperature of our boiler feedwater. The boiler feed pumps are five-stage 4 by 6 in. pumps with 10 in. impellers and designed for 700 GPM and approximately 1300 ft total head. The condensate storage tank is mounted 23 ft above the pump suction. Our required NPSH at 242 °F is 14 ft. We have been unable to raise our feedwater temperature above 256 °F without occasionally *gassing off* a pump, even though the calculated NPSH seems to be great enough.

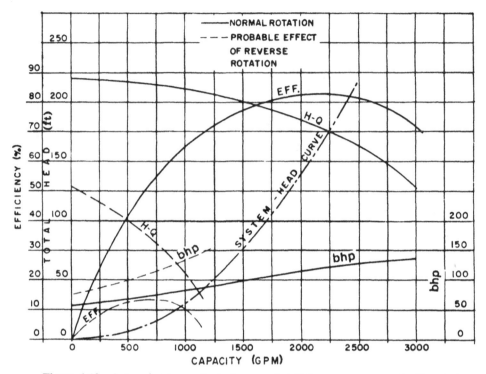

Figure 6.13 Approximate results to be expected if the same pump as in Fig. 6.12 was operated in reversed rotation.

We would be interested in what affects pumping ability as water increases in temperature and some way of estimating "extra" NPSH requirements for additional temperature increments.

Answer. By definition, the net positive suction head required by a pump to handle a certain quantity of water at a given speed is independent of the temperature of this water, as I have discussed it in Question 1.19. The reason for this is that a centrifugal pump requires a certain amount of energy in excess of the vapor pressure of the liquid pumped to cause flow into the impeller, and the required NPSH is the measure of this net energy *over and above* the vapor pressure of the liquid.

Under stable operating conditions, the feedwater in the condensate storage tank of the deaerator is at saturated temperature; that is, it is under a pressure corresponding to the vapor pressure at its temperature. Therefore, whether the water is at 242 or 285 °F, which is the temperature it could reach with 40 psig steam, the required NPSH is not affected. (Actually, there is a very slight temperature effect, as the theoretical required NPSH decreases somewhat with temperature increase. This effect does not become significant until much higher temperatures are encountered.)

The fact that you have been unable to maintain temperatures in excess of 256 °F without steam binding the pump (or flashing) is an indication that you are encountering unstable or transient conditions. This is a frequently encountered phenomenon in steam power plants in which the steam supply to the deaerator is taken from a bleed stage of the main turbine. If this supply is uncontrolled, the pressure at the bleed stage varies almost directly with the load carried by the turbine. Consequently, if a sudden drop in load takes place, the pressure at the bleed stage falls temporarily below the pressure in the deaerator, the check valve in the steam supply line closes, and the deaerator is isolated from any further source of heat until such time as pressures are equalized and the supply of steam from the bleed stage of the turbine becomes again available. In the meantime, cold condensate continues to enter the deaerator, and the feedwater continues to be withdrawn from the storage space. This causes the total heat in the deaerator to be reduced. To maintain equilibrium, some of the feedwater at the surface of the storage space flashes into steam and the pressure in the deaerator "decays" at a rapid rate. This reduction in pressure is transmitted instantaneously to the pump suction. On the other hand, the temperature of the feedwater entering the pump does not begin to decrease until all the feedwater contained in the suction piping between the deaerator and the pump has been evacuated.

I have said that the available NPSH is the difference between the suction pressure and the vapor pressure in feet of water. Since, under steady-state conditions, the pressure in the deaerator is equal to the vapor pressure at pumping temperature, the available NPSH is equal to the static submergence from the water level to the pump centerline less suction piping friction losses. But under the transient conditions I have described, the pressure in the deaerator is reduced below the vapor pressure of the feedwater at the pump suction, and the available NPSH is no longer equivalent to the static submergence less friction losses but is significantly reduced. The exact amount of this reduction depends on a number of factors related to a particular installation, such as the feedwater temperature under initial conditions, condensate temperature after the load reduction, volume of the deaerator storage, and volume of the suction piping.*

Means must be employed to take care of the time lag that exists between the instantaneous reduction of pressure in the deaerator following a

---

*For a complete treatment of this subject, see the Bibliography at the end of this chapter.

sudden load reduction and the ultimate reduction of temperature at the pump suction after the feedwater already in the suction piping has been pumped out into the discharge header. Generally, this consists of adding a certain amount to the NPSH required by the pump under steady-state conditions. The exact value of this factor of safety can be calculated from certain relations developed in papers (1, Jan. 1960) and (2) cited in the bibliography.

Returning to the problem you have presented, it would appear that your installation suffers from a condition similar to the one that I have described, even though you do not indicate that the pressure of the steam supply to the deaerator varies with the load. The fact that you say that you have been unable to raise the feedwater temperature above 256 °F without *occasionally* gassing off a pump implies that you can get satisfactory operation except at certain times. These certain times are occasions when the available supply of 40 psig steam dwindles, the supply pressure falls, and the deaerator pressure follows suit.

I suggest that you investigate this phase of your installation and if you find that pressure variations do occur, that you apply to the installation the analysis outlined in the papers referred to. Should you find that the installation is not safe under the existing conditions, you have several choices available to correct it:

1. You can calculate the maximum pressure-temperature conditions for the deaerator under which no pressure decay can take place or under which the pressure decay is not excessive in the light of the margin available between the available NPSH under steady-state conditions and the required NPSH.
2. You may be able to so alter the deaerator storage volume or the suction piping volume that you will eliminate the unfavorable effect of the steam pressure variations.
3. You may find it most economical to install protective controls that would measure the available NPSH, weigh it against the required NPSH, and, whenever a deficiency arose, admit auxiliary steam from a higher pressure source through a reducing valve. A typical arrangement for such a control is illustrated in Fig. 6.14.

Question 6.17 More on Transient Conditions in Open Feedwater Cycles

Several of your Clinics have dealt with the unfavorable effects of sudden load drops in open feedwater cycles. Is there a simple and short method to check on the adequacy of a proposed installation? What can be done to counteract these effects? I understand that in the 1940s and 1950s antiflash baffling was used in deaerators for this purpose. Why has this practice been discontinued?

Chapter 6

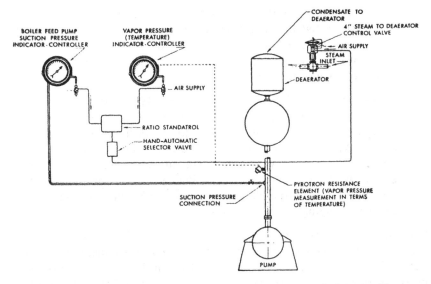

Figure 6.14 Typical arrangement for a protective control that would measure available NPSH, weigh it against required NPSH, and whenever a deficiency arose, admit auxiliary steam to deaerator from a higher pressure source through a reducing valve.

Answer. As I have explained elsewhere, the problem is created by the fact that the moment a significant and sudden drop in load occurs, the pressure at the turbine extraction stage that supplies steam to the deaerator drops proportionally with the load. When this occurs, this pressure falls below the pressure in the heater itself; this causes the intercept valve to close and isolates the heater from the supply of steam. Meanwhile, feedwater continues to be withdrawn from the heater storage and cold condensate continues to be delivered to the heater. In consequence, both the temperature and pressure in the heater are being steadily reduced. However, until all the hotter feedwater in the suction piping has been evacuated, its temperature and vapor pressure remain at the level that existed prior to the load drop. Thus, the available NPSH is reduced by the amount of the pressure reduction in the deaerator. Unless, the installation provides a rather substantial margin between the available NPSH under normal steady-state conditions and the NPSH required by the boiler feed pumps, cavitation will occur at the first stage of the pumps.

The calculation of the necessary margin is a very complex procedure and involves such factors as the volume of the deaerator storage space, the volume of the suction piping, and the values of the enthalpies of the feedwater before the load drop and of the condensate after the drop. The process consists of stipulating that the *actual* rate of pressure decay in the deaerator should not exceed the *allowable* rate of decay for the system.

The allowable rate of decay can be calculated quite simply, on the basis of the following concept:

> If under stable conditions the available NPSH exceeds the NPSH required by a certain number of feet, then under fluctuating conditions the heater pressure must not reduce more than that many feet in the time it takes feedwater just leaving the deaerator to reach the pump suction. Thus, the allowable rate of pressure decay in the heater can be readily calculated for any particular flow by dividing the excess NPSH available (the margin) under stable conditions by the residence time in the suction piping at that flow rate.

As I said, the *actual* rate of decay is more complex and in theory can only be determined by carrying out a series of successive heat balances around the heater with respect to time. But in fact, it is hardly necessary to be that precise as long as an approximate solution is developed that is sufficiently conservative so that it permits the design and operation of steam power plants with an acceptable degree of security.

By equating the *allowable* and *actual* values of the pressure decay, one could then establish a relationship that would ensure such security. The formula thus developed was

$$\text{Minimum } \frac{Q_h}{Q_s} = \frac{(h_{x0} - h_{c2})}{K_h H_x}$$

where

Q_h = volume of feedwater in deaerator storage, in gallons
Q_s = volume of feedwater in suction piping, in gallons
h_{x0} = enthalpy of feedwater in deaerator prior to load drop, in Btu/lb
h_{c2} = enthalpy of condensate to dearator for final conditions after load drop, in Btu/lb
K_h = dh/dp in Btu/lb per ft of absolute pressure, at steam conditions prior to load reduction (see Fig. 6.15)
H_x = available excess NPSH, in ft = available NPSH - required NPSH

I should add that very extensive tests carried out in a steam power plant under actual load drop conditions corresponded very closely both to a "step-by-step" heat balance prediction and to the simplified algebraic equation just given.

Chapter 6

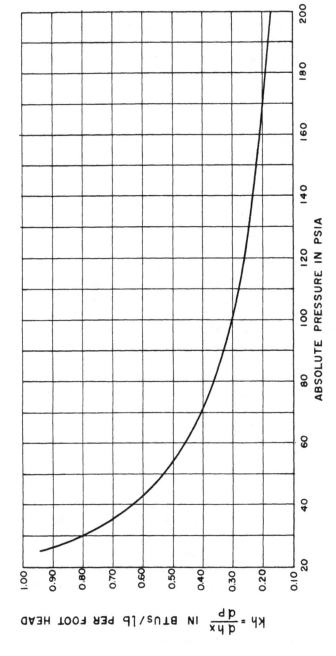

Figure 6.15 Enthalpy change with change of pressure for water.

Let us now examine the solutions available to the steam power plant designer to counteract the unfavorable effects of a load drop. Since the factors involving the overall design of a power plant (h_{x0}, h_{c2}, and K_h) cannot and should not be changed, we can only operate on the values of Q_h, Q_s, and H_x. In other words, to ameliorate a given situation we can only increase the deaerator storage volume, reduce the volume of the suction piping, and/or increase the NPSH margin.

But if we reach the limits that can be economically justifiable in the manipulation of these three variables, we need resort to another approach: it becomes necesary to install protective coontrols that admit either auxiliary steam to the deaerator from a higher pressure stage of the turbine or subcooling cold condensate to the boiler feed pump suction. Such a control is described in Question 6.16 and illustrated in Fig. 6.14. The effect of subcooling is described in Question 1.24 and specifically in Figs. 1.30, 1.31, and 1.32.

Now let me address myself to the question regarding "antiflash baffling" in the deaerator. This solution became quite popular between 1943 and 1955. It consisted of providing baffling immediately below the heating and deaerating section, along with internal piping (see Fig. 6.16) to conduct feedwater directly from the heating element to the outlet nozzle and on to the suction piping. An annular clearance space is provided at the bottom of the storage space between the downcoming pipe and the outlet nozzle, so that any excess of deaerated water over and above the boiler feed pump demand can spill into the storage space. Conversely, any deficiency in feedwater flow can be made up from the storage space. The beneficial effect of the antiflash baffling was claimed to be that lower temperature water was delivered to the suction piping immediately after the load drop.

Unfortunately, later analysis and field tests disclosed that the baffling contributed too little and too late to the security of the system. What was even more disturbing was that the baffling could cause excessive subcooling of the stored feedwater. This, then, robbed it of the possibility of mitigating against an immediate and significant drop in pressure in the deaerator, since no feedwater would flash into steam and reduce the decay rate in the heater. Finally, if a sudden load drop were to take place very shortly after the start-up of a unit, the storage space would be full of quite cold feedwater and the effect I just described would be quite catastrophic. The net result of all this is that most installations incorporating antiflash baffling were ultimately corrected and the baffling removed.

Question 6.18 More About Antiflash Baffling

In your answer to the previous question, you made reference to the fact that antiflash baffling of deaerators may aggravate the ill effects of a sudden load

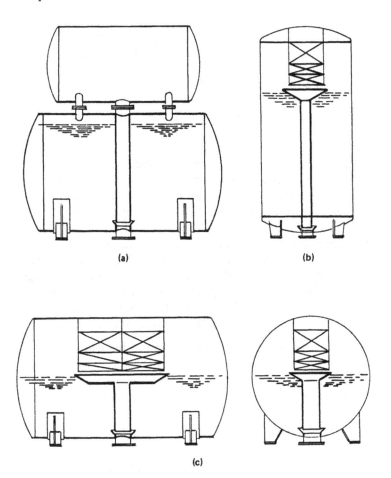

Figure 6.16 Typical arrangements of antiflash baffling.

drop in certain cases. You said that this occurs when the feedwater in the storage space of the deaerator is at a temperature materially below saturation temperature. How can such a situation exist? Isn't the stored water in a deaerator always at, or very close to saturation temperature?

Answer. You are quite right in assuming that under normal operating conditions and after the unit load has been stabilized for an appreciable amount of time, the feedwater in the deaerator storage space is substantially at saturation temperature. Under such conditions, the immediate effect of a sudden load drop is to interrupt the supply of steam from the extraction stage of the main steam turbine to the deaerating section of the feedwater heater. As soon as the pressure in the heater starts dropping, flashing occurs at the surface of the storage space. The pressure in the heater decays slowly—possibly over a time period of 3/4 to 1 1/2 minutes, depending on a variety of factors. This acts to retard the reduction in the available NPSH. But if enough NPSH margin has been provided, the safety of the installation will not be unduly endangered.

There can, however, arise circumstances under which the decay of pressure at the deaerator is not gradual, but rather sudden, and it is this sudden pressure drop at the deaerator that can be most harmful to the installation. These circumstances may best be understood from the description of a case in which such a possibility was first brought to my attention.

As I explained, by 1953 I had become convinced that antiflash baffling could only provide "too little and too late" protection. But since there was insufficient evidence of any danger from this baffling, its use was continued by the industry. A dramatic demonstration of the potential danger took place in a steam power plant completed in Japan in the late 1960s. At that time, steam-electric plants in Japan operated under conditions much more severe than those normally encountered then in the United States:

> Hydroelectric power provided a large portion of the demand.
>> Because these hydroelectric units are of the run-of-the-river type, steam plants are operated to take all load variations.
>
> There was a very sharp and very frequent variation in the load demand.
>
> Electrical storms cause a high frequency of partial or even complete loss of load.

Because of the severity of these operating conditions, the Japanese government was imposing some very stringent tests on all new steam-electric installations before they could be commissioned and incorporated into the overall system. One of these tests required that a new unit be rapidly brought up from a cold start-up to full load and then suddenly tripped out.

It was during such a test that it was determined that the antiflash baffling incorporated into the deaerating heater can be the cause of severe cavitation and potential damage to the boiler feed pumps.

When the unit was brought up to full load rapidly, the temperature of the water in the storage space lagged substantially behind the temperature flowing to the boiler feed pump. (Observations made during later tests showed that by the time the unit had come up to full load, the water in the storage space was about 35 °F below the deaerator saturation temperature.) But since the deaerator was equipped with antiflash baffling, the feedwater in the pump suction line was right up to saturation temperature.

Within seconds after the test trip-out, the boiler feed pumps had lost suction and contact was made between running and stationary parts. What had obviously happened is that with feedwater in the storage space at 35 °F below saturation temperature, the deaerator pressure had to drop almost instantaneously by about 35 psi before flashing in the storage would help moderate the rate of pressure decay. This is equivalent to approximately 80 ft, and since the installation did not provide such an excess in available NPSH, flashing took place at the boiler feed pumps. It is interesting to note that other tests carried out after a long period of operation at full load—such that the stored feedwater was at saturation temperature—indicated that such a violent and sudden pressure decay did not take place after a trip-out.

Conditions at this Japanese power plant did not lend themselves to the exhaustive and rigorous tests that would once and for all demonstrate the exact effects of antiflash baffling. Fortunately, however, there was an installation in the United States that provided ideal conditions for such a test. It comprised two 100 MW units identical in all respects except one: the older unit had no storage tank antiflash baffling, but the latest unit was provided with such baffling. The deaerators were of the tray type, rated for 670,000 lb/hr, and their storage capacity was about 10 min at rated load. Rated design operating pressure was 71 psia. They were installed outdoors and were fully insulated.

Tests were carried out in December 1960 under conditions of normal plant start-up, followed by a period of steady maximum load, and finally with diminishing load. The results of these tests are presented in Fig. 6.17. To provide a clearer comparison between the various temperatures, deaerator pressure readings have been converted to corresponding saturation temperatures on the curves in Fig. 6.17.

It will be noted that there is a very close correlation between the temperatures in the storage space and at the boiler feed pump suction in the case of the deaerator without antiflash baffling; but a substantial time lag exists between the two temperatures in the case of a deaerator with antiflash baffling. This discrepancy persisted for a period of over 5 hr after start-up! In other words, the period of potential danger to the boiler feed pumps extended throughout these 5 hr during which the temperature at the pump suction was higher than the storage tank temperature. Should the load have been suddenly reduced during this period, a substantial drop in deaerator pressure would have taken place before flashing of the stored feedwater could have contributed its cushioning effect and reduced the rate of pressure decay.

It should be noted that the period immediately following start-up of a unit is not the only time when antiflash baffling can introduce a danger.

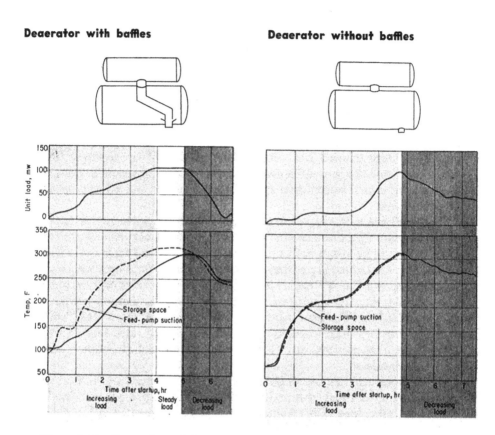

Figure 6.17 Comparison of deaerator storage temperatures with and without anti-flash baffling.

Any time a given unit operates over a wide range of loads, the danger is present. For instance, if a unit has been carrying a reduced load for some time, the deaerator storage will be at a temperature corresponding to this reduced load. If then the load increases rapidly and then drops suddenly, the storage feedwater will not have had time to come up to full saturation temperature. Although the subcooling may be less extensive than if the sudden load drop were to follow a cold start-up, it could still cause trouble. After the results of these tests had been written up, it seemed to me that the subject of anti-flash baffling could be considered as exhausted. And yet, the question refused to move back stage and become a historical curiosity. As a matter of fact, it threatened to heighten in controversial stature and to split power plant engineers into two almost hostile camps. I hesitate to compare this

Chapter 6 529

problem to the controversy that raged in the Middle Ages over the question whether the sun and the stars revolved around the earth or whether we should bow to the inevitable conclusion that the Copernican concept of the universe was to relegate our earth to the insignificant role of one of the planets (and not the biggest one, at that) of a star of the fourth order. But "coexistence" was a word that apparently was missing from the vocabulary of many steam power plant designers, and the controversy continued unabated. Reluctantly, I took pen in hand once more and reentered the fray. Ultimately, facts prevailed and antiflash baffling did become the historical curiosity it deserved to be.

The preceding facts, then, form the basis of my recommendation that any deaerator that is still provided with antiflash baffling should be modified and the baffling eliminated.

Question 6.19 Horizontal Runs in the Suction Piping of Boiler Feed Pumps

I have heard it said that locating a horizontal run in the suction piping to boiler feed pumps too close to the outlet from the deaerator may lead to flashing during the transient conditions that follow a sudden drop in load. Can you explain why this can happen?

Answer. Even where the analysis seems to indicate that the boiler feed pumps themselves are assured of their required NPSH during a reduction in main turbine load, there is no guarantee that their operation will not be interrupted by flashing at some point in the suction piping. The criterion in determining whether this will occur is to consider that the water that left the heater outlet at a saturated condition must pick up static pressure, due to the vertical drop, at a rate at least equal to the pressure decay in the heater after a load drop, or it will flash.

The most adverse conditions are those introduced by locating a horizontal run too close to the heater outlet. A typical case is illustrated in Fig. 6.18. In the comparison of the two installations, we will stipulate that the total length of the piping and the sizes of this piping are the same for both arrangements and further that the volume of the suction piping between the heater outlets and points C and E of the two arrangements, respectively, are the same. If a time interval x is selected such that water having left the heater outlet at the start of the transient conditions will have reached points C and E at the end of this interval, respectively, it becomes apparent that flashing will take place at point C, whereas the pressure at point E will be such that no flashing will occur.

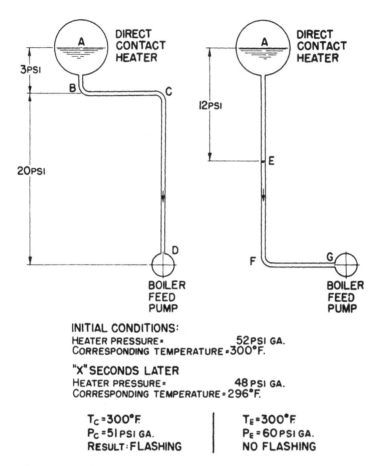

Figure 6.18 Comparison of suction piping arrangements.

Question 6.20 Suction Piping Vibrations

We have a problem on which you could probably give some helpful advice. It is one that recurs with nearly every project, despite our efforts to resolve it. The problem is one of boiler feed suction pipe line vibration.

We try to run this line straight downward from the deaerator outlet connection and to the boiler feed pumps with a minimum of horizontal run. We have even sloped this line to avoid strictly horizontal piping. We keep in mind the factor of thermal expansion and use flexible spring supports where this factor calls for it.

Chapter 6 531

With piping designed in the general manner, we receive reports from start-up operations of vibration of this piping. Often field forces begin at once to install rigid restraints in order that they may proceed with boiler safety valve setting, boiler boiling out, boiler filling, steam pipeline blowing out, and similar operations that generally require low and erratic flow conditions of the boiler feed pumps. The consulting engineers receive complaints and requests for added restraints to hold the piping rigidly. Often such restraints are installed by field forces, entirely neglecting the effects of such on thermal expansion. These restraints reduce the vibration but the cause remains. We would like to formulate design criteria for this piping that would react without vibration to all flow conditions. We realize that this is a very large order. Pumping a flashing liquid is indeed a difficult service for a pump, especially with varying flow conditions and a varying head (floating deaerator pressure). The reaction of many engineers to this problem is to tie down the piping and let the effect continue. I think that there surely must be a better answer to this problem. Your comments on this matter would be greatly appreciated.

Answer. You could have hardly chosen a subject that interests me more or a question that is more difficult to answer conclusively than that dealing with vibration in the suction piping. I am not sure that it is possible to present an exhaustive treatment of this problem, because so many different causes may contribute to what appears to be the same effect.

Nevertheless, I may be in the position to suggest certain areas of exploration that should in many cases serve to pinpoint the problem. I have in mind two specific instances in which I was personally involved and that I can therefore document quite thoroughly. I refer to these two instances as cases A and B.

Case A: Suction Line Vibrations

The installation consisted of three boiler feed pumps, each designed to handle 400,000 lb/hr of 312 °F feedwater against a discharge pressure of 2200 psig. The pumps take their suction from a deaerating heater. They are driven by electric motors through hydraulic couplings at variable speeds. Two pumps feed the boiler, the third remaining on standby service. The three pumps are exact duplicates of three other pumps installed earlier to serve a duplicate main unit in the same station.

Shortly before being placed in service, the individual pumps were all tested with the recirculation line open and no flow to the boiler. Severe vibration was experienced in the piping, at all discharge pressures from 500 to 2000 psi, the amount of vibration increasing with the pressure. It was noticed that the vibration ceased when the line to the boiler was opened, even if flow to the boiler was just nominal. The vibration was most noticeable in the suction piping, and although some discharge piping vibration did occur, it was eliminated after pipe hangers were relocated and secured. The vibration was accompanied by a rather rapid fluctuation of suction pressure, which varied from 40 to 60 psig.

The bypass orifices in the recirculation lines had been rated at 75,000 lb/hr. The first conclusions reached in the field were that the pumps had an unstable head-capacity curve, although no evidence of such an instability had been shown during the shop tests of the pumps. It was therefore decided that increasing the minimum flow through the pumps might eliminate the cause of vibration and pressure fluctuation. Accordingly, a 1/2 in. bypass line was installed around the orifice in order to permit increasing the minimum flow. The valve in this 1/2 in. line was cracked open, passing something of the order of 30 or 50 gpm in addition to the flow through the orifice. Immediately, the vibrations disappeared.

This experiment was considered to be sound circumstantial evidence of the instability of the head-capacity curve. The original orifices were replaced by new ones rated at 250 GPM, in other words of slightly more capacity than the sum of the original orifice rating plus the flow through the 1/2 in. bypass line. To everyone's surprise, the vibrations returned. They again disappeared as soon as the valve in the 1/2 in. bypass line was cracked open.

This definitely disproved the theory of an unstable curve and indicated that some sort of a "resonance" problem existed. Such a conclusion was reinforced when one recollects that with the original 75,000 lb/hr orifices vibration ceased the moment the line to the boiler was cracked open, changing the configuration of the hydraulic circuit.

In addition, it became noted that the vibration would occur only if the pumps had been idle or even drained out prior to a run. If the pumps were operated for 12 to 24 hr on normal operation, that is, feeding the boiler, return to minimum flow conditions did not cause vibration or fluctuation of pressure. The theory was developed that the piping permitted the trapping of some quantity of air that acted as a resonant spring and led to a pulsating condition. After operating for a number of hours, the air would be washed out of the lines and the condition disappeared. Circumstantial evidence indicates the probability of the theory: the original size orifices were reinstalled and the vibration never recurred after the unit was placed in regular service. The pumps have been operating for close to 3 years, and even with no flow to the boiler and all flow limited to the bypass recirculation, there is no sign of distress.

Case B: Suction Line Vibration and Apparent Instability of Operation

Two full-capacity turbine-driven pumps are involved in this installation, either pump delivering full flow to the boiler and the other pump remaining on standby duty. Each pump is designed to handle 1,600,000 lb/hr of 270 °F feedwater against a rated discharge pressure of 2720 psi. Feedwater control is maintained by varying the steam turbine speed.

Two separate and distinct conditions of vibration or instability were observed during the initial stages of operation. The first involved violent piping vibration and pressure fluctuations when the pumps were operating with a closed discharge valve or against a closed check valve, and the flow through the pump was limited to the recirculation through the bypass. It was observed that when such vibrations took place, it sufficed to crack the valve to the boiler just slightly to stop the vibration. These vibrations *did not* recur after the pumps went into regular operation.

The second involved equally or even more serious vibrations that took place whenever the second pump was brought on the line and the pump that was running was backed off manually prior to shutdown. If for instance one pump was handling 1,200,000 lb/hr and the second pump was started and brought on the line to operate in parallel, the two pumps split the load at 600,000 lb/hr each and ran quite smoothly. If at this point one of the pumps was put on automatic control and the second one backed off manually, violent vibrations started when the second pump was reduced to about 400,000 lb/hr. These vibrations were so dangerous that the operators were forced to bring the pump to rest as rapidly as possible. On the other hand, if the bypass on this second pump was opened manually when the flow was down to 500,000 lb/hr, the pump could be brought down slowly without untoward difficulties.

These circumstances were the subject of extensive analysis and discussions that resolved the problem in a manner sufficiently clear to be conclusive.

The clue to the solution lay in the fact that the two types of vibration were apparently caused by completely contrary conditions. In the first case vibration was stopped when the pump delivered even an insignificant amount of water to the boiler in addition to the bypass flow. In the second case, the vibration stopped once the check valve was closed and the only flow was through the bypass. This observation and the similarity of this case with that of case A led us to the conclusion that the first vibrations were due to the presence of air in the system. Conclusive additional proof was made available from the fact that air vents were subsequently installed ahead of the individual pump check valves and this type of vibration disappeared completely.

The cause of the second vibration was explained very readily. Although the pumps have a steady rising characteristic curve expressed in feet head, the heating up that takes place within the pump changes the feedwater temperature (and hence its specific gravity) during the passage through the pump. The effect varies with pump flow, of course being more pronounced as capacity and efficiency are reduced.

This effect is illustrated graphically by the curves in Figs. 6.19 and 6.20. The first shows the rise in specific volume (or decrease in specific gravity) corresponding to the reduction in efficiency with reduced flow. The second illustrates the effect of this reduction in specific gravity on the net pressure generated by the pump and expressed in psi. It will be noted that despite the fact that the *total head in feet* is not unstable, the *pressure in psi* developed by the pump begins to drop off at flows below 800 or 750 GPM. This makes parallel operation at these flows impossible.

This effect, of course, is entirely independent of pump design and occurs with all centrifugal boiler feed pumps. Figure 6.20 shows the type of *total head curve* that would be necessary to develop in order to have a steady rising *net pressure curve*. This, of course, is not possible with a centrifugal pump, and parallel operation in the range where the specific gravity effect causes such a reduction in net pressure is not practical.

If the bypass is opened before the flow drops below 500,000 lb/hr, the check valve closes as soon as the pump that is being slowed down develops less pressure than the pump on automatic control and no longer reopens. The pump is then capable of being brought to rest at leisure and without vibration. It was agreed that whenever the pumps have to be switched, operators would open the bypass on the pump being brought down at 500,000 lb/hr. No further trouble has taken place since.

The conclusions reached were that the pump performance was not and had not been responsible for either the vibrations or the flow fluctuations, and the pumps were given a clean bill of health.

General Comments

I would suggest that you conduct a study to determine whether the type of vibration that has been reported to you has any similarity to the two cases I have outlined.

As to the question of flashing under conditions of sudden load drops, that is an entirely separate matter and one that—today—should present no particular mystery. I have conducted extensive studies on this matter and I believe that there is no longer any reason that a group of boiler feed pumps, the deaerator, and the suction piping configuration should not be so selected and so arranged that flashing difficulties are eliminated under any

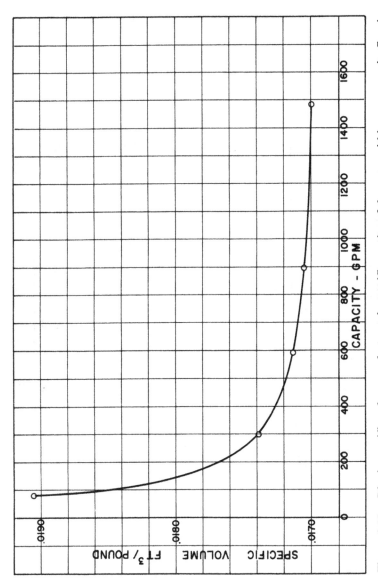

Figure 6.19. Rise in specific volume or decrease in specific gravity of the water within a pump as its flow is reduced and efficiency dropped. Suction conditions: 267F, 175 PSIA, 7000 RPM. Computed from measured hydraulic efficiency at 7000 RPM, using mollier diagram.

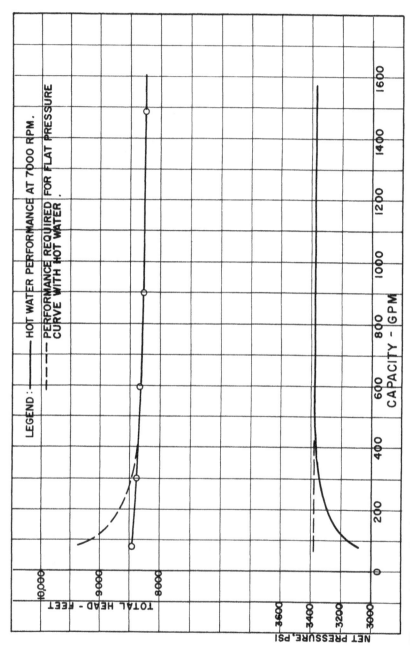

Figure 6.20 Reduction in specific gravity (Fig. 6.19) has its effects upon pressure generated by the pump.

transient condition that can be expected to occur. A discussion of this subject appears in my answers to Question 6.17.

Question 6.21 Effect of Transients in Closed Feedwater Cycles

In many of your Clinics you have discussed the effects of sudden load drop on boiler feed pumps that take their suction from a deaerating heater. What sort of problems are encountered by the boiler feed pumps in a closed feedwater cycle?

Answer. Transient operating conditions in a closed feedwater cycle give rise to entirely different problems than those encountered in an open cycle. They are equally important and must be thoroughly analyzed, in order to avoid the possibility of serious difficulties that could endanger the operation of the entire steam power plant.

Fundamentally, these problems can be classified into two separate groups: (1) the effect of load changes on the hydraulic performance of the boiler feed pumps and (2) the effect of failures of condensate pumps.

Effect of Load Changes on Hydraulic Performance

Unlike the case of an open cycle installation where the boiler feed pumps take their suction from a deaerating heater, in a closed cycle a sudden reduction in main turbine load does not introduce any hazard to the continued operation of the feed pumps since this load reduction is not accompanied by a reduction in available NPSH. Instead, as the turbine load is reduced, the pressure at the boiler feed pump suction rises, but the temperature and the corresponding vapor pressure are reduced after the residence time is elapsed. This is because the pressure generated by the condensate or booster pump increases with a reduction in flow, but the friction losses between the two pumps operating in series are reduced. Therefore, it can be stated that no hydraulic problems arise with a reduction in load.

Likewise, a sudden increase in load should have no ill effects on the hydraulic performance of the boiler feed pumps, assuming that the characteristics of the pumps preceding them in the cycle have been selected with proper consideration in connection with this type of cycle and that the relative effects of acceleration in the boiler feed and condensate pump discharge, respectively, have been considered. In other words, when so-called half-capacity boiler feed pumps are used, it is necessary to analyze the effects of a sudden load increase while a single pump is on the line. Under such conditions the head-capacity of the single pump will intersect the

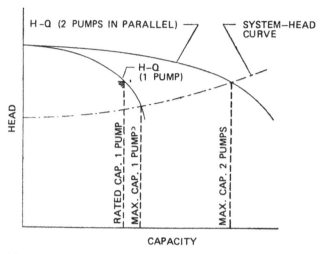

Figure 6.21 Analysis of maximum flows when half-capacity pumps are used.

system-head curve at a flow greater than the design capacity of the pump (see Fig. 6.21). It becomes necessary to ascertain that at such a flow, the pressure supplied by the condensate or booster pump exceeds the prevailing vapor pressure by an adequate margin, so that the available NPSH remains sufficient at all times.

Failure of Condensate Pumps

In a closed feedwater cycle, condensate or booster pumps discharge directly into the suction of the feed pumps, through a series of closed heaters. The discharge pressure of these pumps is selected in such a manner that the suction pressure at the boiler feed pumps exceeds the maximum vapor pressure that may occur at the point of the cycle by a comfortable margin, well beyond the minimum NPSH required by the boiler feed pumps.

Failure of the condensate pumps can obviously occur independently of the conditions prevailing at the boiler feed pumps. Not only can the condensate pumps themselves fail, but their drivers or the power supply to these drivers can fail independently of the power supply to the boiler feed pump drivers. This last condition is obvious when it is considered that generally, when both drivers are electric motors, they will operate on different voltages because of a wide difference in motor size.

In the case of an open cycle, with a direct-contact heater at the suction of the feed pumps, failure of the condensate pumps presents no urgent or immediate problems, as the direct-contact heater contains an adequate storage that will normally permit the boiler feed pumps to continue operating for anywhere from 5 to 15 min or more (depending upon the operating load). This is sufficient time to either reestablish condensate pump operation or to bring the boiler and the unit down without incurring undue hazards and jeopardizing the safety of the equipment.

There is no such safety measure in the case of the closed cycle, and since complete failure of feedwater supply at the suction of the boiler feed pumps cannot be tolerated, some protective means must be provided to avoid such a condition.

In the early days of closed feed cycles, the operating temperatures of the feed pumps in most stations were relatively moderate. Consequently, the minimum permissible suction pressures were likewise moderate. For instance, if the feed pumps handled feedwater at 240°F (vapor pressure = 10.2 psig) and if the minimum NPSH were 25 ft (10.2 psi) the suction pressure could be permitted to fall to 20.4 psig before any harm befell the feed pumps. It was possible then to install a surge tank floating on the suction line and containing a reasonable volume of cold water storage, ready to supply water to the feed pump on failure of condensate supply. The water level could be held at about 50 ft (equivalent to 20.4 psig minimum suction pressure), and a check valve would be installed in the piping from the surge tank to the suction header, as shown in Fig. 6.22.

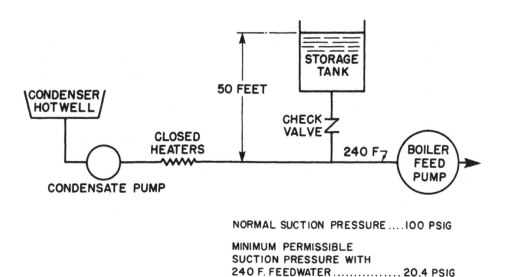

Figure 6.22 Arrangement of surge tank in closed feedwater cycles.

On failure of condensate supply, the suction pressure would drop, until at 20.4 psig the check valve would open and cold water from the surge tank would flow to the boiler feed pump. Although the level in the surge tank would start to fall and the suction pressure would ultimately fall below that required for 240 °F water, the cold water would soon enough reach the boiler feed pump suction and the minimum suction pressure would be reduced by an amount equal to the vapor pressure at 240 °F.

Two separate factors have united to make this solution inapplicable in most cases today. Boiler feed pump capacities have increased so that the minimum NPSH required may be as high as 100, 150, or even 250 ft. But more serious a problem has been introduced by the rapid growth in operating temperatures.

For instance, if the feedwater temperature is 360 °F and the minimum NPSH is 150 ft, the minimum suction pressure becomes 195.8 psig and the water level in the storage tank must be 510 ft above the boiler feed pump centerline, obviously an impractical solution.

Two separate approaches exist to this problem. The first is based on the thought that complete failure of condensate to the boiler feed pump will cause a complete failure of the latter in a matter of seconds and that it is best to cut the losses short. Consequently, failure of suction pressure caused by failure of condensate flow is permitted to stop the boiler feed pumps, letting boiler protective controls take care of the boiler problem.

The condensate pump discharge pressure is then selected to provide some excess over the minimum permissible feed pump suction pressure. If the suction pressure falls below the normal value, but still above the minimum, a pressure switch starts a standby condensate pump in the hope that this will arrest any further decrease in suction pressure. If, however, this maneuver fails to solve the problem, on reduction of the suction pressure to the minimum permissible value, a second pressure switch simply cuts the power supply to the boiler feed pump drives and brings the pumps to rest.

For instance, if the minimum permissible suction pressure is 135 psig, the first pressure switch may be set at 160 psig and the second at 135 psig. The condensate pump pressure is selected to normally provide 180 psig boiler feed pump suction pressure. If for some reason the pressure falls below 180 psig, the standby condensate pump will be started at 160 psig and further decay of the feed pump suction pressure to 135 psig will cut the feed pumps out.

A time delay relay should be incorporated in this second switch to take care of the time lag that will occur in reestablishing the required suction pressure by starting the standby condensate pump. Without such a time delay, failure of the condensate pump could cause the suction pressure to fall rapidly enough to shut down the feed pumps before the standby equipment will have had time to restore pressure. The selection of the time delay that can be allowed is a difficult task, as it is impossible to predict a safe duration for flashing or partial flashing at the feed pump suction. It is considered, however, that something of the order of 5 sec should be sufficient to restore pressure and not excessively long from the point of view of safety under flashing conditions. An automatic reset feature should be incorporated in the pressure switch that causes stopping of the main feed pumps, so that immediate reestablishment of sufficient suction pressure would restart the feed pumps. A selective switch would be used to manually eliminate the automatic reset feature, if desired. However, no such automatic reset should be used with the pressure switch controlling the start of the standby condensate pump.

The second solution is based on the principle that, no matter what happens, it is unthinkable to shut down the boiler feed pumps, on the theory that a flashing condition at the pump suction for a very limited period of time will not necessarily be fatal to the pumps. A cold-water surge tank is located on the suction side of the boiler feed pumps as in Fig. 6.23, with the level of the storage held at the maximum elevation consistent with sound economics. Obviously, this elevation is insufficient to prevent flashing of high-temperature feedwater, should the condensate pumps fail to deliver. The reduction of pressure at the suction will continue until such time that the static pressure of the cold storage water is sufficient to open the check valve and cold water will flow to the pump suction. It is hoped that this reduction in pressure will take place at a sufficiently rapid rate so that flashing and cavitating conditions will prevail at the boiler feed pumps for but a few instants and that the pumps will clear themselves of vapor and start delivering to the boiler again.

The idea is extremely tempting, because it requires no special controls, but I must admit that I have no direct knowledge of such an installation and therefore cannot comment conclusively on the feasibility of the arrangement. Nor am I in the position to submit any valid method of calculating the time element involved. It is doubtful that any steam power plant would be willing to permit carrying out the experiment of willingly flashing an expensive boiler feed pump for the purpose of establishing the permissible duration of such flashing conditions before the boiler feed pump will seize.

A number of modifications of the surge tank arrangement have been considered at some time or another. In all cases, they contemplate the provision of the necessary pressure in a closed surge tank to replace the required—but unobtainable—static elevation. This "supercharging" of the surge tank would require the use of either steam or of an inert gas, such as nitrogen. The latter solution would be cumbersome and has never found favor with power plant designers.

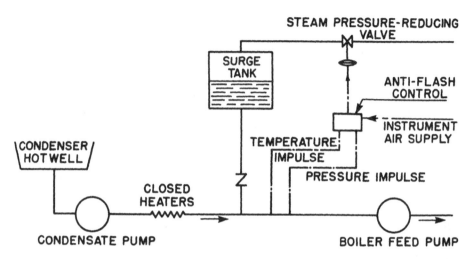

Figure 6.23 Application of "antiflash" control mechanism to the surge tank used as protection in closed feedwater cycles.

Steam blanketing the surge tank to maintain the necessary pressure has a greater appeal. With such an arrangement it might be desirable to provide controls so that the required pressure be made available only when the emergency arises, in order to minimize heat losses. A pressure-reducing valve would be used in a line bringing live steam to the surge tank, two different impulses being available for the control of this valve. In one case, a pressure switch, similar to the switches used to start a standby condensate pump or to shut down the feed pumps, can be incorporated into the system to control the opening and closing of the valve in the steam supply line.

If greater refinement is desired, the valve may be subjected to the impulse from an antiflash control mechanism, as shown in Fig. 6.14. This mechanism measures the difference between the suction pressure and the vapor pressure at the pump suction (hence, the available NPSH) and weighs it constantly against the required NPSH. In case of the reduction of the available NPSH to a value that endangers the operation of the pump, an impulse is generated that, in this particular case, would operate the steam pressure-reducing valve. In order to prevent "hunting," the mechanism should be provided with a "reset" feature, permitting the valve to open when necessary but requiring manual reclosing after normal conditions have become reestablished.

Question 6.22 Effect of Reverse Flow

Will any damage occur to a centrifugal pump if the check valve fails to close when the pump is stopped and the pump runs in the reverse rotation? What is the runaway speed of a centrifugal pump in such a case?

Answer. This is a rather general question, and the answer to it must be qualified, since the results will depend on a number of factors. To begin with, the results will depend on the value of this runaway speed. The latter varies from as high as 175% of rated speed for low-head, large-capacity, high specific speed pumps, such as propeller-type pumps, to less than rated speed for high-head, low specific speed types, such as multistage boiler feed pumps.

If the runaway speed exceeds the speed for which the pump has been designed, mechanical damage may easily occur. Even with reasonable values of reverse rotation speeds, some damage is possible.

Some pumps are provided with sleeve and thrust bearings forced-feed-lubricated by means of an oil pump driven directly from the main pump shaft. When this pump operates in reverse rotation, the oil pump can no longer supply lubrication to the bearings. Instead, it will return to the oil sump all the oil in the discharge pipe adjacent to the oil pump casing and that in its own casing in a few revolutions of the main pump. Although the oil pump must be capable of operating in a semidry condition for sufficient time to prime itself under normal conditions, operation in reverse direction will generally upset the normal thrust balance of the oil pump rotors and contribute to the possibility of heating, galling, and seizing the oil pump.

Besides the damage to the oil pump itself, there is the probability that the main pump will suffer. Obviously, if the rotation is reversed and no lubrication is available, the pump bearings will be damaged unless a separately driven auxiliary oil pump is available to start automatically on failure of oil pressure. Failure at the pump bearings will in itself lead to further damage, as the shaft center of rotation will drop, permitting contact to take place at the close internal running clearances. This can lead to a complete destruction of the pump parts.

Reverse rotation may likewise be harmful to the pump driver in many cases, and careful investigation of this problem is necessary.

If the effect of reverse rotation is found to be hazardous for a particular installation, it is frequently possible to introduce means to prevent it. For instance, with electrically driven centrifugal pumps, a motorized gate valve can be incorporated into the discharge piping, so arranged that the interruption of current to the pump motor starts the gate valve to close. Even though the gate valve may take some time to close fully, the effects of reverse rotation will be minimized, since the head available to cause it is much reduced below the design operating head. After all the net head causing reverse rotation will be equal to the static pressure in the discharge header *less* all friction losses between the header and the pump, and even partial closure of the gate valve will cause appreciable friction loss.

If even this eventuality is to be eliminated, a time-delay relay can be incorporated into the electric circuit so that the electric motor driving the pump is not disconnected until the gate valve has closed fully. Of course, the system must be provided with bypass protection so that the pump is not made to operate against complete shutoff.

If it is the normal practice to hold the pump ready to restart after it has been secured, by keeping the discharge valve open, the motorized gate valve can be made to reopen after the operator has assured himself that the check valve is securely sitting in its seat and closed against reverse flow.

Question 6.23 Reverse Rotation of Vertical Pumps

We have taken bids on several small vertical centrifugal pumps. Examination of the proposals we have received indicates that all the pumps include several screw-type couplings in the shafting. There seems to be no provision for locking these screw-type couplings on the pump shaft. Our installation is such that it is possible to have reverse flow through these pumps and therefore reverse rotation. Is it not necessary to lock these couplings in some manner to avoid unscrewing the couplings?

Answer. No, such a provision is actually not necessary. The hand of thread used in these couplings is such that, during normal operation, the torque exerted by the motor tends to tighten the threads in the couplings (Fig. 6.24). If the motor is stopped and no check valve is available in the pump discharge line, flow through the pump will be reversed, assuming that one or more pumps operating in parallel with the pump in question are kept on the line or that the pump was discharging into a tank or reservoir.

Under these conditions, the pump is transformed into a turbine and becomes the driver, running in reverse rotation to its normal rotation. It produces torque to drive itself and the motor. Now the torque developed by the pump acts in the same direction on the threads as when the motor was driving the pump. Therefore, no tendency exists to unscrew the coupling threads and no special device to lock the couplings on the shaft is required.

On the other hand, if the motor of a vertical shaft pump is not wired correctly, the pump will be started in the wrong direction. In this case, the torque exerted by the motor will act to unscrew the coupling from the shaft, and serious damage may occur. For this reason, motor-disengaging clutches

Chapter 6

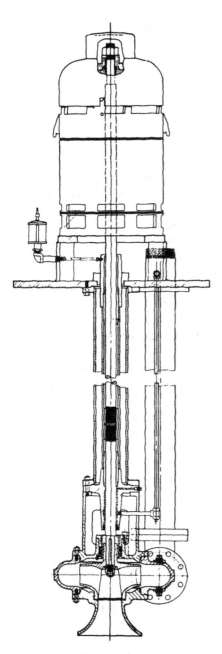

Figure 6.24 Vertical pump with screwed coupling.

are used in some of the hollow-shaft motors intended to drive vertical pumps. At the same time, there is no reason a user would fail to try out motor rotation during the initial installation before the motor has been connected to the pump.

Question 6.24   Zero-Speed Indicator

Do you know of any *zero-speed indicator* that could be applied to boiler feed pumps and that might be tied in with automation so as to prevent trouble caused by check valve failure and the subsequent running of the pumps backward? We experienced such an accident recently with considerable damage to the pumps.

The idea would be to use a zero-speed indicator and tie it in somehow with an automatic valve, so that backflow through the pump would be prevented when the pump is at rest.

Answer. A zero-speed indicator is a readily available gadget. We have furnished electric speed indicators of at least three different makes with our boiler feed pumps. Any one of these can probably be provided with a contact showing that the speed is zero or even that it has fallen to some 30 or 50% of rated and therefore that the pump is coming to rest.

Of course, the problem now is what to do with this impulse. If the discharge gate valve is motorized, the impulse could be applied to energize this valve and close it. But such a valve may take from 20 to 40 sec to come to a fully closed position. It is possible, therefore, that with the pump motor deenergized and the check valve stuck open, the pump will have come to a full stop in a matter of some 5 sec under the turbining action of the reverse flow and will have reversed itself before the motorized valve has closed entirely. Still, the protection afforded to the installation by this arrangement is a bit better than nothing.

Actually, I am much in favor of motorized gate valves in the boiler feed pump discharges. If one is installed, I recommend that it be closed before the motor is tripped out. If an automatic arrangement is desired, it is very simple to install a time-delay relay of a duration just in excess of the motorized valve closure time. Then, when it is desired to bring a pump to rest, the trip switch starts closing the discharge valve and, after it is closed, deenergizes the motor.

Of course, this gives no protection against an accidental motor trip out. But if pump speed indication is also made to provide an impulse to start closing the gate valve, a definite improvement in protection will be obtained.

Bear in mind that proper relay sequence is neccesary for the reverse operation, that is, when the pump is started up. Either the motorized valve must be capable of being opened at reduced speed when the pump is coming up from rest, or the valve must be left closed until the pump comes up to full speed. Remember that if the pump is properly protected by an automatic recirculation bypass, there is no danger in operating it against a closed gate valve.

Question 6.25 Use of Two Check Valves in Series

Instead of using zero-speed indicators and switches to prevent damage in the event of a check valve failure, would it not be simpler to use two check valves in series?

Answer. Indeed, in most cases, a second check valve would be a simpler solution than a zero-speed indicator that would provide an impulse to actuate a motorized gate valve in the pump discharge. I know of several installations in the United States where two check valves in series have been used. In all such cases, one is a piston check valve and the second one a tilting disk valve.

The major objection to the use of a motorized gate valve actuated by a zero-speed indicator is that the impulse is only given when the pump has come to rest and that another 20 to 40 sec will elapse before the gate valve is fully closed. In the meantime, the pump will have gone into reverse rotation, and some damage may take place. Thus, the protection afforded by a zero-speed actuated gate valve is not fully effective, and that is why some design engineers have preferred to go the route of two check valves in series.

There are, however, several possible variations to the motorized gate valve solution that can give reasonable protection. To begin with, a motorized gate valve should always be closed before the driver is deenergized. A time-delay provision is incorporated to assure that the valve is in a fully closed position before the driver is off the line. Of course, this gives no protection against an accidental shutdown of the driver.

Experience has indicated that piston check valves are more prone to sticking than tilting disk check valves. One way to improve reliabilty would be to use a tilting disk check and to replace the gate valve by a flow control valve that can also act as a check valve by having air pressure transmitted to it by a pilot control when the driver is shut down.

All these observations are a good example of the fact that there are always risks involved in the solution of technological problems. The success of engineering solutions is judged by the degree by which these risks are minimized.

Question 6.26 Effect of Reverse Rotation on Boiler Feed Pumps

In Question 6.24 and 6.25 you discussed a zero-speed indicator that one of your readers thought could be applied to boiler feed pumps for the purpose of providing an impulse to actuate a protective device, which, in turn, would prevent reverse rotation. Could you amplify somewhat on your discussion of the problems that arise from sticking check valves and the resultant reverse rotation of the boiler feed pumps?

Answer. Most boiler feed pump installations have two or more boiler feed pumps, of which one or more pumps serve to feed the boiler and one remains idle on standby duty, ready to go on the line whenever necessary. Even when an installation is served by two half-capacity pumps with no standby, there may be times when a single pump is kept running during light conditions so as to conserve power. All pumps are therefore provided with a check valve in the discharge line, and both suction and discharge gate valves are kept open whether the pump is running or not. These gate valves are closed only when it becomes necessary to isolate a pump for inspection or maintenance. The check valve is located between the pump and the discharge gate valve and remains closed as long as the pump is idle, by virtue of the pressure in the discharge header. This prevents reverse flow through the idle pump and reverse rotation.

The use of this check valve permits starting an idle pump by remote control merely by energizing the driver. As soon as the pump reaches a speed at which it develops a pressure equal to the header pressure, the check valve will open, and the pump will start delivering feedwater into the system. Conversely, when the pump is to be shut down, the power supply is cut off. As soon as the pump slows down sufficiently, the check valve closes, preventing reverse flow through the decelerating feed pump.

Unfortunately, the operation of a check valve is not necessarily foolproof. Every once in a while, the check valve will fail to close either because of a mechanical defect or because foreign matter has lodged itself in a critical spot and interferes with the proper operation of the valve. Since the pressure in the header is the discharge pressure of the pump or pumps remaining on the line, an appreciable quantity of feedwater will flow backward through the idle pump and cause it to operate in reverse rotation as a turbine. The reverse speed will depend on the general design conditions of the pump and on the relative opening of the check valve, but it may reach as much as 125% of the rated pump speed. Under these conditions, the damage that can occur to the boiler feed pump and its driver may be very extensive. And, unfortunately, there is no practical mechanical means available to prevent this reverse rotation. Suggestions have been made at various times to incorporate some sort of a brake into the boiler feed pump and driver train. Such a solution would be very cumbersome, expensive, and possibly even less reliable than the check valve whose lack of reliability it would be intended to offset.

Effects of Reverse Rotation

The possible damage to the driver, be it electric motor or steam turbine, could be quite severe. However, this phase of the problem must be examined separately and will not be discussed here. The pump itself should not necessarily suffer from reverse rotation, since even 125% speed is not excessive, especially when it is considered that the pump absorbs rather than develops energy.

But a very serious problem is caused by interference with the proper lubrication of the unit. High-pressure boiler feed pumps are generally provided with sleeve line bearings and a Kingsbury-type thrust bearing, all pressure lubricated. Oil is delivered to the bearings by a gear-type positive displacement service pump, directly driven from the boiler feed pump (see Fig. 4.13). When the pump runs backward, no lubricating oil can be delivered by this service pump. Lack of lubrication will lead to burned-out bearings and then to *indirect* damage to the boiler feed pump itself. Failure to maintain proper alignment of close clearance rotor parts will cause pump seizure and very serious damage. When an integrated oil lubrication system also provides oil to motor bearings, the motor may likewise suffer from lack of lubrication.

In addition, the service oil pump itself may be damaged severely. When the pump operates in reverse rotation, the oil pump will return to the oil sump all the oil in the discharge pipe adjacent to the oil pump casing and that in the discharge casing in a few revolutions of the main pump. The check valve in the oil discharge line will prevent further backflow from the oil system, but air vented in through the passage to the oil pump coupling will help clear the pump of oil, particularly at what is normally the discharge end of the oil pump. An oil pump of this type must be capable of operating in a semidry condition for sufficient time to prime itself under normal conditions, but operation in reverse direction will upset the normal thrust balance of the oil pump rotors. Because of the close clearances necessary in a gear-type oil pump, this unbalance will contribute to the heating, galling, and seizing of the oil pump rotors.

There are two possible approaches to the prevention of damage of high-pressure boiler feed pumps from check valve failure. Positive steps can be taken to prevent reverse rotation by using motor-operated valves or other valve arrangements. Alternatively, adequate lubrication can be provided to the pump and driver bearings by including an auxiliary oil pump.

Solutions Involving Proper Choice of Valves

First, the discharge gate valve can be designed as both a check and gate valve built into one. This combination valve can be powered so that when the main boiler feed pump driver is deenergized, it closes at some predetermined rate and seats tightly.

A simpler solution would be to use a tilting disk check valve, fitted with a dashpot to prevent slamming. Such a valve is probably more reliable and more foolproof against hanging up than a piston-type valve. The valve element in a piston-type check valve fits into a recess in the body when the valve is open. An accumulation of solids may restrict the clearances between the valve body and the valve element itself. Under these conditions, the valve element may hang up and fail to give the desired leakproof closure on shutdown.

The third and probably best solution consists in motorizing the discharge gate valve while leaving the check valve as normally installed. This motorized valve can be closed by means of an interlock whenever the boiler feed pump driver is deenergized. At the operator's leisure, the gate valve can be reopened manually so that the boiler feed pump is again ready for a remote-control start. Before reopening the gate valve, the operator would assure her- or himself that the check valve is properly seated.

Use of an Auxiliary Oil Pump

To repeat, a direct-driven, positive displacement oil pump will not perform its intended function when operating in a reverse direction. In modern installations, a motor-driven auxiliary oil pump is generally included to provide lubrication at start-up and to act as a standby if the direct-driven service oil pump fails to deliver oil for any reason (see Fig. 4.14). (Some of the older installations may not have such an auxiliary oil pump.) This motor-driven auxiliary oil pump is usually interlocked with the main boiler feed pump motor. When the boiler feed pump is started, the auxiliary oil pump comes on the line immediately before the main motor can start up. A pressure switch energizes the main motor only after a predetermined oil pressure is built up in the lubrication system. The time delay in this operation is just a few seconds. When the direct-driven service oil pump comes up to speed with the boiler feed pump, it develops a higher pressure than that developed by the auxiliary oil pump. This higher pressure shuts down the motor-driven auxiliary oil pump by actuating a pressure-operated switch.

In the event of service oil pump failure, pressure in the lubrication system drops to a point at which another pressure switch restarts the auxiliary oil pump. Failure to maintain operating oil pressure with the auxiliary oil pump (as, for instance, in the case of an oil line break) energizes an electric interlock that stops the main boiler feed pump motor and actuates some form of an alarm.

This particular interlock does not provide protection under reverse rotation conditions, since the boiler feed pump would be deenergized. A specific electric interlock can, however, be incorporated to protect the system even against reverse rotation.

The operating sequence would be as follows:

1. Start auxiliary oil pump before main drive is energized.
2. Stop auxiliary oil pump when the regular service oil pump develops sufficient pressure.
3. Restart auxiliary oil pump when the regular service oil pump fails to maintain sufficient pressure.
4. Restart auxiliary oil pump when the main driver is deenergized. The auxiliary oil pump would then be stopped manually after check valve closure is assured.

Note that the final step in this sequence is the important one in the event of reverse rotation.

Question 6.27   Frozen Leak-off Line

I want to depart this once from my accustomed question-and-answer style to relate a series of events that transpired once upon a time in a steam-electric power plant, events that for awhile baffled the combined analytic efforts of the plant personnel and of the manufacturer's engineers. This account should be of interest to utility engineers and station operators and may possibly alert them to potential operating troubles with their equipment. I shall describe the events essentially in the chronological order in which they took place. And although all those involved did learn the cause of the problem we had encountered, I shall disclose the solution at the exact chronological moment that it became evident.

The power plant is question had two 300 MW units, each served by one main shaft-driven* boiler feed pump and by a 35% capacity 4500 hp

---

*The practice of driving boiler feed pumps directly from the main turbogenerator has now disappeared in favor of separate steam turbine drives (see Question 1.53).

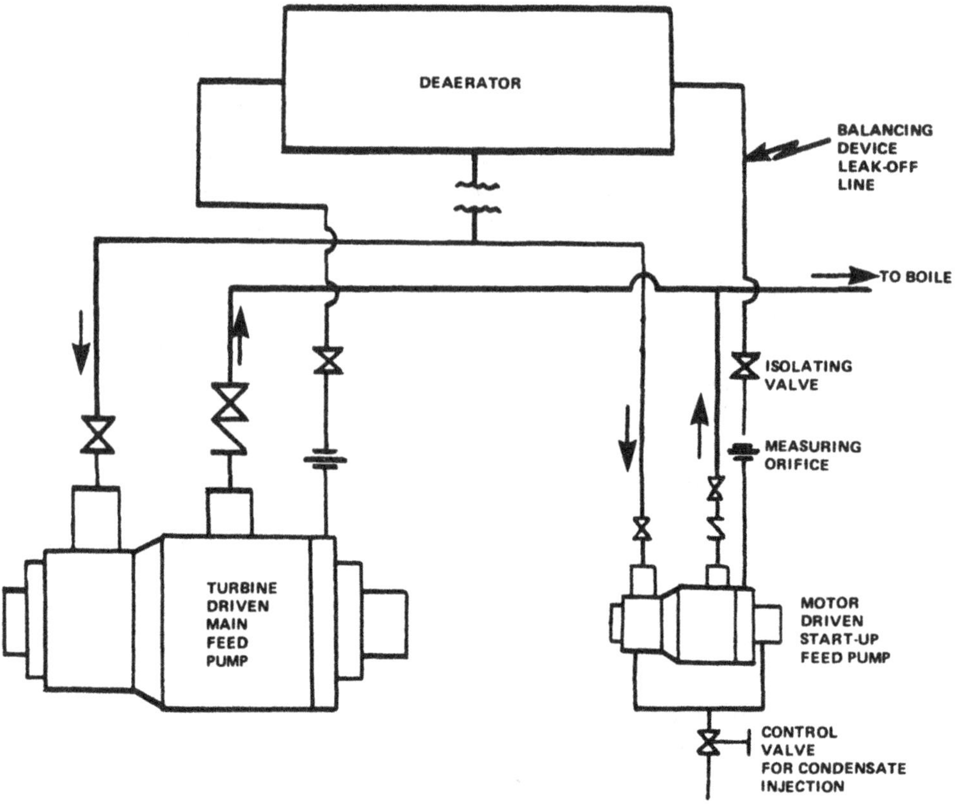

Figure 6.25 Schematic arrangement of boiler feed pumps.

motor-driven start-up pump. Figure 2.21 shows the cross section of the start-up pump and the schematic arrangement of the installation, and relevant piping is shown in Fig. 6.25.

Unit 1 of that steam-electric station had been in operation for some 5 years, and no problem had arisen in that period of time. Several weeks before the incident I am about to describe, the unit was shut down because of some malfunction of the fuel supply. When it came time to restart the unit, it became necessary to place the motor-driven start-up pump on the line so that the boiler could provide steam to roll the main steam turbine and thus energize the main boiler feed pump. It was known that the start-up pump had last been operated successfully some 5 to 6 months prior to these events.

Immediately on starting up, it became evident that the pump was in serious distress. The rotor thrusted over toward the inboard side by as much

Chapter 6

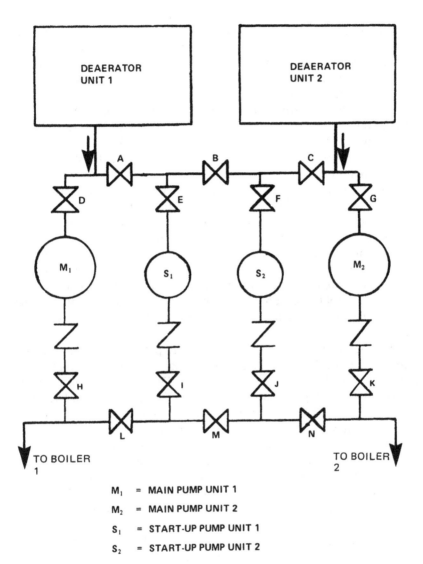

Figure 6.26 Interconnection of start-up boiler feed pumps of two adjacent units.

as 40 mils, destroying the inboard thrust shoes in the Kingsbury bearing, the pump was tripped, and it seized on coastdown.

The dismantling of the pump to examine the damage was quite difficult because one of the areas of seizure was at the condensate injection sealing area where the rotating and stationary parts had become welded to each other. But the inner assembly was finally removed from the casing barrel and sent to the factory for examination. The spare inner element was placed into the pump, and new thrust bearing components were used. Prior to restarting the pump with the new inner element the rotor was turned over by hand. It turned freely; the pump alignment was rechecked and the pump was coupled to the motor, ready to start up.

At about this time word came to the power plant from the factory that the damaged inner element had been dismantled. There was evidence of considerable damage: the balancing disk and stationary disk head had become welded together; one twin volute bushing had broken; the thrust collar was deeply grooved on the inboard side.

When the start-up pump, fitted with the spare inner element was ready, the pushbutton of the starter was pressed and the pump was coming up to speed. But even before reaching full speed there was evidence of rubbing at the seal covers and the operators tripped the pump out. As the pump was coasting down, it seized at what was estimated to have been 300 to 600 RPM.

As soon as word of this second failure reached the factory, an engineer and a technician were dispatched to the scene. It was imperative to find the solution to this problem as soon as possible, since the main unit could not be put on the line until the motor-driven start-up pump was placed successfully in service.

At the same time a suggestion was made to provide an interconnection between units 1 and 2 that would permit the motor-driven start-up pump of any unit to be used in starting up the other unit. This interconection had to be analyzed rather thoroughly, and I shall speak of it in some detail later on.

Fortunately no serious damage was caused to the pump on this second occasion, and it was quickly made ready to start up again. Imagine the discomfiture of the steam power plant and factory personnel when, on this third restart, trouble arose again. This time, the pump did not seize but the diaphragm of the condensate injection control valve ruptured. The pump was tripped out before any damage had occured. This rupturing of the diaphragm proved to be a "godsend," because it saved the pump from a repetition of the first two failures and became the clue to the solution of the problem.

The deaerating heaters of units 1 and 2 from which the main and start-up pumps take their suction are located on the roof of the power plant. The balancing device leak-off lines lead to their respective heaters and pass through the roof and outside before entering the deaerating heaters. The events of which I am speaking took place during an extreme cold spell in the winter of 1971.

The balancing device leak-off line from the start-up pump on unit 1 had frozen solid! When the pump was started up, there was no possibility of flow taking place through it. The condition was equivalent to having closed a valve in that line. As the pump ran up to speed, the pressure in the balancing device leak-off chamber reached almost full discharge pressure instead of being just a few pounds above the deaerating pressure. The only flow past the balancing device was only the small flow that could take place between the condensate injection seal and the shaft sleeve. The rotor became "massively" unbalanced, thrusted over toward the suction, contacted the thrust shoes and burned them out, and contacted at the balancing device and chewed it up. The axial imbalance and the resulting contacts led to radial imbalance before the pump ground to a halt, and of course, contact took place at the wearing rings and the twin-volute bushings, damaging the latter.

"Eureka!" The mystery was solved, but not before having imposed serious delays in the restart of unit 1 and considerable damage to the motor-driven pump. When on the third restart the diaphragm of the control valve ruptured, a path was opened for the balancing device chamber to relieve itself somewhat and thus prevented a third seizure and failure.

To eliminate any further recurrence of this situation, the balancing device leak-off line to the deaerator was cut off and a connection was provided instead right back to the pump suction, with no valve in this short return line. In this manner it became absolutely impossible to impose any excess pressure on the balancing chamber.

To diverge for a moment, I might mention that, in the past, balancing leak-off lines were returned to the deaerator in the majority of cases. The reasoning for this practice is discussed in my answer to Question 3.26. Present practice is to return the balancing leak-off directly to the pump suction and to resist efforts to route it to the deaerator.

Returning to our problem, unit 1 was successfully placed into service with no further incidents.

Let us now come back to the question of interconnecting the motor-driven start-up pumps for units 1 and 2. Not only does such an interconnection give additional backup for the restart of a main unit, but it can permit running both motor-driven pumps if for any reason a main feed pump is out of service. Under such conditions, a unit could probably carry up to 85 or 90% of the load with the two motor-driven pumps running.

However, some precautions are necessary if such an interconnection is used. (See my answer to Question 3.30). The interconnection should include provision for letting each start-up pump take its suction from either deaerator and discharge to either boiler by proper manipulation of certain valves. The arrangement might be as shown schematically in Fig. 6.25. For the purpose of simplification only the valves that have a bearing on our problem have been indicated. For the same reason, minimum flow bypass lines, condensate injection lines, and other parts have been ommited from the sketch.

There are two different ways in which the interconnection can be achieved:

1. The start-up pump on unit 2 can be isolated from its own deaerator and take its suction from unit 1. It would discharge into unit 1.
2. The start-up pump can remain hooked into the deaerator of unit 2 but discharge to the unit 1 boiler.

The first arrangement is probably preferable whenever both start-up pumps are to operate in parallel as standby to the main feed pump on either unit. It does, however, present some problems. These are not unsurmountable but must be given thorough attention.

First, we must remember that the two main units are adjacent to each other but may still be physically separated by appreciable distances. It is necessary to check carefully the NPSH that would be available to a pump taking its suction from the adjacent unit. There is yet another problem: although a start-up motor-driven feed pump is not operated frequently, it is generally kept warmed up, ready to go into operation at a moment's notice in its secondary role of standby pump. It is therefore full of hot feedwater at approximately the temperature of the unit in operation. This problem is also discussed in my answer to Question 3.30.

Table 6.5 describes the positioning of the valves shown in Fig. 6.25 for the various purposes described here.

Question 6.28 Diagnosing Pump Troubles by Type of Noise

Can the type of noise made by a centrifugal pump serve as a clue to the source of trouble?

Answer. Pump noise will very frequently give a definite indication as to the source of trouble to an experienced maintenance person. If a pump produces a crackling noise, it is most likely that the source of trouble will be found at the pump suction. This type of noise is generally associated with *cavitation*. In general this term describes conditions that exist in flowing liquid whenever the pressure at any point falls below the vapor pressure of the liquid at the prevailing temperature. Some of the liquid flashes into vapor, and bubbles of the vapor are carried along with the remaining liquid. Whenever this happens in the suction area of a centrifugal pump or within the entrance of the impeller, the bubbles on proceeding farther into the impeller undergo an increase in pressure and recondense. This process is accompanied by a violent collapse of the bubbles, possible pitting and erosion of the impeller vanes, and a definite crackling noise. Of course, the presence of vapor within the liquid pumped causes a reduction in the pump capacity.

Table 6.5 Setting of Valves for Fig. 6.25

Valve	Normal operation	Using S_2 to start unit 1 from deaerator 2	Using S_2 to start unit 1 from deaerator 1	Using S_1 and S_2 in parallel taking suction from deaerator 1, feeding boiler of unit 1
A	Open	Immaterial	Open	Open
B	Closed	Closed	Open	Open
C	Open	Open	Closed	Closed
D	Open	Immaterial	Immaterial	Immaterial
E	Open	Immaterial	Immaterial	Open
F	Open	Open	Open	Open
G	Open	Immaterial	Immaterial	Immaterial
H	Open	Immaterial	Immaterial	Immaterial
I	Open	Immaterial	Immaterial	Open
J	Open	Open	Open	Open
K	Open	Immaterial	Immaterial	Immaterial
L	Open	Open	Open	Open
M	Closed	Open	Open	Open
N	Open	Closed	Closed	Closed

Cavitation, therefore, is a direct result of insufficient pressure at the pump suction, in other words of operation with insufficient NPSH (net positive suction head).

It is possible to check whether the diagnosis is correct. For instance, throttling the pump discharge will reduce pump capacity and possibly restore pump operation back into a range in which sufficient NPSH is available at the pump suction. Should this eliminate the crackling noise, the diagnosis is correct, and of course, steps to eliminate the trouble are self-evident. They will consist of either increasing the NPSH for the normal range of operating capacities or of replacing the existing impeller with one that can operate with the prevailing NPSH if the latter cannot be altered.

A rumbling noise is generally caused by conditions in the discharge waterways of the casing, either because of operation at part-load capacities when the pump is not hydraulically suitable for such operation or because the pump operates at capacities well in excess of those for which it was designed.

Question 6.29 More on Centrifugal Pump Noise

We have occasionally run into the problem of excessive flow noise or pump noise on centrifugal pumps. What suggestions would you make in designing a job or writing a specification to ensure quiet operation?

Answer. The problem of noise from pumps or in a pumping system is a rather complex one. I shall try to give you some background material on it, but I am not certain that you can "write into a specification" the assurance of a quiet installation. For one thing, background noise has a great deal to do with the relative acceptance of equipment noise levels, and yet it is very seldom that you can predict background noise with a sufficient degree of accuracy at the time the installation is projected.

The other difficulty is that although a pump may in some cases be a source of noise, many of the field problems I have investigated in the past were finally found to be based on system-originated noise, such as excessive velocities in the piping, improperly supported piping or equipment, or operation of the pumps at extremely unfavorable flows. Nevertheless, it is possible to present in a brief form certain considerations that you may find helpful in evaluating different means of combating objectionable noise from pumping systems.

One popular misconception is that higher operating speeds are inherently more conducive to noise. This frequently leads to the specification of an 1800 RPM pump when a 3600 RPM pump may well be a more logical application. It should be remembered that it is not the number of revolutions per minute that would be the factor here, since the total head developed by an impeller is a function of its peripheral speed. The latter remains the same whether a given impeller rotates at 3600 RPM or an impeller of twice the diameter runs at 1800 RPM. On the other hand, it is quite possible that a properly applied 3600 RPM pump will be less noisy than an 1800 RPM pump that has been poorly selected for the actual operating conditions of service.

Actually, much of the prejudice against using 3600 RPM on service for which quiet operation is necessary, such as in buildings, stems from experience with old designs of 1800 RPM models "souped up" to 3600 RPM. This, of course, could easily lead to unsatisfactory operation. Thus, any users who have never tried to apply a 3600 RPM pump on this type of service owe it to themselves to try a *modern design* unit, which is specifically designed for this speed and which should give a satisfactory account of itself.

Of course, where noise is extremely objectionable, special precautions are indicated. Isolation mountings should be used under the pump baseplate and flexible nozzle connections at the piping.

I had commented in Question 6.28 on the fact that the type of noise made by a centrifugal pump can serve as a clue to a source of trouble. Briefly, a crackling noise that *definitely* originates within the pump is a symptom of possible suction difficulties, because this type of noise is generally associated with cavitation. Consequently, it is extremely important to match the pump and system characteristics so as to avoid any possibility of cavitation if this type of noise is to be avoided.

On the other hand, a rumbling noise is generally caused by conditions in the discharge waterways of the casing when the pump operates at capacities either far too low or far in excess of those for which it was designed. Hydraulic losses in a centrifugal pump are least in the range of its best efficiency capacity, and since these losses cause turbulence, a pump operating at or near its best efficiency capacity (see Fig. 6.27) will be considerably quieter than at lower or greater flows. This points to the danger of oversizing or undersizing pumps.

Extremely high liquid velocities in pipes rigidly tied to building structures can produce a noise that may be unjustly attributed to the pump. Care should be exercised in the sizing of piping to give reasonable velocities as well as in providing adequate support. Precautions should be taken to install insulation of piping attachments wherever possible. On the other hand, specifying limiting velocities at the pump discharge will not necessarily eliminate turbulence and noise. This may lead to the selection of an oversized pump and greater noise from operation below the best efficiency capacity. Liquids do not know where the pump stops and discharge piping begins. If the pump is properly sized and lower piping velocities are desirable, an increaser at the pump discharge nozzle will solve the problem. Noise can originate from a motor driver where it comes principally from the fan or the bearings. Fan noise can be reduced to a remarkable degree by the proper design and location of the fan. For instance, where a large fan may have been used to provide more air movement through the motor, it has been frequently found that a smaller fan with a change in baffling will assure satisfactory cooling and will lower the noise level appreciably.

Considerable noise can originate from vibration of a pump or of its driver. A well-designed and balanced unit, with proper alignment, will produce very little if any vibration.

Finally, proper isolation of noise transmission along the pump and driver mounting is very effective in reducing noise. In many cases in which extremely quiet operation is absolutely necessary, such as in air-conditioning cooling installations, the provision of insulating pads below the pump baseplate may be indicated.

Question 6.30 Noisy Jet Pump Causes Concern

Although my pump problem is not an industrial problem, there may be similar situations in industry. My general question is this: Do all jet pumps "sing" when operating with fluid recirculating to the jet?

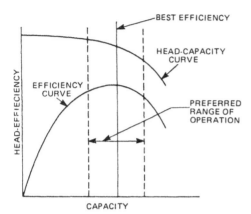

Figure 6.27 Pumps operating at or near their best efficiency capacity will be considerably quieter than when operating at lower or higher flows.

My installation consists of a vertical shaft 3/4 hp centrifugal pump that is pumping river water to an underground lawn-sprinkling system. The lift from the jet to the pump is about 12 ft. Operating as a conventional centrifugal pump, it develops about 30 psig. When valve A (the valve allowing recirculation to take place, as shown in Fig. 6.28) is opened and the jet functions, a pressure of 60 to 80 psig is developed, indicating good operation. But as soon as valve A is opened, a singing noise immediately develops. This gives me concern lest it be a sign of air entering the system that will cause excessive wear to the rotor blades. Inasmuch as the lawn is some 20 to 30 ft in elevation above the pump, the pressure of 30 psig that the pump develops when the jet is not working is insufficient to operate the traveling lawn sprinkler successfully. Therefore it is desired to operate the pump as a jet installation, but one wonders what rate of wear is occuring to the pump rotor.

The system seems tight as evidenced by a static pressure test of the system, using the city water pressure of 60 psig. No leaks could be detected at any fittings at the pump, jet, or connecting lines. Nor could leakage be detected at the permanent seal of the pump rotor shaft. Yet air seems to be entering the system, as evidenced by the observation of air bubbles in the line of a plastic garden hose carrying water from the pump while in operation.

Is it possible for air to be sucked in at the rotating shaft seal during operation of the pump through a path that water will not seek during a static pressure test?

My local pump person says that "all jet pumps sing" when water is recirculated and that there is nothing to be alarmed about. Is this a common situation, and will the wear to the rotor tips be any greater when a water

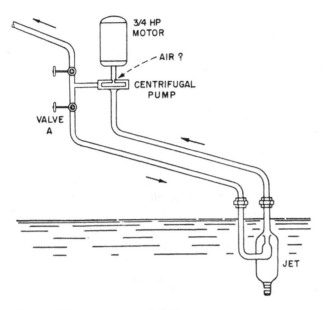

Figure 6.28 Jet pump installation.

stream containing air bubbles enters the chamber than for the rotor of a conventional centrifugal pump that may also suck in air at the shaft but that does not develop an audible noise?

This pump was dismantled last winter and the seal examined. The seal appeared to be in perfect condition.

Your discussions are usually related to industrial problems, but I would appreciate it if you were to comment on this condition.

Answer. First, let me assure you that a pump does not know whether the service on which it is installed is industrial or if it is being used to water a lawn. Thus, this problem could easily be encountered on any other service, as you state.

Unfortunately, however, I have not encountered sufficient numbers of jet installations to be able to verify your pump person's flat statement that "all jet pumps sing." Of course, any hydraulic installation will cause a certain amount of audible noise, and it becomes a question of differentiating between a noise one would normally expect and countenance and a greater or different noise that leads one to suspect that not all is well with the installation.

Noise in itself is not, of course, harmful. But excessive noise is generally a symptom of turbulence or of high velocity. For instance, noise can indicate cavitation in some cases, or in other cases it may be the symptom of turbulence in a pump operating at a capacity much beyond the normal intended flow.

You imply that the pump makes no significant noise when operating as a conventional pump. Since the suction lift is decreased when the pump is operated as a jet pump, the noise is not a sign of cavitation. On the other hand, the pump may be handling a total flow considerably greater than normal design when operating as a jet installation. Since the piping, fittings, and sprinklers remain the same in either case, a discharge pressure of 60 to 80 psig will cause a much greater flow to take place to the sprinklers than when the discharge pressure is 30 psig. This greater flow is further increased by the additional flow through the recirculating valve A. In combination, it is quite possible that the pump handles more water than it was originally intended to handle and that the turbulence leads to excessive noise. Does the noise reduce if you throttle the discharge to the sprinkler system at the second valve (the one above valve A in Fig. 6.28)?

I rather doubt that air is entering the pump chamber at the rotary seal if the system is tight under a static pressure test. As you describe it, the suction pressure becomes approximately 30 psig when the jet is in operation. Thus, the pump casing is under positive pressure and could not have an inward air leak from the atmosphere. As a matter of fact, a slight in-leakage of air is sometimes used to quiet a noisy pump.

As to the air you find in the plastic garden hose, it could well be air originally dissolved in the river water and liberated by the action of the impeller blades as if by an egg beater.

After I had written this, I received a further communication from the same reader as follows:

> Thank you for your recent letter concerning the "singing" of my jet pump. The idea of the "egg beater effect" releasing the dissolved air in the river water had never occured to me. It may well be the reason. I know of no way to eliminate it. It's like a rattle in a car: Once you know the reason for the rattle, we don't worry about it or even bother to fix it. In the case of my pump, I'll use the jet and hope the rotor wear is not excessive.

Further Comments About Noisy Jet Pumps from Another Reader

I would like to comment regarding the "singing" jet pump discussed in Question 6.30. I purchased a 3/4 hp jet pump and installed it without the jet

Chapter 6

to pump from a cistern located approximately at ground level. It was pumping to a pressure tank with the switch set at 20 to 40 lb and ran very quietly.

After about six months, I installed this same pump on a well with the jet set at about 70 ft in the well. It immediately became quite noisy as described in Question 6.30. It was not overloaded, as the reading on an ammeter was slightly below the rated amperes on the motor nameplate.

I do not believe the owner of the aforementioned pump has anything to worry about, as my pump operated for several years making this singing noise without any failure. I later removed it from the well and sold it. To the best of my knowledge, it is still singing and still pumping water.

Question 6.31 Vibration Caused by Operation at Low Flow

Our plant has an installation of two vertical centrifugal pumps intended for general house service. They are designed to handle 4000 GPM each against a total head of 56 ft at 1150 RPM. The service is not intermittent, but the demand varies quite widely, ranging from as low as 2000 GPM. We have noted that whenever the pumps operate at capacities below 2500 GPM, they begin to vibrate quite noticeably, and it becomes impossible to maintain packing in the stuffing boxes for a reasonable length of time. The operation of the stuffing boxes becomes erratic, with spurts of water issuing frequently from beyond the gland. If the glands are tightened to prevent this, the packing begins to smoke, and we have scored the bronze sleeves on several occasions. Our major concern, however, is with the mechanical vibration, as we have already broken a shaft in one of the pumps. We first thought that this could be because of an overload condition at light flows, as horsepower curves on pumps of this specific speed range tend to rise to shutoff. However, as you can see from the performance curve in Fig. 6.29, the H-P curve of these pumps is not rising and is practically a straight horizontal line over the entire operating range. What could be the cause of this vibration, and what remedies would you recommend?

Answer. I believe the difficulties you have encountered are caused by the radial thrust acting on the impeller when the pump is operated at reduced capacities. The type of pump that will have the performance characteristics shown on the curve in Fig. 6.29 is a volute pump. If a volute pump is operated at other than its design capacity, a certain imbalance of the hydraulic forces acting radially on the impeller takes place. The maximum imbalance occurs generally at zero capacity and is reduced as rated capacity is approached. This imbalance creates a radial load on the pump shaft. I

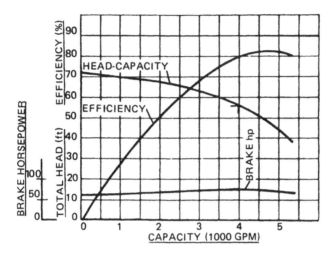

Figure 6.29 Performance curves for the 4000 GPM centrifugal pumps discussed in Question 6.31.

have known cases of shaft failures to take place when part-load operation was not contemplated at the time that the pump was ordered, and therefore the designer did not incorporate a pump shaft suitable for maximum loading over the entire operating range of the pump.

Although this radial thrust is a function of the head generated and its effect should not normally be excessive in a pump designed for 56 ft, the problem is aggravated here, since a vertical pump is involved and since an overhung design has to be used without a bearing below the impeller.

A means used to combat the effects of radial thrust in volute-type pumps is the use of a dual volute. The effect of radial thrust in single-volute pumps and that in dual-volute designs is illustrated in Figs. 6.30 through 6.33. It will be seen that the resultant forces at part-capacity operation are balanced even though they are not eliminated. Had the designer been acquainted with the requirement to operate this pump at widely changing capacities, a dual volute may have been incorporated in your pump. Unfortunately, it is not practical to do so now unless you are willing to spend the money for replacement casings (always assuming that the manufactuer can produce such a casing). There is, however, a means available to you that will eliminate this difficulty. I note that the power consumption is hardly changed whether you pump 1000 or 4000 GPM. If you were to install a bypass that would permit you to handle 4000 GPM in this pump, regardless whether the demand is 4000 or 1000 GPM, the ill effects of the radial thrust

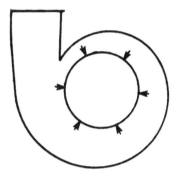

Figure 6.30 In single-volute pumps, casing pressures are uniform at the design capacity and there is no radial reaction.

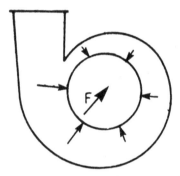

Figure 6.31 Pressures are not uniform at reduced capacities in a single-volute pump, giving rise to radial reaction F.

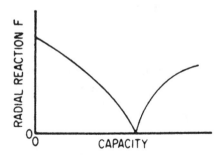

Figure 6.32 The magnitude of the radial reaction F decreases from shutoff to design capacity and then increases again with overcapacity in single-volute pumps. (With overcapacity, the reaction is roughly in the opposite direction from that with part capacity).

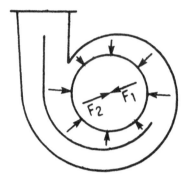

Figure 6.33 Although in a double-volute pump pressures are not uniform at part-capacity or overcapacity operation, the resultant forces F_1 and F_2 for each 180° volute section oppose and balance each other.

will disappear. The slight increase in power consumption when less demand exists is a reasonable price to pay for the elimination of the vibration, longer life of packing and sleeves, and freedom from the danger of shaft breakage. The excess flow over the demand can be dumped back to the source from which the pump takes its suction.

Question 6.32 More About Operation at Reduced Flow

We have a 10-in. double-suction pump which takes suction from a river and provides cooling water for 10 heat exchangers connected to a header from the pump discharge. The pump operates satisfactorily when all 10 heat exchangers are on the line, but when we cut out five of the heat exchangers, the pump becomes very noisy, and the thrust bearing becomes hot. Examination of the pump impeller shows serious cavitation damage at the discharge of the impeller vanes. The cause may be cavitation from low suction pressure, but we do not think so as the manufacturer's curve shows we have 5 ft more head than required.

Answer. You are correct that the problem does not result from low suction pressure. All pumps exhibit a condition at low flows that we refer to as *recirculation*. Some pumps are worse than others, and the severity of the symptoms is dependent on the speed and diameter of the impeller. Recirculation is a turbulent reversal of a portion of the flow at the discharge of the impeller. The result is cavitation-like damage at the discharge tips of the vanes and a disturbance of the rotational flow patterns on each side of the impeller between the impeller shrouds and the casing walls. When the pump

Chapter 6

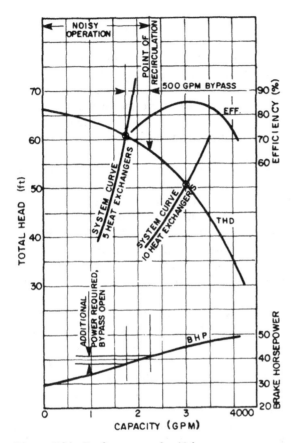

Figure 6.34 Performance of a 10 in. pump.

operates at or near its design capacity, the rotational flow patterns on each side of the impeller are symmetrical and impose no side thrust on the impeller. At low flows, however, the rotational flow patterns are no longer symmetrical, and a pressure differential exists between the two sides of the impeller. The result is an end thrust on the bearing.

Figure 6.34 shows the performance of a typical 10-in. pump. I have shown an assumed system curve for 10 heat exchangers and for 5 heat exchangers on the line to illustrate the problem. Also shown is the low-flow condition at which recirculation starts. With 5 heat exchangers on the line, the operating capacity of the pump is reduced from 3000 to 1750 GPM, 500 GPM below the recirculation capacity.

Since the problem is one of operation at a flow less than the recirculation flow, one solution is to bypass sufficient capacity from the discharge back to the suction to prevent the pump from operating in the recirculation zone. The size of the bypass can be determined only after the head and capacity at which noise starts are known. This can be done by first throttling the pump until the noise appears. Determine the total head at this point from the suction and discharge gauge readings. Then determine the capacity corresponding to this total head from the manufacturer's performance curve.

The next stop is to determine the pump capacity when the minimum number of heat exchangers are in operation. This can be done by determining the total head from gauge readings and the capacity from the manufacturer's performance curve. The difference between the two capacities is the flow that should be bypassed. In the example shown in Fig.6.34, the bypass capacity is 500 GPM. To prevent bypassing when the pump is operating at design capacity, a pressure-operated valve should also be installed that will open at the pressure corresponding to the point of noisy operation on the pump head-capacity curve.

Bypassing is the best solution to the problem as it eliminates both the noise and excessive end thrust at low flows. Operation with a bypass, however, requires additional power, as shown in Fig. 6.34. If the additional power consumption is an uneconomical solution, the noise level and damage to the discharge tips of the vanes can be reduced by the introduction of air into the suction of the pump. The air acts as a cushion to prevent the sudden collapse of the cavitation vapor bubbles and suppresses the noise and reduces the damage to the metal surfaces of the impeller.

To determine the amount of air required, throttle the pump to the lowest flow at which the pump will operate on the system curve. Then bleed air into the suction of the pump through a petcock until the noise is reduced to an acceptable level. Lock the petcock in position to prevent tampering with the air supply. A pressure-operated valve controlled by the pump discharge pressure could also be incorporated into the air line if it is desired that air be introduced only at low flows, not during normal operation. A word of caution: Free oxygen in the water may result in accelerated corrosion of the tubes of the heat exchangers. Before this method is adopted as a permanent solution, the tubes should be examined periodically for evidence of corrosion.

The hot thrust bearing, however, is another problem, and the introduction of air will not reduce the thrust. Obviously a larger bearing would be helpful. Many bearing housings can be bored out for larger bearings. If this is not possible, a larger bearing housing would have to be adapted to the bearing bracket.

Chapter 6

Question 6.33 Standby Boiler Feed Pump Common to Two Units

Our power plant has two 15,000 kW units operating at 450 psig. The two units operate independently, except for a tie-in at the boiler feed pumps, as shown in Fig. 6.35. A motor-driven boiler feed pump supplies all the feedwater required for each one of the units. A common standby boiler feed pump is arranged for turbine drive. This pump is started automatically on failure of either of the two motor-driven pumps to develop sufficient pressure. Since it may have to take its suction from either one of the two deaerating heaters, its suction valves are kept closed. One of these is opened by the impulse that, after a suitable time delay, starts the turbine-driven pump.

Obviously, the pump is not expected to be operated very frequently. But the very first time this pump started owing to an emergency about 6 months after its installation, the pump seized and was seriously damaged at

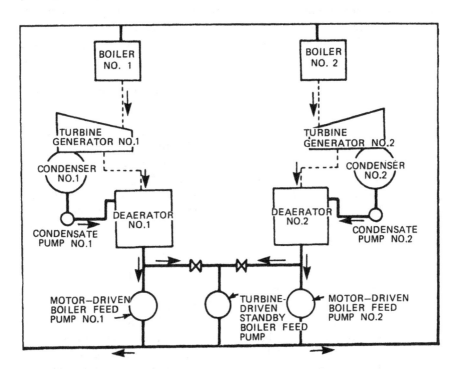

Figure 6.35 Standby turbine-driven pump is tied in with two otherwise independent units.

all its internal running joints. What could have caused this failure, and what can be done to avoid its repetition?

Answer. The reason for installing two valves in the suction lines to the turbine-driven pump is that the pressure in the two heaters and the temperature of the feedwater at the suction of the two individual pumps may be unequal if the two units are operating at different loads. Thus, the suction valves are normally closed, and one of them (depending on which of the two motor-driven pumps loses pressure) opens a fraction of a minute before the pump is put on the line.

This arrangement satisfies the requirement that the two suction lines will never be interconnected lest unequal pressures in the two deaerating heaters result in flashing and steam binding, but it also acts to cause difficulties when the turbine-driven pump is started up.

As is stated in the question, this pump is apparently never operated under ordinary circumstances and is held only for emergencies. When such an emergency does not arise for 6 months after the pump has been installed, the pump is not in condition to be operated for the simple reason that leakage of water through the stuffing boxes will have drained the pump casing down to the level at which no further leakage can take place and the pump is air bound. This is undoubtedly what happened when the pump seized on start-up.

There are several means available to avoid the recurrence of this type of failure:

1. The pump could be operated manually at least once a week, say over the weekend. The suction valve to one or the other heater would be opened and a high point on the casing vented just in case enough feedwater has escaped at the stuffing boxes.
2. The pump can be always kept connected to one of the heaters. Thus, 50% of the time it would be connected to the one unit for which it must replace the motor-driven feed pump. In the event that it is the other pump that requires replacement, the standby pump can operate temporarily, taking its suction from the other heater. It can be switched into its proper unit manually after the emergency has been taken care of.
3. A bypass can be installed around the check valve in the discharge of the turbine-driven pump with an orifice in this bypass (Fig. 6.36). This orifice should be selected to pass a nominal flow of feedwater—say 3 to 5 GPM—under the pressure differential between the discharge header pressure and the pump internal pressure. To prevent building up pressure within the casing, a second bypass or

Chapter 6

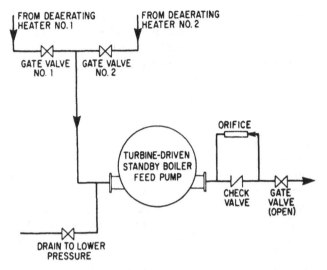

Figure 6.36 Hookup permitting circulation through standby pump even while both suction valves are closed.

drain should be installed between the pump suction and some lower pressure reservoir, such as the surge tank for condensate into which the deaerating heaters overflow. In this manner, the pump casing will always be kept full of water and primed, ready to start. This arrangement will also accomplish the purpose of a warm-up line.

The bypass around the check valve need not be closed after the pump has been started, since the flow through it will reverse, and its presence will have no effect on the operation of the pump. The drain line from the suction to a lower pressure reservoir can be closed manually, at leisure, after the pump has been started.

Question 6.34 Comments on Question 6.33

The piping arrangement shown in Fig. 6.36 is inherently hazardous if care is not taken to prevent misoperation of certain valves. If, by some misoperation, gate valves 1 and 2 were closed, the gate valve in the line entitled "Drain to lower pressure" were closed, and the discharge gate valve of the pump were open, it would be possible for the boiler feed discharge header pressure to back up through the pump and be imposed upon the entire pump and its suction piping.

I would suggest that a small relief valve be installed on the pump suction line to warn of excess pressure in the suction line and to relieve this pressure. I would also suggest that a gate valve be installed in the little orifice bypass line for more reliable isolation of the pump for maintenance if the gate valve in its discharge line could not be closed off tight and should leak slightly. A possible substitution for the relief valve would be to lock open the gate valve in the drain to lower pressure.

Answer. I shall not deny that certain hazards exist from misoperation of the system shown in the piping diagram in Fig. 6.36. Unfortunately, misoperation hazards will probably always be with us until the day that a steam power plant is built that is completely automatic in its operation and in which no human hand or human judgment is permitted to interfere with a prefabricated programming. Even then, I fear, the plant will not be 100% free of random failures; witness the record compiled in the launching of our rocket-powered satellites! I have frequently said that there is *always* a risk to operating a steam power plant. Once we admit this to ourselves and devote our efforts to *reducing* this risk rather than to *eliminating* it, we shall make progress. Otherwise, we shall be chasing an elusive and unattainable will-o-the-wisp.

And so it is in this case. If we assume that through misoperation the gate valve in the drain line is closed, it is obvious that pressure would build up in the pump except for whatever relief is afforded through leakage past the stuffing boxes. Your other two assumptions are actually not necessary, since they are facts: Gate valves 1 and 2 are *closed* and the discharge gate valve is *open*, not through misoperation but rather through intent, as indicated in the body of the question. Thus, the closing of the drain valve would be a major error in operation.

As to the provision of a relief valve, I believe that it is optional rather than mandatory. I know of installations where such relief valves are not used merely on the theory that infrequently operated relief valves may be as hazardous as the misoperation against which they are intended to provide protection. For instance, this is most often true of the balancing device leak-off line of high-pressure boiler feed pumps that returns to the deaerating heater at the pump suction. This line is always provided with a gate valve to permit isolating the pump for inspection or complete dismantling. Closure of this valve through misoperation would cause a major catastrophe to the pump, and yet few such leak-off lines are protected by relief valves against such an emergency. Thus, for my part, I would prefer locking the drain valve in the open position while the common spare pump is out of operation.

As to the valve in the small jumper line around the check valve, I agree entirely with you with regard to its necessity. I had merely oversimplified the diagrammatic sketch. In the system drawings supplied to customers for this portion of the auxiliary hookup, the isolating valve is always shown. It is, however, kept in the locked open position until such time that a pump is to be opened for inspection.

Question 6.35 Short Ball Bearing Life

We have two centrifugal pumps on boiler feed service. They are two-stage pumps designed for 300 GPM and 650 ft at 3560 RPM driven by electric motors. The pumps are fitted with grease-lubricated ball bearings with water-cooled jackets. We find that the life of the bearings is extremely short, and on occasion we have had to replace the bearings after only 6 weeks of operation. We would like to know what investigation program we should instigate to search out the source of this trouble. The only symptom we have so far uncovered is that some water has been found on occasion within the bearing housing.

Answer. The manufacturers of ball bearings have done a great deal of research on the life expectancy, load limits, design, and materials of ball bearings. This research has resulted in quite accurate methods of selecting the right ball bearing for any given application. We can assume, therefore, that the pump manufacturer has selected the ball bearings for this particular application for a long life and for loads well within the allowable limit.

We must therefore search for possible sources of trouble in the manufacturing process itself, in the installation of the unit, or in the operation procedures. The occasional presence of water in the bearing housing eliminates the possibility of a porous bearing housing casting, since in the latter case cooling water would be constantly seeping in. But this symptom has a definite significance, as will be seen later.

We can list an imposing array of possible causes for these difficuties. It will remain for you to check each one of these causes, the remedy for which is in most cases self-evident. It is entirely possible, incidentally, that more than one cause is contributing to the short life of these ball bearings.

1. There may be misalignment between the bearing housings and the bore of the casing or of the bearing brackets (Fig. 6.37). This may be the result of an accumulation of inaccuracies in manufacture or of the growth of the castings in service.
2. You may have been using incorrect bearing replacements. This is especially possible in the case of double-row thrust bearings, which should be of the so-called *matched* type (Fig. 6.38). If individual

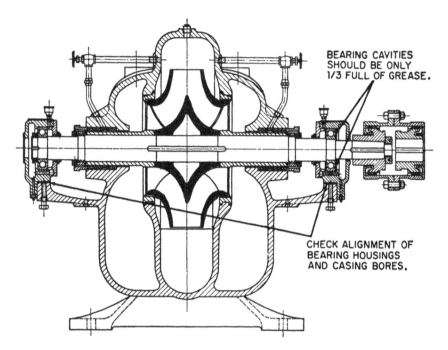

Figure 6.37 Possible points of misalignment between bearing housings and bore of casings or bearing brackets.

single-row bearings are purchased and used in pairs, the possible mismatching of the pairs can frequently lead to difficulties.
3. Replacement bearings may be improperly mounted on the shaft. In the process of installing, it is best to heat the bearing slightly in an oil bath to expand its inner race. If the bearing is forced on the shaft, an arbor press should be utilized, or a tubular sleeve can be employed. In the latter case, make certain that the blows against the sleeve are gentle and applied alternately on opposite sides of the sleeve held against the inner race (Fig. 6.39). Uneven pressure on the inner race may distort it and preload the bearing unevenly, shortening its life very materially.
4. Check the doweling of the pump feet to the base. If the two inboard feet are doweled, the other end of the pump is free to allow for thermal expansion. Doweling at both ends will prevent this expansion and distort bearing alignment.

Figure 6.38 When double-row thrust bearings are used, be sure they are of the so-called *matched* type. (Courtesy *New Departure*.)

5. Excessive casing strains may be caused by misalignment of suction or discharge piping. The piping should be disconnected at the pump flanges, and if there is any misalignment, it should be corrected.
6. Troubles may be caused by misalignment between the pump and its driver. The coupling alignment must be as nearly perfect as possible when both pump and driver are at *operating temperature*. This is very important because even though a flexible coupling will compensate for slight misalignment, it is not designed to operate as a universal joint.
7. Check your method of lubrication. An antifriction bearing should be lubricated sparingly. In this case, with grease lubrication, the bearing housing should not be packed more than one-third full (Fig. 6.37). Excess lubrication leads to churning and overheating.
8. One of the most frequent sources of ball bearing failure is excessive cooling. This is particularly true in the case of pumps handling hot liquids. The inner race expands under the action of the heat conducted along the shaft, and the outer race contracts from the effect of extremely cold cooling water in the cooling jackets. The result is that the balls are squeezed, so to say, and much overloaded. Distortion and flattening of the individual balls can often be observed in such cases. There should not be

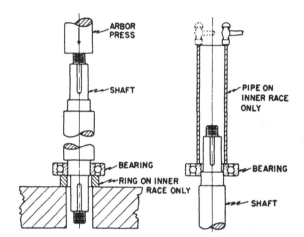

Figure 6.39 Arbor press (left) is best for mounting ball bearings on shaft. If press is not available, a length of pipe and a hammer can be used as shown at right.

too large a temperature differential between the two races, and cooling-water flow should be throttled until a definite temperature rise is noted at the outlet of the water jackets.

As a mater of fact, this particular cause should be investigated first in your installation. The presence of water in the bearing housings is probably due to the condensation of atmospheric moisture caused by the *refrigerating* action of excessive cooling-water supply. This condensation can in itself be causing serious difficulties, since it can produce small rust spots on the balls and destroy their useful life.

9. Suction conditions should be thoroughly investigated to determine whether flashing or even partial cavitation is taking place. Vaporization in the first stage will result in an abnormal axial thrust, which can severely damage the thrust bearing. Even though this effect does not directly act on the line bearing of the pump, a damaged thrust bearing will set up conditions that reflect on shaft alignment and start a chain of circumstances that affects both bearings ultimately.

10. Operation of the pump at abnormally low flows may cause excessive overheating of the pump. Check whether the pump is protected by a proper recirculation arrangement to eliminate this possible source of trouble.

This is quite an imposing list of potential hazards. It points out very forcibly the fact that equipment requires careful manufacturing, installation, and operation in order to give satisfaction. But this is true of all equipment. Unfortunately, it is not practical to design mechanical equipment that does not require attention to details in these three areas.

Question 6.36   More About Short Bearing Life

One of our small two-stage, axially split casing pumps has had a series of mysterious thrust bearing failures. The pump is designed for 700 U.S. GPM capacity and 800 ft total head at 3560 RPM. It ran quite satisfactorily for about 4 years, at which time the thrust bearing failed. After this bearing was replaced, the new bearing failed again after only 3 months. Since then, we have had to replace the thrust bearing at even shorter intervals. The pump itself has never been dismantled, as nothing else seems to be wrong with it, other than it now appears to have lost some capacity. Because the pump had been selected with sufficient margin over the operating requirements, this loss of capacity is insufficient to indicate the necessity for overhauling the internal clearances.

Could you indicate the possible cause or causes of the short life of this thrust bearing?

Answer.   I am afraid that I cannot agree fully with you regarding the decision of postponing an overhaul of the internal clearances. Today's high cost of energy is forcing us to reexamine our past practices with respect to the evaluation of increased power consumption versus the cost of overhauling equipment, such as pumps. I suspect that you will find that restoring the pump internal clearances at this time will prove to be a most cost-effective maintenance procedure. There is one more reason that you may find it advisable to examine the pump internals, and this reason will become apparent further on in my answer.

Your case differs in one respect from that discussed in the preceding question and that is the steady decrease in the MTBF (mean time between failures) of the thrust bearing. I note that you made no comments about the inboard bearing, and I assume that it has not been subject to constantly decreasing intervals between failures if it has even failed at all. This would be a definite sign that the conditions that cause the thrust bearing failure were being steadily aggravated.

Thus, although I suggest you examine carefully the various causes for bearing failures I have listed in my answer to the preceding question, I think that the culprit is the excessive wear at the internal clearances. This excessive wear causes the impellers to handle a higher flow than that discharged by

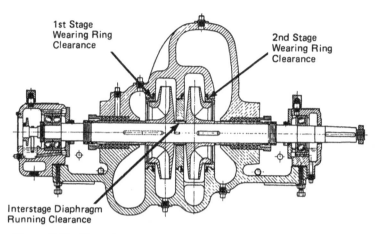

Figure 6.40 Two-stage pump.

the pump. If the suction conditions were marginal to begin with, cavitation of the first stage impeller may now be taking place. This can lead to an upset of the axial thrust balance if, as I imagine, the two impellers of your pump are placed back to back (see Fig. 6.40). The unbalanced axial thrust may exceed the rated capacity of the thrust bearing. This will significantly shorten the bearing life.

But even if cavitation does not take place, the excessive wear at the running clearances can upset the axial thrust balance of the two back-to-back impellers very materially. If you examine Fig. 6.40, you will notice that there are three such close running clearances: two of them are at the wearing rings, and the third one is at the interstage diagragm. As wear at these clearances takes place, there is a net loss of capacity, as illustrated in Fig. 6.41, which shows the head-capacity curves of a two stage pump designed for the conditions of service that you mention in both new and worn states.

It can be assumed that the impellers of this back to back two stage pump are of equal diameter and produce the same total at all flows. Thus the unbalanced area of each single suction impeller is subjected to the same pressure difference, and the unbalanced axial thrusts of the two impellers act in opposite directions and balance each other.

Presumably, as the pump wears, each wearing ring clearance increases essentially by the same amount and the leakage past each wearing ring increases likewise by the same amount. Therefore, the increased clearances at the wearing rings should have no effect on the axial thrust of the pump and therefore should not be the cause of the premature failures of the thrust bearing.

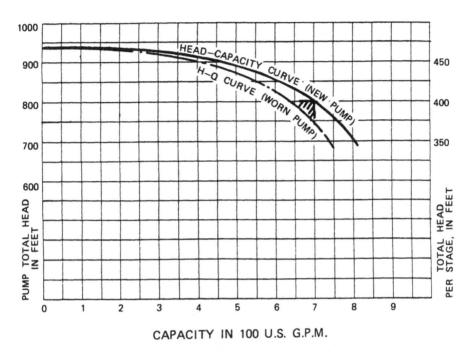

Figure 6.41 Performance of a two-stage pump.

This is not, however, true in the case of the third clearance joint, that at the interstage diaphragm. As this particular clearance is enlarged and the flow past it increases, the two impellers no longer handle the same flow.

Let us assume, for instance, that the net flow discharged by the pump happens to be 600 U.S. GPM, which of course is also the net flow entering at the suction. If the leakage past the interstage diaphragm has increased by, say, 100 GPM, the second-stage impeller handles a total of 700 GPM, of which 600 GPM is discharged by the pump and 100 GPM flows past the clearance joint into the discharge area of the first stage. In turn, the first-stage impeller handles only 600 GPM, but 700 GPM flow to the suction of the second stage from the first stage volute. The total heads produced by the two impellers are no longer equal. Their difference can be determined by an examination of the curves in Fig. 6.41 using the head-capacity curve of a worn pump:

Stage	Capacity (U.S. GPM)	Stage (ft)
First	600	415
Second	700	375

There is now a difference of 40 ft or 17.3 psi between the total heads generated by the two impellers. This difference in pressure acting on the unbalanced areas of a single-suction impeller will create an unbalanced axial thrust imposed on the thrust bearing. We can make an approximate estimate of this thrust. Assuming an unbalanced area of about 36 sq. in., the force will be of the order of 625 lb. This does not appear to be such a large force, but it may be sufficient to cause premature bearing failure, particularly if any one or more of the other possible causes I have listed are also operative and create a marginal situation to begin with.

I should add that I have encountered premature bearing failure caused exactly by the phenomenon of excessive interstage running clearance wear on many occasions. As a matter of fact, in one particular case, the wear had led to an increase in leakage of almost 250 U.S. GPM in a 1000 GPM pump. The resulting difference in the pressure between the heads generated by the two stages reached an equivalent pressure of about 50 psi, and the unbalanced axial thrust exceeded 3000 lb. In all such cases, the restoration of the internal clearances at suitable intervals eliminated these premature failures. I am reasonably sure, on the basis of this experience, that when you dismantle your pump you will find considerable wear at the interstage joint and that after you rebuild the pump, you will not only save on the power consumption but you will also return to a more normal bearing life.

Question 6.37 Rapid Wear of Wearing Rings

We have an installation of several multistage pumps on severe service that we find necessary to overhaul about every 2 or 3 years. The pumps have *in-line* single-stage impellers and a hydraulic balancing device. Sleeve-type bearings are used in these pumps and a Kingsbury-type thrust bearing.

On several occasions, we have noticed that, following overhaul and return to service, damage to the wearing rings and to the impeller hubs has manifested itself very rapidly, within the first few days. We have checked very thoroughly for ring concentricity and horizontal split gasket thickness and type, in short for all the possible sources of difficulty suggested by the pump manufacturer, but with no success in isolating the source of our particular trouble. Could we have overlooked something?

Answer. This is one of the most difficult types of questions to answer, especially since the pump manufacturer possesses considerably more information about the design of this pump, the materials used, the type of service (which you term as "severe" without further description), and so on. But it might be worthwhile investigating one particular point that is frequently overlooked and into which I once ran some years ago in a pump of this general description.

One of the checks necessary after such a pump has been reassembled in the field is that for endwise or axial clearance. This is accomplished by "shoving" the rotating element axially as far as it will go in one direction and measuring the amount by which it can be moved back in the opposite direction. But it is extremely important that the rotor be moved over gently. Otherwise, the impeller hubs will strike hard in the wearing rings, and the hub corners will turn up, developing burrs. The burred corners of the impellers start catching and ripping the rings, and this is all that is necessary to start trouble.

Whether this could have been the source of your troubles, I cannot surely say. But even if the cause is different, this warning may save you from damage to the wearing rings and impeller hubs on this or similar equipment.

Question 6.38 Casing Wear

About 3 years ago we installed a small single-stage centrifugal pump designed for 250 GPM and a head of 225 ft to handle raw river water supply to our plant. We recently found that the pump casing was worn well beyond our expectations. Is raw river water considered to be severe service, or is 3 years of operation an unreasonably short life for this pump?

Answer. Unfortunately, the term "raw river water" is hardly a fully descriptive definition, since there are many types of river water, ranging from clear pure water to an extremely acid water because of contamination of the river or muddy, sandy, and gritty water. Obviously, all this will materially affect the life of any pump that may be installed.

It is also difficult to determine from your question exactly in what manner the casing of this pump has become badly worn. I do not know whether it is a question of corrosion, of erosion, or a combination of both. The water handled by this pump may contain an excessive amount of gritty and sandy matter. In this case, you may expect a certain amount of erosion at the wearing joints and possibly at the casing tongue. If the water is strongly acid, galvanic corrosion can readily be expected, especially if the pump has a cast-iron casing and bronze impeller and rings. The failure from this cause can be accelerated through intermittent use of the pump if the acid water is allowed to stand in the pump without completely filling it. This generally leads to greater corrosion at the line of demarcation between the water in the pump and the air above it.

Thus, to determine whether the pump materials are suitable for the particular water that will be handled by the pump, it is imperative to determine exactly what kind of water is pumped, both from the point of view of its chemical analysis and from that of its content of abrasive materials in suspension.

If a standard fitted pump (cast-iron casing and bronze fittings) is used on slightly corrosive water, it is possible to lengthen the pump life by painting the casing with an acid-resisting paint. There are several such paints available on the market, and I have known them to be used in pumps handling mine waters, which are generally more corrosive than average river waters. Pumps should then have an inspection at frequent intervals, say once a year. At such times, the paint should be touched up after the casing has been scoured with a wire brush wherever the initial application has worn off.

Question 6.39   Wear Caused by Sand in River Water

In our power plant we have four vertical intake pumps that pump raw water from the river to the open water reservoir in the plant area. These pumps handle sand-laden water as we have not been able to prevent the entry of the sand into the intake.

Basic details of these pumps are as follows:

Type: Vertical, single-stage, open-type impeller and single diffuser
Capacity of each pump: 1885 m^3/hr (8300 GPM)
Total dynamic head: 24.4 m (80 ft)
Speed: 985 RPM (50 cycle current)

Due to the sand-laden water, pump suction bells, impellers, wearing rings, diffusers, and other parts are wearing very fast. Bronze and high-chrome steel impellers, cast-iron suction cones, diffusers, and high-chrome wearing rings have been tried, but the results are not satisfactory.

We would request your comments and suggestions, particularly in regard to the following:

1. Are there any better materials for the pump wearing parts that will overcome abnormal and rapid wear due to sand-laden water? If available, kindly indicate the likely supplier of such materials.
2. Is the use of a two-stage pump of closed impeller design at 1460 RPM suitable for this service? (Incidentally, we have an offer for such a pump.)

Answer

1. Material selections for pumps handling abrasive particles entrained in the liquid vary between either of two extremes. In one, very

hard materials, such as austenitic manganese steels or martensitic white irons, are used to provide maximum resistance to high-impact erosive forces. This approach is most commonly used in dredge pumps and heavy-duty gravel pumps, in which the particles handled are relatively large and hard. The other extreme, generally preferred when the solids are quite fine, as in the case of most river sand, is to use pumps having their principal parts either molded of or coated with natural rubber or neoprene. This approach would appear to be better suited to your needs, and I would suggest you contact the pump manufacturer involved regarding procurement or adaptation of parts as necessary.
2. I definitely recommend against the use of higher speed pumps, which are likely to suffer even more serious wear problems than you already have. Conversely, experience suggests that lower speed pumps, even with equal heads per stage, will probably give better service. As to closed versus open impeller designs, the latter are usually easier to deal with when rubber or neoprene coatings are applied to otherwise conventionally designed components.

Question 6.40 Galvanic Corrosion

Several of our centrifugal pumps on cold-water service have shown a considerable amount of corrosion of the casings, which are made of cast iron. The impellers are bronze and seem to be untouched. The water handled by these pumps has a pH varying from 6.2 to 6.5. We have been told that bronze-fitted pumps are not suitable for our conditions. What is the reason for this corrosion, and why is bronze unsuitable, since it is not attacked?

Answer. The pH of a liquid is a measure of the relative acidity or alkalinity of this liquid. A solution with a pH of 7 is neutral. pH values above 7 (up to 14) indicate alkalinity, and values below 7 (down to 0) indicate acidity.

The corrosion that has taken place is apparently due to galvanic action. It will occur whenever a metal is electrically connected to another metal in the presence of an electrolyte (in this case acid water with pH of 6.2 to 6.5). The amount of corrosion will depend on the conductivity of the electrolytic solution, the proximity between the surfaces of the dissimilar metals, and the difference in electric potential, that is, the distance between the two metals in question in the galvanic series.

The galvanic series presented in Table 6.6 gives an approximate idea of the interrelation of various metals when they are in the presence of an electrolytic solution.

Table 6.6 Galvanic Series of Metals Commonly Used in Pump Construction[a]

Corroded end (anodic):
Zinc
Iron
Chromium iron
Chromium nickel iron
Tin
Lead
Brasses
Bronzes
Nickel-copper alloys
Copper
Protected end (cathodic):

[a]Note: The series is not complete, as it includes only those materials most commonly encountered in centrifugal pump construction. In addition, it does not include the various stainless steels used for pump impellers, wearing rings, or sleeves. The position of these metals in the series cannot be definitely fixed, as it has been known to change depending on the exact nature of the electrolyte. It is safe to state, however, that with a weak electrolyte such as slightly acid or slightly alkaline water, they will occupy a position in the iron and chrome iron group and that the galvanic action between these metals is almost negligible.

Cast iron being anodic to bronze, a continuous flow of iron ions from the casing takes place toward the bronze impeller. It deposits on the impeller in the form of an iron oxide and is washed away by the flow of water. In the meantime, the surface of the casing that loses the iron becomes graphitized. The relatively soft surface left exposed is readily washed away by the jetting action of the water. In addition, there is a definite electrochemical action set up between the graphitized surface and the underlying or adjacent cast iron, thus accelerating the loss of iron in the affected parts.

Because of this galvanic corrosion, it is preferable to use *all-ferrous* materials when the water has a pH of 6.2. If the casings involved have not been damaged beyond repair, it would be advisable to replace the bronze impellers with stainless steel, which is closer to cast iron in the galvanic series.

Question 6.41 Corrosion of Cast-Iron Casings of Boiler Feed Pumps

We have two separate steam power plant installations of 15,000 kW operating at a throttle pressure of 625 psig. The boiler feed pumps serving this system are designed for a discharge pressure of 800 psig, two full-capacity electric motor-driven pumps being used in each case, with one pump running and the second pump serving as standby. In the case of the first installation we used cast-iron pump casings; the second unit was provided with pumps having 5% chrome steel casings.

In the 18 years since the first installation was started, we have had to rebuild the pumps several times, repairing the cast-iron casings with ni-rod welding in areas where corrosion-erosion had taken place. Needless to say, the 5% chrome steel casings, which have seen 15 years of service, have had no need of repair. As a matter of fact, one of these pumps has never been opened. By observation of the pump performance we know that the time has again come to carry out casing repairs. Probably the best solution would be to replace the casings with 5% chrome steel. On the other hand, we believe that this expense may not be quite justified, as this particular 15,000 kW unit will be decommissioned in another 8 or 10 years. Still we are concerned that after so many repairs of the casings, one more repair may not give us sufficient life and may even have a relatively short duration since so much of the original material has had to be replaced with ni-rod. What would you recommend in this case?

Answer. Without examining the cast-iron casing pumps after they have been opened, I find it difficult to predict whether the repairs you will be carrying out will be as short-lived as you fear. On the other hand, it would appear that you have had to open and repair these pumps at intervals more frequent than every 8 or 10 years, since you state that you have repaired them "several times" in the past 18 years. Thus, we may assume that you will have to repair the casings now and at least one more time before the decommissioning of the unit.

On the other hand, you state that the 5% chrome steel casing pumps in the second unit have had no need of casing repairs. Presumably one of the pumps was opened up at least once—possibly to replace internal wearing parts, such as wearing sleeves, bushings, or sleeves.

What I would recommend would be to reduce the amount of maintenance to the minimum possible by swapping two of the pumps between the two units, after once more repairing the cast-iron casings. In this way each unit will be served by one pump with 5% chrome steel casing to be used as the main pump and by one standby pump with cast-iron casing. Such an arrangement will probably serve you quite well until you decommission the first unit, after which the 5% chrome steel casing pump from the first unit can be returned to service in the second unit.

It is more than likely that the repaired cast-iron casing will have enough life to last the next 8 to 10 years if it is run only for a few hours each week or each month to assure the operators that the pump is available for standby service and ready to run in an emergency.

Whether the two 5% chrome steel casing pumps need be rebuilt internally or not can be readily established by running a simple performance test and comparing their present performance with that when they were new.

Question 6.42 Pumps Handling Seawater

We have an installation of centrifugal pumps handling seawater with about 15 ft suction lift. The pump casings are cast iron, and the pumps are bronze fitted. Considerable wear has taken place in a very short time in the casing near the impeller suction area. Can this wear be ascribed to a form of cavitation? Could you give us some general comments on the materials that should be used in centrifugal pumps handling seawater?

Answer. It is most likely that the casing wear has been caused by electrolytic action and graphitization rather than by cavitation. You do not state whether the seawater is relatively clean or contaminated as generally all harbor waters are. The following remarks will give you a general idea of the behavior of different materials in handling seawater.

1. *Standard fitted pumps.* Thousands of these pumps, with cast-iron casings and bronze fittings, are in use handling seawater. However, they generally give trouble under conditions in which the seawater is contaminated, as in the case of harbor waters. Failures are caused by the galvanic action between the bronze and the cast-iron, leading to serious corrosion near the bronze rings on the cast-iron casings and the eventual deterioration of the casings beyond the possibility of repairs. Another area that can be affected markedly is the casing volute tongue, where velocities are high, and the flow carries away corroded and graphitized material, constantly exposing new material to attack.

Of course, high suction lifts can aggravate this condition, because dissolved gasses are released in the water and accelerate the corrosive attack. When the seawater contains sand in suspension, the wear on the rings has been known to be rapid enough so that they have to be replaced every 6 to 12 months in order to keep the pump up to capacity.

2. *All-iron pumps* avoid the electrolytic action between the bronze parts and cast-iron casings. Unfortunately, however, under some operating conditions, the cast-iron parts are graphitized, and then the impeller gives a very short life. More trouble is encountered after the impeller replacement than with the original equipment, because the new cast-iron impeller is put into service in an old partially graphitized casing, setting up an even more effective galvanic cell. The new cast iron is anodic to the cathodic graphitized

surface, and the replacement impeller has been known to last only a small fraction of the lifetime of the original impeller.

3. *Chrome-stainless steel fittings.* The use of 13% chrome stainless impellers, rings, and sleeves has generally resulted in considerable improvement and eliminates the graphitization of cast iron impellers. The stainless steel is also more resistant to the effects of cavitation. If the pump is operated constantly, the 13% chrome stainless tends to avoid the serious pitting that has been observed in tests under quiet, unagitated conditions. For this reason, intermittent service is less favorable to long life than operation on constant duty.

I remember one particular instance concerning an 8 in. single-stage double-suction pump that had to be repaired approximately every year and that had been fitted first with bronze impellers and then cast-iron ones. It was reequipped with 13% chrome stainless steel impeller, rings, and sleeves. The last I heard of it, the pump had given uninterrupted service for 4 years and was still going strong.

4. *Chrome-nickel stainless steels.* When such materials as 20 chrome-24 nickel stainless are used instead of 13% chrome steels for the pump fittings, even less galvanic action is encountered between the fittings and the cast-iron casing. In addition, the high chrome-nickel content stainless steels have the advantage of being completely impervious to seawater action under any conditions. Thus, if a pump is shut down frequently, the 13% chrome stainless steel parts would eventually pit, but the high chrome-nickel content steels would not.

5. *An all-bronze pump* should give the longest life for seawater conditions, and I know of pumps lasting over 20 years handling seawater and even harbor water. Of course, not all bronzes are equally suitable for this service.

To get maximum life from a pump and to minimize the influence of the saltwater in combination with a suction lift, it is advisable to use a so-called *zincless bronze*, as, for instance, an alloy of 89% copper, 10% tin, and 1% lead. On the other hand, the type SAE-43 bronze, which is made up of 53 to 62% copper and 38 to 47% zinc and which is frequently used for propeller wheels in saltwater is completely unsuitable, especially in proximity with cast-iron because of the rapid dezincification that will occur.

In general, one must realize the difficulty of making very definite statements as to the possible wear expected on a pump. The wear and ultimate life of a pump handling seawater and operating under high suction lift will depend on many different factors, such as the relative freedom from contamination of the water, its acidity, and whether the pump operates regularly at its normal head and capacity or whether it is at times operated at very reduced or abnormal capacity. Irregular operation may make a difference on the wear of as much as 4:1 when compared with the regular operation of the pump.

Question 6.43 Further Discussion of Materials for Seawater Pumps

One of your earlier centrifugal pump clinics dealt with pumps handling seawater. Although it did refer to chrome-nickel stainless steel pumps, it stated that "an all-bronze pump should give the longest life for seawater conditions." Of course, you qualified this by stating that not all bronzes are equally suitable for this service and that zincless bronzes should be used.

Would not *all-316 stainless steel* pumps be preferable to all-bronze pumps on seawater service? I am refering specifically to vertical turbine pumps.

Answer. I must admit that my answer to Question 6.42 was slightly ambiguous. There is no question that an all-316 stainless steel pump will withstand the action of seawater better than any other metal (other than possibly some exotic and highly expensive metal, such as titanium). Note that one of the paragraphs in that clinic states". . . the high chrome-nickel content stainless steels have the advantage of being completely impervious to seawater action." Thus, one can infer even in the light of the statement you quote ("an all-bronze pump should give the longest life for seawater conditions") that such stainless steels are superior to all-bronze pumps.

Nevertheless, you are right in questioning the wording of that particular centrifugal pump clinic. In addition, the paragraph in question referred to materials such as 20 chrome-24 nickel stainless steels and made no mention of 316 stainless. Yet the latter, although not equal to the higher chrome-nickel series, is superior to all-bronze.

It should be remembered that at the time the original article appeared, the use of 316 stainless steel was considered to be almost prohibitive in cost. It is only in more recent times that there has been so little difference in cost between all-bronze and all-316 stainless steel pumps.

A note of caution is necessary: if pumps remain immersed during extended periods of idleness and if the water is essentially stagnant, corrosion of 316 stainless may occur from oxygen starvation. This effect is aggravated by warm water (above about 85 °F) and/or high concentrations of marine life. Where such conditions exist, it would be better to use 316 L.

Question 6.44 Corrosion of Bronze-Fitted Cast-Iron Bowl Vertical Turbine Pumps

We have a problem about which we would like to ask your advice. We have a number of vertical turbine pumps fitted with bronze impellers, stainless steel shafts, and cast-iron bowls. When the pumps were removed after less than 4000 hr of operation, severe corrosion was found in the cast-iron bowls. The analysis of the water in the wells indicates pH values from 6.7 to 6.8. It also contains some free carbon dioxide.

What are the reasons for this corrosion, and what would be an economical remedy for it? Could vitreous enameling of the bowls extend the life of the bowls?

Answer. The corrosion is related to the fact that your pumps contain dissimilar metals in the presence of an acid solution. The presence of carbon dioxide accelerates this corrosion. As a matter of fact, as the pressure increases, the amount of carbon dioxide going into solution will also increase. Thus, in the last stages of the pumps the pH will be even lower than the 6.7 your analysis shows, and with this higher acidity I expect that the corrosion you have found is more severe in the last stages of the pump than in the first stages. In the presence of acid water, a galvanic action is set up between the bronze impellers and the cast-iron bowls. The cast iron is anodic to the bronze, the electrochemical reaction causes an electric current, and small metal particles flow from the cast iron (anodic, corroded metal) to the bronze (cathodic, protected) end of the electrochemical cell.

The injection of a neutralizing material into the water (such as either sodium carbonate or sodium hydroxide) would be helpful by increasing the pH of the water. This, however, may not be too practical.

Vitreous enameling of turbine pump bowls is employed for better efficiency and to provide some protection against erosion from abrasive particles in the water. It is not particularly useful for corrosion resistance as it cannot be applied to machined fits, and these would still be exposed to corrosion. Furthermore, enameling of existing pump bowls would not be practical because the process requires firing at temperatures around 1400°F. This would distort the machined parts of the bowls.

You may try applying sprayed epoxy coatings to your existing bowls in an attempt to lengthen their life. However, the only certain solution is to replace them with bronze bowls. An all-bronze pump should give you very satisfactory life even when handling water with the characteristics you describe.

Question 6.45 Unequal Damage to the Two Sides of a Double-Suction Impeller

We recently had occasion to inspect a 12 in. single-stage double-suction pump after only 6 months' service. Much to our surprise we noted severe erosive wear in the inlet area of one side of the impeller, but no wear is apparent on the other side. The wear seems to be similar to that produced by cavitation. Why should there be such a wear pattern? Could it be caused by some improper foundry procedure, with one side of the impeller having some imperfection created by the casting process, which made one side of the impeller more vulnerable to cavitation attack than the other side?

The pump is designed to handle 4000 GPM of cold water against a total head of 180 ft at 1750 RPM. The performance curve furnished with the pump indicates that at this capacity of 4000 GPM, the pump requires 20 ft of NPSH, but the installation provides 22 ft NPSH. This margin appears to be ample to prevent cavitation.

Answer. I don't think that the problem you have described can be attributed to the casting process. Instead, the symptoms are indicative of a rather frequent field problem resulting from the configuration of the suction piping, I refer to the use of an elbow immediately at the suction flange of a double-suction pump, with the elbow in a plane parallel to the pump shaft. (See Fig. 6.42B.)

When liquid flows through an elbow, both higher pressures and higher velocities occur on the outside of the turn. Incidentally, this is not a contradiction of Bernoulli's law, which applies only to individual streamlines. The phenomenon of this pressure and velocity distribution is caused by a conversion of streamlines as a result of inertia.

Thus, if your installation includes such an elbow, as I suspect, there will be a higher capacity flowing to one side of the impeller than to the other. Although I cannot predict the exact flow distributions between the two sides of the impeller, for the sake of establishing the effect of such an uneven distribution, let us assume that when the total flow is 4000 GPM, 2200 GPM will flow to one side of the impeller and the second half handles only 1800 GPM. Now let us examine the relation between flows and NPSH required. Figure 6.43 shows the NPSH plotted both against the total flow and against the theoretical one-half flow entering each side of the impeller. Although, as you say, the required NPSH at 4000 GPM is 20 ft (or at 2000 GPM per side), this NPSH becomes a little over 24 ft at a flow of 2200 GPM per side. This is in excess of the 22 ft available, and the pump will cavitate on that side of the impeller (not on the other side), creating the symptoms you have described.

The preferred arrangement for an elbow at the suction is that shown in Fig. 6.42A, that is, with the elbow in a plane at right angles to the pump shaft. If for whatever reason your installation cannot accommodate such an arrangement, it is necessary to provide a section of straight piping between the elbow and the pump suction flange of a length of at least 5 to 10 pipe diameters.

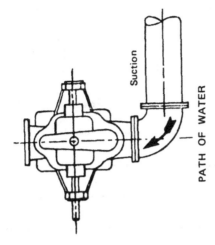

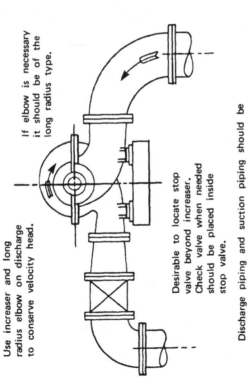

Figure 6.42 Suction elbow on double-suction pump.

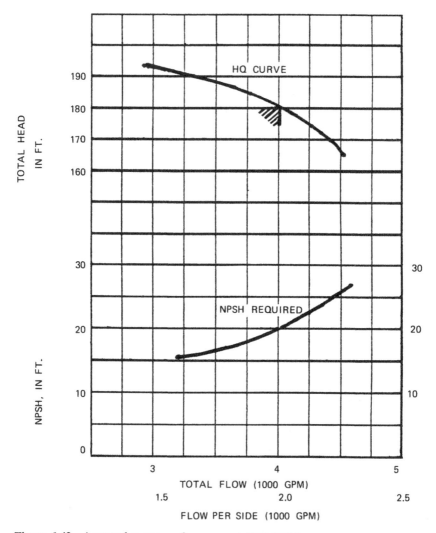

Figure 6.43 Assumed pump performance at 1750 RPM.

Question 6.46 Leakage at Boiler Feed Pump Gasket

Our power plant is served by three axially split casing multistage boiler feed pumps. The pumps have given quite satisfactory service. We have, however, one problem that we would like to eliminate. When the boiler feed pumps have to take cold water whenever there is a sudden shutdown of hot water from the heaters, there is a certain amount of leakage at the bolts holding the horizontal casing flanges. Can you suggest the cause of this leakage and a solution that would eliminate it?

Answer. I would like first to comment on the sudden change in temperature of the feedwater that you mention. You do not describe the type of feedwater cycle used in your plant. Of course, sudden load variations will lead to changes in feedwater temperature. However, the rapidity of the temperature changes and the range of these changes vary depending on whether an open or a closed cycle is employed.

An open cycle is one that embodies a direct-contact heater at the suction of the feed pumps; in the closed cycle all the heaters are of the closed type, and the condensate pump delivers feedwater to the boiler feed pump suction directly through one or more of these closed heaters.

The reduction of temperatures in an open cycle is not so drastic or so rapid as it is in a closed cycle. For instance, if the feedwater temperature is 300°F, a complete trip out of the main turbine will be followed by a reduction in temperature to about 220°F, because the direct-contact heater is generally provided with means to prevent operation under vacuum. These means consist of a pressure *pegging* valve that opens to admit auxiliary steam to the direct-contact heater when the main supply from the turbine extraction stage is lost. The reduction in temperature from 300 to 220°F may take as long as 5 to 10 min because of the large amount of heat stored in the storage section of the heater.

In the case of a closed cycle, a complete trip out will bring the feedwater temperature right down to that of the condensate, since no bleed steam is made available under these conditions to the individual closed heaters. Thus, the feedwater temperature may drop to as low as 100 or 120°F. Furthermore, the temperature change is quite abrupt because it takes place as soon as all the hot water in the line between the condensate pump and the boiler feed pump has been evacuated and replaced by cold water. This may take as little as 1 or 2 min or less.

The fact that your boiler feed pumps are of the axially split casing type implies that they are designed for a discharge pressure that does not exceed 1250 psi, as radially split casing barrel-type pumps would be used for higher pressures. Consequently the main turbine pressure probably does not exceed 850 psi. The majority of installations operating at 850 psi or lower generally employ an open feedwater cycle, so that I am puzzled by your reference to drastic temperature changes.

Now as to the cause of the leakage at the bolts: When a pump is operating in its normal load range or, at least, within a narrow range of temperatures, the pump casing and the bolts reach a uniform temperature.

Since both have the same coefficient of thermal expansion, they expand equally, and the casing gasket is maintained under uniform compression to prevent leakage.

If the casing becomes cooled abruptly, it contracts more rapidly than the bolts, since the latter are not in intimate contact with the casing where they pass through drilled holes in one-half of the casing flange. Since the casing studs cannot immediately follow the contraction of the flange thickness, the compressive load on the gasket is temporarily relieved, and leakage may occur until the studs also cool down to the casing temperature.

Assuming that drastic temperature changes cannot be eliminated, the only solution lies in the selection of a gasket material having sufficient recoverability characteristics to follow changes in flange thickness with a minimum loss of residual stress in the gasket. Such materials do not have the tensile strength or temperature resistance of gaskets normally preferred for axially split casing multistage boiler feed pumps. They are, however, suitable for boiler feed service, and in this case their high recoverability characteristics would seem to take precedence over the other properties.*

Question 6.47 Shaft Sleeve and Packing Wear

Our boiler feed pumps are provided with packed stuffing boxes cooled with circulating water. We are experiencing considerable excess of leakage and rapid sleeve wear. We have lately been using copper foil-type packing, but this has not improved the situation.

In addition, we have the problem of incrustation of the internal water passages, water-cooled outer covers, and gland sleeve cooling-water passages. Filtered water was being used for cooling purposes. After switching over the supply to condensate from outlet of condensate pumps, the problem has actually been overcome. Although the problem has been solved for the present, we request your comments on this problem for our information and future guidance.

Answer. Most packing being sold for this type of service should be capable of running up to 12 months without difficulty. My suspicion is not that the packing is inadequate but rather that packing installation practices may need to be upgraded.

Stuffing-box maintenance primarily consists of packing replacement. Although this sound simple, it must be done correctly, or pump operation will not be satisfactory. The procedure to be followed in repacking a stuffing box is described in detail in Question 5.18.

---

*See also the answers to Question 5.13 and 5.14.

Packing removed from a stuffing box being repacked should be examined in order to obtain as much information as possible on the cause of packing wear. Often, correctable operating conditions or inadequate packing procedures are revealed by this examination. For a detailed analysis of symptoms, see Question 5.19.

As to the incrustation in the cooling-water passages, I am afraid that if the water used for cooling contains considerable minerals that may be deposited in cooling passages, the only solution is to do what you are doing, that is, to use condensate for cooling purposes. Remember that this is not a loss of water because the condensate can either be returned back to the condenser or rejoin the stream of the condensate being delivered to the deaerating heater.

Question 6.48   Source for Condensate Injection Sealing

We have an installation with an open feedwater cycle, that is, with the boiler feed pumps taking their suction from a direct-contact deaerator. The boiler feed pumps are provided with condensate injection sealing at the stuffing boxes. We are concerned over the possibility that condensate pump delivery may be interrupted while the main boiler feed pumps may continue to operate for some significant period of time by drawing down on the deaerator storage. Under this condition, we would not have adequate condensate injection supply to the stuffing boxes. Is there a means available to arrange injection sealing without recourse to the condensate pump discharge?

Answer. There will always be some risks in the operation of a steam power plant. We can reduce these risks—sometimes at quite a cost—but we can never hope to eliminate risk completely.

Two possible arrangements could be used to provide an alternative source of condensate injection:

1. Use the pump discharge flow, passing it through a pressure-reducing orifice and then through a heat exchanger to cool it down to around 100 °F.
2. Install a small standby condensate injection supply pump, possibly driven by a dc motor, to provide a temporary supply of condensate injection until the regular condensate pumps have been restored to service or until the main feed pumps have been shut down and secured if the interruption of condensate is to be of long duration.

Either arrangement can be made automatically responsive to failure of condensate pump discharge pressure below some predetermined minimum.

Question 6.49 Possible Contamination of Feedwater at Condensate Injection Seals

In one of your earlier clinics you commented on the question of the contamination of the feedwater by oxygen caused by the use of condensate injection seals. As I remember it, you pointed out that the extent of this contamination is negligible. What does concern me, however, is the fact that the drains from that portion of the condensate injection flow that is ultimately returned to the condenser may be almost saturated with oxygen. Will this fact create any problems in the ability of the condenser to provide condensate with 0.01 mL/L or less, which is the guarantee usually made for deaerating condensers?

Answer. You are quite correct in recalling my statements. I had used an example in which the total feedwater flow was 3000 GPM and the inflow from the condensate injection through worn labyrinths had reached 10 GPM. I had assumed a nondeaerating condenser, with the condensate containing 0.03 mL/L of dissolved oxygen. The mixing of such condensate with 3000 GPM of deaerated feedwater would cause an increase in oxygen concentration of only 0.0001 mL/L, which is certainly negligible. I had added that if we were dealing with a deaerating condenser in a closed feedwater cycle, the condensate injection would have exactly the same oxygen concentration as the feedwater entering the pump.

Matters are slightly different when it comes to the condensate injection drains returned back to the condenser, but still there should be no problem here.

Water thrown off the high-speed shaft into the ventilated collection chamber (see Fig. 6.44) may reach as much as 75% saturation before it reaches the drain pipes (see Fig. 6.45). Assuming that this drain piping has been selected with ample margin and therefore is running only partially full and that this piping is of considerable length, the oxygen content of the drains could approach 90% of saturation at the traps. Of course, if as in many cases the drains are not returned directly to the condenser but rather to an open tank where all other drains are returned before being pumped to the condenser, the condensate injection drains will probably approach 100% of saturation.

Since the saturation level at 100 °F and atmospheric pressure is about 4.7 mL/L, the amount of oxygen in the condensate injection drains appears to be an appreciable amount. On the other hand, these drains are a rather small proportion compared with all the other returns and the makeup that is generally fed to this open tank. The deaerating capacity of a modern

Chapter 6

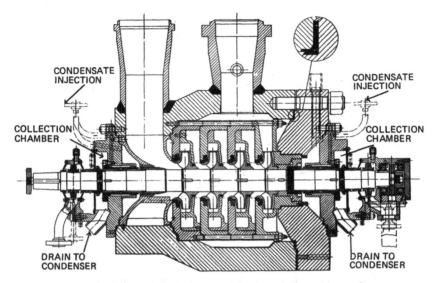

Figure 6.44 Section showing condensate injection sealing construction.

deaerating condenser is greatly in excess of the requirements covering the sum of all drains and of the maximum amount of makeup, and you need not concern yourself with the fact that condensate injection drains may be saturated with oxygen.

Question 6.50 Maintenance Problems with Mechanical Seals

I am searching for a remedy for our heating system. We have a number of pumps with carbon mechanical seals. We also have fine sand in this system, and this cuts the seals rapidly. I would appreciate any suggestions you might make to solve this problem.

Answer. The increased use of mechanical seals on many pumping applications is obviously an indication of the fact that mechanical seals have a number of advantages over conventional stuffing-box packing for these applications. In addition to their most important advantage of reducing leakage to the point at which to all intents and purposes it is eliminated, they reduce wear at the shaft or shaft sleeves, reduce stuffing-box friction losses, and—when properly applied—reduce maintenance. The first and the last of these advantages, that is, reduction or elimination of leakage and

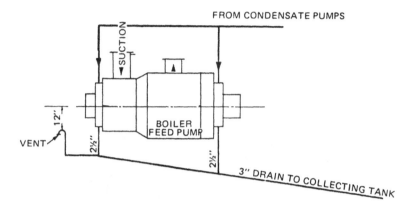

Figure 6.45 Drains from condensate injection seal collection chambers.

reduction of maintenance, are the two basic reasons so many pumps on circulating service in heating systems are furnished with mechanical seals.

One must realize, however, that these advantages can only be obtained if the seals are applied properly, installed in an adequate manner, and made to operate under favorable conditions. In your particular case I can assume that the seals have been selected with full knowledge of the normal expected conditions of service as to pressure and temperature. I shall also assume that they have been installed in the pumps with the necessary care. The only difficulty that is apparent from your question is that they do not operate under favorable conditions because of the fine sand present in your system.

I shall not inquire as to the reasons the system is not clean. If it were not for the fine sand, your system could become contaminated with mill scale from the piping or with rust, and therefore I shall assume that you have concluded that cleaning out the system completely is not practical.

The presence of foreign matter in the liquid being pumped is probably one of the most prevalent causes of premature failure of mechanical seals. The basic principle of mechanical seals involves the use of two perfectly flat surfaces, one stationary and one rotating, separated by a thin liquid film that prevents contact between the faces. These faces are generally lapped to a flatness of within two light wavebands (about 0.00002 in.). Any roughness of these surfaces caused by foreign matter abrasion will destroy the effectiveness of the mechanical seal. Therefore, it is imperative to prevent the intrusion of foreign particles into the general area of the seal faces, lest they work their way between the faces and damage them through abrasion.

Chapter 6

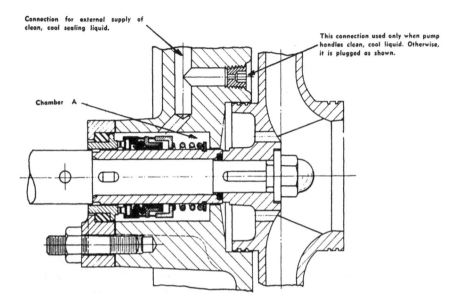

Figure 6.46 Single-stage pump with single mechanical seal.

If you refer to Fig. 6.46, which shows a typical centrifugal pump equipped with a single mechanical seal, you will note that the liquid pumped fills the chamber A adjacent to the seal. Therefore, any foreign particles present in this liquid can work their way to the seal faces. How, then, if the liquid pumped always contains foreign matter, is it possible to supply clean liquid to chamber A to prevent this? First, this clean liquid must come from some external source, since the liquid pumped is not clean. Second, it must be provided at some pressure higher than that which prevails in the chamber—say at 15 to 20 psi higher pressure.

This clean liquid will form an isolating barrier between the seal and the liquid handled by the pump, and a portion of the injection liquid will flow into the pump proper. To prevent an excessive amount of this clear *isolating* liquid from flowing into the pump, it becomes necessary to locate a bushing at the bottom of the stuffing box and to maintain the minimum possible clearance between the stationary bushing and the rotating shaft.

If the amount of clear liquid that will enter the pump must be held to a very minimum—that is, to practically nothing—it becomes necessary to use a double seal, such as illustrated schematically in Fig. 6.47 and shown mounted in a pump in Fig 6.48.

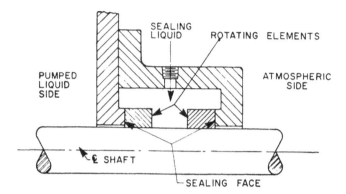

Figure 6-47 Schematic arrangement of double seal.

Which of the two arrangements is best to use in your particular case will depend on a number of factors. To begin with, I do not know whether you have a source of clean water available for this injection or if it is practical to supply it to each pump in question. If the answer is negative, it will become necessary to use the water being pumped, passing it first through a microfilter that can eliminate particles down to about 5 or 10 μm in size. This water can be taken directly from the pump discharge, taken through the filter, and led to the chamber adjacent to the seal.

If you use a single seal and a throttling bushing at the bottom of the stuffing box, you may have as much as 1 GPM of water flowing into the chamber (depending on the size of the pump, the pressure differential, and the internal clearance at the bushing). The filter size has to be adequate for a somewhat greater flow that will take place when the bushing is worn and the clearance increases. Also, if the fine sand is present in appreciable quantity, the filter will clog up after a while and will have to be attended to.

On the other hand, if the pump stuffing box has sufficient room to permit the installation of a double mechanical seal, there will be no significant flow into the chamber between the two seals, and the injection sealing will essentially be merely a pressurizing system that will maintain a pressure in this chamber some 15 to 20 psi higher than in the pump next to the inner seal. The flow into the chamber will be of the order of a few drops per hour. The filter can be much smaller and will probably never clog up during the lifetime of the installation, the injection water just barely seeping through the filter.

To sum up, then, if you can fit a double seal into your pumps, do so. Provide clean filtered water to the chamber between the two seals, and they will last a long time. If you cannot use a double seal, make sure that there is a

Chapter 6

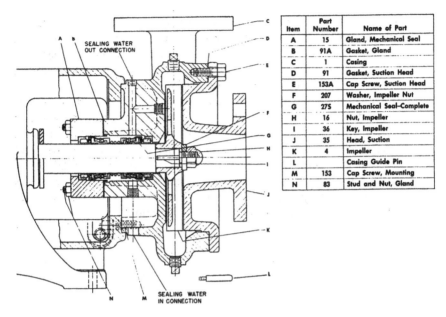

Item	Part Number	Name of Part
A	15	Gland, Mechanical Seal
B	91A	Gasket, Gland
C	1	Casing
D	91	Gasket, Suction Head
E	153A	Cap Screw, Suction Head
F	207	Washer, Impeller Nut
G	275	Mechanical Seal–Complete
H	16	Nut, Impeller
I	36	Key, Impeller
J	35	Head, Suction
K	4	Impeller
L		Casing Guide Pin
M	153	Cap Screw, Mounting
N	83	Stud and Nut, Gland

Figure 6.48 Single-stage pump with double mechanical seal.

throttling bushing at the bottom of the pump stuffing box and inject clean filtered water ahead of the seal. You will need somewhat more water than with the double seal, but you will still get adequate life from your equipment.

Question 6.51 Centrifugal Separators

It is with great interest that we have read your article in which you answered a question concerning abrasives in the form of sand that shorten mechanical seal life to a significant degree. We certainly agree with your approach to this problem but would like to also suggest that the use of centrifugal separators should be considered since they are being used more and more lately to combat just such contamination problems as you are referring to.

Answer. Indeed, you are quite right in calling my attention to the growing use of centrifugal separators, and I believe that I shall do a service to users of centrifugal pumps and of mechanical seals if I give some information on these separators. As a matter of fact, we have used a number of these separators in connection with several water-lubricated bearing pumps and have found them quite satisfactory.

The development of the centrifugal separator was probably stimulated by the fact that some form of filtration is absolutely essential if absolutely clean sealing water is to be introduced into the mechanical seal cavity. Although excellent fiters are available for this purpose, they all suffer from the fact that they ultimately get clogged up, and unless they are frequently attended and backwashed or otherwise cleaned out, sealing water supply will be interrupted to the great detriment of the mechanical seal.

The operating principle of the centrifugal separator is based on the fact that if liquid under pressure is introduced tangentially into a vortexing chamber, centrifugal force will tend to make it rotate in the chamber, creating a vortex. Particles heavier than the liquid in which they are carried will tend to hug the outside wall of the vortexing chamber, and the liquid in the center of the chamber will be relatively free of foreign matter. Such a separator is shown mounted on a pump in Fig. 6.49 and its action is demonstrated in Fig. 6.50, where a clear plastic model is shown in operation.

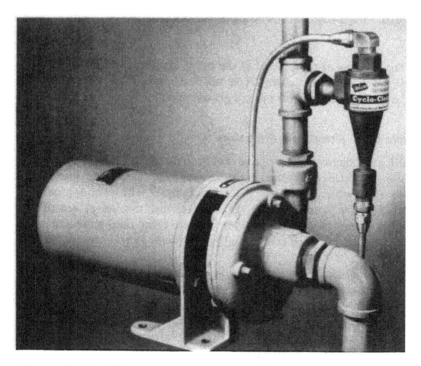

Figure 6.49 Centrifugal separator mounted on a pump to supply clean sealing liquid to the mechanical seal.

Chapter 6

Figure 6.50 Illustration of the principle of centrifugal separators.

Liquid piped from the pump discharge enters the feed tap and is rotated inside the cone-shaped bore at a velocity dependent on the available pressure differential. The heavier particles remain close to the tapered wall of the cone (see the spiral line in Fig. 6.50) and flow downward to the discharge outlet at the bottom of the separator. They are piped back to the suction side of the pump along with a small amount of liquid if this liquid is not to be lost. If the liquid can be disposed of, the outlet can be connected to a waste drain. The clean liquid rises up through the outlet at the top and center of the separator and is then piped to the housing for flushing the seal faces of the mechanical seal. If conventional packing is used for the pump, the same arrangement can be used to provide clean sealing liquid to the lantern gland or seal cage.

To be effective, a centrifugal separator must be applied to an installation where the abrasive particles have a higher specific gravity than the transporting liquid and where there is enough pressure differential to provide an adequate supply of clean liquid. About 50% of the liquid supplied at the feed tap is eliminated with the particles at the bottom outlet, and the remaining 50% is available as clean sealing liquid.

Sand that will pass through a #40 sieve will be 100% eliminated in such a separator with supply pressures as low as 20 psi. A typical tabulation of the effect of supply pressure and particle size on particle elimination is given in Table 6.7. Note that the centrifugal separator is not 100% efficient. No manmade apparatus ever is. But it is very obvious that a significant portion of the smallest particles is eliminated and that all large particles, such as sand, are eliminated through its use. There is no question but that a mechanical seal supplied with sealing liquid that has gone through a centrifugal separator will last longer than one that takes raw sealing liquid supply.

Table 6.7 Effectiveness of Cyclone Separators

Particle size (μm)	Differential pressure (psi)	% particle elimination
15	20	95.6
15	50	97.3
15	100	99.4
5	20	88.3
5	50	92.6
5	100	95.5

Question 6.52 Short Life of Mechanical Seals

Not every problem with which I have been involved can be recounted in question/answer form. In some cases, the facts underlying the problem are difficult, if not impossible, to ascertain from correspondence. This is particularly true when the person reporting the problem is not aware that some specific circumstances have a significant bearing on the sequence of events that create the problem. As a result, such circumstances are not reported, and only a visit to the site of the installation and a most thorough investigation of the facts can possibly lead to a solution.

In this clinic I shall recount such an instance. I was first alerted to the problem by a telephone call that described the symptoms in rather general terms. My questions failed to elicit a plausible clue to the source of the problems and therefore to a possible solution. It was obvious that an on-site investigation was required.

Case of the Short-Lived Seals

The equipment in question consisted of two 10-in. double-suction single-stage pumps installed in the basement of a large office building and used on air-conditioning condenser circulating service in conjunction with the cooling towers on the roof of the building. The pumps were furnished with mechanical seals at the stuffing boxes.

From the very start of operation of these pumps the mechanical seals had given poor service and had been replaced three times a year in one pump and four times in the second.

On-site inspection disclosed that the pumps were installed some 300 ft below the sump of the cooling tower, with a total of about 450 ft of suction piping, as there were several horizontal runs. Deducting the suction piping friction losses from the 300 ft static head, suction pressure at the pumps was approximately 110 psig.

A visit to the cooling tower on the roof revealed that the water circulated was very dirty; there was approximately a 1 in layer of "muck" on the bottom of the cooling tower sump floor. The system had apparently never been cleaned because, said the maintenance man, no one had told him that it needed to be cleaned.

There was very little mystery now with respect to this problem. Mechanical seals *will not operate satisfactorily* if exposed to dirty liquids. The foreign matter will wear the faces of a mechanical seal very rapidly and will collect around and under the O rings and the springs.

Dirty water can be particularly bad when a pump is started up. When the pump is not running, sediment in the water contained in the suction piping will settle in each of the horizontal runs. When a pump is started, slugs of concentrated muck will go through the pumps and aggravate the situation at the mechanical seals.

It was imperative to keep the dirty water away from the seals. Cleaning the system every year or even every 6 months would not be the answer. Elimination of all the dirt as a continuous process would be very expensive if the water were to be maintained free of foreign matter to a degree suitable for satisfactory seal life.

The most economical and logical solution was to use city water pumped into the seal chamber to exclude dirt from the vicinity of the seals. Referring to Fig. 6.51, it becomes necessary to provide clean liquid from an external source at a pressure higher than that which prevails in the seal chamber. The excess pressure should be about 15 to 20 psi and must provide flow whether the pump is running or not. When the pump is running, the pressure in the seal chamber is the suction pressure, or 110 psig. When the pump is idle, suction pressure will correspond to the 300 ft static head from the cooling tower sump, or 130 psig. The supply of clean water should be about 150 psig, which is higher than available city water pressure and requires a small booster pump.

The best means of providing this is to use a small water-sealing package unit similar to that illustrated in Fig. 6.52. A tank is equipped with a float valve that maintains the water level in the tank. A small close-coupled pump is mounted directly on the tank and maintains the supply of clean water at the stuffing-box seals of the battery of pumps it serves. Incidentally, this system

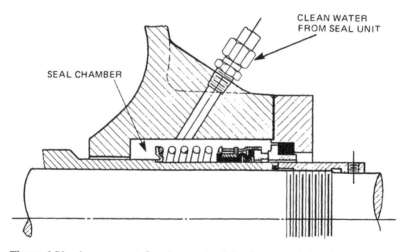

Figure 6.51 Arrangement for clean water injection to seal chamber.

Figure 6.52 Water-sealing supply unit.

does not require any extra city water because the supply of water to the seals (approximately 2 GPM per seal) reduces the quantity of makeup water to the system. In this particular case, total makeup was 200 GPM.

Question 6.53 Acid-Handling Chemical Pumps

We have a large number of frame-mounted end-suction single-stage chemical pumps ranging in size from 1" to 4", installed in an area where acids are handled. A considerable number of bearing failures have taken place because of acid vapors or spillage entering the bearing housings through the Klozure seals.

The pumps have external slinger rings mounted on the shaft, but they are ineffective because they merely sling the chemicals to the top of the body yoke from where they trickle back down to the shaft Klozure seal. We tried installing deflectors to eliminate trickling, but this did not prove entirely satisfactory. We have talked to a number of people about improved seals but never received satisfactory answers.

I would like to know if an air purge system could be installed to pressurize the bearing housing, thereby creating a positive pressure to prevent foreign materials from entering the housing. Of course, the pressure would be only a few inches of water, which could be controlled by an air purge meter.

My main concern is whether the pressurization of the bearing housing would in any way affect proper lubrication of the bearings. Your comments on the practicability of this type of installation will be greatly appreciated.

Answer. Bearing failure caused by the entrainment of acid vapors through the Klozure seals is not a normal or frequent occurrence. Tens of thousands of chemical pumps are installed in acid service without encountering such difficulties, since they are generally designed for this type of service. On the other hand, statistics by themselves can never effect a cure for an actual problem such as you describe. Therefore, some steps are indicated in your case.

In principle, if the bearing housing is pressurized just slightly, proper lubrication of the bearings will still take place. Exactly how much pressurization will be required must be established by experiment, because surface tension and capillary action at the Klozure seals may still permit the entrainment of some acid contamination into the bearing housing. On the other hand, excessive pressurization may cause some difficulty.

No modification need be made to the pumps if grease-lubricated bearings are involved (see the bottom part of Fig. 6.53) other than to connect the supply of air and to arrange control of air admission by means of an air purge meter.

Chapter 6

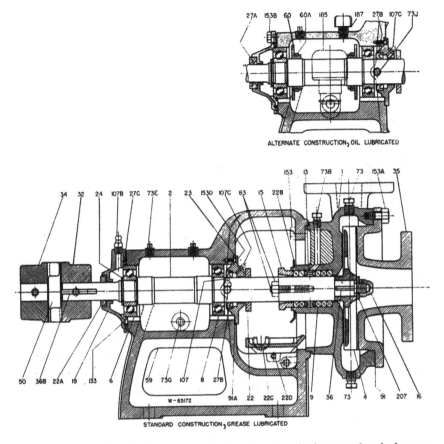

Figure 6.53 Section of a frame-mounted, end-suction single-stage chemical pump.

If the pump bearings are oil lubricated (top of Fig. 6.53), it is generally the practice to use a constant-level oiler. This device operates on the basis of a quantity of reserve oil stored in a bottle above the operating level of the oil (Fig. 6.54). The oil in the bottle is prevented from coming out since the opening to the bottle is below the operating oil level, and as long as air cannot get into the bottle, oil cannot come out. However, when the operating level drops, the opening to the bottle is uncovered, and air is allowed into it. Reserve oil is thereby released until the operating level rises to cover the opening again. In doing so, the operating level is maintained practically constant.

Therefore, if your pumps are provided with constant-level oilers, these must be removed if you pressurize the bearing housings, as oil may be

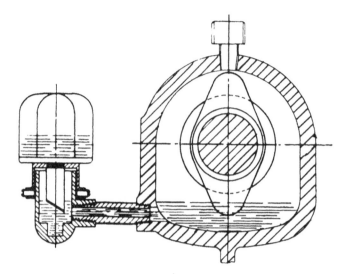

Figure 6.54 Constant-level oiler.

prevented from entering the housing or may even be blown out of the bottle reservoir if the pressurization is high enough. A sight gauge glass must be installed instead, with top and bottom connections. Oil will have to be added periodically on a scheduled basis and of course more frequently than if a constant-level oiler is retained.

A more sophisticated approach is to use the so-called *oil-mist* systems. These are combined filter-regulator-lubricators that supply clean and thoroughly lubricated air into the bearing housing at constant pressure.

There are several manufacturers who supply such oil-mist systems. Generally, one system supplies all the pumps in a given location. I have heard of an installation where some 2000 points are served from 28 oil-mist lubricating units.

Question 6.54 How to Limit Bearing Abrasion

I have a serious pump problem that appears to defy solution. Any comments relative to suggested methods of improvement would certainly be appreciated.

In essence, we use 2 and 3 in. vertical sump pumps of the type illustrated in Fig. 6.55. Approximately 100 of these pumps are installed to handle a very abrasive slurry containing approximately 10% maximum solids. There is no appreciable chemical attack. Pump lengths vary from 6 to 15 ft.

Chapter 6

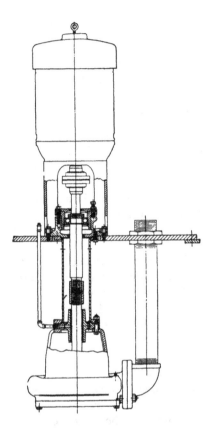

Figure 6.55 Vertical sump pump.

We have been experimenting with a pump in an effort to determine the best shaft and sleeve bushing material. These are the points of failure. The pumps are fitted with 316 stainless steel columns, cast-iron housings, and impellers. All run at 1750 RPM.

We find that we have never gotten over 700 hr operating time (the pumps are on a 24 hr/day service). This was with crushed glass-Teflon sleeve bushings and cold-rolled shafting (1 1/8 in.) with about 0.5 GPM flushout water to each of three bushings.

After this period, the pump began to vibrate excessively and had to be pulled. Examination showed that one of the crushed glass-Teflon sleeve bushings had "disappeared" and that the other two were badly pitted, with mud particles embedded. The cold-rolled steel shaft was severely nicked down at each sleeve bushing location.

We have tried the following materials with not so much success as above:

1. Brass bushings (Brinell 55) with cold-rolled shafting
2. Urethane polymer bushings with cold-rolled shafting
3. Cast-iron bushings with cold-rolled shafting
4. Copper-graphite bushings with cold-rolled shafting

Our latest experiments have been with Stoody "6" bushings (400 to 500 Brinell hardness number) and cold-rolled shafting. We found little wear on the bushings but great wear on the shafting. Service life was about 200 hr.

We are contemplating using Stoody "6" bushings with Metco-sprayed molybdenum hard-surface shafting and, in theory, hope to make a miniature crushing roll out of each bushing and shaft area and flushing out the resultant finer particles. Normal clearances of 0.015 to 0.020 in. are used on these materials. Operating temperatures are about 65 °C.

Can you furnish us with some suggestions? Is our approach toward a *crusher roll* correct? How rigid must the vertical pump and motor be on the tank mounting flanges to prevent minute vibrations that may ultimately increase, causing *whipping* at the foot of the shaft? Evidence lends itself to the belief that these vertical motors are not mounted rigidly enough. What is the effect of oversize shafts toward whipping?

Answer. In my opinion, the major source of the difficulties you have been encountering lies in the fact that the pumps you describe are essentially water pumps designed for sump dewatering service, not process pumps. They are therefore far from being rugged enough to withstand the severe service to which they are applied. Although the material selection has been upgraded to improve the potential life of such parts as the pump columns and the bearing bushings, the sole effect is to lengthen the life of these parts slightly over what it would be with standard construction, for instance, with standard bronze bushings. The mechanical inadequacy is hardly improved.

The major problem seems to be the fact that the shafting appears to be too light and probably "whips" to a considerable degree. I do not know the critical speed of these pumps or therefore the magnitude of deflection, but from all appearances this is a "light" construction with a considerable tendency to accelerate wear in the bearings regardless of the materials used there. Even if clear, filtered flushing water is supplied to the bearings, slight unbalancing forces and shaft vibrations send the shaft to one side or another. All the clearance exists on the opposite side, and dirt particles get in to cut the bushings or the shaft.

It appears to me also that the clearances of 0.015 to 0.020 in. that you mention (and which I assume are diametral clearances) are too liberal to permit

Chapter 6

the bushings to act as good steady bearings. A diametral clearance of 0.004 to 0.005 in. would be the maximum that could be relied upon to give adequate bearing action.

Over the years, there have been three schools of thought on how to minimize the rate of cutting at such bearing surfaces:

1. Make very hard surfaces.
2. Provide an elastic surface on one element that permits depressing the foreign particles into the material during the rubbing process.
3. Exclude *any* foreign particles from the bushing area.

The first method has seldom been applied to vertical sump pumps because it is expensive and because it only reduces the rate of cutting. Even when the additional cost of the materials may be justified because of the severity of the service, as in your case, I doubt that alone this method will be satisfactory, since it may not provide an acceptable life to the parts. The crushing roll effect has its limitations.

The second method, that is, of providing an elastic surface, involves the use of a liberal jet of clean water and a lot of longitudinal grooves into which the foreign particles may be dumped and washed away after being "rolled" by rotation to the grooves. This is the *cutless rubber bearing* solution that has frequently been used for wet-pit pumps handling water with some sand particles. Whether it would be practical in this case is hard to say, but I doubt that it would be effective with the concentration of mud that you encounter. Furthermore, experience has shown that cutless rubber bearings are a failure unless properly flushed down before the pump is started.

The third solution, especially if combined with the hard materials of the first one, is the one that I would recommend. The complete exclusion of mud particles from the bushings, however, would require in my opinion considerably more than the 0.5 GPM flushing water that you mention even if the clearances are reduced to the 0.004 or 0.005 in. that I recommend. Although one could calculate the amount of water required, given the diameters, the clearances, and the grooving, I believe that a figure of 2 to 3 GPM is more nearly the required quantity. Actually, rather than a specific flow, it is necessary to provide a sufficient differential pressure to ensure flow in the desired direction and accept whatever flushing flow this will give you. With a liberal flow flushing into grooved or tapered bushings, it should be possible to assure elimination of abrasive particles within the bearings.

Now as to materials: Assuming that shaft size cannot be increased but that clearances of the order of 0.005 in. can be provided and that you will supply clear filtered water to flush mud particles, the choice that you contemplate is in the right direction. Stoody "6" bushings with either Metco-sprayed shafting or Colmonoy, Stellite, or ceramic surfaces on the shaft should result in a considerable improvement in pump life. I have also found that Graphitar 39 is an effective material for bearings to use in conjunction with hard sleeve materials in this type of pump.

Of course, if heavier shafts can be incorporated in the pump construction, you will decidedly get further improvement of the pump life between overhauls.

The motor mounting must be quite rigid and, of course, "square" so as to avoid vibration on the one hand and a "cocking" effect of the shaft running in the bearings on the other. Proper alignment assumes greater importance where accelerated wear at the bearings can lead to the entrance of abrasive foreign particles into the clearances and further acceleration of this wear that leads to ultimate failure.

I shall certainly be interested in learning what improvements are obtained from the use of closer clearances, adequate supply of clean flushing water, and harder materials.

Question 6.55 Iron Compound Deposit on Impellers

Our new 25,000 kW unit is served by two 100% capacity boiler feed pumps that are duplicates of pumps installed in an older station. Recently we noted a very rapid increase in power input to these feed pumps, and in our concern over the possibility of premature wear, we had them opened up. No internal leakage was reported, nor were the clearances markedly increased. The rotor was cleaned up and the pump reassembled. After reassembly, the pump power consumption returned to normal. What could have been the cause of the increase in power we noted, and why did this disappear after the pumps were opened up? We did not replace any of the wearing parts.

Answer. Although you do not mention the appearance of the pump parts when the pump was opened up, you do state that "the rotor was cleaned up." I imagine that your maintenance mechanics found the impellers heavily coated with oxides and brushed them or sandblasted them, depending on the degree of adherence of the oxides to the impellers.

Caution should be exercised in the interpretation of field test results that show evidence of premature reduction in the pump net capacity or of a rapid increase in power consumption. The first conclusions reached generally are that internal clearances have increased to the point that a complete overhaul is required. Another possibility contemplated is that serious internal leakage is taking place between the stages. However, in a number of cases, these conclusions prove to be erroneous.

Although the use of chromium steels for pump parts and for pump casings of boiler feed pumps is quite effective in resisting corrosion-erosion attack, there still remains the problem of oxide formation in equipment located upstream of the feed pumps. Various iron compounds are formed in deaerating heaters, feedwater lines, heater drip systems, and so on. These compounds are quite readily deposited on the internal walls of the impeller waterways as well as on the casing walls, external impeller walls, and other surfaces.

Normally, the reduction of the flow areas due to these deposits is negligible. However, a few extreme cases have in the past come to my attention, where up to 0.040 or even 0.060 in. buildup of dark brown oxide had formed on each internal impeller wall in impellers of approximately 3/4 in. width. The reduction in area in such cases amounted to between 10 and 16%. Obviously, such reduction in area is accompanied by an equivalent reduction in pump capacity.

The pump efficiency suffers doubly: first from the reduction in capacity and second from the fact that the resulting profile of the impeller becomes distorted from its ideal design shape and the hydraulic efficiency of the impeller and of the pumps is unfavorably affected.

The iron compounds deposited on impellers vary from very soft consistencies that can be removed with a wire brush to extremely hard, brittle compounds that generally adhere quite strongly to the impeller walls and that need to be sandblasted. The latter compounds are dangerous for another reason: because of the hardness of any broken-off particles. If these lodge themselves within close internal clearances, they may cause damage to the pump.

Unfortunately, there is no ready means available to predict the possibility of such a buildup, nor is there any assured and safe method developed to remove it without dismantling the pump. On the other hand, there are a number of feedwater-treating methods to reduce or eliminate this condition once it has been detected. Although the problem of feedwater treatment is extremely complex and each case should be thoroughly analyzed on its own merits, I shall list a few of the methods that I have seen employed. In one particular case, a program of dosing the makeup water with a slight amount of ammonia was used. Addition of concentrated boiler water to the deaerator is frequently successful. Since dosing with sulfite for oxygen elimination frequently leads to iron compound buildup, a different method of oxygen elimination will generally reduce fouling of the pump internals.

But once such fouling has been detected, the proper procedure is to institute a thorough study of the problem by a feedwater treatment specialist.

Question 6.56 Elbow Erosion in Bypass Piping

Our boiler feed pumps are provided with recirculation bypasses for thermal protection at light loads. The pumps are designed for 400 GPM and 900 ft total head and are driven by 150 hp 3560 RPM motors. The recommended minimum flow is 20 GPM, and it is provided by means of an individual bypass line from each pump ahead of the check valve and returning to the deaerating heater at the pump suction. This line has in it a 7/32 in. orifice 4 in. long drilled in a piece of stainless steel. Because of the expense of an automatic control, we have installed a manually controlled valve to open and close the bypass. The orifice is followed by an elbow in the return line. We find that the high velocity of the water issuing from the orifice erodes the elbow in less than 2 months. What can be provided to eliminate this difficulty?

Answer. The description of the installation indicates that great care has been taken to protect the pump against overheating when operating at light loads. By rule-of-thumb methods, the amount of bypass provided would limit the temperature rise to 15°F, which is generally the recommended value. However, the piping of the recirculation line can readily be improved. The orifice should first be followed by a length of straight pipe of at least 12 to 18 in. After this, a tee should be provided, as shown in Fig 6.56, to lead back to the heater. The tee should then be followed by another length of straight line, terminating in another tee or pipe coupling. The end of this should be fitted with a stainless steel plug.

It is this stainless steel plug that will take the brunt of the high-velocity stream coming through the orifice. The plug should have a very long life. At worst, it may require replacement, but it is cheaper and simpler to replace it than an elbow in the recirculation line.

Question 6.57 Noise in Boiler Feed Pump Recirculating Lines

Can you suggest or refer me to material regarding any solutions found effective for eliminating noise in high-pressure boiler feed pump recirculation lines?

Answer. On several occasions I have searched the literature on the subject of noise in recirculation lines, and so far I have found no reference of any significant value. Should you or any reader come across any such reference, I would myself appreciate hearing about it.

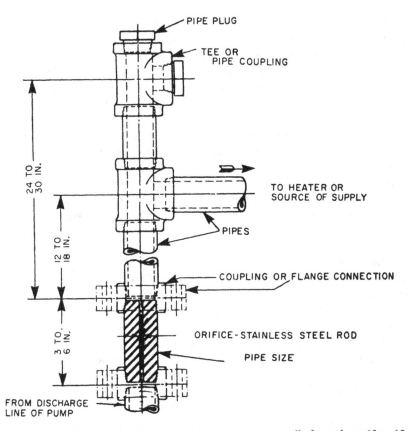

Figure 6.56 Suggested orifice and piping arrangement calls for at least 12 to 18 in. of straight run between orifice and a tee takeoff to the heater plus another run of straight pipe to a second tee fitted with a stainless steel pipe plug.

The entire subject of noise in or about hydraulic machinery, piping, and valves is an extremely complex subject. In the case of the recirculating lines, noise can be traced to three separate areas:

1. The pressure-reducing orifice installed in the line to establish a specific flow under the prevailing pressure differentials
2. The control valve, which is generally of the open-or-closed type, although modulating valves are now being used for larger installations to replace both the orifice and the valve
3. The piping itself

To a considerable degree, noise is a function of flow velocities and of abrupt changes of velocities or of flow direction. Let us then examine the range of velocities encountered in the recirculating system. A typical arrangement for this system is illustrated in Fig. 3.27. Since the recirculation bypass returns to the direct-contact heater at the pump suction (Fig. 3.27 illustrates an open feedwater cycle), the pressure drop through the system is equivalent to the net pressure generated by the boiler feed pump, which in turn is a function of the power plant operating steam pressure. For a 2400 psi unit, the pump net pressure at the capacity at which recirculation flow has to be provided may be of the order of 2500 psi (for variable-speed operation) or as high as 3000 psi (for constant-speed operation).

Friction losses through the piping may be of the order of 25 to 50 psi. An open-or-closed control valve will have from 50 to 100 psi pressure drop. The remaining pressure drop (2400 to 2900 psi) must take place through the pressure-reducing orifice. To accommodate such high pressure drops and avoid an excessively rapid erosion, multiple pressure-reducing orifices are used, similar to the type illustrated in Fig. 6.57. Obviously, even though the pressure drop takes place across several orifices, the velocities at each orifice are extremely high, and it is to be expected that a certain amount of noise is unavoidable.

Appropriate insulation will decrease the noise somewhat. Another advisable approach is to locate the orifice at such a point in the recirculation line that it will be farthest from areas where operators are likely to find themselves.

During the early 1940s, Public Service Electric and Gas Co. of New Jersey installed a recirculation system in one of its generation plants where the pressure-reducing orifice was replaced by a very long pipe (of several hundred feet), the function of which was to create a friction loss equivalent to that of the orifice itself. The velocities in the piping were, of course, of a considerably higher order of magnitude than in normal recirculating piping. Nevertheless, it was thought that less erosion would take place than in a conventional multiple pressure-reducing orifice and that there would be a significant reduction in noise level.

I have also heard of another installation where some difficulties were encountered with recirculation control and where the pressure-reducing orifice was replaced by a long length of small-diameter piping. At the same time, the amount of minimum recirculation flow was increased somewhat. The pumps involved in that installation were not of my company's manufacture so that I am not in position to say whether the change from a pressure-reducing orifice to the length of piping or the increased recirculation flow can be credited with correcting the difficulty.

I might add that since the number of installations where small-diameter piping is used to provide the necessary pressure drop is almost negligible, the

Chapter 6   619

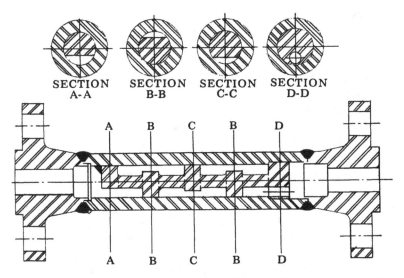

Figure 6.57 Multiple pressure-reducing orifices designed to guard against erosion from fluid flow experience high velocities and attendant noise.

conclusion to be reached is that the noise problem is not considered as a very serious one by steam power plant designers or operators.

Question 6.58 Excessive Vibration in a Vertical Pump Installation

We have recently installed a vertical volute pump with the motor mounted on a separate floor 20 ft above the pump floor elevation. The motor is connected to the pump through intermediate shafting with a support bearing at the midpoint between the pump and motor. At initial start-up the bearing vibrated in excess of 0.030 in., and this vibration could not be reduced even after careful alignment of the shafting. The intermediate bearing is a roller bearing mounted in a housing and supported at the end of an I beam embedded in a concrete wall. What is the cause of this excessive vibration, and what changes could be made to reduce the vibration?

Answer. From your description of the bearing support and the excessive vibrations you have experienced, the cause of the problem is undoubtedly a resonant frequency of the bearing support. This phenomenon can be described as follows.

Any flexible integral structure exhibits a progressive series of natural frequencies from its fundamental or first mode of vibration up to infinity. The simplest structure is the cantilever, such as a simple I beam or channel supporting an intermediate line shaft bearing at one end and built into concrete at the other. Any beam held rigidly at one end will vibrate in a definite and predictable series of frequencies that will vary directly as the square root of the modulus of elasticity and the moment of inertia and inversely as the square root of its weight and length cubed. The reason for this is as follows: If the beam is struck at its free end, it will deflect a certain distance because of the kinetic energy imparted. As it travels through this deflection, the kinetic energy is converted into an energy of position or potential energy. Its magnitude of deflection is thus dependent on the kinetic energy available, and when the conversion is complete, the beam will stop and reverse its direction to restore its original position of zero strain. When the beam returns to its original position of zero strain, however, its potential energy has been converted back into kinetic energy, and this energy carries the beam in the opposite direction and at the same amplitude.

Internal friction will eventually dissipate this energy in the form of heat, but if an external force at the same frequency is imposed upon the beam, the internal and external energies are in phase, and the deflection amplifies and results in severe vibration. If the rotational speed of the shaft is not quite in phase with the support, a beat will be set up. A beat is always an indication that the vibration results from a near resonant condition.

The static deflection at the end of a cantilever beam of length L, weight W, moment of inertia I, and modulus of elasticity E is

$$\delta = \frac{WL^3}{3EI}$$

The frequency of vibration in cycles per second is

$$FCPS = \frac{1}{2\pi} \sqrt{\frac{3EIg}{WL^3}} = 3.13 \sqrt{\frac{3EI}{WL^3}}$$

Converting to frequency in cycles per minute, which is equal to the rotational speed of the shaft in RPM, we obtain

$$FCPM = 187.7 \sqrt{\frac{3EI}{WL^3}}$$

The weight W is an equivalent weight that is approximately equal to one-third of the weight of the beam plus the weight of the bearing.

Chapter 6 621

You do not indicate in your question the speed of the pump or the dimensions of the cantilever beam supporting the bearing. To illustrate the problem, however, a typical installation is shown in Fig. 6.58. We have assumed a pump speed of 1200 RPM and an I beam weighing 5.7 lb/ft with a moment of inertia about the neutral axis through the web of 0.46 in.[4]. The natural frequency of the bearing support can be calculated as follows.

Assume that the bearing is mounted at the end of an I beam 36 in. long. The weight of the beam is 17.1 lb. The weight of the bearing and housing is 10 lb. Then

$$W = \frac{17.1}{3} + 10 = 15.7$$

$$E = 29 \times 10^6$$

$$L = 36 \text{ in.}$$

$$FCPM = 187.7 \sqrt{\frac{3 \times 29 \times 10^6 \times 0.46}{15.7 \times 36^3}} = 1385 \text{ CPM}$$

As can be seen, a natural frequency of 1385 CPM would be too close to an operating speed of 1200 RPM for vibration-free operation. The structure should be redesigned with additional stiffening to raise the natural frequency to at least 50% above the operating speed.

Using this simple relation, you can determine the natural frequency of your bearing support. If your calculated natural frequency is not at least 50% higher than the pump speed, it will be necessary to install bracing from the beam to the wall to apply the additional stiffness required.

Question 6.59 Effect of Frequency Reduction on Capability of Steam Power Plant Pumps

Our power plant is isolated from any other electric system. Recently, our load has grown to the point that we have practically no reserve over the peak demand, and a new unit will be added shortly. In the meantime, occasions arise when the load exceeds the installed capability and causes a reduction in frequency for as long as 1 hr or 1 1/2 hr. Obviously, this reduction in frequency has the effect of slowing down all electric-driven auxiliaries. I am primarily concerned with the boiler feed pumps as I suspect that this reduction in speed might render them incapable of maintaining the required rate of feeding. Can you indicate a method whereby the effect of frequency

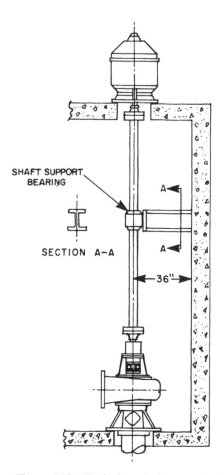

Figure 6.58 Typical vertical pump dry-pit installation.

variation on the capability of the boiler feed pumps may be predicted, so that we can find out whether we are in serious trouble or rather danger from that quarter?

Answer. Because a centrifugal pump is a velocity machine, a change in speed will affect the head-capacity curve of the pump. Since all velocities in the pump impeller and in the casing *for similar points on the characteristic curve* will vary in direct proportion to the peripheral velocity, it becomes very easy to calculate the effect on the head-capacity curve. The pump capacity, which is a direct function of the velocities, will vary directly as the operating speed. The total head, which is a function of the square of the peripheral speed, will vary as the square of the operating speed. Finally, since the power consumption varies as the product of the head and the capacity, the power will vary as the cube of the operating speed.

We can therefore prepare a tabulation of factors by which we can multiply the capacity and head points of a pump curve so as to predict the pump characteristics at any frequency other than the rated frequency. If the latter is 60 cycles, these factors will follow Table 6.8.

You can use these factors to prepare a family of head-capacity curves for your boiler feed pumps for any frequency between 60 and 55. (I earnestly hope that you are not faced with anything as drastic as such a reduction.) A typical example is given in Fig. 6.59, in which a typical performance curve of a boiler feed pump is expressed in terms of percentages of its rated conditions at 100% speed (or 60 cycles) as well as at lower frequencies. Thus, for instance, with the frequency reduced to 56 cycles, the rated point of 100% capacity and 100% head will have been reduced to 93.3% capacity and 87.1% head.

Whether frequency reduction will impair the ability of your boiler feed pumps to deliver sufficient feedwater to the boiler will depend in part on the extent of frequency reduction and in part on the margin that has been included over and above the system demand in selecting rated conditions of the pumps. To illustrate this point, I have assumed in constructing Fig. 6.59 that 8% has been added to the maximum capacity requirements and that the pump head has been selected arbitrarily at 2% above the system at the rated capacity so established. Thus, at the normal maximum demand, the pump capacity must be 92.5% of rated and the head 96.5% of rated.

If this is the case, and assuming that the pump is not worn, the frequency could drop to 58.3 cycles without preventing the pump from delivering the full maximum demand flow to the boiler. If, however, the frequency were to drop to 56 cycles, the head-capacity curve would intersect the system-head curve at 76% of rated capacity or 82.2% of the maximum demand.

Table 6.8 Effect of Frequency Reduction

Frequency	Capacity multiplier	Head multiplier
60	1.00	1.00
59	0.983	0.967
58	0.967	0.935
57	0.95	0.903
56	0.933	0.871
55	0.917	0.84

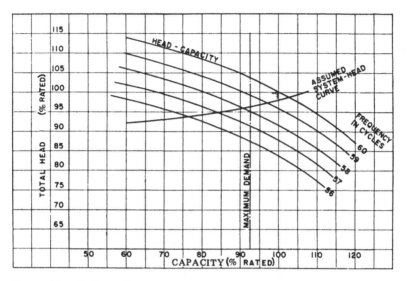

Figure 6.59 Effect of frequency variation on relatin between H-Q and system-head curve.

Of course, if the pump is worn appreciably and internal leakage has reduced its effective capacity appreciably below its rating, the effect of frequency reduction will be more severe in restricting the ability of the pump to deliver sufficient flow to the boiler. This consideration leads to the recommendation that if your pumps have not been reconditioned for a long time, renewing internal clearances may well be indicated at this junction.

Another suggestion may be in order if your installation includes spare equipment. Assuming, for instance, that your unit is served by three pumps of which two are running for full load and one is a spare, running this spare pump whenever you are expecting frequency reduction may get you over the hump until your new unit is available. Figure 6.60 illustrates the effect of running three such pumps under reduced frequency conditions. You will note that even were the frequency to drop to 55.5 cycles, the three pumps will still deliver the maximum demand. Of course, it may be wise to start the spare pump in advance of the anticipated frequency reduction, lest the starting current impose further difficulties on your plant once it is loaded beyond its normal capacity.

Question 6.60 Overload of Motor Drivers

What can be the cause of overload of an electric motor driving a centrifugal pump?

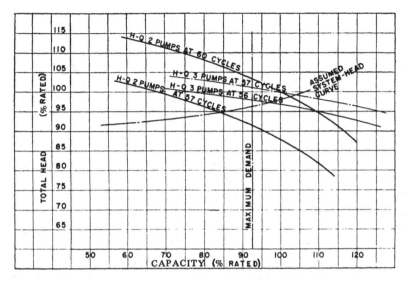

Figure 6.60 Operation with two and three pumps in parallel at varying frequencies.

Answer. This question illustrates very convincingly the classic observation that "any given cause can have only one effect, but any given effect can have a multitude of causes." You will presently see that the list of causes that can lead to the overload of a pump driver is quite extensive. To simplify our analysis, we shall assume that the pump, as built, met its guaranteed conditions of service insofar as capacity, head and efficiency are concerned. We shall further assume that the size of the driver has been so selected that it is not overloaded when the pump is operated at its rated conditions of service. We can now start examining all the circumstances that can lead to motor overload. These circumstances can be very conveniently subdivided into two groups:

1. Hydraulic or system problems
2. Mechanical problems

Hydraulic or System Problems

Let us first dispense with some possible causes that in this particular case probably cannot apply. For instance, if the pump were to be operated at a speed higher than the rated speed, the power consumption would increase significantly. But since you have referred to an electric motor driver—presumably a squirrel cage induction motor—this probability of operation at a higher speed is eliminated. The possibility that the pump may have been connected to a higher speed motor than originally intended is rather farfetched. This, however, still leaves us with a large number of possibilities.

1. The pump may be operated at flow conditions quite different from those contemplated originally. This circumstance may arise either because the system-head curve or the actual capacity requirements differ substantially from the preliminary calculations. Whether the overload occurs at capacities in excess or well below the rated conditions will depend on the type of the pump in question, that is, on its specific speed.

If, for instance, we are dealing with a low specific speed pump of, say, 1550, the horsepower curve increases with an increase in flow, as shown in Fig. 4.27. Thus, if the system-head curve had been overestimated, when the pump is installed and operated in the system, its head-capacity curve will intersect the system-head curve at a capacity in excess of the design or rated capacity and the motor may be overloaded.

On the other hand, if we deal with a high specific speed pump, whose characteristics are ilustrated in Fig. 4.29 (specific speed = 10,000), the bhp increases with a reduction in capacity. For instance, were this pump to be operated at 50% of its rated capacity by means of throttling the discharge, the brake horsepower would increase to almost 140% of the rated horsepower. If the motor size has not been selected with this end in view, it will be overloaded.

2. The motor may be running in the wrong direction. This can easily happen if the unit has not been tested for rotation at the time of the installation. Figure 6.13 illustrates what happens to the pump performance under this condition. You will note that not only is there a marked deterioration in the pump head-capacity curve, but the efficiency is reduced drastically and the power consumption runs well above the normal power curve.

3. The liquid characteristics may deviate from those at the time the pump and its driver were selected. It is quite obvious that if the specific gravity or the viscosity of the liquid are much higher than originally contemplated, the pump brake horsepower will be increased to the point where the motor is overloaded.

Mechanical Problems

Here again, we may be facing a large number of possible causes.

1. One possibility is that the pump impeller has been mounted backward. The performance of a pump in which this has happened is shown on Fig. 6.12. I should add that this circumstance can only arise in the case of a double-suction impeller in a single-stage pump or with single-suction impellers of a multistage pump. If the pump has been tested in the manufacturer's test lab, such an event is improbable. It can then occur only if for

whatever reason the pump has been dismantled in the field and then reassembled.

If, on the other hand, the size of the pump precludes running a shop test, it is conceivable although improbable that the impeller mounting has been reversed during the assembly of the pump at the factory. The diagnosis of such a problem is not difficult. If the motor is checked for rotation and runs in the proper direction, the pump performance when the impeller is mounted incorrectly can be detected by examining the resulting head-capacity curve.

2. If a significant amount of foreign matter becomes lodged in the impeller waterways, the pump efficiency will suffer and the power consumption may exceed the rating of the driving motor.

3. Rubbing between the rotating and stationary parts of a pump, for whatever cause, will of course increase the power consumption. This rubbing may be caused by misalignment of pump and driver, by a bent shaft, or by incorrect reassembly of a pump during maintenance and replacement of internal parts.

4. Excessive wear of the parts at the internal running clearances will generally lead to increased power consumption. Increased clearances cause increased internal leakage and the impeller is forced to handle a much higher flow than is being discharged by the pump into the system.

5. Improper conditions at the stuffing boxes will frequently increase power consumption. This is particularly true for small pumps for which the mechanical losses at the stuffing boxes represent a significant portion of the total power consumption. The packing may have been improperly installed, or it may be inadequate for the actual operating conditions; the glands may have been tightened excessively. Anyone of these problems will increase the friction losses at the packing.

It remains for the operator of a pump to analyze each one of the possibilities that I have listed, starting with the most probable cause, and to correct the problem that leads to motor overload.

Bibliography

1. Karassik, I. J., "Steam Power Plant Clinc," a series of articles in *Combustion*, containing a number of items dealing with transient operating conditions in the operation of boiler feed pumps: Oct. 1958, May 1959, June 1959, Nov. 1959, Jan. 1960, Feb. 1960, May 1960, July 1960, Jan. 1962, June 1962, May 1963, April 1965, Jan. 1968, Sept. 1968, Sept. 1973, and Dec. 1973.
2. Karassik, I. J., G. E. Bosworth, and W. D. Elston, "Centrifugal Boiler Feed Pumps under Transient Operation Conditions," preprint of Paper 53-F-32, ASME Fall 1953 Meeting, Rochester, N.Y.

3. Karassik, I. J., E. Zabel, and R. A. Neal, "Tests Strengthen the Case against Anti-flash Baffling in Deaerators," *Power*, Aug. 1961.
4. Liao, G. S., "Protection of Boiler Feed Pump against Transient Suction Pressure Decay," Paper 73-WA/PWR-1, presented at the Winter Annual Meeting of the ASME, Nov. 11-15, 1973.
5. Liao, G. S., "Analysis of Drain Pumping System for Nuclear Power Plants under Transient Turbine Loads," published in ASME Paper No. 75 PWR-A.
6. Liao, G. S., "Protection of Drain Pumps against Transient Cavitation," published in ASME Paper No. 75-WA/PWR-4, July 1976.
7. Liao, G. S., and P. Leung, "Analysis of Feedwater Pump Suction Pressure Decay under Instant Turbine Load Rejection," Paper 71-WA/PWR-2, presented at the Winter Annual Meeting of ASME, Nov. 28-Dec. 2, 1971.
8. Strub, R. A., "Reduction of the Suction Pressure of Boiler Feed Pumps as a Result of Sudden Load Drops," *Sulzer Technical Review*, March 1960.

Index

A

Abbreviations, 99-101
Abrasion of bearings, 610-614
Abrasive fluids or slurries, 66, 197
Absolute pressure, 29
Accuracy of testing, 399-407
Affinity laws, 7-13, 155-156, 415, 426
Air entrainment, 55, 87-92, 271-280, 420, 492, 496-497, 499-500, 502, 510, 513-515
Air leakage into pumps, 18, 121, 127, 240, 310, 474, 493-497, 505-509, 560, 562
Air pockets, 240, 494-497, 516
Alignment, 195, 244-249, 329, 614
All-bronze pumps, 587-589
All-iron pumps, 586
Altering design capacity, 5-7, 11-16, 113-118
Altitude,, effect of, 33-34, 485
ANSI Standards, 36, 98, 236-237, 324
Anti-flash baffling, 520, 524-529
Antifriction bearings (*See also* Bearings), 172, 193-197
API Standards, 36, 87, 98, 219-220
Application of centrifugal pumps, 1-189
Arbor press, use of, 475, 576
ASME Boiler Code, 163, 383
Atmospheric moisture, effect of, 154, 320, 360-363, 366
Atmospheric pressure, 23, 29, 33-34, 119, 121, 127, 241-242, 485
Attainable efficiencies, 183-188
Automatic control of auxiliary services, 360-363
  of recirculation by-pass, 282-289, 294-298
Automatic priming system, 142
Automatic start-up, 338-342, 360-363, 373-382, 427-428
Automation of Boiler Feed Pump Operation, 373-382, 394-398, 428, 546-547
Auxiliary oil pumps, 363-368, 428, 543, 550-551
Available NPSH (*See* NPSH)
AWWA Test Specifications, 155
Axially split cashings, 172, 205, 222, 224, 450, 455-466, 592-594
Axial clearances, 262, 395, 581

629

Axial thrust, 16-18, 192, 215-217, 219-225, 252-260, 261, 303, 576, 578, 580

B

Back-to-back impellers, 16, 220-225, 231
Back wearing rings, 214-217
Backwashing, 316
Balancing axial thrust, 16-18, 220-225
Balancing device, 220-225, 231, 298-305, 310, 377, 401, 408-409, 439, 551-556
Balancing device leak-off, 221, 285, 298-305, 310, 312-313, 325, 551-556, 572
 calibrated orifice for, 221, 439-440
 check valve in, 301-302
 from lean-oil pump, 303-305
Balancing holes in impellers, 214-217
Balancing impellers, 22
Balancing radial thrust, 563-566
Ball bearings (*See also* Bearings), 172, 193-197
 excessive cooling of, 360, 471, 563, 575
 failures of, 573-580, 610-614
 life of, 573-580, 610-614
 mounting of, 573-574, 576
Barometric pressure, 23, 29, 33-34, 119, 121, 127, 241-242, 485
Barrel-type pumps, 303-304, 319, 327, 330, 443, 451, 476
Baseplates, 235, 243-249
Basic rules for maintenance and operation, 369-373
Bearings
 abrasion, 610-614
 antifriction, 172, 193-197
 ball, 172, 193-197
 bowl, 197
 cone discharge, 197
 cooling, 360-363
 failures of, 191-192, 196, 608-610

[Bearings]
 Kingsbury, 197, 303-304, 364-365, 543, 549, 553-555
 lubrication of, 196, 232, 244, 362-369, 428, 473, 543, 549-551, 575, 608-610
 moisture in, 154, 320-322, 360-363, 366, 573, 576
 short life of, 573-580, 610-614
 sleeve, 193-197, 401, 549
 suction case, 197
 temperature, 420-421
 thrust, 18, 254, 256, 259, 573, 577-580
 types of, 193-196
 water-lubricated, 172, 197, 199-200, 259, 343, 349-351, 386, 420-424
Bedplates and pump supports, 235, 243-249
Best efficiency point, 15-16
Bid evaluation, 181-183
Boiler circulating pumps, 179-181
Boiler feed pumps, 144-151, 152-153, 160-177, 190, 202-205, 220-225, 229-231, 240, 263-267, 282-298, 305-318, 318-321, 323, 325, 327, 343, 347-355, 382-385, 387-398, 401-412, 420-435, 443-445, 466-467, 476-479, 504-508, 517-542, 548-556, 569-576, 584-586, 592-597, 614-619, 621-624, 627-628
 automation of, 373-382, 394-398, 428, 546-547
 capacity of, 152-153
 condensate injection sealing, 190, 263, 299, 380, 426-427
 flashing of, 347-349, 433-435
 lubrication, 363-369
 materials for, 202-205
 minimum flow, 282-298, 351-355, 373-382
 minimum operating speed, 349-351
 NPSH, 44-45, 46-50, 57-61, 298, 381, 389, 391, 396-397, 517-530, 537-542
 overhauling, 436-440, 584-586

Index 631

[Boiler feed pumps]
  recirculation by-pass (*see* Recirculation by-pass)
  seizure of, 314, 348-349, 392-394, 424, 551-556, 569
  spare parts, 443-445
  spare or standby, 285, 289, 305-314, 332, 387, 427-428, 433-435, 548, 556, 569-573
  starting of, 309, 311, 314-315, 335-336, 374-379
  stopping of, 306, 374-379
  stuffing box leakage, 263-267
  testing for rotation, 327-330
  transient conditions, 49-50, 225, 240, 299, 317-318, 389-394, 410-412, 517-530, 537-542
  warm-up, 302, 310-313, 331-333, 382, 421-422, 556
Booster pumps, 145, 164-165, 170-174, 179-180, 302, 349, 389, 391, 433-435
Bottom suction, 166-167
Boxes, stuffing (*See* Stuffing boxes)
Brake horsepower, 5-7, 15, 21, 115, 128-129, 150-151, 334-335, 352, 355, 384-385, 441
  curves, 115-356, 399
Break, operation in the, 48, 61-69, 152, 334
Brine, for stuffing box sealing, 270-271
Brine pumps, 153-154
By-pass, recirculation (*See* Recirculation by-pass)

C

Can pumps (*See* Vertical can pumps)
Capacity
  changing, 5-7, 11-16, 113-118, 333-334
  of condensate pumps, 151-153
  vs. impeller diameter, 5-6, 9-13, 16-18, 119, 121, 241

[Capacity]
  loss of, 105-106, 494-497, 515-517, 615
  reducing, 333-334
  vs. speed, 6-11, 121, 243
Capacity required, 3
  total, calculation of, 130-133
Casing, bolt elongation, 458-460
  bolt stresses, 464-465
  flanges, 205-206, 456-466
  gaskets, 205, 456-466
  maintenance, 205-206, 468, 583, 585
  materials, 202-204, 237-238, 581-589
  volute tongue, 468
  wear, 581-582
Cast-iron casings, 202, 205, 237-238, 581-589
Cast-steel casings, 202-205, 237-238
Cavitation, 23-25, 28, 47-48, 50-51, 57, 59, 63-65, 68, 70-75, 79-83, 84, 91, 113, 119, 152, 154, 159, 192, 208, 210, 241, 279, 334, 342-345, 359, 386, 389, 392-394, 412, 508, 513-515, 541, 556, 578, 590
  causes of, 92-94, 119
  coefficient, 24-26, 28, 71-75, 156
  noise and, 51, 80, 119, 489, 508, 513-515, 556, 558
Central automatic-priming system, 142
Centrifugal separators, 601-604
Characteristic curves, 5, 7, 11
Check chart of pump problems
  causes, 486-489
  symptoms, 483-484
Checking rotation, 327-331, 516-517, 544-546
Check valves, 38, 134, 136, 286-287, 289, 302, 317-318, 338-339, 347, 491-492, 498, 533-534, 543-550
  in balancing device leak-off line, 301-302
  by-pass around, 570
  in reverse flow, 543
Chemical analysis, 3

Cleaning piping, 314-317
Clearances, 93, 105, 161, 172, 184-188, 204, 222, 225, 227, 314-315, 343, 347, 349-351, 405-407, 429-431, 436-439, 440-442, 446, 447-456, 577-580
  measuring of, 450-456
Close-coupled pumps, 235
Closed circuit, operation in a, 121-127
Closed discharge, operation with, 14, 27, 94-97, 134-144, 280-282, 289, 334-335, 412-415, 511, 533
Closed system, 121-127
Cold brine, 153-154
Cold starts, 331-333
Cold water injection, effect of, 59-61
Compression of water, isentropic, 402-403
Condensate injection sealing, 190, 299, 380, 553-554, 595-597
Condensate pumps, 61-65, 151-153, 165, 170-177, 347, 385-387, 416-420, 537-542, 595
Condensate booster pumps, 170-174
Condenser, 284, 314
Condenser circulating service, 177-179
Conditions of service, 1-5
Constant discharge pressure, 138-140
Constant level oiler, 609-610
Construction, pump, 190-234
Contamination by oxygen, 504-508, 596-597
Continuous service, 4, 357-360
Coolers, oil, 360-363
Cooling water, for bearings, 360-363
  for stuffing boxes (*See also* Sealing water for stuffing boxes), 360-363
Coriolis effect, 273, 277-279
Corrosion, 91-92, 232, 319-320, 446, 479, 581-582, 584-586
  galvanic, 197, 468, 581-582, 589
Corrosion-erosion, 202-205, 230-231, 437, 581-582, 615
Couplings, flexible, 200-203
  flexible disk, 201-203

[Couplings, flexible]
  gear type, 201
  grid type, 201
  limited end-float travel, 200-202
  quick-disconnect, 424
  rigid, 256
  screw type, 328-329, 544-545
Critical speed, 612, 619-621
Critical temperature, 48
Cryogenic pumps, 92-94
Curves (*See* Characteristic curves; Efficiency curves; Head-capacity curves; Performance curves; Power curves; System-head curves; Test curves)
Cutdown of impellers, 5-6, 9-13, 16-22, 117, 121, 127, 241, 243-244, 334, 388, 425-426
Cyclone separators, 601-604

D

Data sheet, 1-2
Deaerator (or deaerating heater), 44-45, 46-50, 57-61, 152-153, 164, 172, 240, 283-284, 290-293, 298-302, 306-314, 317-318, 410-412, 435, 505, 517-537, 554-556, 570-573, 615
Deep-well pumps (*See* Vertical turbine pumps)
Deflection, 145, 228-230, 343, 347, 424-425, 620-621
Diagnostic chart of pump problems
  causes, 486-489
  symptoms, 483-484
Dial indicator, use of, 450-454
Dirty water, 506, 581-583, 597-608
Discharge conditions, 3, 107-109
Discharge head, 29, 107-109
Discharge nozzles, 235-238
  velocities, 36-38
Discharge piping, 107-109
Discharge pressure for boiler feed pumps, 160-164

Index

Discharge pressure, loss of, 339
Discharge, throttling of, 121, 134, 243, 333-334, 388, 503, 509-513, 557, 562, 568
Discharge valves, 27-28, 305-314, 334-335, 543-550, 571-572
Disconnecting pump and driver, 327-331, 351, 423-424, 516
Disengaging clutches, 544
Disk horsepower, 15, 20-21, 98, 425
Dismantling pumps, 301, 350, 398-401, 423, 436-440, 492, 516
Dissolved air or gas, 87-92, 359, 492, 562
Documentation of field troubles, 480-485
Double-casing pumps, 220-225, 331-333, 382-383
Double-extended shaft, 210, 213-214
Double-suction impellers, 18-19, 38, 75-76, 78, 191-192, 206-210, 515-517, 589-592, 626-627
Double volute, 563-566
Double wearing rings, 225-228
Dowelling, 248-249, 574
Drains, 479
Driver overload, 94-97, 342-343, 437
Drives (*See* Electric motors; Turbines, steam)
Drooping head-capacity curve, 133-135, 498, 532, 536
Dry pit pumps, 4, 206, 210

E

Eccentric reducers, 238-240
Efficiency, pump, 6, 14-17, 70-71, 151, 154, 183-188, 213, 241, 352, 355-356, 401-407, 426, 440-442, 516, 615
Efficiency curves, 14, 107, 162, 241
Ejectors, 493-494
Elbows, 38, 240, 249-250, 590-592, 616

Electric motors, 200-201, 248-249, 321-322, 343-344, 425-428, 543, 624-627
  bearings, 200-201, 254, 256, 259
  hollow-shaft, 262
  incorrect wiring of, 327-331
  overload of, 94-97, 342-343, 437, 624-627
  rotation, 327-331
  starting controls, 330, 338-342
Electrolytic corrosion (*See* Galvanic corrosion)
Elevation above sea level, effect of, 33-34, 485, 489-491
Elongation of casing bolts, 457-462
  of shafts, 260-262
End-float travel, 200-202
End play in vertical turbine pumps, 262
End suction pumps, 235-237
Entrance losses, 101
Entrained air or gas, effect of, 55, 351, 513-514
Erosion (*See also* Corrosion-erosion), 202-205, 229-231, 291, 293, 446, 581-583, 590, 616
Evaluation of bids, 181-183
Excess capacity, operation at, 557
Excessive cooling of bearings, 360, 362, 471, 575-576
Excessive suction lift, 485, 492, 508
Exit losses, 101
Expansion joints, 249-252
Expansion of shaft, 260-262
Extended storage of pumps, 319-323
Extra-deep stuffing boxes, 193
Eye diameter, effect on NPSH, 33-36

F

Feedwater cycles:
  closed, 164-166, 170-174, 537-542, 593
  open, 152-153, 164-166, 171, 313, 517-537, 593, 595
  split pumping, 166-170, 172

Feedwater treatment, 615
Feeler gauges, use of, 450-451
Field tests, 22-32, 53, 398-415
Field troubles, 480-628
Fittings, losses in, 109-113
Flanges, casing, 205-206, 456-466
  raised face, 237-238
Flashing, 291, 293, 347-349, 391-394, 541
Flat head-capacity curves, 13, 128-130, 158-159
Flat system-head curves, 158-159
Flexible couplings, 200-203
Float control, 66-67, 267-268, 493, 503, 509-514
Flow, minimum permissible, 143, 188, 282, 298, 351, 354, 359, 385-287, 428-433
  reverse, 338, 543-544, 546-547
Flowmeter control of recirculation bypass, 283, 285, 294-298
Fluid drive (*See* Hydraulic couplings)
Flushing piping, 314-316, 423
Foot valves, 259, 338, 492-493, 500
Forced-feed lubrication, 363-369, 543, 549-551
Foreign matter in liquid pumped, 3, 217, 228, 314-317, 319, 325, 343, 350, 382-383, 407, 410, 423-424, 426, 548, 597-608, 613, 627
Forty basic rules for maintenance and operation, 369-373
Foundations, load on, 252-258
Frame-mounted pumps, 231-232, 235-237
Frequency of overhauls, 436-439, 443, 589
Frequency reduction, effect of, 621-624
Friction losses, 4, 37, 41, 44, 101-103, 107-113, 114-117, 128, 154, 240-243

G

Galling at running joints, 299
Galvanic corrosion, 197, 468, 581-582
Galvanic series, 583-584
Gas, entrained, effect of, 55, 87-92, 271-280, 420, 492, 496-497, 499-500, 502, 510, 513-515
Gaskets, casing, 205, 332-333, 456-466, 592-594
Gauge pressures, 24, 29-32, 42-44
Green liquor service, 65-69
Gritty water, 506, 581-583, 597-608
Grouting, 245-248

H

Head-capacity curves, 6-11, 13-15, 20-21, 22-25, 48-49, 55, 63, 67-68, 81, 89-90, 98-99, 104-107, 109-113, 115-144, 158-159, 161-162, 333-334, 336, 342, 344, 347, 383, 388, 399-400, 408-409, 415, 490, 503, 510, 515, 532, 536-538, 578-579, 622-624, 626-627
  drooping, 133-135, 498, 532, 536
  flat, 13, 128-130, 158-159
  steep, 13, 128-130, 158-159
Head, discharge, 3, 108-109, 121
  static, 3, 101, 114, 123, 128, 132, 134, 333, 510
Head, total, 5-12, 16, 25-26, 29-32, 70-71, 74-75, 90, 94-95, 97, 101, 104-107, 113-114, 117, 128, 130-132, 134, 145-146, 151, 156, 352, 355-356, 588, 623
  velocity, 29-31, 40-43, 102
Heater drain pumps, 152, 173-174
Hollow-shaft electric motors, 329, 546
Horizontally split casings (*See* Axially split casings)
Horizontal pumps,

Horsepower:
  brake, 5-7, 21, 334-335, 352, 355, 384-385
  water, 5-7, 334-335, 384-385
Hydraulic balancing devices (*See* Balancing devices)
Hydraulic coupling drive, 141-144, 149, 160, 172-174, 176-177, 181, 322-323, 349
Hydraulic Institute,
  NPSH corrections, 46-57
  for hot water, 46-51
  for hydrocarbons, 51-57
  specific speed limitations, 72-78
  Standards, 11, 23-26, 28, 32, 46, 49, 51-57, 72-76, 98, 100, 156, 238-239, 418, 509
  Test code, 24, 26, 28, 32
Hydraulic losses, 15
Hydrocarbons, 51-57, 355-357
Hydrostatic test, 461

I

Impellers, balancing of, 22
  closed vs. open, 583
  cutdown of, 5-6, 9-13, 16-22, 117, 121, 127, 241, 243-244, 333-334, 388, 425-426
  cutting and trimming of, 5-6, 18-22
  diameter change, effect of, 5-6, 9-13, 16-22
  double-suction, 18-19, 73, 76, 78, 191-192, 206-210, 515-517, 589-592
  eye diameter of, 33-36
  iron deposit on, 614-615
  materials, 468, 581-584, 589
  mounting on shaft, 248, 331, 475-476
  reversed, 498, 515-517
  shapes, 70-71
  single-suction, 76, 77, 206-210, 214-217

[Impellers, balancing of]
  vane tips, 6, 18-22
  width, 13-17, 114
Increaser at discharge, 36-38
Independent seals, 506, 606
Inducers, 97-98
Information required to select pumps, 1-5
Initial start-up, 314-316, 336-337, 349
Injection of sold water at suction, 57-62
Inner assembly, 144
Inquiries, preparation of, 1-5
Inspection, 172, 302, 336, 436-439, 450-454
Installation, 327-435
Instruction books, 247
Instrumentation, 28, 373-382, 394-398, 437, 439-440
  of balancing device leak-off, 221, 439-440
Intakes for wet-pit pumps, 272, 275, 277, 325-326
Intagible factors, 181-183
Interchargeability, 194
Intermittent operation, 357-360
Intermittent service, 4, 179, 587
Internal clearances, 93, 105, 161, 172, 184-188, 204, 222, 225, 227, 314-315, 343, 349-351, 392-393, 405-407, 436-439, 446-456
Internal leakage, 15, 105-106, 204-205, 228, 400, 405-410, 577-579, 624
Internal recirculation, 76, 79-80, 82-88, 184, 188, 351, 359, 566-568
Internal water-lubricated bearings, 172, 197, 199-200, 230, 343, 349-351, 386, 420, 422-424
Interruption of power supply, 141, 543
Iron deposit on impellers, 614-615
Isentropic compression of water, 402-403
Isolation mounting, 558

J

Jet pump, 493-494, 559-563

K

Kigsbury-type thrust bearings, 197, 303-304, 364-365, 543, 549, 553-555
Kinetic energy, 29-32, 38-44

L

L-type wearing rings, 450-451
Lantern rings (*See* Seal cages)
Leakage, at casing, 457, 592-594
  internal (*See* Internal leakage)
  at stuffing boxes (*See* Stuffing box leakage)
Leakage losses (*See* Internal leakage)
Leak-off (*See* Balancing device leak-off)
Level control, 61-69, 267-268, 493, 503-504, 509-514
Limitations of space and weight, 4
"Limited end-float" couplings, 200-202
Limiting pump capacity, 115-118
Liquid, nature of, 3, 190-191, 626
Load, sudden reduction of, 49-50, 225, 240, 299, 318, 410-411, 519-528
Location of installation, 4, 33-34, 485
Loss of capacity, 105-106, 494-497
  after starting, 494-497
Loss of discharge pressure, 339
Loss of prime, 338-342, 491-494, 510
Loss of suction, 338-342, 491-494
Low flow, operation at, 79-83, 92-94, 563-568, 576
Lubricating oil
  temperatures, 364-369
  viscosity, 364-369

Lubrication, bearing, 197, 303-304, 363-365, 543, 549, 553-555, 575, 608-610
  systems, 244, 363-369, 543, 549-551

M

Magnetic drive, 141, 149, 177-178, 181
Maintenance, 369-373, 436-479
  intervals, 436-439, 443, 589
  tools, 442-443
  of wearing rings, 225-228, 447-456
Materials for pump parts, 5, 202-205, 227, 260-262, 446-447, 468, 581-589, 610-614
Measuring internal clearances, 450-456
Mechanical losses, 15
Mechanical seals, 190-192, 263, 270, 427, 597-608
Metering, 101-102, 104
Minimum flow, 143, 188, 282, 298, 351, 354, 356, 359, 385-387, 428-433, 532
Minimum operating speed, 349-351
Minimum submergence, 271-279
Misalignment (*See also* Alignment):
  of internal parts, 573
  of piping, 249, 482, 575
  of pump and driver, 482-575
Mixed-flow pumps, 327-329
Modulating control valves, 179, 285-289, 297
Moisture in bearings, 150, 320-322, 360-363, 366, 573, 576
Molybdenum disulfide, use of, 475-476
Monitoring pump performance, 373-382, 394-412, 420-421
Monitoring shaft deflection, 425
Motors (*See* Electric motors)
Mounting impellers on shaft, 248, 331, 475-476
Multistage pumps, 16-18, 155, 184, 220-225, 228-230

Index

N

Nameplates, 106
Narrow and wide impellers, 13-17, 114
Noise, 51, 80, 489, 508-509, 513-515, 556-563, 616-619
Nonreversing ratchets, 329
Nozzle sizes, 36-38
NPSH (net positive suction head):
   available, 23-26, 38-44, 45-57, 59-69, 84-87, 98, 111-113, 119-121, 145, 154, 159-160, 17?, 242, 345-347, 349, 357, 381, 396, 420, 428-429, 485, 490-491, 503-504, 510, 517-524, 537-542, 590
   effect of temperature on, 44-45`
   extra, for transient conditions, 49
   field test for, 22-28
   vs. impeller eye diameter, 33-36
   insufficient, 381, 397, 485, 557
   required, 22-28, 33-36, 45-57, 59-61, 61-69, 70-94, 98, 111-113, 119-121, 129, 145, 159-160, 173-175, 186, 241, 344, 346, 357, 389, 391, 399, 420, 428-429, 431, 485, 490-491, 503-504, 517-524, 538-542, 590
   and submergence control, 18, 61-69, 152
   velocity head and, 28-32, 38-40

O

Oil coolers, 360-363
Oil mist lubrication, 608-610
Oil pumps, 321, 363-369, 543, 549-551
Oil viscosity, 364-368
Opening for inspection, 172, 302, 336, 436-439, 450-454
Operating speed, 6-11, 69-79, 144-149, 190, 196, 208, 349-351, 400, 407-409, 421-424, 485, 489-491, 621-624
   minimum, 349-351

Operation of pumps, 327-435
   in the "break," 48, 61-69, 152, 334, 345, 347, 491, 510-511
   in a closed circuit, 121-127
   with closed discharge, 14, 27, 94-97, 134-144, 280-282, 289, 334-335, 338, 412-415, 511, 533, 544
   continuous, 357-360
   at excess capacity, 342
   at higher speeds, 485, 489-491, 558
   intermittent, 357-360
   at low flows, 79-83, 92-94, 563-568, 576
   in parallel, 127-133, 135-138, 150, 177-179, 338, 344-347
   at reduced head, 115
   at run-out capacity, 344-347
   in series, 127-131, 167-168, 172, 210-214, 342-343
   at shutoff, 94-97, 134-144, 280-282, 289, 334-335, 338, 412-415, 511, 533, 544
   unattended, 138, 192, 338-339, 373-382
   at variable speed, 3, 8, 101, 135-144, 149, 160, 172-174, 176-169, 181
Optimum efficiencies, 70-71, 183-188
Orifices:
   metering, 102, 104, 221, 439
   pressure-reducing, 284, 286, 291-294, 616-619
Overfilling impeller vanes, 21-23
Overhaul, 161, 436-439
Overload of drivers, 94-97, 342-343, 359, 624-627
Oversizing pumps, 118-121
Oxygen contamination, 60-61, 504-508

P

Packing for stuffing boxes, 190-193, 264, 267, 270, 427, 443, 468-470, 472, 473-474, 476, 563, 594-595, 597, 604

Parallel operation, 127-133, 135-144, 150, 177-179, 338, 344-347, 534, 556
  of centrifugal and reciprocating pumps, 150
Performance characteristics, 70, 123, 398-399
Performance curves, 104-107, 121-130
Performance tests, 22-32, 154-158, 204-205, 383-385, 437
Peripheral speed of impellers, 5-6, 146
pH values, 3, 197, 203-204, 583-584, 589
Piping:
  discharge, 36-38, 107-109, 249-252
  suction, 36-38, 249-252
Pitometers, 104
Power consumption, 5-11, 94-97, 128-130, 334
Power curves, 98-99
Power supply, interruption of, 141
Press fit, 475-476
Pressure, loss of, 339
Pressure energy, 38-44, 251
Pressure gauges, 28, 30-32, 41-42, 415
Pressure-reducing orifices, 284, 289-294, 616-619
Pressure switch, 339-340, 364-366
Price, 181-183
Prime, loss of, 338-342, 491-494
Priming, 142, 335-342
  automatic, 142, 336-342
Pumps, application, 1-89
  construction, 190-234
  location of, 4, 33-34, 485
  specifications (*See* Specifications)
Pump-out vanes on impellers, 217-220

Q

Quenched stuffing boxes, 362

R

Radial thrust, 3, 192, 351, 358, 415, 563-566
Radially split casings, 222, 451-454, 593
Raised-face flanges, 237-238
Range of operating capacities, 3, 79-80, 359
Reciprocating pumps, paralleling centrifugal pumps with, 150
Recirculation by-pass, 121, 124-126, 134, 143, 179, 280-298, 306-307, 310-312, 335-336, 351-355, 385-387, 428-433, 511, 531-533, 544, 568, 616-619
  automatic control, 282-289, 294-298
  to drain, 280-282
  elimination of, 351-355
  erosion in, 616
  modulating control valves, 179, 285-289
  temperature rise control, 282-285
Recirculation control valves, 179, 282-298, 431-433
Recirculation line, common, for several pumps, 289-290
Reducers at suction, 36-38, 238-243
Reducers, eccentric, 238-243
Reducing capacity, 5-22, 113-118, 121-127, 333-334
Reducing impeller diameter, 5-6, 9-13, 16-22, 117, 121, 127, 241, 243-244, 334, 388, 425-426
Reliability, 181-183
Relief valves, 280-282, 303-307, 312, 572
Remote control, 360, 548
Renewing internal clearances, 405-407, 436-444, 447-450
Repair parts, 194, 196
Repairs, 445-450
Required NPSH (*See* NPSH)
Reversed impeller, 515-517
Reverse flow, 338, 347, 543-550

Index

Reverse rotation, 306-307, 327-331, 347, 498, 516-517, 544-551, 626
Rigid coupling, 256
Rotation, checking motor, 327-331
Rules for operation and maintenance, 369-373
Runaway speed, 543, 548
Runout capacity, 344-347

S

Screens in piping, 314-317
Screw-type couplings, 328-329, 544-545
Sea level, effect of elevation above, 33-34
Sea water, pumps handling, 586-588
Seal cages, 235, 360-363, 469, 474, 493, 495, 506-508, 604
Seal rings (*See* Seal cages)
Sealing water for stuffing boxes, 360-363
Seals, mechanical (*See* Mechanical seals)
Seizure of pump internals, 229, 347-350, 392-394, 551-556
Self-priming pumps, 335, 493
Series operation, 127-133, 167-168, 172, 210-214, 342-343
Series units, 210-214
Service conditions, 1
Setscrews, 328
Sewage effluent for stuffing box sealing, 270-271
Sewage pumps, 267
Shaft, breakage, 3, 467, 563
  deflection, 145, 228-230, 343, 347, 424-425, 620-621
  double-extended, 210-213
  elongation, 260-262
  materials for, 260-262
  runout, 395
  span, 228-230
  straightening, 466-467
  for vertical pumps, 260-262, 612, 614

Shaft nuts, 327-328, 330
Shaft sleeves, 230-231, 474-475, 594-595
Shims, 246-248
Shrink fit, 331, 452
Shutoff, operation at, 14, 94-97, 134-144, 280-282, 334-335, 338, 412-415
Shutoff head, 14, 94-97, 133-138, 412-415, 497
Side discharge, 166-167
Sigma (cavitation constant), 24-26, 28, 71-75, 156
Single-stage pumps, 184, 210, 213-214
Single-suction impellers, 76-77, 191-192, 206-210
Single volute, 351, 563-566
Single wearing rings, 225-228
Siphon, 107-109
Sleeve bearings, 193-197
Sound (*See* Noise)
Space and weight limitations, 4
Spare pumps (*See* Standby pumps)
Spare and repair parts, 144, 443-445
Specific gravity, 3, 33, 36, 151, 168, 383-385, 534-535, 626
Specific heat, 355-357
Specific speed, 55, 69-76, 94-97, 106-107, 146-148, 183-189, 327, 337, 351, 359, 406-407, 415-418, 426, 563, 626
Specific Speed Limit Charges, 72-78
Specifications, 1-5, 193-194, 202-205, 557-559
Speed, operating, 6-11, 69-79, 144-149, 190, 196, 208, 349-351, 400, 407-409, 421-424, 485, 489-491, 621-624
  minimum, 349-351
Speed changes, effect of, 6-11, 154-158, 485, 489-491
Speed controls, 135-144
Stable curves, 133-135
Stainless steel fitted pumps, 202-205, 587-588

Standard fitted pumps, 581-582, 586, 615
Standards of Hydraulic Institute, 11, 23-26, 28, 32, 46, 49, 51-57, 72-76, 98, 100, 156, 238-239, 418, 509
Standard lines of pumps, 4-5, 114-115, 194-195, 213
Standby pumps, 161, 180, 204, 285, 289, 305-314, 336, 360-369, 427-428, 556, 569-573
Start-and-stop operation, 512-513
Starting pumps, 331, 336-337, 339, 344, 367, 374-379, 427-428
  cold, 331-333
Starting torque, 195
Static head, 4, 101-103, 128-130
Static pressure, 41-43
Steam turbine drive, 135-138, 148-150, 160, 165, 169, 344, 533
  parallel operation, 135-139
Steep head-capacity curves, 14, 128-130, 158-159
Steep system-head curves, 128-131
Stopping pumps, 306, 339, 344, 374-379
Straightening pump shafts, 466-467
Strainers, 314-319, 410, 475, 498
Stresses of casing bolts, 464-465
Stuffing boxes, 18, 129, 170, 190-193, 263-271, 310, 360-362, 468-470, 473-475, 488, 493, 505-508, 594-595
  extra deep, 193
  glands, 263-267, 362-363, 468-470, 472
  leakage, 263-267, 469-470
  vs. mechanical seals, 190-192
  packing, 190-193, 468-470, 473-474, 594-595
  pressure, 190-192, 217
  quenched, 267, 362, 470
  seal cages, 235, 360-363, 469, 493, 495, 506-508
  sealing water, 360-363
  water-flooded, 231-234

Subcooling, effect of, 57-62
Submergence (*See also* Suction head), minimum, 271-279
Submergence control (*See also* Operation in the "break,"), 18, 61-69, 152
Submergence of suction piping, 495
Suction:
  single vs. double, 206-210
  throttling of, 85, 333-334
Suction conditions, 3, 69-94
Suction head (*See also* NPSH), 3, 28-32, 33
Suction lift, 3, 28-32, 33-36, 70-73, 119, 154, 241, 338
  excessive, 485, 489-491, 508
  measurement, 28-32
Suction lines, cleaning, 314-317
  relief valves, in, 305-307
Suction, loss of, 338-342, 491-494
Suction nozzle velocities, 36-38, 240-243
Suction piping, 36-38, 49, 238-243, 495, 522, 524, 529-537
Suction pressure, 44-45, 47, 217
Suction pressurization, 45-46, 158-160
Suction recirculation, internal (*See* Internal recirculation)
Suction specific speed, 69, 75-79, 84, 87, 156, 428-430, 490
Suction strainers, 314-319, 410, 475, 498
Suction valves, 305-307, 309-312, 336, 569-573
Sudden load reduction, 49-50, 225, 240, 299, 317-318, 517-530, 534, 537
Suppression tests, 80, 84-88, 90
Symbols, 99-101
System-head curves, 8, 63, 67, 101-107, 109-113, 115, 119-121, 124-127, 128-134, 136-138, 161, 333-334, 342, 344-345, 347, 408, 503, 510, 515-517, 538, 567-568, 623-624, 626

Index 641

[System-head curves]
 flat, 128-131
 steep, 128-131

T

Tail pump, 61
Temperature, effect of:
 on brake horsepower, 150-151, 383-385
 on NPSH, 26, 57-61
 on stuffing boxes, 190-191
Temperature rise, 93-94, 281-285, 292-294, 298, 304, 335-336, 351-358, 401-404, 412-415
 of bearing cooling water, 360-362, 365, 471
Temperature-rise control of recirculation by-pass, 282-285
Test Code of Hydraulic Institute, 23-26, 28, 32
Test curves, 104-107
Testing motors for rotation, 327-331
Tests, pump, 22-32, 154-158, 204-205, 383-385, 409-412, 437
Thermal expansion, 260-262, 452-454
Thoma-Moddy cavitation parameter, 24-26, 28, 71-75, 156
Throttling discharge, 121, 333-334, 388
Throttling suction, 26, 333-334
Thrust, axial, 16-18, 176, 192, 215-217, 219, 220-225, 252-261, 303, 576, 580
 radial, 3, 192, 351, 358, 415, 563-566
Thrust bearings, 18, 254, 256, 259, 573
Tools, pump, 442-443
Top discharge, 235-237
Torque, starting, 195
Total head, 5-12, 16, 25-26, 28-32, 70-71, 74-75, 90, 94-95, 97, 101, 104-107, 113-114, 117, 128, 130-132, 134, 145-146,

[Total head]
 151, 156, 352, 355-356, 558, 623
 measurement, 28-32
Transient operating conditions, 49-50, 225, 240, 299, 317-318, 389-394, 410-412, 517-530, 627-628
Turbines, steam, 135-138, 148-150, 160, 165, 169, 519, 521, 524
Turning gear, 421-424
Twin volute, 563-566

U

Unattended operation, 138, 192, 338-339, 373-382
Underfiling impeller vanes, 21-23
Unstable curves, 133-135, 532, 534
Unwatering reservoirs, 499-501
Unwatering tanks, 501-504

V

Vacuum service, 232
Valves, fittings, 109-113
Vapor pressure, 23-24, 26, 36-46, 50-56, 59, 63-64, 93-94
Variable speed operation, 7-8, 101, 135-144, 149-150, 342
Velocity head, 29-31, 40-43, 102, 251
Venting centrifugal pumps, 240, 336, 495-497, 516, 570
Venturi meters, 104
Vertical can pumps, 172, 386, 416-420
Vertically split casings (*See* Radially split casings)
Vertical propeller pumps, 327-328
Vertical pumps, 206-210, 318-319, 325-328, 338, 386, 544-546, 563, 610-614, 619-621
Vertical turbine pumps, 153-155, 197-200, 252-262, 279-280, 328-329, 588-589

Vibration, 80, 245, 247, 347-349, 359, 395-398, 466, 471, 489, 530-537, 563, 611, 619-621
Viscosity, 98-99, 154, 364-368, 626
Volume of steam to volume of water, 50-52
Volutes, 563-566
Volute tongue, casing, 468
Vortex formation, 271-279, 495, 499-500, 502-503, 510

W

Warm-up, 302, 310-313, 331-333, 350, 395, 421-422
Water:
　dirty, 506, 581-583, 597-608
　　removing from basin, 499-501
　river, 582-583
　sealing (*See* Sealing water for stuffing boxes)
　isentropic compression of, 402-403
Water-cooled bearings, 360-363, 573-576
Water-cooled stuffing boxes, 360-363
Water hammer, 258
Water horsepower, 15, 334-335, 352, 355, 384-385, 440-441

Water-lubricated bearings, 172, 197, 199-200, 230, 343, 349-351, 386, 420, 422-423, 424
Water-quenched glands, 267, 362, 470
Water-sealing unit, 267-269, 606-608
Water-works service, 101-107, 138-142
Wear, 106, 119, 161-162, 215, 229-230, 405-409, 577-580, 580-583, 594-595
Wearing rings, 93, 105, 343, 423, 447-456, 476-479, 580-581
　double, 225-228, 448-449
　flat type, 450-451
　L-type, 450-451
　materials for, 227
　single, 225-228, 449
Weight and space limitations, 4
Wet-pit pumps, 3-4, 210, 272, 318-319, 325-326
　intakes, 272, 275, 277, 325-326
Whirlpool action (*See also* Vortex formation),
Wiring, incorrect, 327-331, 544
Witness test, 399
Wound-rotor motors, 141

Z

Zero-speed indicator, 546-547